D1806277

154 Topics in Current Chemistry

Carbohydrate
Chemistry

Editor: J. Thiem

With contributions by
D. R. Bundle, G. Descotes, J. Gigg, R. Gigg,
W. Klaffke, B. Meyer, J. Staněk, Jr., T. Suami,
J. Thiem

With 33 Figures and 21 Tables

Springer-Verlag Berlin Heidelberg New York
London Paris Tokyo Hong Kong

This series presents critical reviews of the present position and future trends in modern chemical research. It is addressed to all research and industrial chemists who wish to keep abreast of advances in their subject.

As a rule, contributions are specially commissioned. The editors and publishers will, however, always be pleased to receive suggestions and supplementary information. Papers are accepted for "Topics in Current Chemistry" in English.

ISBN 3-540-51576-3 Springer-Verlag Berlin Heidelberg New York
ISBN 0-387-51576-3 Springer-Verlag New York Berlin Heidelberg

Bookbinding: Lüderitz & Bauer, Berlin
2151/3020-543210 — Printed on acid-freepaper

Guest Editor

Prof. Dr. *Joachim Thiem*
Organisch-Chemisches Institut,
Westfälische Wilhelms-Universität Münster,
Orléansring 23, D-4400 Münster

Editorial Board

Table of Contents

Synthesis of Oligosaccharides Related to Bacterial O-Antigens

David R. Bundle

Division of Biological Sciences National Research Council of Canada Ottawa, Ontario, CANADA K1A 0KR6

Table of Contents

The O-polysaccharides of bacterial lipopolysaccharides are in general regular, periodic polymers with diverse structures that contain, in many instances, comparatively rare monosaccharides. The oligosaccharides that constitute the repeating units of these polysaccharide antigens provide a demanding challenge in terms of glycoside synthesis, an objective which is particularly important since these structures act as antigenic determinants which are valuable markers of bacterial infection. Advances in glycoside synthesis together with the ancillary techniques of chromatographic separation and high resolution NMR spectroscopy have permitted rational synthesis of such oligosaccharides to be planned, successfully completed and have in a limited number of instances, even allowed small polysaccharides, representative of the O-polysaccharide, to be synthesized via polymerization of synthetic repeating units. The most intensive synthetic efforts have focused on the O-antigens of *Salmonella*, *Shigella*, and *E. coli*, although increasing attention is being given to the synthesis of antigenic determinants of other Gram-negative pathogens, including those of *Brucella*.

Topics in Current Chemistry, Vol. 154
© Springer-Verlag Berlin Heidelberg 1990

1 Survey of Synthesized and Structurally Defined O-Antigens

1.1 Introduction

The lipopolysaccharides (LPS) of most Gram-negative bacteria, pathogenic for humans and animals, carry an O-polysaccharide component. Morphologically the colonies of such organisms have a smooth rather than rough appearance. The latter are often associated with LPS that lack this structural component and, as a consequence, the designations smooth (S) and rough (R) indicate the presence or absence of an O-polysaccharide [1]. During the course of bacterial infection, a humoral and cellular response is mounted most often with specificity for antigenic determinants of the O-polysaccharide and, ultimately, immunity to reinfection by the causative bacterium can be established [2]. The antibodies produced to unique cell-surface antigens constitute diagnostic markers by which the infectious agent can be subsequently identified. Consequently, these antigens or their fragments may be used both as diagnostic markers of specific infectious agents and possibly may even serve as vaccines in special circumstances [2]. The lipopolysaccharide molecule itself, due to its toxic lipid A component, is unsuitable for this purpose [3] and hence synthetic variants of the O-polysaccharide antigen become attractive candidates as chemically defined antigens for diagnostic and prophylactic purposes [2].

The advances that have led to the almost routine synthesis of tri- and tetrasaccharides began in the early 1970s and included a new approach to α-glycosides, the halide-ion catalyzed glycosylation [4], and the introduction of improved heavy metal catalysts such as silver trifluoromethanesulfonate (triflate) [5]. Numerous other technical and conceptual advances punctuated the intervening period and reference to these developments are to be found in the cited literature of several reviews [6—9].

1.2 Lipopolysaccharide Structure

A general architecture of LPS has been established based in large part upon extensive studies of *Salmonella* LPS [10]. The picture of LPS that has emerged and been confirmed in general detail for most LPS has three well-defined regions (Fig. 1). The O-specific antigen (also called somatic antigen or O-polysaccharide) is a polysaccharide generally consisting of from 5 to 40 repeating units that may contain between

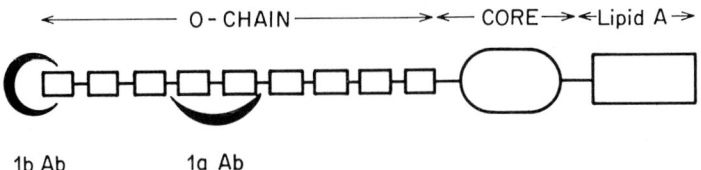

Fig. 1. Schematic representation of the three distinct regions of LPS. The repating unit structure of the O-polysaccharide may be recognized by antibodies with specificities for determinants associated with the terminal non-reducing residues (antibody type 1 b) or internal sequences often spanning more than one repeating sequence (cf. Refs. [72, 73])

one to seven monosaccharides [3, 10]. This component is biosynthesized by addition of single repeating units to the reducing end of the growing O-polysaccharide and then completed polysaccharide is translocated to a specific monosaccharide close to or at the non-reducing terminus of an oligosaccharide core, which is composed of approximately ten monosaccharides. This core is biosynthesized, prior to O-chain attachment, by sequential introduction of monosaccharides to the third region, Lipid-A, a hydrophobic segment which anchors the assembly in the outer membrane of the Gram-negative bacterial cell wall.

It should be noted that the biosynthetic assembly involves polymerization of a specific repating unit structure, which is termed the *biological repating unit* as opposed to the *chemical repating unit* [11]. The latter term refers to the repeating unit structure elucidated by analytical techniques, which rely on chemical fragmentation of the polysaccharide and thus reflects the relative resistance of inter-residue glycosidic bonds to cleavage. Consequently, structural analysis of O-polysaccharide rarely yields information on the biological repeating unit, a structural feature that is generally difficult to establish. The antigenic specificity of chemical repeating units can differ appreciably from those of the biological sequence [2], a fact which may have considerable relevance to the planning of synthetic targets and strategies.

1.3 *Salmonella* O-Polysaccharides Structures

The structures of the O-antigens of *Salmonella* serogroups A, B, D, and E were amongst the first to be studied structurally [11] and biosynthetically [12, 13]. These serogroups also comprise by far the largest clinically important group of *Salmonella* responsible for gastrointestinal infections [2].

A linear sequence of Man-Rha-Gal is to be found in the repeating units of all O-polysaccharides belonging to these serogroups, A–E. The repeating units of serogroups A, B, and D are closely related (Table 1) and share a common linear Man-Rha-Gal sequence, which is branched at O-3 of the mannose residue by a 3,6-dideoxyhexose, the stereochemistry of which defines the A, B, and D_1 serogroups. These tetrasaccharide repeating sequences are presented together with the original literature reporting the structural elucidation in Table 1. In this regard it is to be noted that the erroneous determination of the configuration of rhamnopyranose as β in structures belonging to serogroup B was never corrected in the original literature and this was perpetuated in several reviews. In both the serogroup A and D polysaccharides, this linkage was correctly assigned as an α-L-rhamnopyranose despite the difficulty of unambiguously identifying this feature. A similar error was made for the equally problematic mannopyranosyl linkage but this was corrected following the results of enzymatic and biosynthetic studies [17, 19]. Thus, it is now clear that in all antigens of the serogroups A, B, and D_1 both mannose and rhamnose residues are involved as α linkages [20]. The exception is the serogroup D_2 where the mannose residue not only adopts the β configuration but also is substituted by galactose at O-6 instead of O-2. Elaboration upon these serogroup A, B, and D structures by the action of a phage-induced glucosyl transferase, introduces branching on the galactose residue of the main chain at either position 4 or 6 [10]. Since this modification occurs after polymerization, not every repeating unit is glucosylated [10, 14]. Despite this additional level of complexity

David R. Bundle

Table 1. *Salmonella* serogroups A, B, and D

Salmonella Group (O Facotrs)	Biological Repeating Units	References to Structural Studies	References to Synthetic Work
S. paratyphi (1, 2, 12)	**Group A** [2]-α-D-Man$p$(1 → 4)-α-L-Rha$p$(1 → 3)-α-D-Gal$p$[1] 　　3　　　　　　　　　　　　　　4 　　↑　　　　　　　　　　　　　　↑ 　　1　　　　　　　　　　　　　　1 　α-D-Par$p$　　　　　　　　α-D-Glc$p$	14	26–28, 33, 34, 37
S. typhimurium (1, 4, 5, 12)	**Group B** [2]-α-D-Man$p$(1 → 4)-α-L-Rha$p$(1 → 3)-α-D-Gal$p$[1] 　　3　　　　　　　　　　　　　　4/6 　　↑　　　　　　　　　　　　　　↑ 　　1　　　　　　　　　　　　　　1 　α-D-Abe$p$　　　　　　　　α-Glc$p$	15–19	30–32, 34–36, 38
S. bredeney (1, 5, 12)	2-*O*-Ac-α-D-Abe$p$		40
S. typhi (9, 12)	**Group D$_1$** [2]-α-D-Man$p$(1 → 4)-α-L-Rha$p$(1 → 3)-α-D-Gal$p$[1] 　　3　　　　　　　　　　　　　　4 　　↑　　　　　　　　　　　　　　↑ 　　1　　　　　　　　　　　　　　1 　α-D-Tyv$p$　　　　　　　　α-D-Glc$p$	19, 21	24, 28, 29 33, 34, 37
S. strasbourg (9), 12$_3$, 46)	**Group D$_2$** [6]-β-D-Man$p$(1 → 4)-α-L-Rha$p$(1 → 3)-α-D-Gal$p$[1] 　　3　　　　　　　　　　　　　　4 　　↑　　　　　　　　　　　　　　↑ 　　1　　　　　　　　　　　　　　1 　α-D-Tyv$p$　　　　　　　　α-D-Glc$p$	21–23	25, 39

the principal antigenic specificity involves the 3,6-dideoxyhexose moieties, for which the term *immunodominant* was originally introduced [11]. Thus, most synthetic approaches address this major antigenic feature [32–36], although attempts, principally by Kotchetkov and co-workers [39], have been made to incorporate the second branch point involving α-glucosylation of the main chain galactose residue (Table 1). Serogroup B structures are known in which *O*-2 of the 3,6-dideoxy-α-D-*xylo*-hexopyranose residue carries an *O*-acetyl group [14]. This structural feature radically alters the antigenic specificity (O-factor 4 changes to O-factor 5) of antibodies directed toward the branching residue but the intrinsic difficulty of introducing and maintaining an *O*-acetate throughout de-protection stages has limited attempts to synthesize determinants with O-factor 5 specificity [40].

Much of the current knowledge of the biosynthesis of bacterial O-polysaccharide assembly was derived from early studies of the polysaccharides of *Salmonella* serogroup E [13]. This group consists of a series of four closely related structures belonging to the serogroups E$_1$ → E$_4$ (Table 2). The basic main chain sequence of three monosaccharides, Man-Rha-Gal, seen in the D$_2$ serogroup is the structural element of all four polysaccharides; however, in serogroups E$_1$, E$_2$ and E$_3$, E$_4$ the variation oc-

4

curs by changes in the configuration and position of glycosidic linkages rather than via branching by distinct hexose residues. The polysaccharides of serogroups E_1 and E_2 are amongst the simplest of the *Salmonella* O-polysaccharides and, consequently, attempts to prepare synthetic polysaccharides involved these structures [48–50]. The work which formed the basis for the approach is referenced in Table 2 and covers the synthesis of tri- and tetrasaccharide determinants of the E_1–E_4 structures prepared for the most part as biological repeating units [50–56]. *Salmonella* membrane preparations have also been used to polymerize trisaccharide repeating units, prepared using glycosyl transferase and nucleoside diphosphate sugars [58].

Table 2. *Salmonella* serogroup E

Salmonella Group (O Factors)	Biological Repeating Units	References to Structural Studies	References to Synthetic Work
S. anatum (3, 10)	Group E_1 [6)-β-D-Man$p$(1 → 4)-α-L-Rha$p$(1 → 3)-α-D-Gal$p$(1]	41–43	34, 50, 51, 54, 58–61, 64
S. newington (3, 15)	Group E_2 [6)-β-D-Man$p$(1 → 4)-α-L-Rha$p$(1 → 3)-β-D-Gal$p$(1]	41, 42, 44	48–51, 54, 58, 60, 62, 63
S. minneapolis (3, (15), 34)	Group E_3 [6)-β-D-Man$p$(1 → 4)-α-L-Rha$p$(1 → 3)-β-D-Gal$p$] 4 ↑ 1 α-D-Glc$p$	41, 45	55
S. senftenberg (1, 3, 19)	Group E_4 [6)-β-D-Man$p$(1 → 4)-α-L-Rha$p$(1 → 3)-α-D-Gal$p$(1] 6 ↑ 1 α-D-Glc$p$	46, 47	56, 57

Fragments of the basic repeating unit structures of the serogroups A–E have been prepared [59–63], in one case as an allyl glycoside, which was subsequently co-polymerized with acrylamide to provide a highly active antigen [61, 62]. The synthesis of a disaccharide representative of the *Salmonella* serogroup C antigen in which abequose is linked α-1,3 to α-L-rhamnopyranose has also been reported [64].

1.4 *Shigella* O-Polysaccharides

Shigella flexneri LPSs comprise a large family of interrelated O-polysaccharide structures, the simplest of which belongs to the Y variant polysaccharide (Table 3). This has a tetrasaccharide repeating unit containing three L-rhamnose residues and

Table 3. *Shigella flexneri* O-antigens

Shigella flexneri O-Factor	Biological Repeating Unit	References to Structural Studies	References to Synthetic Work
—: 3, 4	Variant Y structure [2)α-L-Rhap(1 → 2)α-L-Rhap(1 → 3)α-L-Rhap(1 → 3)β-D-GlcNAcp(1]	65, 68	74, 80, 83, 88
—: 7, 8	Variant X structure [2)α-L-Rhap(1 → 2)α-L-Rhap(1 → 3)β-D-GlcNAcp(1] 3 ↑ 1 α-D-Glcp	66, 67	84–87
V: 7, 8	5a structure [2)α-L-Rhap(1 → 2)α-L-Rhap(1 → 3)α-L-Rhap(1 → 3)β-D-GlcNAcp(1] 3 ↑ 1 α-D-Glcp	66, 67	84–87
V: 7, 8	5b structure [2)α-L-Rhap(1 → 2)α-L-Rhap(1 → 3)α-L-Rhap(1 → 3)β-D-GlcNAcp(1] 3 ↑ 1 α-D-Glcp	66, 67	84–87
I: 4	1a structure [2)α-L-Rhap(1 → 2)α-L-Rhap(1 → 3)α-L-Rhap(1 → 3)β-D-GlcNAcp(1] 4 ↑ 1 α-D-Glcp	66, 67	81
I: 6	1b structure [2)α-L-Rhap(1 → 2)α-L-Rhap(1 → 3)α-L-Rhap(1 → 3)β-D-GlcNAcp(1] 2 4 ↑ ↑ Ac 1 α-D-Glcp	66, 67	81

I: 6

```
[2)α-L-Rhap(1 → 2)α-L-Rhap(1 → 3)α-L-Rhap(1 → 3)β-D-GlcNAcp(1]
              2                                  4
              ↑                                  ↑
              Ac                                 1
                                              α-D-Glcp
```
66, 67

II: 3, 4

2a structure
```
[2)α-L-Rhap(1 → 2)α-L-Rhap(1 → 3)α-L-Rhap(1 → 3)β-D-GlcNAcp(1]
                                                 4
                                                 ↑
                                                 1
                                              α-D-Glcp
```
59, 60

II: 7, 8

2b structure
```
[2)α-L-Rhap(1 → 2)α-L-Rhap(1 → 3)α-L-Rhap(1 → 3)β-D-GlcNAcp(1]
                                  4
                                  ↑
                                  1
                               α-D-Glcp
```
66, 67

III: 6, 7, 8

3a structure
```
[2)α-L-Rhap(1 → 2)α-L-Rhap(1 → 3)α-L-Rhap(1 → 3)β-D-GlcNAcp(1]
                  3
                  ↑
                  1
               α-D-Glcp
```
66, 67

III: 3, 4, 6

3b structure
```
[2)α-L-Rhap(1 → 2)α-L-Rhap(1 → 3)α-L-Rhap(1 → 3)β-D-GlcNAcp(1]
   3
   ↑
   1
α-D-Glcp
```
66, 67

IV: 3, 4

4a structure
```
[2)α-L-Rhap(1 → 2)α-L-Rhap(1 → 3)α-L-Rhap(1 → 3)β-D-GlcNAcp(1]
                                                 6
                                                 ↑
                                                 1
                                              α-D-Glcp
```
66, 67

IV: 6

4b structure
```
[2)α-L-Rhap(1 → 2)α-L-Rhap(1 → 3)α-L-Rhap(1 → 3)β-D-GlcNAcp(1]
                  2                              6
                  ↑                              ↑
                  Ac                             1
                                              α-D-Glcp
```
66, 67

7

2-acetamido-2-deoxy-D-glucose and is the basis for a series of elaborations involving α-glucosylation and O-acetylation [65–68]. The serological classification and identification of specific O-factors is less well established than the Kauffman-White scheme [69] for *Salmonella*. Recently, the basis of the previously proposed O-factor scheme for *Shigella flexneri* [70] has been questioned [71–73]. Nevertheless, the structural details of these polysaccharides are well established and several extensive synthetic approaches have been mounted, centered mainly upon the simplest Y variant repeating unit [74–83, 88] — and more recently extended to include the X, 5a, and 5b structures [84–87]. In addition, structural variations designed to probe features of the antibody combining site have been the subject of recent studies in this area [89–91].

A synthetic *S. flexneri* Y antigen containing ten repeating units has been prepared [88].

The *S. flexneri* serotype 6 structure is in fact unrelated to the other *S. flexneri* repeating units and should be classified separately [20]. One of the two proposed structures [92, 93] has been the target of a synthetic study which provided a branched tetrasaccharide structure [94] and separate work gave a disaccharide monoacetate structure [95].

The O-polysaccharides of *Shigella dysenteriae* are structurally very distinct from those of the *S. flexneri* [20] and have been the subject of synthetic studies in which a branched pentasaccharide of the serotype 2 antigen was prepared [96].

1.5 *E. coli* and *Klebsiella* O-Polysaccharides

Although the O-antigens of these organisms provide a rich variety of complex structures, the synthesis of antigenic determinants of only a few structures has been attempted. The synthesis of the 3,6-dideoxy-α-L-*xylo*-hexopyranose-containing structure of *E. coli* 0111 has been attempted but only to the trisaccharide level of complexity [97]. A very elegant synthesis of *E. coli* 075 antigen has been reported [98] and the oligomannose determinant of the 09 antigen [99] was also successfully prepared.

Even fewer structures corresponding to *Klebsiella* O-antigens have been synthesized although elements of the O-7 LPS have been prepared [100].

Finally, a synthesis of the determinant of the enterobacterial common antigen which is not a true O-antigen has been reported [101].

1.6 Other O-Polysaccharides

The O-antigens of *Brucella* species are homopolymers of α-1,2-linked 4,6-dideoxy-4-formamido-D-mannopyranose residues in which there may be α-1,3 linkages [102, 103]. The *Yersinia* 0:9 antigen is a similar 1,2-linked homopolymer but contains no 1,3 linkages. A related structure *N*-acylated by 3-deoxy-L-glycero-tetronic acid in place of the formate group is produced by *Vibrio cholerae* Inaba and Ogawa strains [104]. These structures have been the subject of synthetic work [105–107].

O-Antigens from *Aeromoanas salmonicida* [108] and *Pseudomanas* [109] have also been the subject of synthetic efforts [110, 111].

2 Synthesis of *Salmonella* Oligosaccharides

2.1 *Salmonella* Serogroups A, B, C, and D

Chemical synthesis of disaccharide elements of these oligosaccharides marks the beginning of systematic synthetic studies of the antigenic determinants of bacterial O-antigens [24–31]. The 3,6-dideoxyhexoses, which are the immunodominant mono-saccharides of each serogroup, were an intrinsic synthetic challenge since their synthesis in a form suitable for use as glycosyl donors was a prerequisite to successful synthesis. Early synthetic approaches used in the identification of the 3,6-dideoxy-hexoses [112–114] were judged unsuitable for subsequent incorporation into synthetic strategies for disaccharide synthesis. Recently, *p*-nitrobenzoylation of 3,6-dideoxy-hexoses synthesized from 1,4-lactones [115, 116] shows that the reducing mono-saccharide may indeed be a useful starting point for synthesis. Numerous syntheses of the 3,6-dideoxyhexoses have been reported [112–114, 117–123, and references cited therein] but the most frequently adopted routes for subsequent exploitation in glyco-sylation reactions have used a common methyl hexopyranoside precursor in which the 3-deoxy function is first introduced followed by deoxygenation at C-6, or starting from a 6-deoxyhexose the deoxygeneration at C-3 is the final step in the preparation of starting material.

   Typical examples of these two approaches are to be seen in syntheses of 3,6-dideoxy-D-*arabino*-hexose [29, 33, 121] and 3,6-dideoxy-D-*xylo*-hexose [35, 118, 122]. Methyl 4,6-*O*-benzylidene-α-D-glucopyranoside (*1*) is converted in a one-flask reaction and

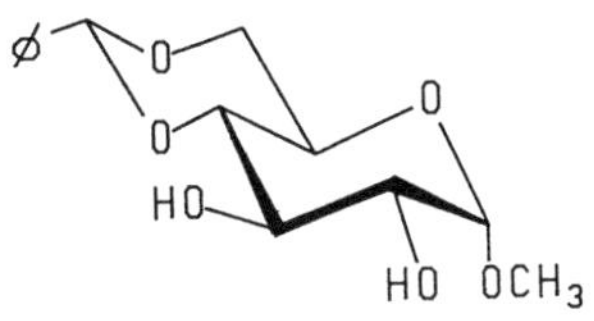

1

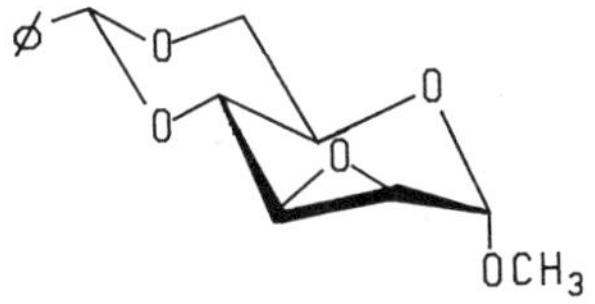

2

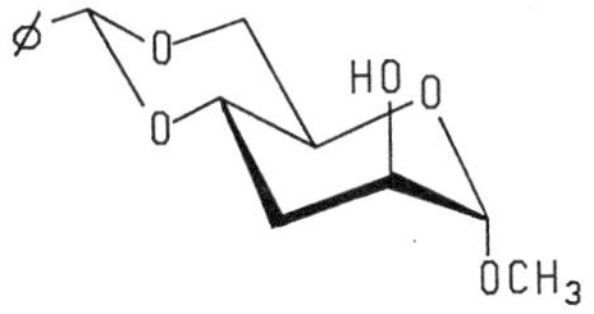

3

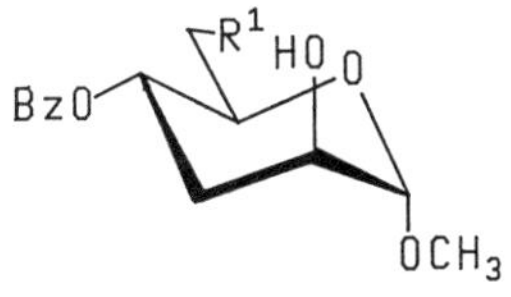

R¹ = Br *4*
R¹ = H *5*

high yield to the 2,3-anhydro-compound (*2*) via the 2-*O*-tosylate and reduction by LiAlH₄ affords the 3-deoxy-mannopyranoside (*3*) [124] from which a 6-bromo-deoxy derivative (*4*) is easily prepared by the Hanessian-Hullar reaction [125]. Reduction affords methyl 4-O-benzoyl-3,6-dideoxy-α-D-*arabino*-hexopyranoside (*5*) [29, 33]. Methyl 6-deoxy-α-D-galactopyranoside (*6*) may be generated by either the deoxy-genation of 1,2:3,4-di-*O*-isopropylidene-α-D-galactopyranose followed by a Fischer

David R. Bundle

glycosylation [35] or from methyl 2,3-di-*O*-benzoyl-4,6-*O*-benzylidene-α-D-galacto-pyranoside [122]. Subsequent formation of the *R* and *S* 3,4-*O*-benzylidene acetals (*7a* and *7b*) followed by *N*-bromosuccinimide actel opening gave the 3-bromo-deri-

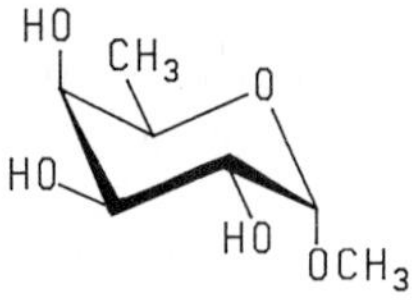

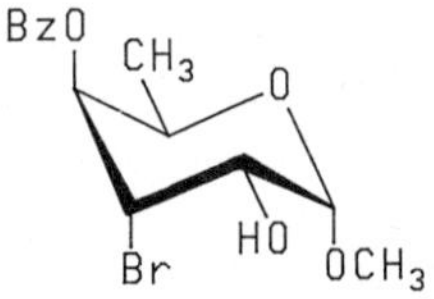

6 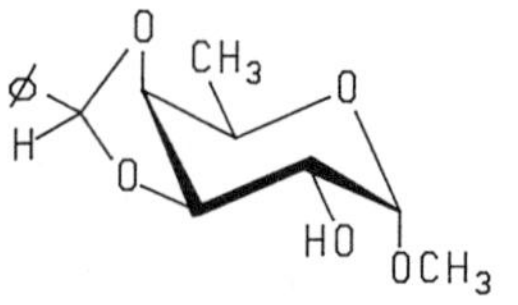   7a   8
           7b

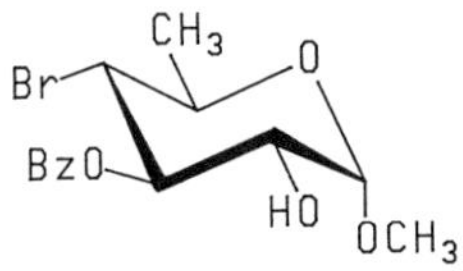

9 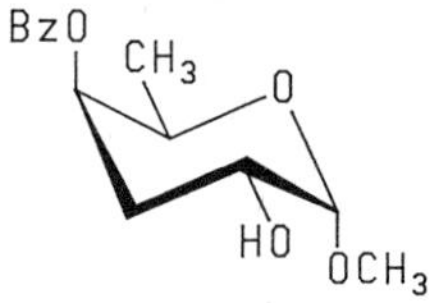   10

vative (*8*) and, in the case of the α-series (cf. Ref. [35, 126]), also methyl 3'-*O*-benzoyl-4-bromo-α-D-glucopyranoside (*9*). Reduction provides the target monosaccharide glycoside (*10*). Severyl convenient syntheses of paratose, 3,6-dideoxy-*ribo*-hexose, using the aforementioned approaches, have been described [32, 33, 37], and as well direct conversion of methyl β-D-glucopyranoside to the 3,6-dichloro-3,6-dideoxy-allopyranoside by chlorosulfation [120] or the bromo analogue with triphenylphos-phine, bromine, and imidazole [122] are practical routes, together with recent modi-fication of the chlorosulfation approach [119].

Finally, as already noted, simple and elegant lactone chemistry [115, 116] starting from the readily available 1,4-galactonolactone (*11*) and HBr/HOAc yields crystalline

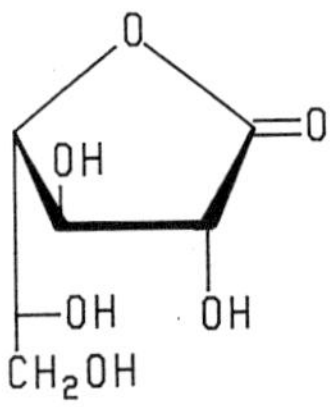

11 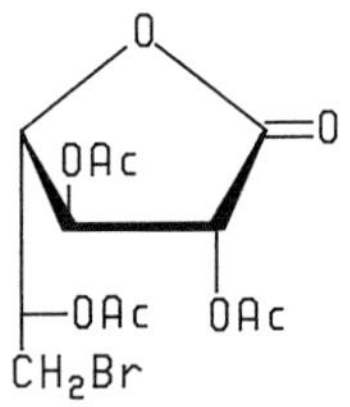   12 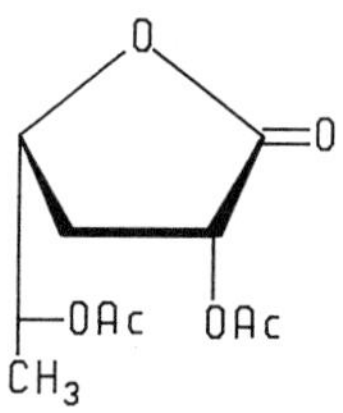   13

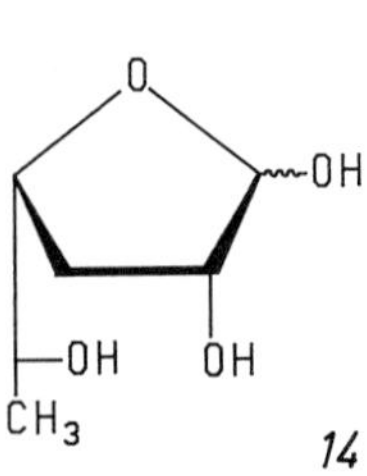

14

10

6-bromo-6-deoxylactone (*12*) from which the 3,6-dideoxy-1,4-lactones (*13*) are obtained by hydrogenolysis in the presence of triethylamine, via a β-elimination reaction that provides a 2,3-unsaturated lactone, which is reduced to *13*. The deacetylated 3,6-dideoxy-1,4-lactone is then reduced under controlled conditions to give the 3,6-dideoxy-D-*xylo*-hexose (*14*). This compound was utilized further as its 2,4-di-*O-p*-nitrobenzoyl glycosyl bromide [36]. Since acetylated lactones undergo this β-elimination, tyvelose may in principale be prepared by HBr/HOAc bromination of the 3-deoxy-D-*arabino*-hexono-1,4-lactone obtained from either D-*manno*- or D-*glucono*-1,4-lactones [115, 116].

The branching 3,6-dideoxyhexoses are exclusively α-linked to the main chain mannose residues of the O-polysaccharide and for paratose and abequose this requires the formation of a 1,2-*cis*-glycopyranoside bond [8]. Thus, glycosyl halides with the *ribo* and *xylo* configuration are required with a nonparticipating group at O-2. In the case of tyvelose, 3.6-dideoxy-D-*arabino*-hexopyranose, glycosyl halides with participating protecting groups lead to the desired 1,2-*trans*-glycosidic bond formation. Thus, 2,4-di-*O*-benzoyl α-D-*arabino*-hexopyranosyl bromide [29] or chloride [33, 34, 37] have been used in the synthesis of the serogroup D₁ O-factor 9 oligosaccharides. Generally, abequose and paratose methyl glycopyranosides have been dibenzylated or the respective 4-*O*-benzoates may be monobenzylated at *O*-2 employing neutral- or acid-catalyzed benzylation condictions [30, 32, 33, 35–37]. Direct conversion of the protected methyl pyranosides to glycosyl halides has employed hydrogen chloride or hydrogen bromide [27, 31], dichloro- or dibromomethyl methyl ether [33, 35] or chloro- or bromotrimethylsilane [34, 37]. Alternately, the less direct aqueous hydrolysis of the methyl glycoside, followed by formation of the 1-*O*-nitrobenzoates, has been recommended to yield cleaner glycosyl bromide preparations [34].

Employing glycosyl donors of the above type and a simple monosaccharide acceptor such as methyl 2-O-benzyl-4,6-O-benzylidene-α-D-mannopyranoside or the

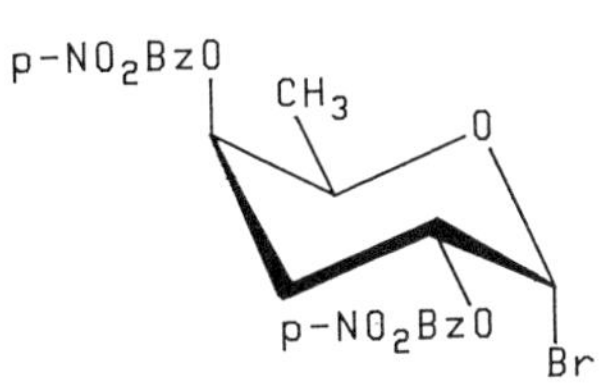

16

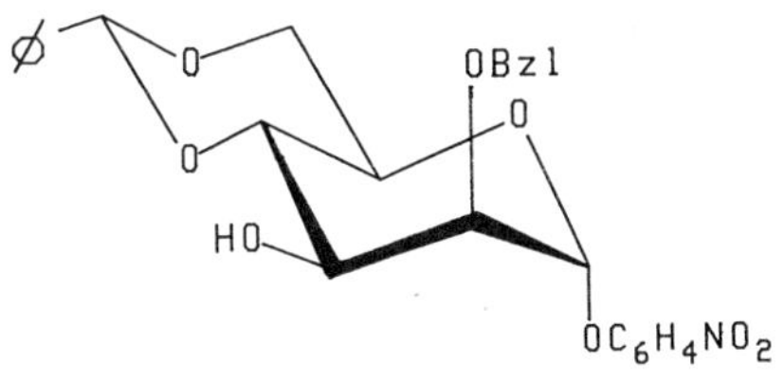

15

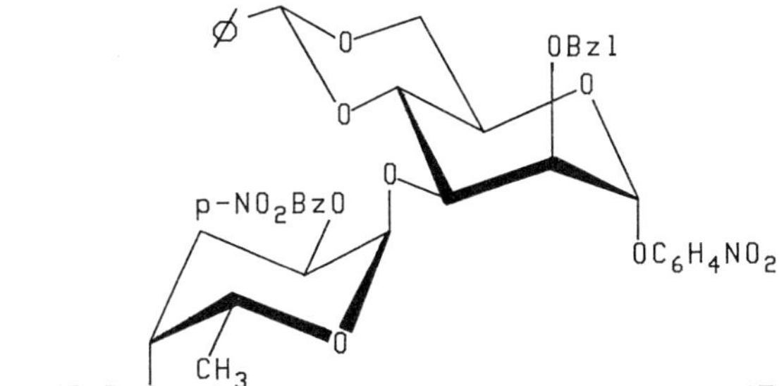

17

David R. Bundle

corresponding *p*-nitrophenyl glycoside *15*, various disaccharides have been prepared by Garegg's group [24–32]. These authors noted that 2,4-di-*O*-*p*-nitrobenzoyl-α-D-*xylo*-hexopyranosyl bromide (*16*), which was more readily available than the corresponding 2,4-di-*O*-benzyl derivative, reacted with the selectively protected mannoside (*15*) in the presence of mercuric cyanide to give almost exclusively the α-linked disaccharide (*17*). Recent systematic studies of the synthesis of related disaccharides

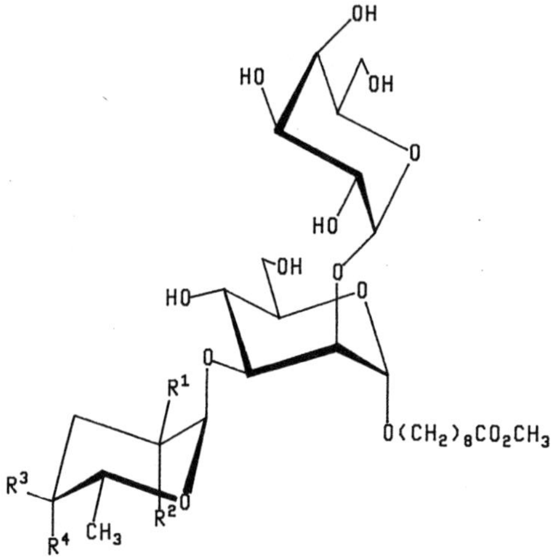

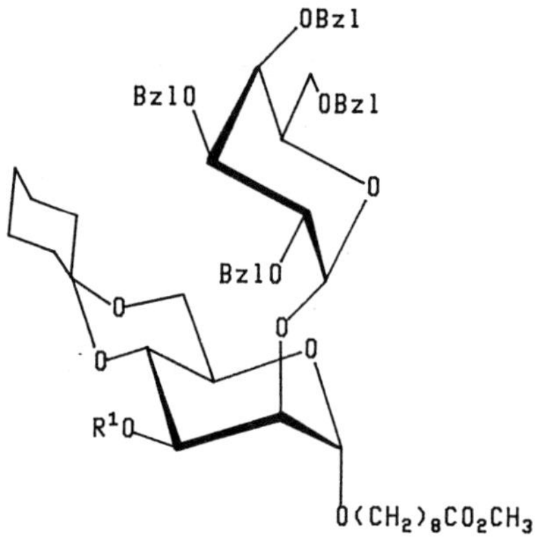

R^1 = R^4 = OH, R^2 = R^3 = H   *18*
R^1 = R^4 = H, R^2 = R^3 = OH   *19*
R^1 = R^3 = OH, R^2 = R^4 = H   *20*

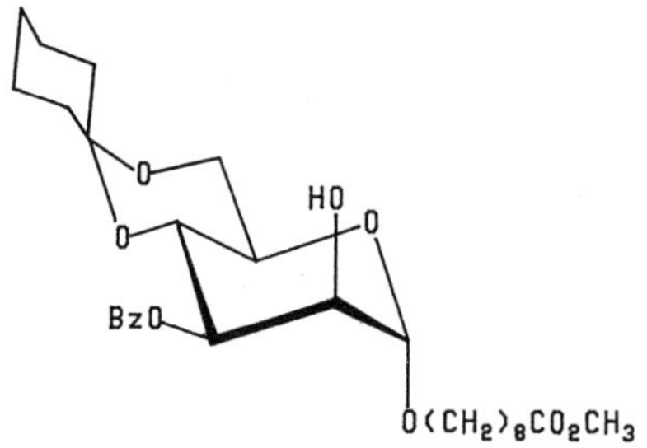

R^1 = Bz   *21*
R^1 = H   *24*

22

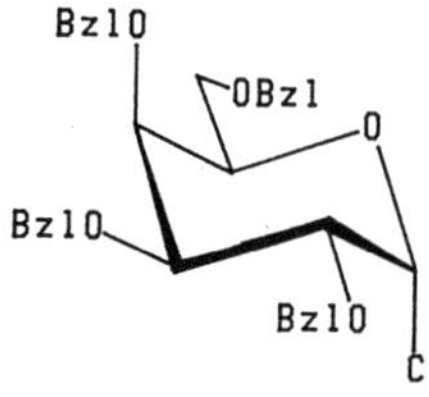

23

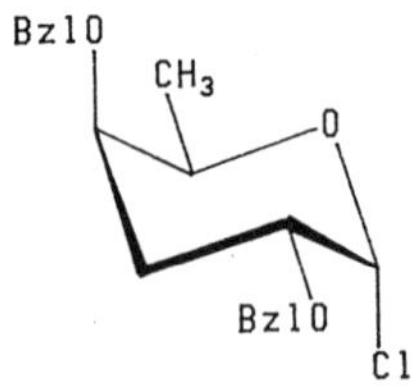

25

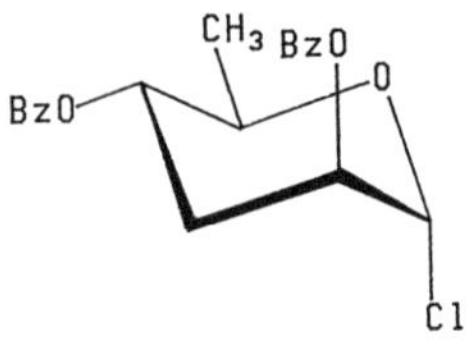

26

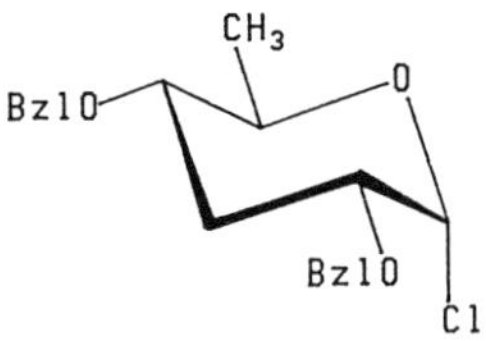

27

12

using toluene nitromethane as solvent gave poor stereoselectivity with $\alpha:\beta$ ratios close to $1:1$ [36]. Syntheses of tri- and tetrasaccharide [33–35, 38] involving this linkage have, therefore, employed non-participating groups at O-2 of the glycosyl halides derived from abequose and paratose.

Three branched trisaccharides (18, 19, and 20) in which the mannose residue was glycosylated by each of the three 3,6-dideoxyhexose isomers, were synthesized from a common disaccharide precursor (21). The latter was constructed from 8-methoxyl-carbonyloctanol 3-O-benzoyl-4,6-O-cyclohexylidene-α-D-mannopyranoside 22 and tetra-O-benzyl-α-D-galactopyranosylchloride (23). Transesterification of disaccharide 24 leads to the selectively protected disaccharide 21 which was reacted with the ap-

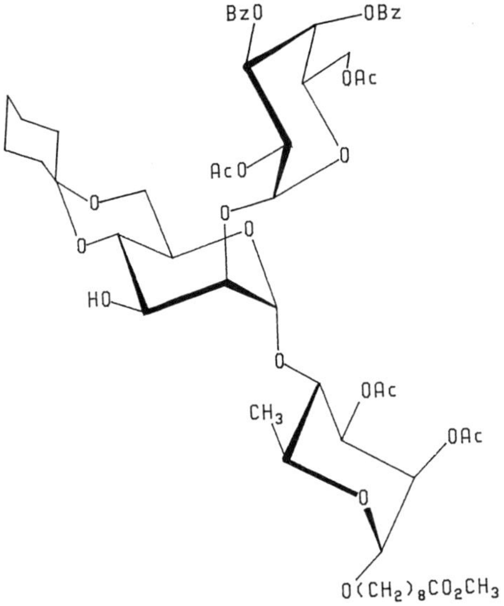

$$28$$

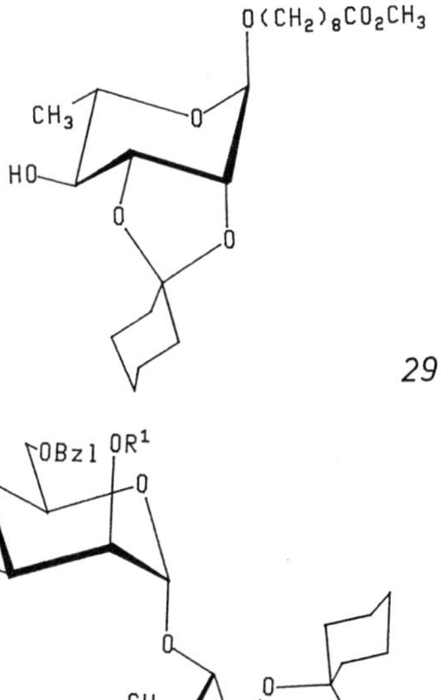

$$29$$

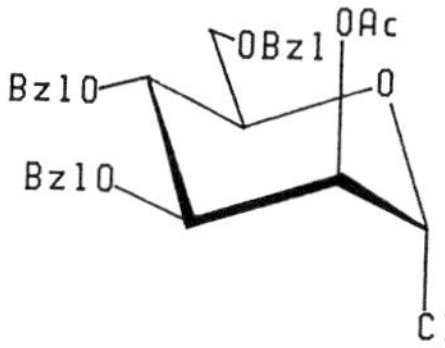

$$30$$

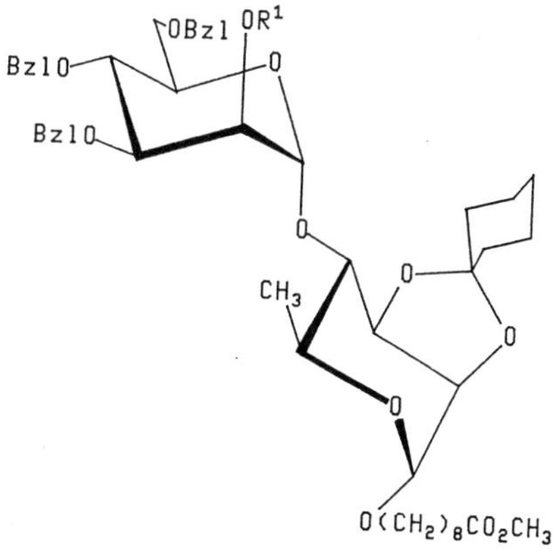

$$R^1 = Ac \quad 31$$
$$R^1 = H \quad 32$$

13

David R. Bundle

propriate derivatives of abequose *25*, tyvelose *26*, and paratose *27* to yield the fully protected branched trisaccharides of serogroups B *18* [35], D₁, *19*, and A *20* [33]. Deprotection was performed by hydrogenolysis in acetic acid and transesterification where appropriate. A similar sequence of glycosylation was used by Bock and Meldal

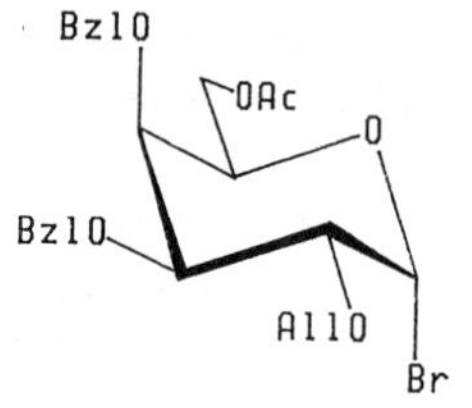

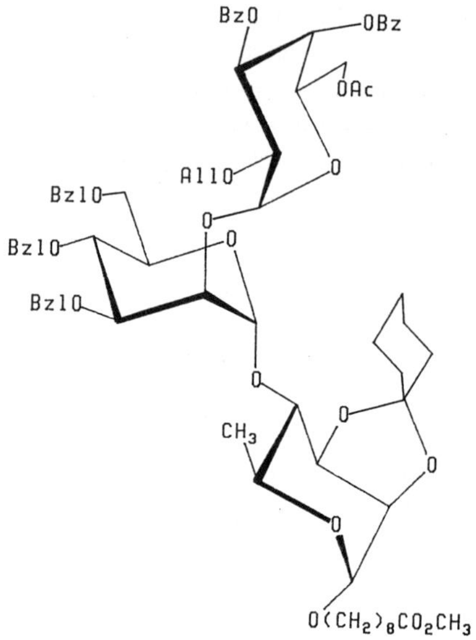

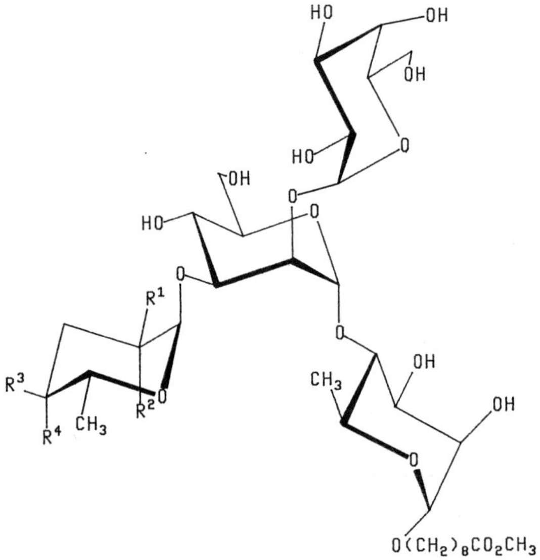

$$R^1 = R^2 = R^3 = H, \ R^4 = OH \qquad 35$$
$$R^1 = OH, \ R^2 = R^3 = R^4 = H \qquad 36$$

[36–38] in the construction of a series of branched tetrasaccharides. The common trisaccharide precursor *28* was synthesized from 8-methoxycarbonyloctanol 2,3-O-cyclohexylidene-α-L-rhamnopyranoside (*29*) and 2-O-acetyl-3,4,6-tri-O-benzylα-D-mannopyranosyl chloride (*30*) which gave disaccharide (*31*). Selective deprotection of the mannose residue *31 → 32* followed by silver triflate promoted glycosylation at *O*-2 of mannose by 6-*O*-acetyl-2-*O*-allyl-3,4-di-*O*-benzoyl-α-D-galactopyranosyl bromide (*33*) gave the linear trisaccharide *34*. Removal of allyl and cyclohexylidene groups, acetylation and hydrogenolysis of the benzyl ether groups gave a tri-hydroxy mannose derivative, which was converted to the trisaccharide alcohol *28* by acetalation with 1-ethoxycyclohexene. In addition to the three tetrasaccharides corresponding to chemical repeating units of the serogroup A, B, and D_1 O-antigens [37, 38], a variety of derivatives, deoxygenated in the branching residue, were synthesized [37]. These included the 2,3,6-trideoxy-α-D-*threo*-hexopyranosyl (*35*) and 3,4,6-trideoxy-α-D-*erythro*-hexopyranosyl (*36*) branched tetrasaccharides. A point of interest during the deprotection of these acid-sensitive tetrasaccharides and the natural abequose isomer was the inclusion of ethylene glycol in acetic acid to catalyze the hydrolysis of the 4,6-*O*-cyclohexylidene acetal [37, 38] under mild hydrolytic conditions, thereby avoiding the loss of the acid labile 3,6-dideoxyhexose or trideoxyhexoses. Conventional aqueous hydrolysis of the acetal caused considerable loss of the dideoxy- or trideoxy α-D-hexopyranosyl residues.

Linear tetrasaccharides representing the biological repeating units of the serogroup A, B, and D_1 polysaccharides were synthesized from a common linear trisaccharide precursor (*37*), which was in turn built by sequential 1,2-*trans*-glycosidations. The synthetic scheme, in contrast to the usual practice, employed benzoate and acetate esters as persistent blocking-groups for hydroxy functions, and benzyl ethers and chloroacetates for temporary protection. Since this strategy demanded hydrogenolytic steps, the functionality of the aglycone used for subsequent coupling of the deblocked oligosaccharide to protein was converted from a *p*-nitrophenyl galactopyranoside to the *p*-trifluoroacetamidophenyl galactopyranoside (*38*) at the outset of the synthesis. The rhamnose building unit *39* used for the first glycosylation step contained a 4-*O*-chloroacetyl group, the selective removal of which from the resultant disaccharide *40* gave the glycosyl acceptor *41* for chain elongation with 2-*O*-acetyl-tri-*O*-benzyl-α-D-mannopyranosyl bromide (*30*). Hydrogenolytic cleavage of the benzyl ethers of *42* and introduction of a 4,6-*O*-benzylidene group gave the trisaccharide precursor *37*, from which the three target structures were prepared by glycosylation with the appropriate 3,6-dideoxyhexopyranosyl halides [34].

The synthesis of a pentasaccharide corresponding to the reported structure for the *Salmonella strasbourg* repeating unit used a block synthesis [39]. This strategy employed a trisaccharide, tyv-man-rha used in an orthoester glycosylation reaction with a disaccharide, Glc-Gal to yield a branched pentasaccharide in which the glucose residue is the branching residue responsible for O-factor 12_3 in the natural antigen. The weakness of this synthetic approach was the acid lability of the 3,6-dideoxyhexopyranosyl(1 → 3)mannopyranose linkage when the trisaccharide precursor *43* was converted to the glycosyl bromide *44* prior to synthesis of the corresponding orthoester *45*. Reaction of this orthoester with benzyl 2,6-di-*O*-benzyl-4-*O*-(2,3,4-tri-*O*-benzyl-6-*O*-benzoyl-α-D-glucopyranosyl)-β-D-galactopyranoside *46* under conditions established for orthoester glycosylations gave a 17% yield of the penta-

David R. Bundle

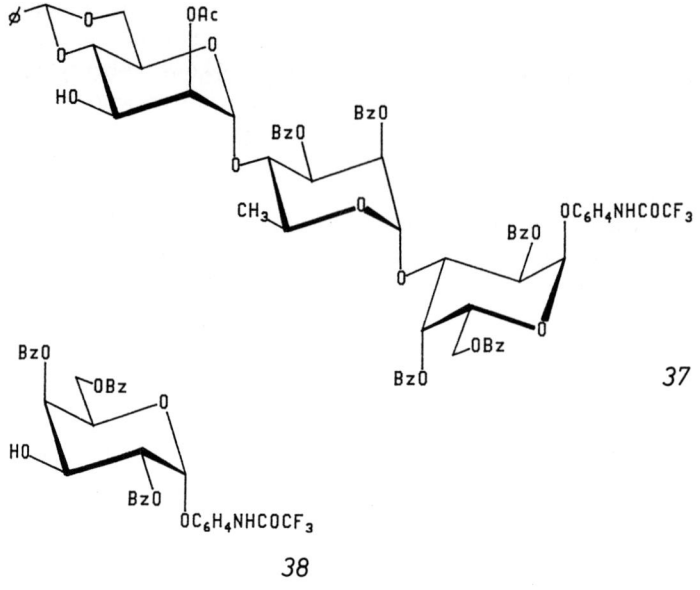

37

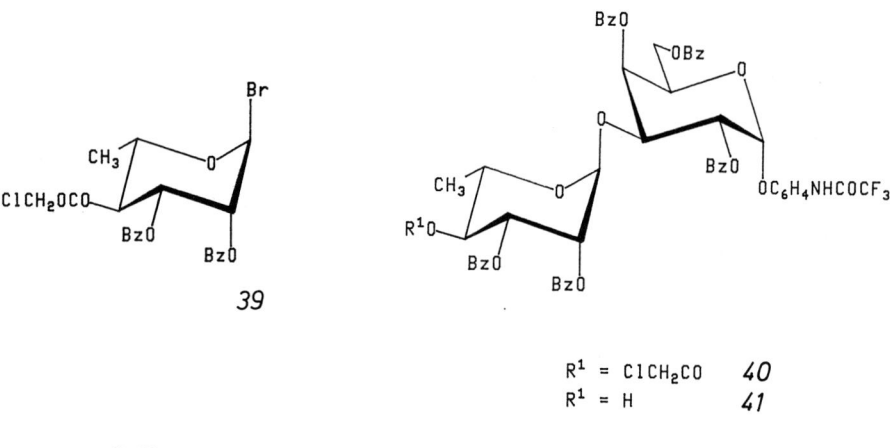

38

39

$R^1 = ClCH_2CO$ *40*
$R^1 = H$ *41*

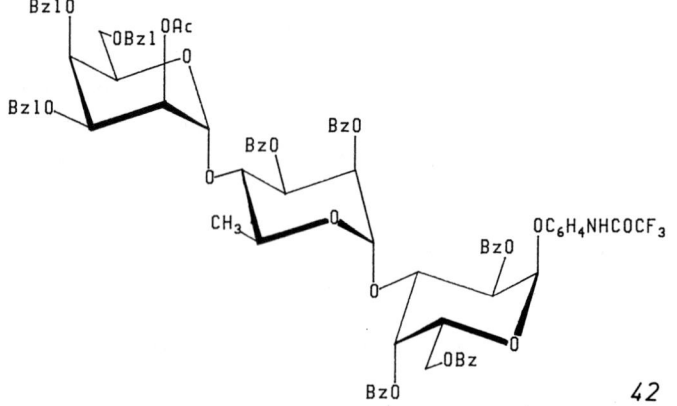

42

16

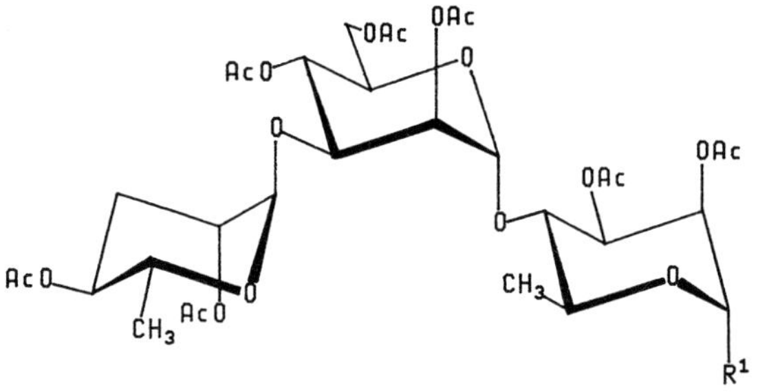

R = Ac *43*
R = Br *44*

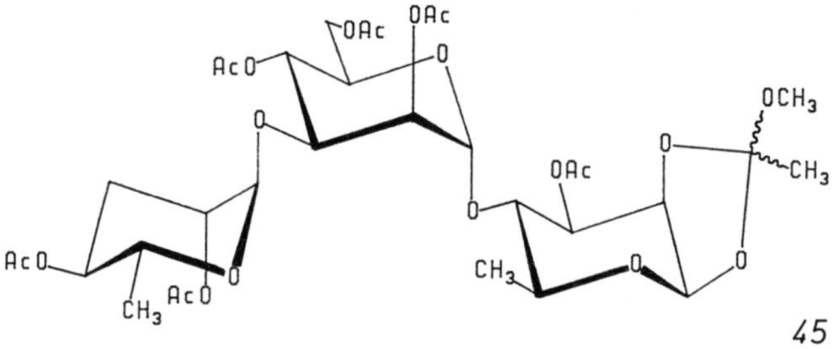

45

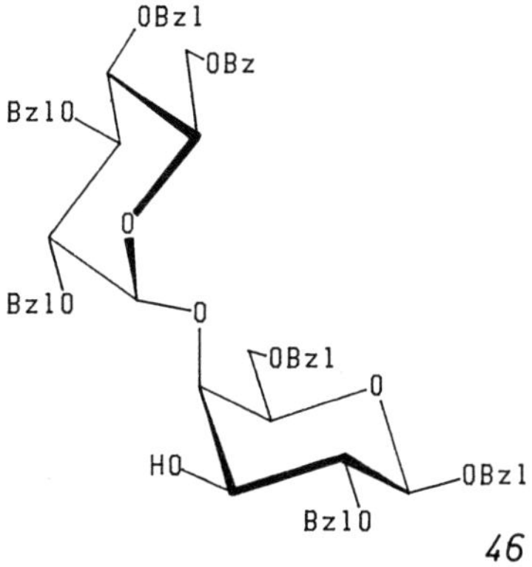

46

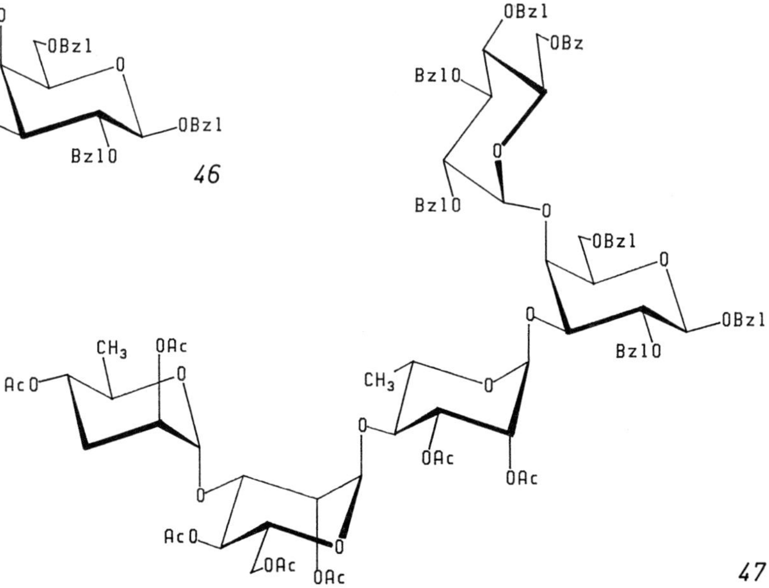

47

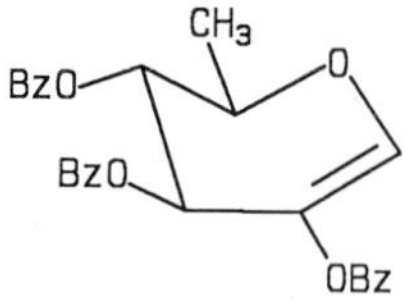

David R. Bundle

48

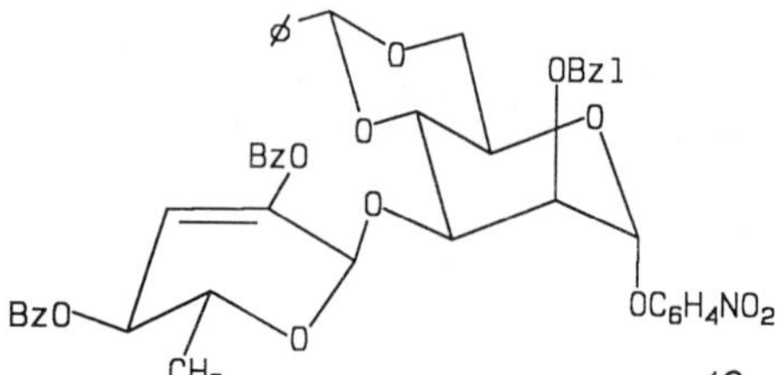

49 a

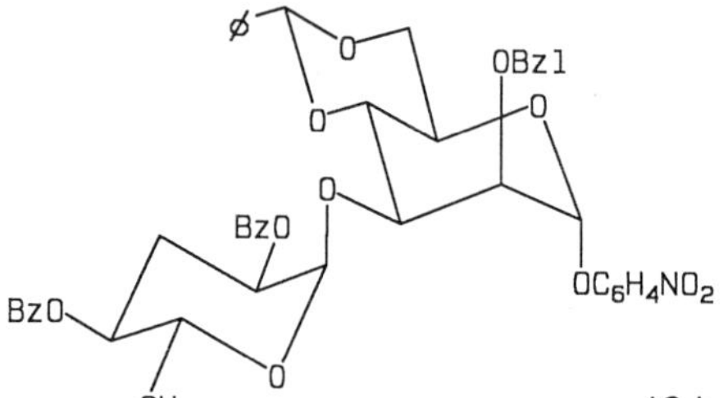

49 b

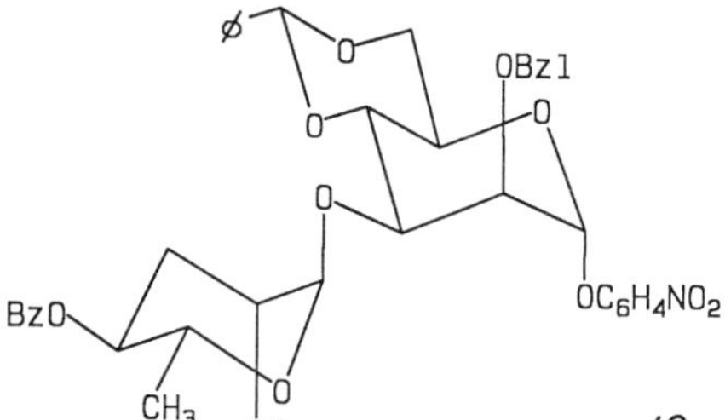

49 c

saccharide, which on deblocking was shown to be a mixture of the desired pentasaccharide 47 and a tetrasaccharide devoid of tyvelose residues.

Starting from 1,5-anhydro-2,3,4-tri-O-benzoyl-6-deoxy-D-*arabino*-hex-1-enitol (*48*),
an unusual approach to the synthesis of disaccharides of serogroup A and D has
applied the allylic rearrangement glycosylation procedure of Ferrier [127] to obtain
an α-D-*erythro*-hex-2-enopyranosyl residue in the disaccharide 49a. Reduction of the
2-enopyranose double bond gave both the paratose (*ribo*) 49b and tyvelose (*arabino*)
49c products [28].

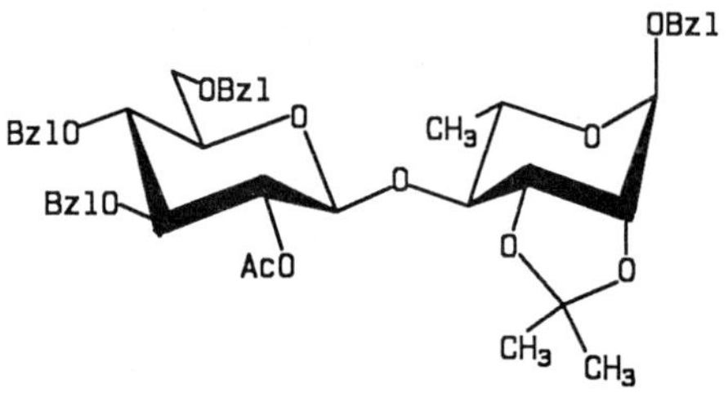

50 a

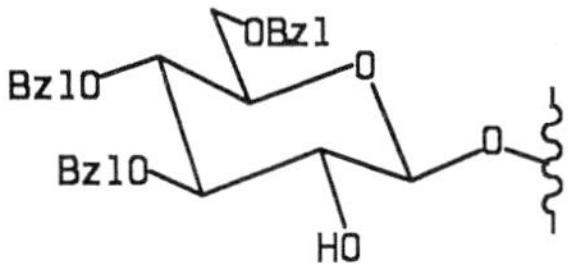

50 b

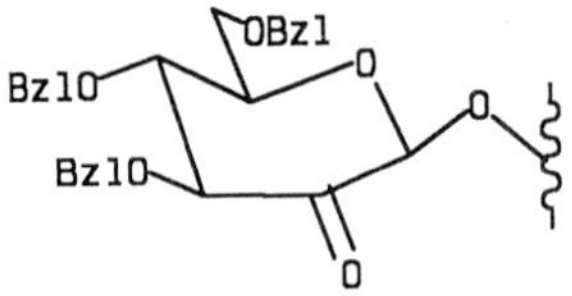

50 c

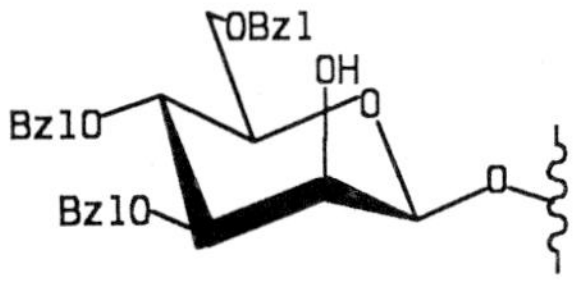

50 d

18

The immunodominant feature of O-factor 5 is a 2-O-acetyl-3,6-dideoxy-α-D-*xylo*-hexopyranosyl residue. So far no successful attempts to incorporate this feature into larger oligosaccharides have appeared. Immunochemical studies using a *p*-amino-phenyl glycoside of this monosaccharide, prepared by partial acetylation and chromatographic resolution of the resulant mixture, have been reported [(*40*)].

2.2 *Salmonella* Serogroup E

The structures of O-polysaccharide repeating units of the serogroup E are a closely related set of four subgroups, all possessing a linear trisaccharide element, man-rha-gal (Table 2). The four subgroups E_1–E_4 are generated by variation of the anomeric configuration of the linkage between galactose and mannose and also by branching α-D-glucopyranosyl residues at either O-4 or O-6 of the galactose unit.

Since the mannose residue is exclusively β-1,4-linked to α-L-rhamnose this disaccharide has been prepared by one of two methods. A method based upon epimerization at C-2 [136] and elaborated by Garegg's group [128] was used to create a β-glucopyrano-side *50a* by 1,2-*trans* glycosylation of benzyl 2,3-O-isopropylidene-α-L-rhamnopy-ranoside using 2-O-acetyl-3,4,6-tri-O-benzyl-α-D-glucopyranosyl chloride. After de-O-acetylation and oxidation of the alcohol *50b* to a β-*arabino*-hexopyranosidulose *50c*, reduction gave a resolveable mixture of β-mannopyranoside *50d* and the corresponding β-glucopyranoside ([51], cf. Ref. [59]). This disaccharide was converted to a mannosyl-rhamnosyl halide and tested for reactivity with three selectively protected derivatives of galactose, 1,2:5,6-di-O-isopropylidene-galactofuranose, 1,2-O-isopropylidene-4,6-O-ethylidene-α-D-galactopyranose *51* and benzyl-2,6-di-O-acetyl-β-D-galactopyranoside. The 4,6-O-ethylidene derivative *51* was found to be the most effective [51–54], taking into account deprotection of *52*, and the separate and eventual utilization of the resultant trisaccharide *53* for polysaccharide synthesis [48–50]. The second method employed to reach *53* used sequential chain extension starting from *51*. In this case the β-mannopyranosyl residue was introduced via the 4,6-di-O-acetyl-2,3-O-carbonyl-α-D-mannopyranosyl bromide first reported by Gorin and Perlin [129] and prepared by an improved procedure in this work [52].

The first synthesis of a biologically specific heteropolysaccharide made use of trisaccharide *53*, which was used to synthesize oligomeric forms of the *Salmonella newington* O-polysaccharide with molecular weights in the range 2000–5000 Daltons. The principle of the method was to construct a 1,2-O-cyanoethylidene derivative 55 from the glycosyl halide *54*. The CN-*exo* and *endo* isomers or their mixtures were effective and stereospecific glycosylating reagents and in the presence of triphenyl-methylium perchlorate and trityl ether, 1,2-*trans*-glycosides were formed stereospecifically [49]. Thus, when a bifunctional trisaccharide (*55*) with a trityl ether at the site of chain extension and a 1,2-O-cyanoethylidene group at the reducing terminus is reacted under these conditions, the chemical polymerization of the building units is accomplished. The deprotected polysaccharide possessed serological activity appropriate to its polymeric nature. The glycoside linkage established between the galactose and mannose residues by polymerization was demonstrated to be exclusively β.

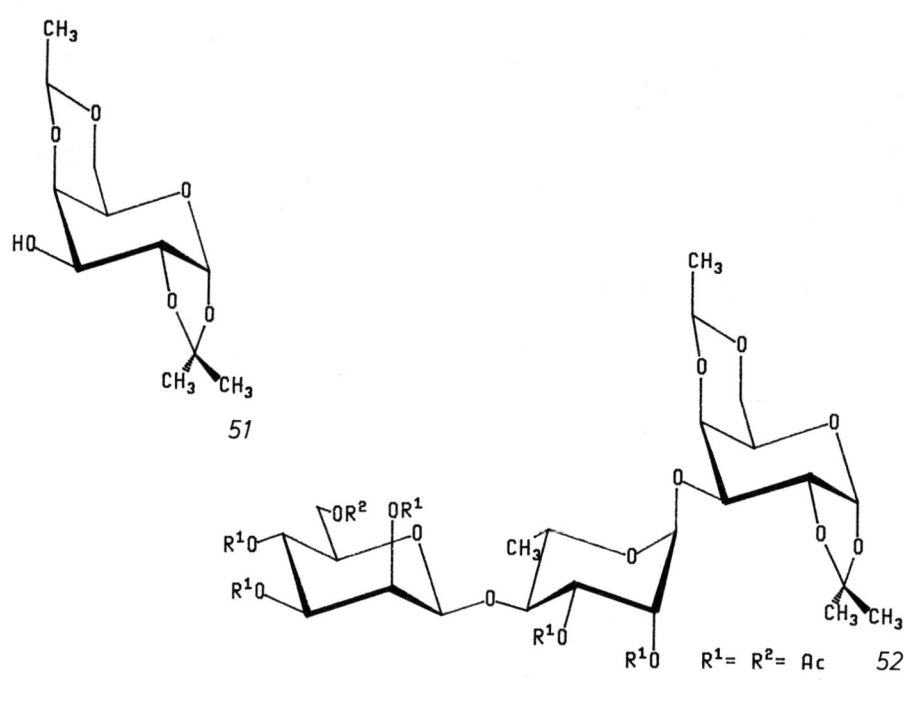

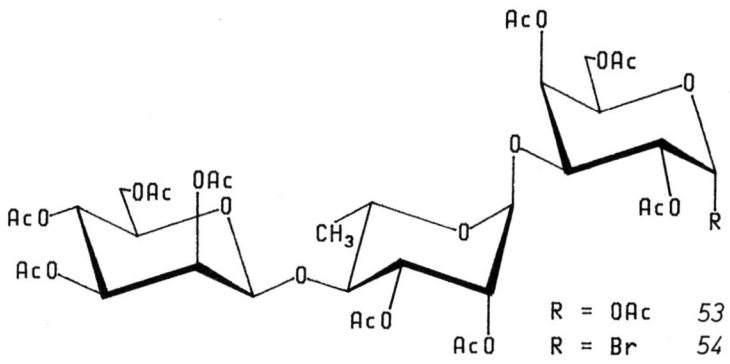

Block syntheses of hexa- and nonasaccharide have been recorded using conventional glycosyl halide block synthesis but based upon chemistry used to synthesize the trisacharide building units [50].

Repeating units corresponding to the O-antigens of *Salmonella muenster* and *Salmonella minneapolis* serogroup E_3 [55] and *Salmonella senftenberg* serogroup E_4 [57] have both utilized the disaccharide unit 56 as the glycosyl donor to, respectively, Glc(1 → 4)Gal 57 or Glc(1 → 6)Gal 58a or 58b disaccharide alcohols. The synthetic schemes utilized the orthoester method to establish the α-L-Rha(1 → 3)-D-Gal linkage of the target branched tetrasaccharides.

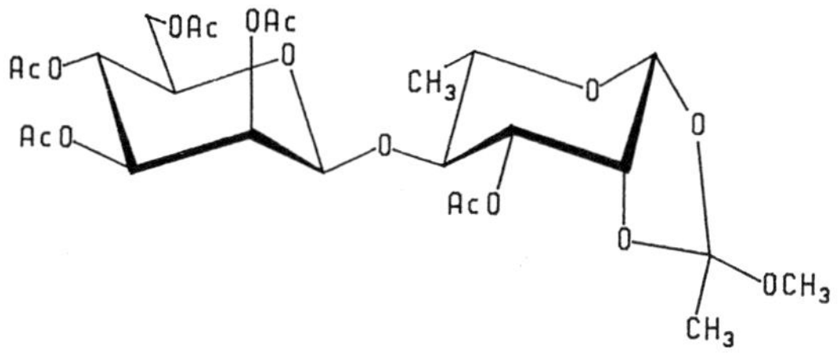

56

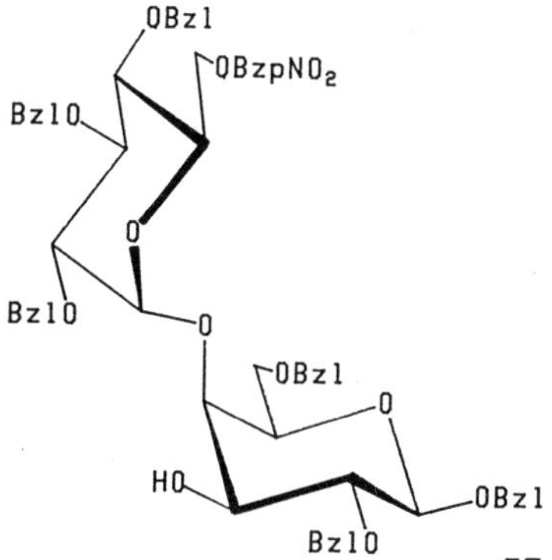

57

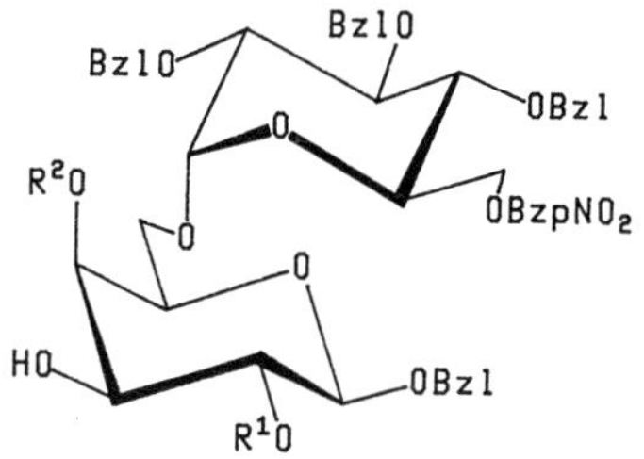

R¹ = R² =Bzl   58a
R¹ = Ac ,R² = H   58b

3 Synthesis of *Shigella* Antigens

A considerable synthetic effort has been expended upon the O-antigens of the *Shigella flexneri* repeating units, especially those of variant Y and X, and the serogroup 5a and 5b structures. A notable synthesis of the *Shigella dysenteriae* structure has also been accomplished.

3.1 *Shigella Flexneri* Variant Y

The simplest structure of the *Shigella flexneri* O-antigens is that of variant Y, which is a linear tetrasaccharide (Table 3). All permutations of frame shifted di- and trisaccharides, together with three of the four possible tetrasaccharide sequences, were synthesized for this structure [74–79]. The synthesis of the tetrasaccharide sequence

Rha→Rha→Rha-GlcNAc that corresponds to the biological repeating unit is shown and incorporates the essential elements of the synthetic approach adopted in the synthesis of the entire series of these structures [78]. The glycosyl donor 2-*O*-acetyl-3,4-di-*O*-benzyl-α-L-rhamnopyranosyl chloride (*59*) provides the structural unit

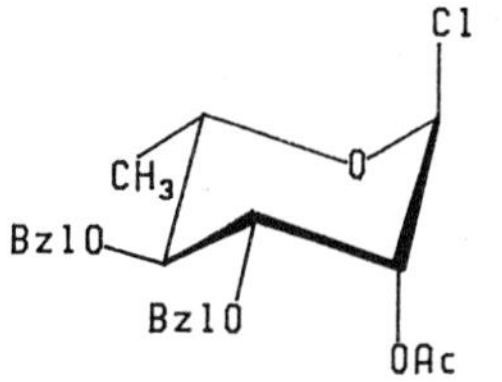

59

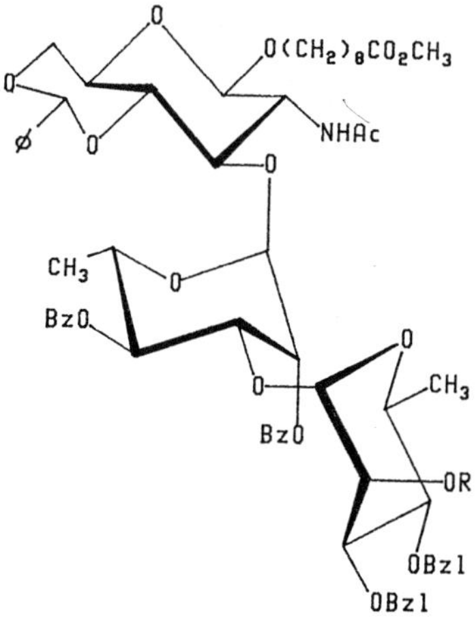

60

R = Ac   *61*
R = H   *62*

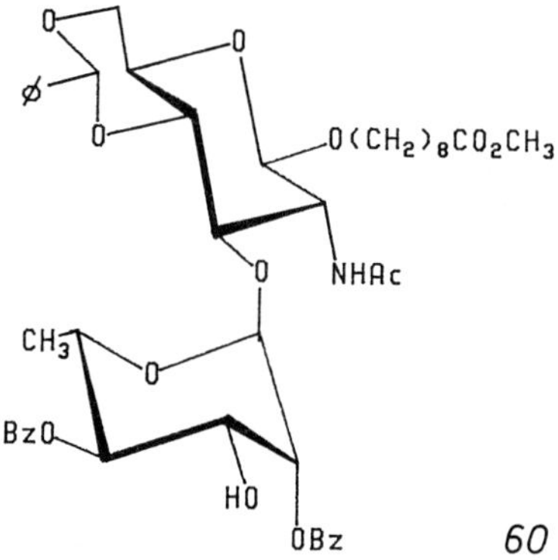

R =

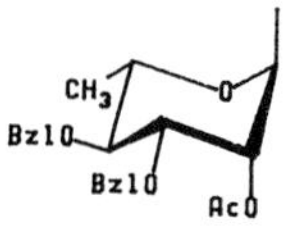

63

which becomes the glycosyl acceptor for chain extension at *O*-2'' following glycosylation to give disaccharide *60*. Transesterification of this 2''-acetate can be achieved selectively in the presence of benzoate groups of trisaccharide *61* [76]. Ultimately, the tetrasaccharide *63* was obtained following glycosylation of *62* by *59*. It was observed during attempts to prepare the disaccharide segment β-D-GlcNAc(1 → 2)α-L-Rha that glycosylation of the *O*-2 position of rhamnose residues by derivatives of D-glucosamine, to give either of the frame-shifted tetrasaccharide sequences Rha-Rha-GlcNAc-Rha [78] or GlcNAc-Rha-Rha [77] was most effectively accomplished by 3,4,6-tri-*O*-acetyl-2-deoxy-2-phthalimido-D-glucopyranosyl bromide [130]. Acetochloroglucosamine [131] and its 1,2-oxazoline [132] derivative, both of which are well-known glycosylating intermediates, were ineffective reagents for chain extension at this site [74].

Sequential introduction of monosaccharide building blocks was employed during the early phase of studies on *Shigella flexneri* variant Y, but more recently effective use has been made of oligosaccharide block synthesis [79, 80, 81, 83]. The key intermediate for the synthesis of penta- to heptasaccharides was identified as the rhamnose trisaccharide *64*, in which the reducing rhamnopyranose residue possessed a nonglycosylated *O*-2 atom, thereby rendering feasible the incorporation of a participating

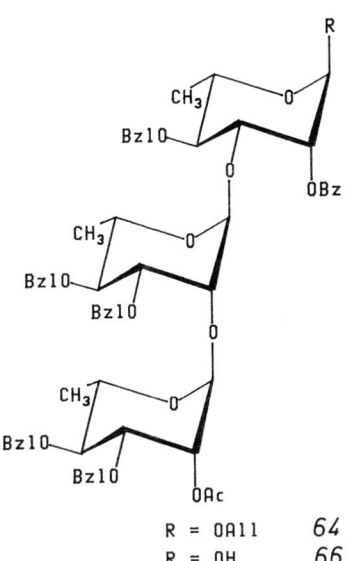

R = OAll   64
R = OH   66
R = Cl   67

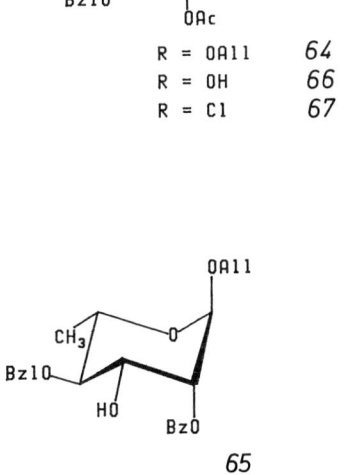

65

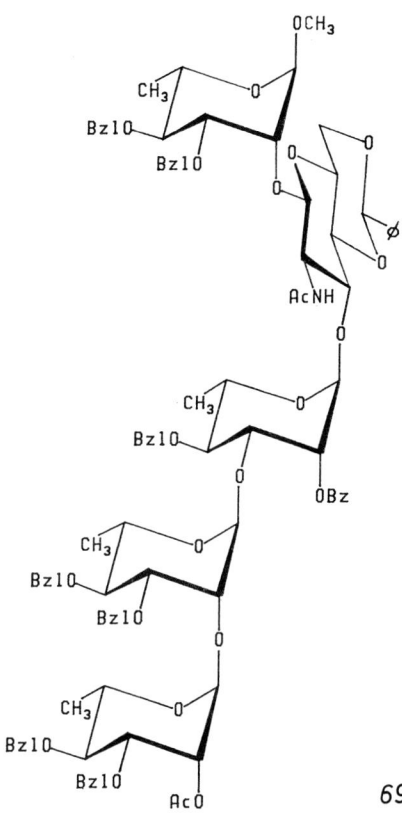

69

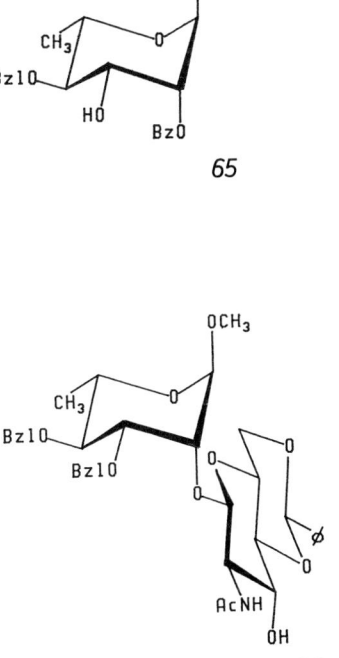

68

23

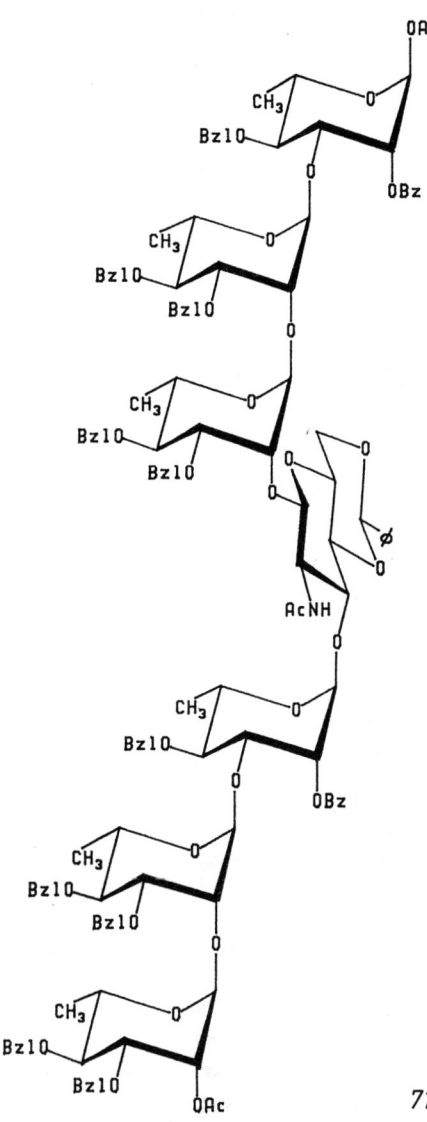

70

71

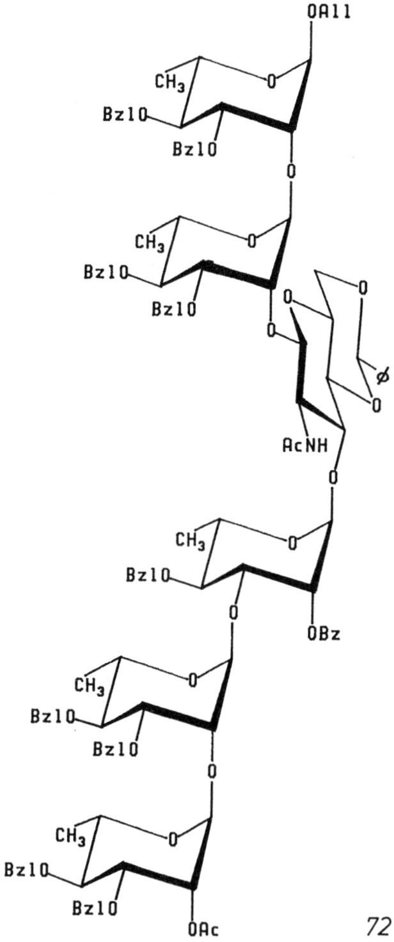

72

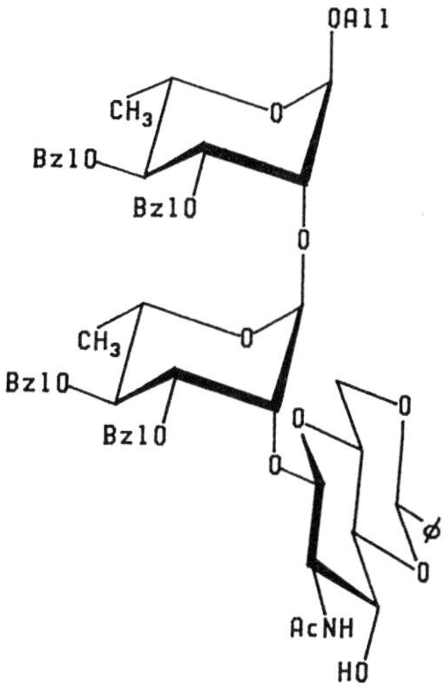

73

ester function, which would provide adequate stereo-control over the glycosylation reaction to ensure the exclusive formation of a 1,2-*trans*-glycosidic bond [82]. The allyl 2-*O*-benzoyl-4-*O*-benzyl-α-L-rhamnopyranoside (65), together with the previously exploited rhamnopyranosyl chloride (59), provided in two glycosylation steps the key trisaccharide 64. Transesterification of the 2'-*O*-acetyl moiety prior to the second chain extension to give 64 used acid catalysis rather than the base-catalyzed reaction used in the earlier studies. As noted by Byramova et al. [133], these conditions are less likely to result in loss of material due to competing transesterification of benzoate esters. The trisaccharide allyl glycoside 64 was reacted to give the reducing trisaccharide 66 which reacted smoothly under rigorously anhydrous conditions with the Vilsmeier-type reagent ($[Me_2^+ N=CHCl]Cl^-$) to provide the glycosyl donor molecule 67. This reacted with disaccharide acceptor 68 under silver triflate promotion to give the pentasaccharide 69 which was deprotected in the usual manner to give the penta-saccharide glycoside that corresponds to the biological repeating unit plus one residue (Rha-Rha-Rha-GlcNAc-Rha).

25

Using this strategy and the key trisaccharide donor molecule *67*, both hexasaccharide *70* and heptasaccharides *71* have been prepared [83]. In the latter case, an allyl glycoside obtained after selective de-*O*-acetylation of *64* is used as a glycosyl acceptor for tri-*O*-acetyl-2-deoxy-2-phthalimido-β-D-glucopyranosyl bromide [130]. The tetrasaccharide obtained was selectively deprotected in a manner which ensured removal of only acetate and phthalimido groups. Following introduction of the 4,6-*O*-benzylidene group the selectively blocked tetrasaccharide *70* was reacted with the trisaccharide glycosyl donor *67* to give the protected heptasaccharide *71*. Synthesis of the hexasaccharide *72* employed a related strategy but used the trisaccharide acceptor *73*.

Employing the cyanoethylidene derivative approach [49] to polymerize synthetic repeating units Byramova et al. [88] have achieved a synthesis of a *S. flexneri* Y polysaccharide containing ten repeating units. Although the polymerized sequence did not correspond to the biological repeat, good immunological activity was reported.

3.2 *Shigella Flexneri* Serogroups 5a, 5b and Variant X

The synthesis of antigenic determinants of the variant X and serogroup 5 lipopolysaccharides exploited a multifunctional rhamnose intermediate *74a* or *74b*, which possessed a persistent blocking group at *O*-4 and distinct non-persistent protecting groups at *O*-2 and *O*-3 [86]. In addition, the methyl glycoside was capable of being converted to a glycosyl halide so that *74* served as a source for both potential glucosyl donor *75* or one of two acceptors *76* or *77*. Thus, the possible combinations of these intermediates with 2-*O*-acetyl-3,4-di-*O*-benzyl-α-L-rhamnopyranosyl chloride *59* and 3,4,6-tri-*O*-acetyl-2-deoxy-2-phthalimido-β-D-glucopyranosyl bromide [130] allowed for branching to be incorporated at either the rhamnose *a* or *b* residue (residues *a* and *b* refer to the first and second rhamnose residues of the biological repeating unit Table 3). In the case of the variant X or serogroup 5a structures, branching by a single α-D-Glc residue occurs at *O*-3 of rhamnose residue *a* and *b*, respectively, while branching glucose residues at both sites provides the 5b antigen (Table 3). It was decided to synthesize the 5b pentasaccharide *78*, which contains all residues and glycosidic linkages involving branching vicinal to the main chain extension sites. Glycosylation of *76* by the corresponding glycosyl chloride *75*, gave the disaccharide *79*. Transesterification of *79* provided the disaccharide alcohol *80*, which was converted to the linear trisaccharide *81* by reaction with 3,4,6-tri-*O*-acetyl-2-deoxy-2-phthalimido-β-D-glucopyranosyl bromide [130]. Exploratory work on the synthesis of the branched X

R = Ac	*74a*	
R = Bz	*74b*	
R = H	*76*	

	75

R = Ac	*77a*
R = Bz	*77b*

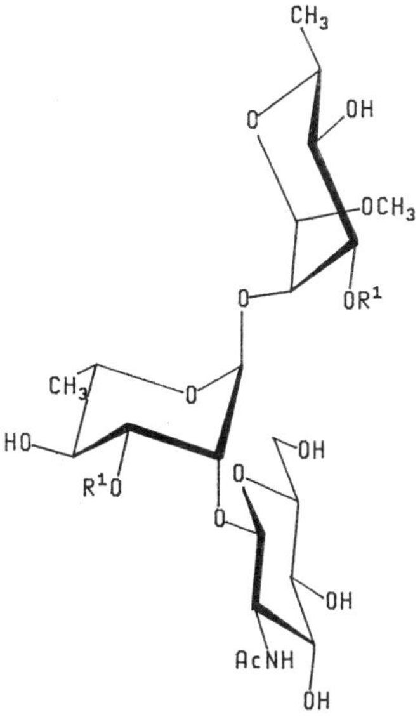

R^1 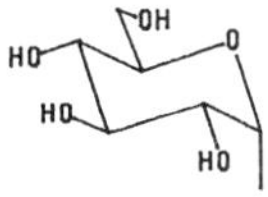 =   **78**

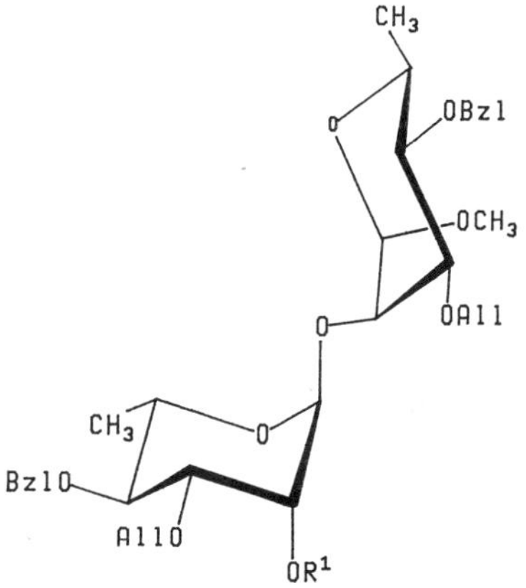

R^1 = Ac   **79**
R^1 = H   **80**

R^1 =   **81**

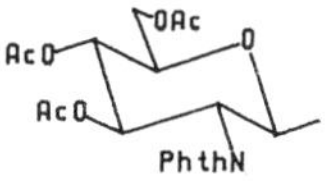

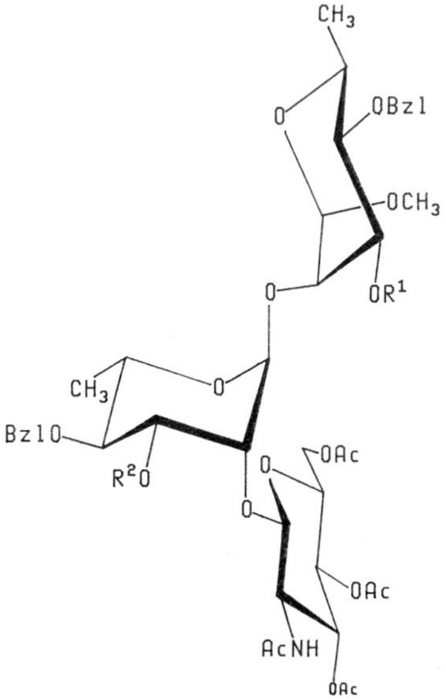

R^1 = R^2 = H   **82**
R^1 = H, R^2 = All   **83**

R^1 = R^2 =   **84**

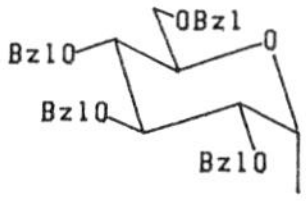

R^1 = 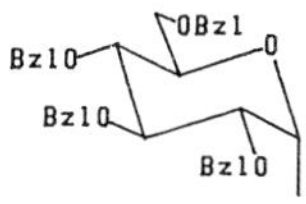 R^2 = All   **85**

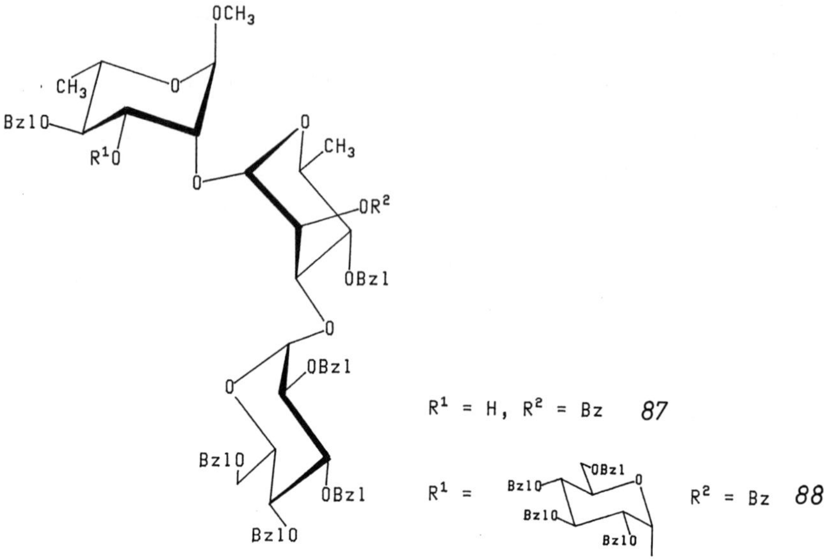

OCH3
CH3
BzlO
R1O
O
CH3
OR2
OBzl
OBzl
BzlO
OBzl
BzlO
R1 = H, R2 = Bz 87
R1 =
BzlO OBzl BzlO BzlO
R2 = Bz 88

OCH3
CH3
BzlO
OBzl
BzlO
BzlO
BzlO
CH3
OR1
OBzl
R2O
R1 = Ac, R2 = H 86
R1 = Ac, R2 =
BzlO OBzl BzlO BzlO
88
R1 = H, R2 =
BzlO OBzl BzlO BzlO
89
R1 =
AcO OAc AcO PhthN
R2 =
BzlO OBzl BzlO BzlO
90

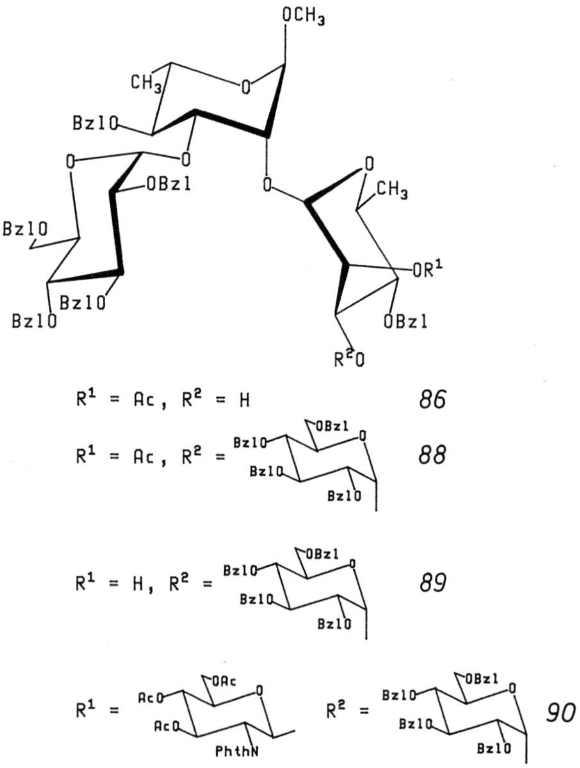

trisaccharide, β-D-GlcNAcp(1 → 2)-[α-D-Glcp(1 → 3)]-α-L-Rhap had established the necessity of converting the phthalimido to an acetamido group for efficient intro-duction of the branching α-D-Glc residue [86]. Therefore, trisaccharide *81* was con-verted to its acetamido derivative prior to removal of the 3- and 3′-allyl groups to give the diol *82*. The mon alcohol *83* was obtained as a side product of the de-allylation reaction. Glycosylation of *82* and *83* by tetra-*O*-benzyl-α-D-glucopyranosyl bromide then gave, respectively, the fully protected derivatives of the 5b pentasaccharide frag-ment *84* and the 5a tetrasaccharide sequence *85*. The relatively inefficient glycosylation of the diol *82* in a stereospecific fashion resulted in poor yields of *84* indicating that a sequential glycosylation of the 3- and 3′-hydroxyl group would be a preferred ap-proached to the doubly branched structure *78*.

An approach of this type was reported independently by Gomtsyan et al. [87]. These authors used either a branched trisaccharide *86* [84] or a linear trisaccharide *87* [85] in reaction with tetra-O-benzyl-α-D-glucopyranosyl bromide to prepare the tetrasaccharide *88*. This was converted to the alcohol *89* and glycosylated by 3,4,6-tri-*O*-acetyl-2-deoxy-2-phthalimido-β-D-glucopyranosyl bromide [130]. In agreement with similar reactions reported by Wessel and Bundle [86], the yield of desired product *90* was very low when silver triflate was the promoter. However, conditions that employed mercuric cyanide and mercuric bromide in acetonitrile gave the penta-saccharide *90* in good yield. In general, both groups attempting the synthesis of the branched X, 5a and 5b structures experienced poor selectivity for α-glucosylation reactions with tetra-*O*-benzyl-α-D-glucopyranosyl bromide as the glycosyl donor [86, 87].

3.3 *Shigella Dysenteriae*

The pentasaccharide repeating unit of *Shigella dysenteriae* serotype 2, a branched pentasaccharide containing exclusively α-glycosidic bonds, was synthesized by Paul-sen and Bünsch [96]. This demanding synthesis was accomplished by a 3 + 2 block synthesis, and the scheme adopted illustrates many interesting themes of current approaches to complex glycoside synthesis. The branched trisaccharide which eventu-ally becomes the glycosyl bromide *91* was constructed sequentially, building up from a 2-azido-2-deoxyl-1,6-anhydrogalactose residue (*92*) and introducing first the α-D-GlcNAc residue, as its azido derivative, at *O*-3 followed by introduction at *O*-4 of a 2-azido-2-deoxy-galactopyranosyl unit (93). Controlled acetolysis was used to open the 1,6-anhydro-ring and the resultant 1-*O*-acetate (94) was reacted with ti-tanium tetrabromide to give *91*. This glycosyl donor was used to prepare a tetra-saccharide by reaction with the monosaccharide *95* or the disaccharide acceptor *96* reaction welt the full repeating unit *97*.

4 Synthesis of Antigenic Determinants of *E. coli*

The colitose-containing repeating unit of the *E. coli* 0111 polysaccharide antigen has been the subject of some synthetic work. The 3,6-dideoxyhexose was synthesized from

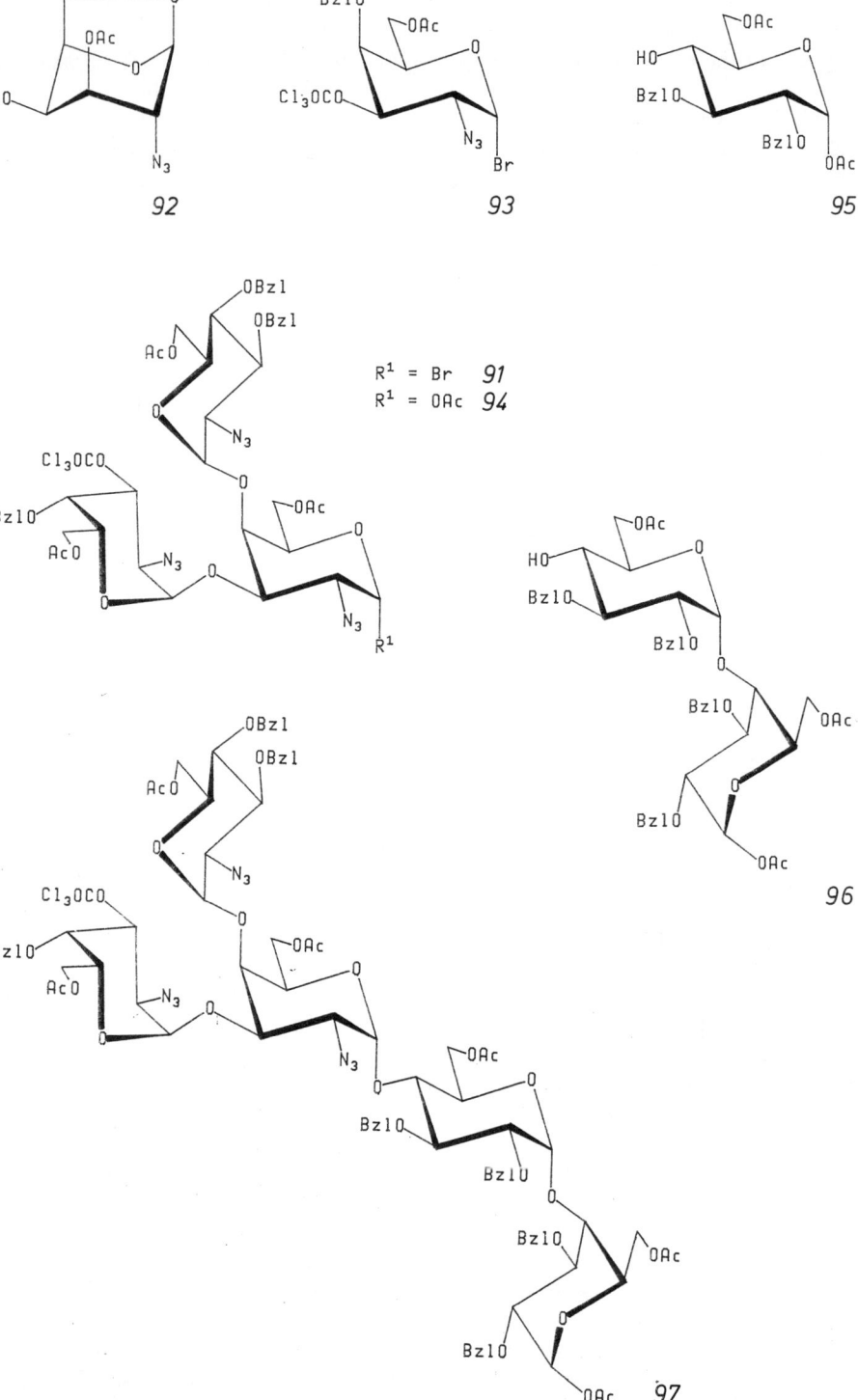

92

93

95

R¹ = Br 91
R¹ = OAc 94

96

97

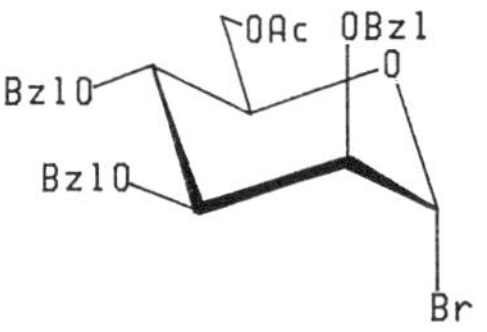

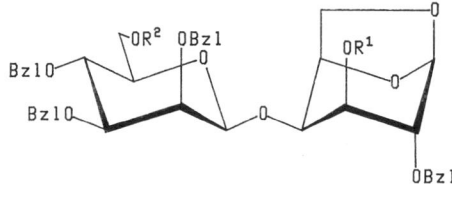

98

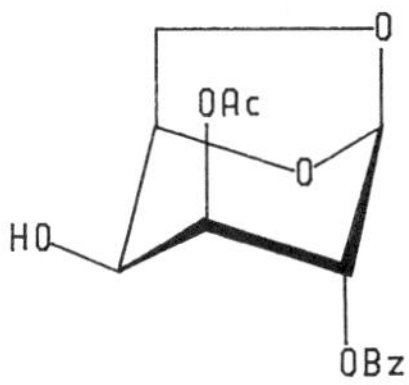

99

R¹ = R² = Ac   **100**
R¹ = R² = H   **101**
R¹ = H, R² = Bz   **102**

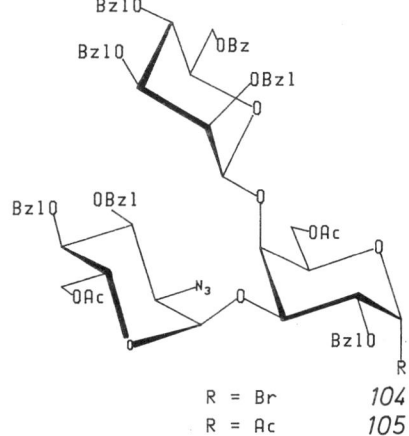

R = Br   **104**
R = Ac   **105**

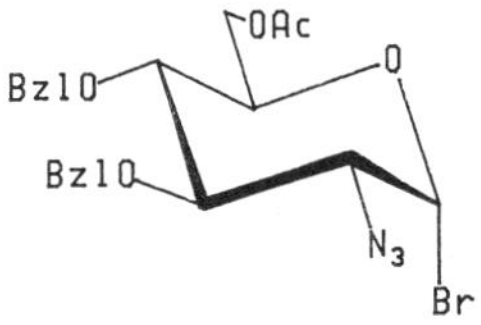

103

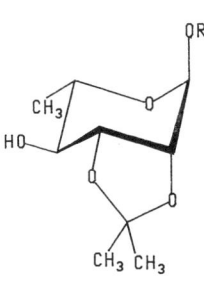

R = $C_6H_5CH_2$ **106**
R = $CH_2(CH_2)_7CO_2CH_3$ **107**

L-fucose [126] and subsequently used to prepare a branched trisaccharide. This trisaccharide α-L-Col(1 → 3)[α-L-Col(1 → 6)]-α-D-Glc forms a crucial element of the pentasaccharide repeating unit (structure: see Ref. [20] p 301) and was synthesized as an 8-methoxycarbonyloctyl glycoside [97].

The branched tetrasaccharide repeating unit of *E. coli* 075 was synthesized as both the reducing oligosaccharide and its 8-methoxycarbonyloctyl glycoside. The synthetic approach was to construct a branched trisaccharide and in a block synthesis glycosylate the forth residue, creating in the process an α-D-Gal(1 → 4)α-L-Rha linkage [98]. The glycosyl donor for the block synthesis, which contained two linages that require special care to construct, namely a β-D-mannopyranosyl and an α-D-glucosaminyl linkage, was prepared by first generating the β-mannose bond. The mannopyrannosyl bromide *98* with a non-participating benzyl group at *O*-2 was coupled to the 1,6-anhydro-galactose derivative *99* in the presence of silver silicate catalyst (silver zeolite has been reported to accomplish similar results [137]). The acetyl groups were removed from the disaccharide *100* and the primary hydroxy group of *101* was selectively benzoylated to give *102*, which was glycosylated by the 2-azido-2-deoxy-glucopyranosyl bromide *103* using a mixture of mercuric cyanide and mercuric bromide. Model studies in support of this synthesis established that the order of reactivity of glycosyl halides increases according to the ring substituents present on the glycosyl halide in the sequence *O*-acetyl < *O*-glucosyl < *O*-benzyl. The order of selectivity toward α-glycoside formation was the reverse. On the other hand, this selectivity could be improved by decreasing the reactivity of the glycosyl halide or by decreasing the nucleophilicity of the hydroxyl groups of the alcohol component, in the case of highly reactive halides [98]. Accordingly, the moderately active catalyst $Hg(CN)_2/HgBr_2$ was used with the trisaccharide glycosyl halide *104* (obtained from *105*) and the selectively protected benzyl rhamnoside *106* or the 8-methoxycarbonyl-octyl rhamnoside equivalent *107*. In this way both reducing and functionalized repeating unit were obtained [98].

5 Synthetic *Brucella* Oligosaccharides

The structures of the classical A and M antigens of *Brucella* were recently shown to be hompolymers of 4,6-dideoxy-4-formamido-α-D-mannopyranose [102, 103]. The A antigen was an α-1,2-linked polymer, while the M antigen was also linear but was composed of a pentasaccharide repeating unit of four α-1,2- and one α-1,3-linked residues.

Synthesis of oligosaccharides with the A structure ranging from di- up to pentasaccharide have been completed. A common intermediate (*108*) was used to prepare either glycosyl donor *109* or glycosyl acceptor *110* molecules. The intermediate *108* was in turn derived from *111*, synthesized from D-mannose in 9 steps [105, 106]. These two intermediates (*110* and *109*) allowed efficient synthesis of a trisaccharide by employing silver triflate as the promotor in sequential glycosylation reactions. The 2′-*O*-acetyl group of the disaccharide being easily removed to provide an alcohol for the second glycosylation step [105, 106].

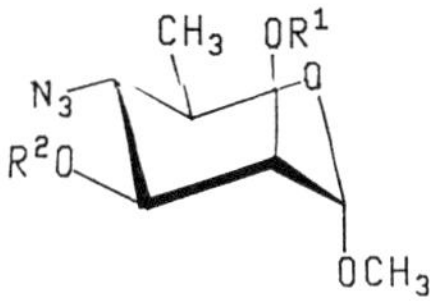

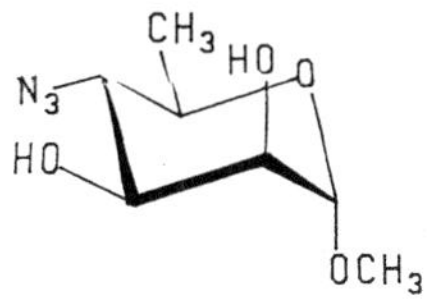

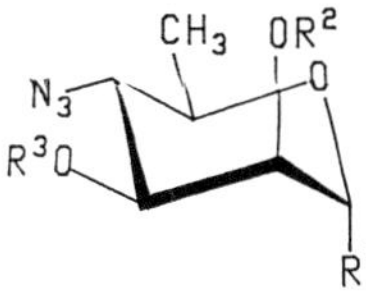

$R^1 = Ac$, $R^2 = Bzl$ *108*
$R^1 = H$, $R^2 = Bzl$ *110*
$R^1 = Ac$, $R^2 = H$ *119*

111

$R^1 = Cl$, $R^2 = Ac$ *109*
$R^1 = SEt$, $R^2 = Ac$ *112*
$R^1 = R^2 = Ac$ *113*
$R^1 = SEt$, $R^2 = H$ *114*
$R^1 = Br$, $R^2 = Ac$ *115*

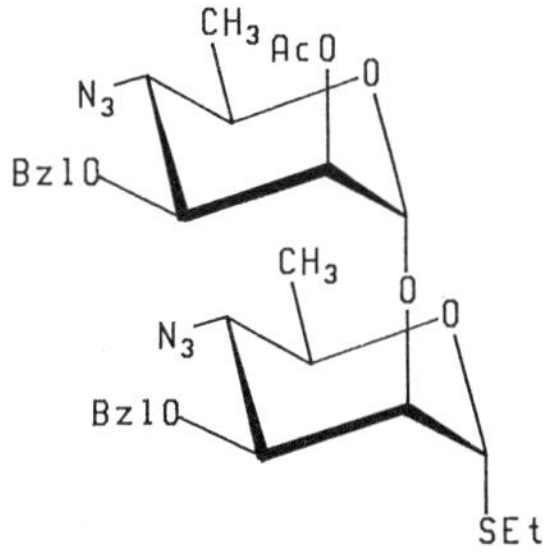

116

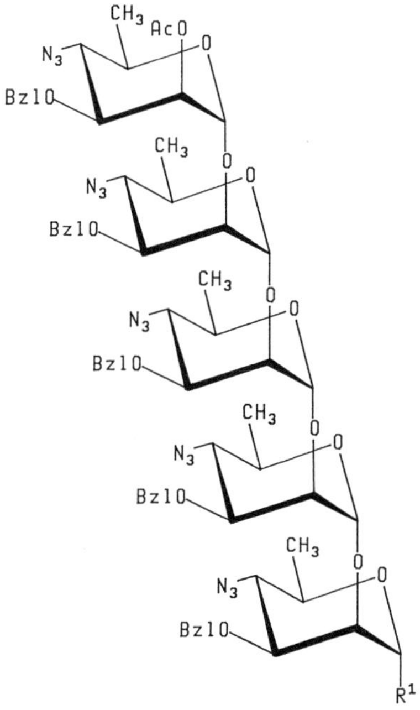

$R^1 = OCH_3$ *117*
$R^1 = O(CH_2)_8CO_2CH_3$ *118*

In order to extend the synthesis to higher oligosaccharides, the S-ethyl-thioglyco-side *112* was prepared from the diacetate *113*. This unit then served as a monomeric building unit or, following transesterification to *114* and silver triflate-mediated glyco-sylation by the bromide *115*, derived from *112* by reaction with bromine, a disaccharide building unit or, following transesterification to *114* and silver triflate mediated glyco-S-ethyl-glycosides and mono- or disaccharide acceptors, odd or even number oligo-mers were constructed, containing 3, 4, and 5 monosaccharide residues. In addition,

33

the aglycone could be varied easily to give either methyl glycoside *117* or 8-methoxyl-carbonyloctyl glycosides *118*. The 1,3-linkage present in the M antigen [103] could also be prepared using a derivative such as *119*. In this way a full range of A- and M-type sequences were prepared [107].

6 Future Prospects

Chemical synthesis can now be used with great effect in the preparation of complex oligosaccharides in the range of tri- up to heptasaccharides. In particular cases, practical syntheses of larger oligomers may be possible, although in general the obstacles to effective syntheses rapidly increases for oligomers larger than a pentasaccharide. Thus, deca saccharides and higher target structures appears at present to depend upon the development of new methodologies.

A radically different conceptual approach to oligosaccharide synthesis will be required if larger oligomers are to be chemically synthesized in an efficient manner. Therefore, it appears that enzymatic synthesis holds the greatest promise for the elaboration of such large oligomers and small polysaccharides. First steps have already been reported in this area as applied to O-polysaccharides [58]. Since the enzyme complex responsible for the construction and polymerization of repeating units is membrane bound, it seems certain that progress in this area will be slow. It is encouraging to note, however, that the genes responsible for the synthesis and assembly of one bacterial O-polysaccharide, that of *Vibrio cholerae* have been cloned and expressed in *E. coli* [135]. Potential access to the genes coding for enzymes that perform O-polysaccharide synthesis is thus within sight. Chemical synthesis of biological repeating units in combination with enzymatic polymerization by bacterial polymerase may be a first step toward taylored, complex polysaccharides produced by enzymes.

7 References

1. Hitchcock, PJ, Leive L, Mäkelä PH, Rietschel ET, Strittmatter W, Morrison DC (1986) J Bacteriol 166: 699
2. Lindberg AA, Wollin R, Bruse G, Ekwall E, Svenson SB (1983) Am Chem Soc Symp Ser 231: 83
3. Luderitz O, Galanos C, Lehmann V, Mayer H, Rietschel ET, Weckesser J (1978) Naturwissenschaften, 65: 578
4. Lemieux R, Hendricks KB, Stick RV, James K (1975) J Am Chem Soc 97: 4056
5. Hanessian S, Banoub J (1976) J Am Chem Soc Symp Ser 39: 36; (1977) Carbohydr Res 53: C13
6. Igarashi K (1977) Adv Carbohydr Chem Biochem 34: 243
7. Lemiueux R (1978) Chem Soc Rev 7: 423
8. Paulsen H (1982) Angew Chem Int Ed Engl 21: 155
9. Schmidt RR (1986) Angew Chem Int Ed Engl 25: 212
10. Jann K, Westphal O (1975) in: Sela M (ed) The Antigens, vol 5, Academic Press, p 1
11. Luderitz O, Staub AM, Westphal O (1966) Bact Rev 30: 192
12. Osborn MJ (1969) Ann Rev Biochem 38: 501
13. Robbins PW, Wright A (1971) in: Weinbaun G, Kadis S, Ajl SJ (eds) Microbial toxins, Vol IV: Bacterial endotoxins Academic Press New York p 351

14. Hellerqvist CG, Lindberg B, Samuelsson K, Lindberg AA (1971) Acta Chem Scand 25: 955
15. Hellerqvist CG, Lindberg B, Svensson S, Holme T, Lindberg AA (1968) Carbohydr Res 8: 43 (1969) Carbohydr Res 9: 237
16. Hellerqvist CG, Larm O, Lindberg B, Holme T, Lindberg AA (1969) Acta Chem Scand 23: 2217
17. Nikaido H (1968) Advan Enzymol 31: 77
18. Hellerqvist CG, Larm O, Lindberg B, Lindberg AA (1971) Acta Chem Scand 25: 744
19. Hellerqvist CG, Hoffman J, Lindberg B, Pilotti Lindberg AA (1971) Acta Chem Scand 25: 1512
20. Kenne L, Lindberg B (1983) Aspinall GO (ed) Bacterial polysaccharides, in: The polysaccharides, vol 2 Academic Press, New York p 287
21. Hellerqvist CG, Lindberg B, Svensson S, Holme T, Lindberg AA (1969) Acta Chem Scand 23: 1588
22. Nghiem HO, Bagdian G, Staub AM (1967) Eur J Biochem 2: 392
23. Hellerqvist CG, Lindberg B, Pilotti A (1970) Acta Chem Scand 24: 1168
24. Boren HB, Garegg PJ, Wallin NH (1972) Acta Chem Scand 26: 1082
25. Garegg PJ, Wallin NH (1972) Acta Chem Scand 26: 3892
26. Alfredsson G, Garegg PJ (1973) Acta Chem Scand 27: 556
27. Garegg PJ, Gotthammer B (1977) Carbohydr Res 58: 345
28. Ekborg G, Garegg PJ, Josephson S (1978) Carbohydr Res 65: 301
29. Ekborg G, Garegg PJ, Gotthammar B (1975) Acta Chem Scand Ser B29: 765
30. Eklind K, Garegg PJ, Gotthammar B (1976) Acta Chem Scand Ser B30: 300
31. Eklind K, Garegg PJ, Gotthammar B (1976) Acta Chem Scand Ser B30: 305
32. Garegg PJ, Hultberg H, Norberg T (1981) Carbohydr Res 96: 59
33. Bundle DR, Iversen T (1982) Carbohydr Res 103: 29
34. Garegg PJ, Norberg T (1982) J Chem Soc Perkins Trans I: 2973
35. Pinto MB, Bundle DR (1984) Carbohydr Res 133: 333
36. Bock K, Meldal M (1983) Acta Chem Scand Ser B37: 629
37. Bock K, Meldal M (1984) Acta Chem Scand Ser B38: 255
38. Bock K, Meldal M (1984) Acta Chem Scand Ser B38: 71
39. Kochetkov NK, Torgov VI, Malysheva NN, Shashkov AS (1980) Tetrahedron 36: 1099
40. Stellner K, Westphal O, Meyer H (1970) Liebigs Ann Chem 738: 179
41. Robbins PW, Uchida T (1962) Biochemistry 1: 323
42. Robbins PW, Uchida T (1965) J Biol Chem 1: 375
43. Hellerqvist CG, Lindberg B, Lönngren J, Lindberg AA (1971) Carbohydr Res 16: 289
44. Hellerqvist CG, Lindberg B, Lönngren J, Lindberg AA (1971) Acta Chem Scand 25: 939
45. Uchida T, Robbins PW, Luria SE (1963) Biochemistry 2: 663
46. Staub AM, Girard (1965) Bull Soc Chem Biol 47
47. Hellerqvist CG, Lindberg B, Pilotti A, Lindberg AA (1971) Carbohydr Res 16: 297
48. Betaneli VI, Ovchinnikov MV, Backinowska LV, Kochetkov NK (1980) Dokl Akad Nauk SSSR 251: 108; (1980) Chem Abstr 93: 204942P
49. Kochetkov NK, Betaneli VI, Ovchinnikov MV, Backinowsky LV (1981) Tetrahedron 37 Suppl 1: 149
50. Dmitriev BA, Nikolaev AV, Shashkov AS, Kochetkov NK (1982) Carbohydr Res 100: 195
51. Kochetkov NK, Dmitriev BA, Malysheva NN, Chernyak AYa, Klinov EM, Bayramova NE, Torgov VI (1975) Carbohydr Res 45: 283
52. Betanelli VI, Ovchinnikov MV, Bachinowsky LV, Kochetkov NK (1980) Carbohydr Res 84: 211
53. Kochetkov NK, Dmitriev BA, Nikolaev AV (1977) Izv Akad Nauk SSSR Ser Khim 2578; (1978) Chem Abstr 88: 152884w
54. Kochetkov NK, Dmitriev BA, Nikolaev AV, Bairamova ME (1977) Izv Akad Nauk SSSR Ser Khim 1609; (1977) Chem Abstr 87: 201939h
55. Kochetkov NK, Torgov VI, Malysheva NN, Shashikov AS, Klimov EM (1980) Tetrahedron 36: 1227
56. Kochetkov NK, Malysheva NN, Torgov VI, Klimov EM (1977) Izv Akad Nauk SSSR Ser Khi 654; (1977) Chem Abstr 87: 53506x
57. Kochetkov NK, Malysheva NN, Torgov VI, Klimov EM (1977) Carbohydr Res 54: 269

David R. Bundle

58. Kochetkov NK, Shibaev VN, Druzhinina TN, Gogilashvili LM, Danilov LL, Torgov VI, Mattsev SD, Utkina NS (1982) Dokl Akad Nauk SSSR 262: 1393; (1982) Chem Abstr 97: 106402a
59. Ekborg G, Lönngren J, Svensson S (1975) Acta Chem Scand Ser B29: 1031
60. Kochetkov NK, Klimov EM, Torgov VI (1976) Izv Akad Nauk SSSR Ser Khim 165; (1976) Chem Abstr 85: 332965
61. Kochetkov NK, Dmitriev BA, Chernyak AYa, Pokrovskii VI, Tendetnik YY (1979) Bioorg Khim 5: 217; (1979) Chem Abstr 90: 187248k
62. Kochetkov MK, Dmitriev BA, Chernyak AYa, Levinsky AB (1982) Carbohydr Res 110, C16
63. Garegg PJ, Hultberg H, Lindberg C (1980) Carbohydr Res 83: 157
64. Hultberg H, Garegg PJ (1979) Carbohydr Res 72: 276
65. Kenne L, Lindberg B, Petersson, K, Romanowska E (1977) Carbohydr Res 56: 363
66. Kenne L, Lindberg B, Petersson K, Katzenellenbogen E, Romanowska E (1977) Eur J Biochem 76: 327
67. Kenne L, Lindberg B, Petersson K, Katzenellenbogen E, Romanowska E (1978) Eur J Biochem 91: 279
68. Carlin NIA, Lindberg AA, Bock K, Bundle DR (1984) Eur J Biochem 139: 189
69. Kauffmann F (1966) The bacteriology of enterobacteriaceae, 2nd ed, Munksgaard, Copenhagen
70. Edwards PR, Ewing WH (1972) Identification of enterobacteriaceae, Burgess Publ Co, Minneapolis
71. Carlin NIA, Lindberg AA (1986) Infect Immun 53: 103; (1987) 55: 1412
72. Carlin NIA, Gidney MAJ, Lindberg AA, Bundle DR (1986) J Immunol 137: 1377
73. Carlin NIA, Bundle DR, Lindberg AA (1987) Immunol 138: 4419
74. Bundle DR, Josephson S (1979) Can J Chem 57: 662
75. Bundle DR, Josephson S (1979) J Chem Soc Perkin Trans 1: 2736
76. Josephson S, Bundle DR (1980) J Chem Soc Perkin Trans 1: 297
77. Josephson S, Bundle DR (1979) Can J Chem 57: 3073
78. Bundle DR, Josephson S (1980) Carbohydr Res 80: 75
79. Wessel HP, Bundle DR (1983) Carbohydr Res 124: 301
80. Bundle DR, Gidney MAJ, Josephson S, Wessel HP (1983) Am Chem Soc Symp Ser 231: 49
81. Forsgren M, Norberg T (1983) Carbohydr Res 116: 39
82. Pinto BM, Morissette DG, Bundle DR (1987) J Chem Soc Perkin Trans 1: 9
83. Pinto BM, Reiner KB, Morisette DG, Bundle DR (1989) J. Org. Chem. 54: 2650
84. Backinowsky LV, Gromtsyan AR, Byramova NE, Kochetkov NK (1984) Bioorg Khim 10: 79; Chem Abstr 100: 210297m
85. Backinowsky LV, Gromtsyan AR, Byramova NE, Kochetkov (1985) Bioorg Khim 11: 254; (1985) Chem Abstr 103: 105237f
86. Wessel HP, Bundle DR (1985) J Chem Soc Perkin Trans 1: 2251
87. Gomtsyan AR, Byramova NE, Backinowsky LV, Kochetkov NK (1985) Carbohydr Res 138: C1
88. Byramova NE, Tsvetkov YE, Backinowsky LV, Kochetkov NK (1985) Carbohydr Res 137: C8
89. Bundle DR in: Lark DL (ed) Protein-carbohydrate interactions in biological systems Academic Press, London, p 165
90. Norberg T, Oscarsson S, Szönyi M (1986) Carbohydr Res 156: 214
91. Bundle DR, Hanna R, Wessel HP (manuscript in preparation) Canad J Chem
92. Katzenellenbogen E, Mulczyk M, Romanowska E (1976) Eur J Biochem 61: 191
93. Dmitriev BA, Knirel YA, Sheremet OK, Shashkov AA, Kochetkov MK, Hofman IL (1979) Eur J Biochem 98: 309
94. Paulsen H, Kutschker W (1983) Liebigs Ann Chem 557; (1983) Carbohydr Res 120: 25
95. Garegg PJ, Norberg T, Konradsson F, Svensson SCT (1983) Carbohydr Res 116: 308
96. Paulsen H, Bünsch H (1981) Tetrahedron Lett 22: 47; (1981) Chem Ber 114: 3126
97. Iversen T, Bundle DR (1982) Can J Chem 60: 299
98. Paulsen H, Lockhoff O (1981) Chem Ber 114: 3079; (1981) Chem Ber 114: 3102; (1981) Chem Ber 114: 3115
99. Ogawa T, Yamamoto H (1985) Carbohydr Res 137: 79
100. Pozsgay V, Nanasi P, Neszmilyi A (1981) Carbohydr Res 90: 215
101. Paulsen H, Lorentzen JP (1986) Carbohydr Res 150: 63

102. Caroff M, Bundle DR, Perry MB, Cherwonogrodzky JW, Duncan JR (1984) Infect Immun 46: 384
103. Bundle DR, Cherwonogrodzky JW, Perry MB (1987) Biochemistry 26: 8717
104. Kenne L, Lindberg B, Unger P, Holme T, Holmgren J (1979) Carbohydr Res 68: C14
105. Bundle DR, Gerken M (1987) Tetrahedron Lett 28: 5067
106. Bundle DR, Gerken M, Peters T (1988) Carbohydr Res 174: 239
107. Peters T, Bundle DR (1987) J Chem Soc Chem Commun 1648
108. Shaw DM, Lee YZ, Squires MJ, Luderitz O (1983) Eur J Biochem 131: 633
109. Horton D, Riley DA, Samreth S, Scheveitzer MG (1983) Am Chem Soc Sym Ser 231: 21
110. Paulsen M, Lorentzen JP (1985) Tetrahedron Lett 26: 6043
111. Horton D, Samreth S (1982) Carbohydr Res 103: C12
112. Fouquey C, Lederer E, Luderitz O, Polonsky J, Staub AM, Stim S, Tinelli R, Westphal O (1958) CR Acad Sci Paris 246: 2417
113. Fouquey C, Polonsky J, Lederer E, Westphal O, Luderitz O (1958) Nature (London) 182: 944
114. Westphal O, Stirm S (1959) Liebigs Ann Chem 620: 8
115. Bock K, Lundt I, Pedersen C (1981) Acta Chem Scand B35: 155
116. Bock K, Lundt I, Pedersen C (1979) Carbohydr Res 68: 313
117. Copeland C, Stick RV (1977) Aust J Chem 30: 1269
118. Baer HH, Astles DJ (1984) Carbohydr Res 126: 343
119. Bundle DR (1979) J Chem Soc Perkins I: 2751
120. Williams EH, Szarek WA, Jones JKN (1971) Can J Chem 49: 798
121. Svensson S (1968) Acta Chem Scand 22: 2737
122. Eklind K, Garegg PJ, Gotthammar B (1975) Acta Chem Scand B29: 633
123. Classon B, Garegg PJ, Samuelsson (1981) Can J Chem 59: 339
124. Rembarz G (1960) Chem Ber 93: 622
125. Hanessian S, Plessas NR (1969) J Org Chem 34: 1035
126. Bundle DR, Josephson S (1978) Can J Chem 56: 2686
127. Ferrier RJ, Prasad N, Sankey GH (1969) J Chem Soc C: 587
128. Boren HB, Ekborg G, Eklind K, Garegg PJ, Pilotti A, Swahn CG (1973) Acta Chem Scand 27: 2639
129. Gorin PAJ, Perlin AS (1961) Can J Chem 39: 2474
130. Lemieux RU, Takeda T, Chung BY (1976) Am Chem Soc Symp Ser 39: 90
131. Horton D (1966) Org Syn 46: 1
132. Khorlin AY, Shul'man ML, Zurabyan SE, Privalova IM, Kopaevich YL (1968) Izv Akad Nauk SSSR Ser Khim 2094; (1969) Chem Abstr 70: 58169y; Zurabyan SE, Antonenko TS, Khorlin AY (1971) Carbohydr Res 15: 21
133. Byramova NE, Ovchinnikov MV, Backinowsky LV, Kochetkov NK (1983) Carbohydr Res 124: C8
134. Lonn H (1985) Carbohydr Res 135: 105
135. Manning PA, Heuzenroeder MW, Yeadon J, Leavesley DI, Reeves PR, Rowley D (1986) Infect Immun 53: 272
136. Antonakis K (1969) Bull Soc Chim Fr 122
137. Garegg PJ, Ossowski P (1983) Acta Chem Scand B37: 249

Synthetic Saccharide Photochemistry

Gérard Descotes

University Lyon I Laboratoire de Chimie Organique 2 Unité Associée CNRS 463
69622 Villeurbanne, France

Table of Contents

Topics in Current Chemistry, Vol. 154
© Springer-Verlag Berlin Heidelberg 1990

Synthetic photochemical applications to carbohydrates are described by photosubstitution, photo-addition, photoizomerisation, photoreduction, and photoxidation reactions. The use of photoremovable protecting groups in osidic syntheses is also evoked.

The recent development of such radical reactions using photoinitiators is mainly due to their regio and stereoselectivities interpreted in terms of stabilized conformations of carbohydrate radicals by stereoelectronic factors. The preferential α-attack of anomeric radicals is a well documented example of such transformations.

Some new methodologies using single electron transfer reactions are also indicated in this review.

1 Introduction

1.1 Introductory Remarks

In recent years, the introduction of modern instrumentation has changed the orientation of research in the photochemistry of carbohydrates. After the first studies of the physical properties of irradiated reaction mixtures, identification of photoprotation of research in the photochemistry of carbohydrates. After the first studies of the physical properties of irradiated reaction mixtures, identification of photoproducts was then undertaken and finally, an increasing use of free radical reactions including photochemical ones was applied to the synthesis of complex organic nds.

This use of free radical reactions offers an alternative to ionic reactions which are well known in carbohydrate chemistry. The neutral conditions and mild work-up of these phototransformations are often of great interest in the synthesis of multifunctional molecules.

The knowledge of photoreaction mechanisms and of the structures of intermediate radicals is also of particular interest in the understanding of many structural modifications of sugars. For instance, radicals which are not solvated and less susceptible to steric factors adopt specific conformations determined by stereoelectronic effects which can explain some regio and stereoselectivities of many photochemical reactions in carbohydrates.

The purpose of this article is to briefly review those synthetic photochemical reactions with mono- and oligosaccharides used as specific substrates or as chiral synthons in synthesis. This article follows the excellent general and complete review published in 1981 by R. W. Binkely [1] in the "Advances in Carbohydrate Chemistry and Biochemistry".

The adopted plan follows the different types of photochemical reactions [substitutions, additions, rearrangements, reduction, oxidation, photolabile protections] with the presentation of some typical synthetic examples and of some mechanistic aspects.

1.2 Abbreviations

Ac	acetyl
AIBN	azobisisobutyronitrile
Ade	adenine
Bn	benzyl
Bu	butyl
Bz	benzoyl
DCNB	*p*-dicyanobenzene
DME	dimethoxyethane
DMF	*N,N*-dimethylformamide
DMSO	dimethylsulfoxide
Et	ethyl
HMPA	hexamethylphosphoric triamide
LAH	lithium aluminum hydride

LDA	lithium diisopropylamide
Me	methyl
Mol	molecular
Ms	methanesulfonyl
NBS	*N*-bromosuccinimide
P	phenanthrene
PCC	pyridinium chlorochromate
Ph	phenyl
Piv	pivaloyl
Tf	triflyl
TFA	trifluoroacetic acid
THF	tetrahydrofuran
TMS	trimethylsilyl
Tr	trityl
Ts	*p*-toluenesulfonyl
Ura	uracile

2 Photosubstitutions

Photochemical substitutions of carbohydrates can be differentiated by describing the carbon atom which is transformed in the reaction. This deals with reactions where the photochemical event is the breaking and the creation of a bond in a carbohydrate moiety:
— at the anomeric or proanomeric position
— at non-anomeric positions.

2.1 Photosubstitutions at the Anomeric or Proanomeric Carbon

2.1.1 Anomeric Hydrogen Photoabstraction

The anomeric hydrogen photoabstraction leading to substitution products is generally obtained by the creation of new bonds such as:
— carbon-halogen bonds
— carbon-oxygen bonds

2.1.1.1 Carbon-Halogen Bond Formation

The photobromination with *N*-bromosuccinimide [NBS) of aldopyranosyl and furanosyl derivatives can efficiently take place at non-anomeric positions to give 5- or 4-bromo derivatives, respectively (Sect. 2.2.1.1). However, other β-substituents on the anomeric carbon such as halogens, nitrile and other electron withdrawing groups can change the regioselectivity of this substitution to yield only 1-α-bromo 1-deoxy sugars.

This *regioselectivity* results from the "*capto dative effect*" of intermediate radicals which are stabilized by the presence of one electron-donor and one electron-acceptor substituents [2]. The *stereoselectivity* of α-bromosubstitution is in agreement with the theory of *"kinetic anomeric effect"* for the preferential axial hydrogen abstraction at the anomeric center of heterocyclic acetals and carbohydrates [3–7]. This preferential cleavage can be explained by the presence of an antiperiplanar relationship between C–H bond and a non-bonding electron pair on the ring oxygen [3–7].

These new compounds such as *2* [8, 9] derived from *1* are of great interest for synthetic purposes because of the double functionality of the *"pseudoanomeric"* center (Scheme 1).

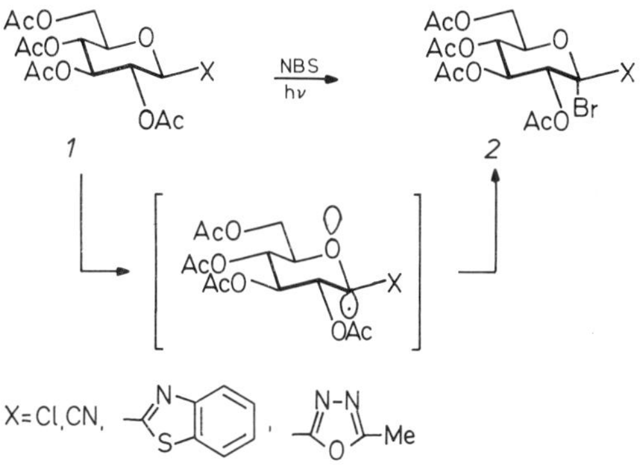

X=Cl,CN,

Scheme 1

In the case of 1-halogeno 1-deoxy sugars, the competition of photobromination at C-5 or C-1 depends on the halogen and its configuration. The regioselectivity of the reaction at C-1 decreases from anomeric chlorides to fluorides, and the α-derivatives are less reactive or inert for the anomeric substitution.

"Proanomeric" carbons substituted by similar withdrawing endocyclic groups are photobrominated with higher regio and stereoselectivities. Thus, the presence at C-2 of oxo or oximino functions increases the radical reactivity at the C-1 center. High yields of α-bromo substituted carbohydrates *4* [10, 11] are obtained from *3*. This photobromination process is useful for the synthesis of di- and trisaccharides with central β-D-mannose, α-D-glucosamine and β-D-mannosamine as basic units [12–13] such *5* (Scheme 2). The favourable position of the carbonyl group is very influential since similar photobrominations of 4-uloses result in substitution at the C-5 position confirming the importance of captodative effects in the carbohydrate moiety.

The anomeric bromination can result from *photosubstitution of other positions* in the ring moiety. Very recenly, bromination of fully protected 1-2-*O*-benzylidenated pyranoses with bromotrichloromethane and U.V. light or NBS yields 2-*O*-benzoyl glycosylbromides which may be converted in situ to glycosides and disaccharides in good yields (Scheme 2). These reaction conditions are compatible with protective groups like esters (acetates, benzoates, tosylates) and silyl ethers.

2.1.1.2 Carbon-Oxygen Bond Formation

The anomeric hydrogen abstraction can be obtained by an intramolecular "Barton type" reaction using hydroxyalkylglycosides in presence of iodine and mercury oxide under irradiation. This methodology is used for the synthesis of anomeric spiroorthoesters [*6* and *7* in scheme 3).

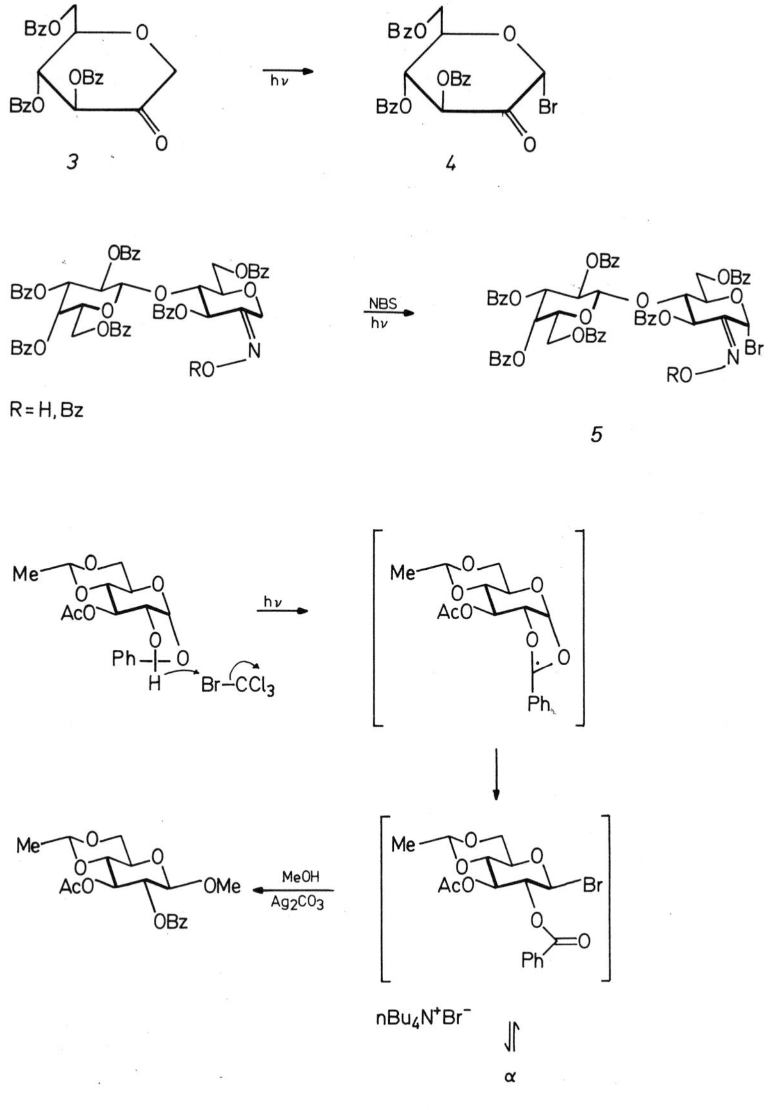

Scheme 2

This photocyclization proceeds in a "one pot" reaction with photolytic cleavage of the intermediate carbone-mercury bond and subsequent hydrogen abstraction through a probable hypoiodite intermediate 6 [15a]. With secondary alcohols (R = H), the reaction is less efficient than with primary ones (R′ = H), and mainly undergoes a retention of configuration at the anomeric carbon [15b] by abstraction of the α-anomeric hydrogen.

Both reactions are of interest for the synthesis of analogs of a new class of antibiotics called "*ortho*somycins" [16].

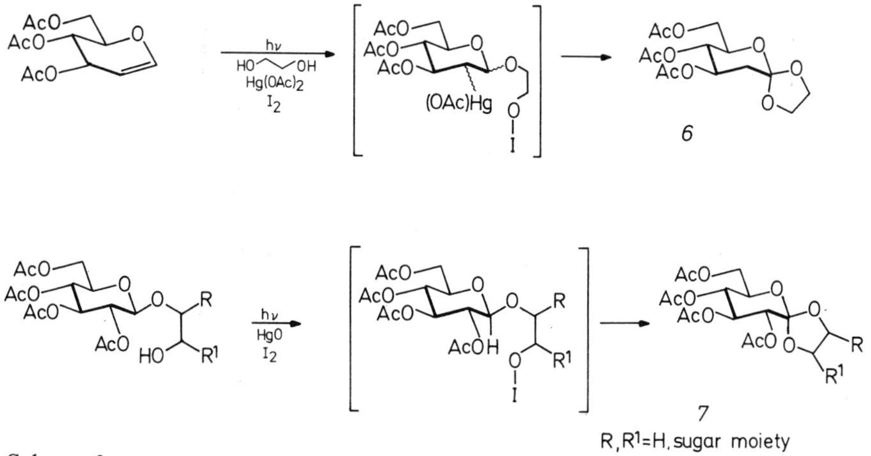

R,R¹=H, sugar moiety

R,R^1 = H, sugar moiety

Scheme 3

2.1.2 Anomeric Carbon-Heteroatom Cleavage

The main photocleavages of anomeric carbon-heteroatoms bonds are classified as:
— carbon-halogen photolysis
— carbon-oxygen photolysis

2.1.2.1 Carbon-Halogen Cleavage

The photochemistry of 1-deoxy 1-halogenosugars has been developed in recent years to create new bonds at the anomeric carbon such as:
— carbon-hydrogen or deuterium bond
— carbon-carbon bond

Anomeric Halosugars.

Metal hydrides reduce 1-α-anomeric halosugars into 1-β-deoxy sugars with inversion of configuration by a SN_2 mechanism. On the contrary, the use of Bu_3SnD under photochemical conditions yields deoxy sugars with retention of configuration with a stereoselectivity of 90% according to Scheme 4 [17a].

If both α and β-halogenosugars *8a* and *8c* are reduced to the same anomeric mixture with the predominant deuterio compound *10*, an easier cleavage of the axial anomeric carbon-bromine bond is observed with shorter times of photoreactions.

In comparison, the stereoselectivity of the radical reduction slightly decreases if the anomeric center is substituted by an electron withdrawing substituent (CN) using AIBN as radical initiator [17b]. The corresponding radical is supposed to be more planar than pyramidal to explain the lower degree of stereoselectivity for the reduction process of *8b* into *9*.

C-Glycosylation

The preceding photoreduction of 1-halosugars with tin hydride under photolytic conditions can be carried out in the presence of electron-poor alkenes to form a

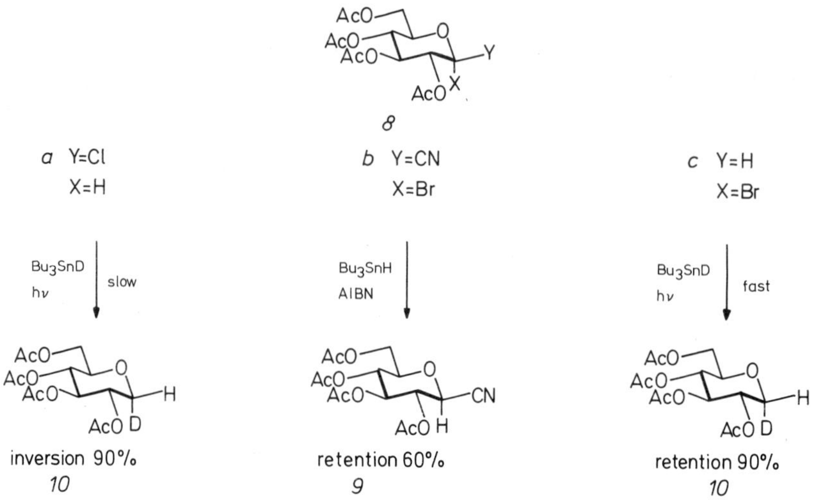

Scheme 4

carbon-carbon bond coupling according to Scheme 5. This photosubstitution occurs with α-bromo or β-phenylselenosugars *11* and *12* to give *13* with a similar stereoselectivity in favour of the axial arrangement of the *C*-glycosidic bond [18, 19]. These stereoselectivities are of interest for the formation of axial C–C bond which is difficult to create at the anomeric center of sugars. This radical reaction has been extensively applied for the carbofunctionalization of carbohydrates and an "Organic Synthesis" procedure has recently appeared for the preparation of *C*-glycosides [18 b].

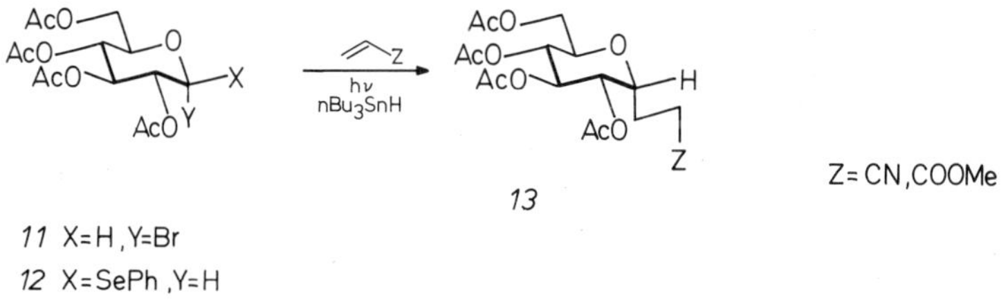

Scheme 5

According to ESR measurements, the intermediate glycosyl radical exists in the boat-like conformation [20], which was previously observed in some photoproducts of spirocyclization (Sect 2.2). This preferred conformation of the pyranosyl radical results from the stabilizing interaction of the single occupied *p* orbital with the σ*-Lumo of the adjacent β-C-OR bond (Scheme 6).

Using this methodology, α-*C*-dissacharides [21] and α-*C*-mannosides such *14* are synthesized by radical intermediates which can adopt chair conformations (Scheme 7).

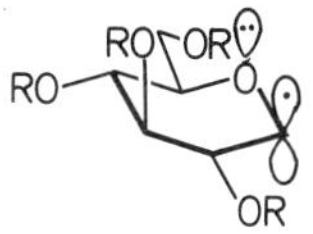

Scheme 6

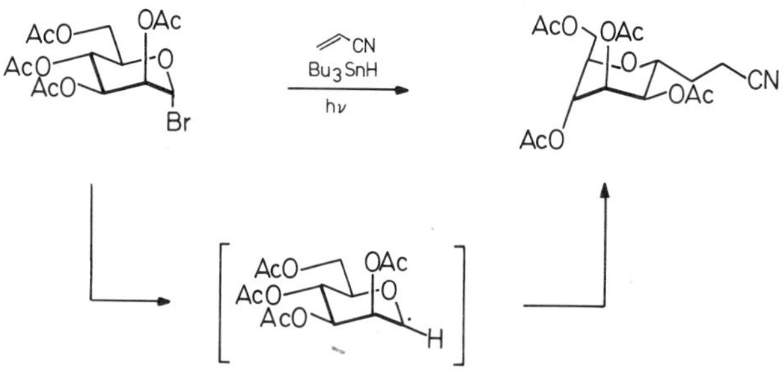

Scheme 7

This anomeric stabilization of radicals is also observed using halonitrosugars such as 1-C-nitroglycosyl halides [22] *15*. Captodative stabilization of the alcoxy nitro radicals explains the radical-chain substitution with mild nucleophiles such as malonate or nitroalkane anions to form *16* (Scheme 8).

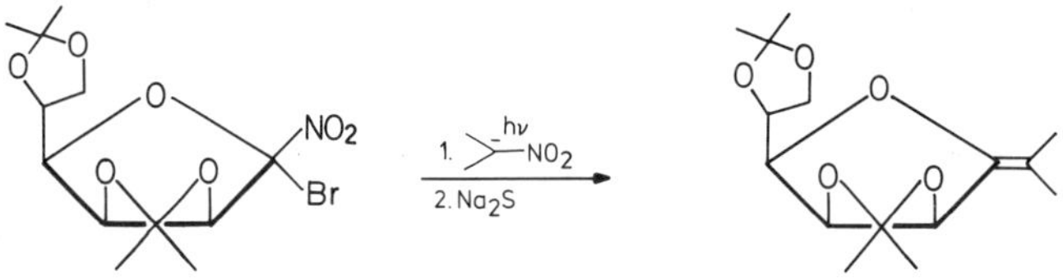

| 15 | 16 |

Scheme 8

2.1.2.2 Carbon-Oxygen Cleavage

Using a binary sensitizing system (phenanthrene P/DCNB: *p*-dicyanobenzene) in acetonitrile solution, *O*-aryl glycosides are transacetalized with alcohols after generation of aromatic radical cations [23]. According to kinetic anomeric effects, the α-side attack of nucleophiles to cyclic oxocarbenium ions follows scheme 9.

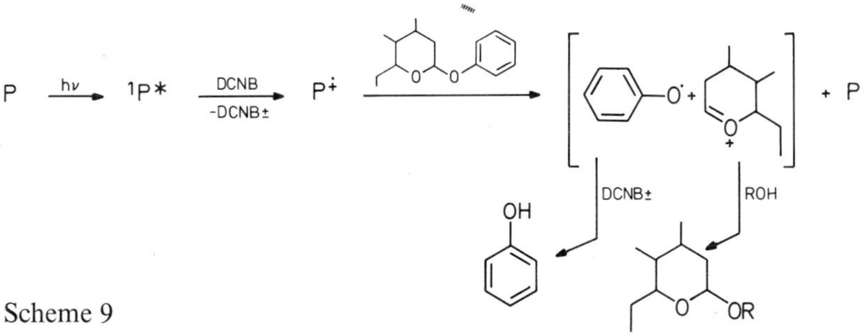

Scheme 9

47

This transacetalization seems useful for intra or intermolecular glycosidation after photoirradiation of glucosides *17* in the same preceding conditions via a one electron exchange mechanism: 1,6-anhydro sugar *18* or 2-deoxyglycoside *19* [α:β = 55:45) are obtained in good yields (Scheme 10).

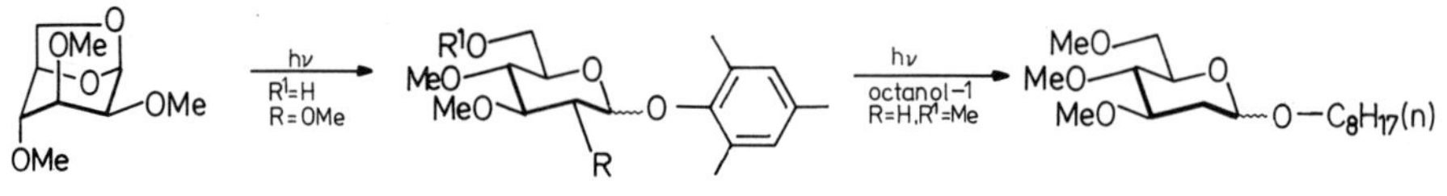

Scheme 10

2.2 Photosubstitutions at Non-Anomeric Carbons

2.2.1 Non-anomeric Hydrogen Photoabstraction

2.2.1.1 Photobromination

The most efficient photosubstitution reaction, with many applications in synthesis, is the photobromination method developed by FERRIER's group [24]. The substitution of H-5 is regioselective and stereoselective for pentopyranosides [25], hexopyranosides [26], 1–6 anhydrosugars [27], uronic acids [24, 28] heterocyclic derivatives from deoxyinosones [29], nucleosides [30, 31], etc... using bromine or *N*-bromosuccinimide in carbon tetrachloride under irradiation.

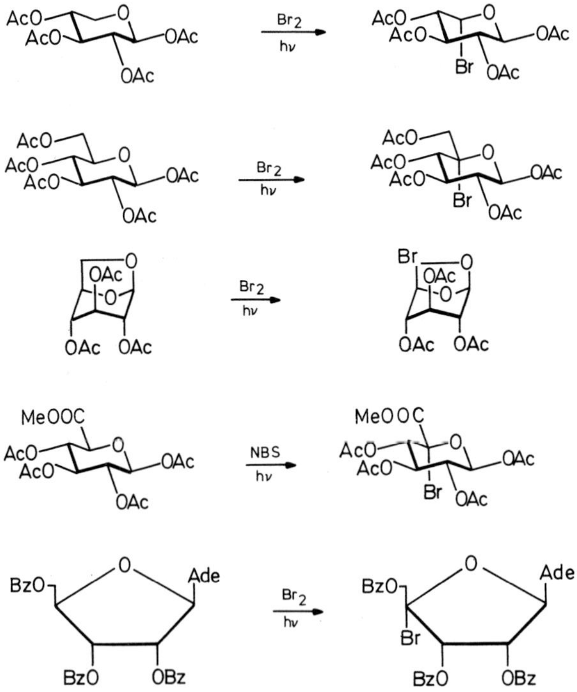

Scheme 11

48

All these transformations are obtained in good yields and 5-bromo derivatives are often crystalline and easy to separate. By-products can be bromides or dibromides with halogen substituent at C-1 and C-5, but the *regioselectivity* of photobromination at C-5 results from the easier formation of tertiary radicals. The α-bromination confirms the *stereoselectivity* of the substitution reaction with the preferential abstraction of axial H-5.

Photobromination with bromine occurs at C-4 with 1-*O*-acetyl-2,3,5,6-tetra-*O*-benzoyl-β-D-glucose *20* or D-galactose *21* giving the same mixture of 4-monobrominated compounds *22* + *23* from which the D-galacto epimer *22* can be isolated in high yields [30a]. These results confirm the formation of the same intermediate radical indicated in Scheme 12.

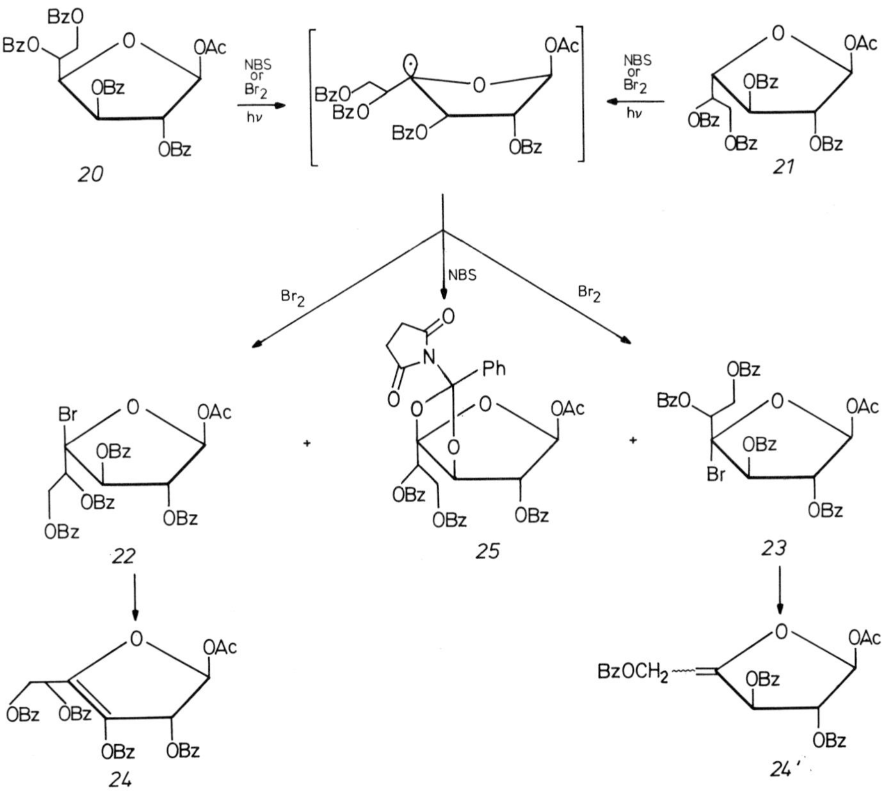

Scheme 12

Nethertheless, more complex mixtures are obtained with the same compounds when treated with NBS (formation of *ortho*amide *25* in good yields).

2.2.1.2 Applications in Synthesis of C-4, C-5, C-6 Photobrominated Carbohydrates

The facile photobromination at C-4 or C-5 position of different furanose and pyranose derivatives has many applications for further modifications of carbohydrates, mainly by elimination and substitution reactions.

Gerard Descotes

a) Different examples of *eliminations* [30] are given in Scheme 12 showing the synthesis of *endo 24* and *exo 24'* cyclic unsaturated sugars.

In the pyranose series, the 6-deoxy hex-5-eno-pyranose derivatives obtained from C-5 photobrominated carbohydrates reacted with mercury (II) salts by an intramolecular aldol process to give β-hydroxy-cyclohexanones according to Scheme 13.

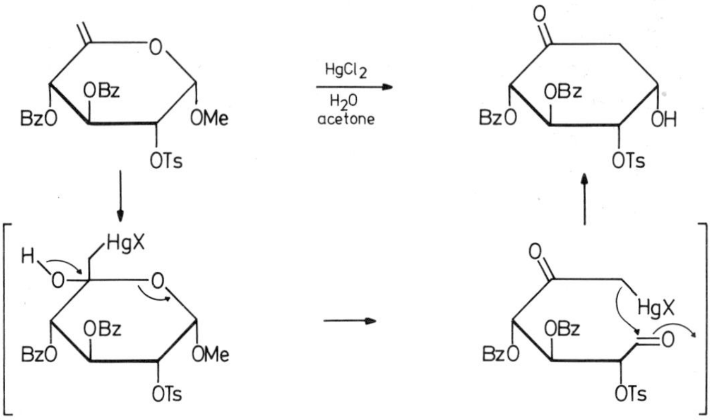

Scheme 13

These reactions were applied to the synthesis of phenol derivatives [32], aminoglycoside antibiotics [33] etc. . . and are of great interest in other synthetic applications.
b) In the field of *substitution* reactions, a more recent application of photobromination reactions is described for the synthesis of L-iduronic acid derivatives from D-glucuronic acid analogs *26*. The C-5 photobrominated product *27* is reduced with tri-n-butyltin hydride to give a mixture of starting material and the L-*ido*pyranuronate *28* by a supposed rapidly interconverting radical [29] (Scheme 14).

The functionalization at C-6 of 1,6-anhydro sugars using the previously described photobromination process allows the synthesis of chirally deuterated hydrosy methyl groups in the glucopyranose series [34] (Scheme 15). Similar sequences can be used in the *galacto* [35] and *ribo* series [36].

2.2.2 Non-anomeric Carbon-Heteroatom Cleavage

Non-anomeric positions can be regioselectively transformed by radical substitutions of halosugars or sulfur derivatives to yield mainly *C*-branched sugars by creation of new C–C bonds.

2.2.2.1 *Halosugars and Unsaturated Sugars*

Photolysis of iodosugars in alcohol solution is of great interest to obtain deoxy sugars, but in the presence of tri-n-butyltin hydride and poor electron alkenes, equatorial carbon-carbon bonds are obtained with a high stereoselectivity at C-4 position [39] from *29* forming mainly *30*.

At C-2 position, carbohydrate radicals were formed starting from unsaturated sugars by acetoxymercuration followed by reduction of the carbon-mercury bond

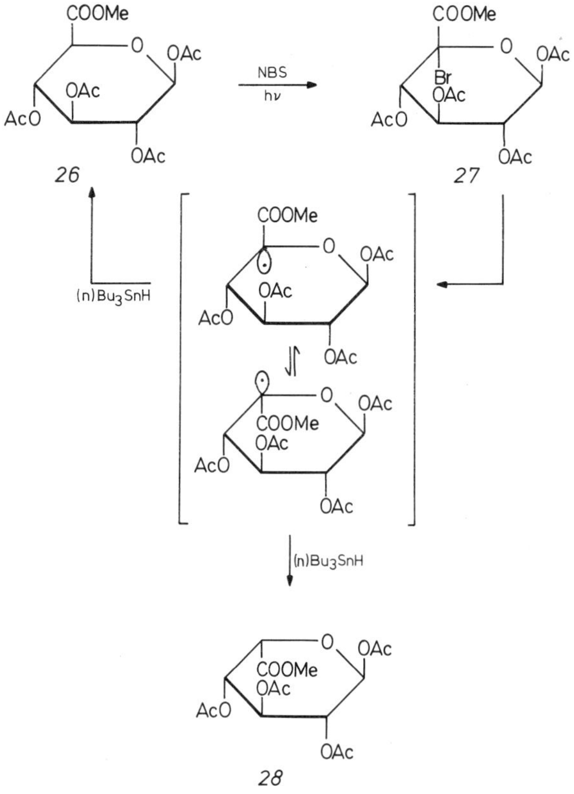

Scheme 14

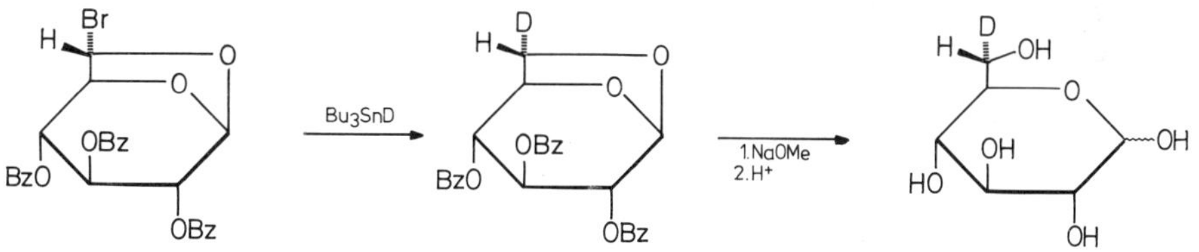

Scheme 15

with tin hydride in the presence of an electron-poor alkene [40]. For this reaction, axial attack of the intermediate predominates if vicinal substituents are both axial [5] (Scheme 16) to afford a mixture of *31* + *32* (4:2).

2.2.2.2 Sulfur Derivatives

Similar homolysis of xanthates derived from *galacto* compounds *33* by a tin radical can form deoxy sugars or, in the presence of acrylonitrile, *C*-branched sugars at C-3 such as *34* [37, 38] according to Scheme 17.

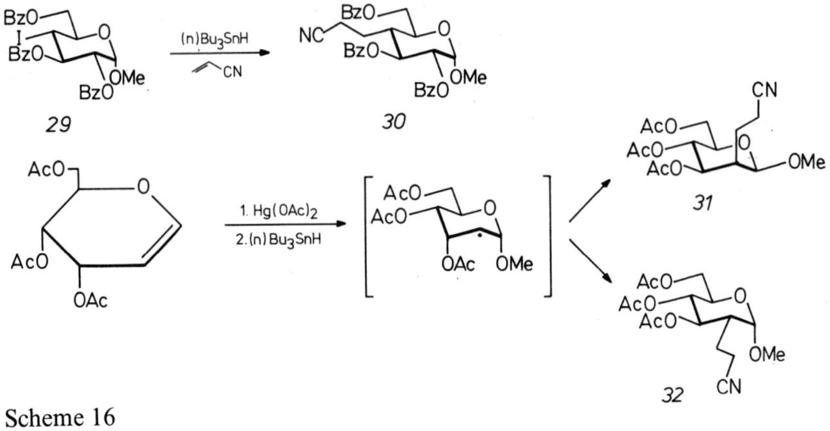

Scheme 16

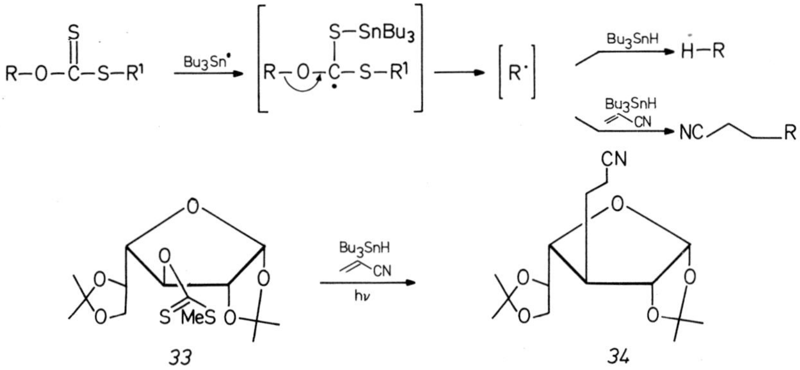

Scheme 17

Radical chain reactions can take place via organotin reagent without hydrogen donors such as allyltin [38, 39]. This methodology has been applied to synthesize some important chiral precursors of natural products (for example pseudomonic acid). Thus, the pentose derivative 35 affords the compound 36 with retention of configuration via the carbohydrate radical R· at C-4 position [Scheme 18].

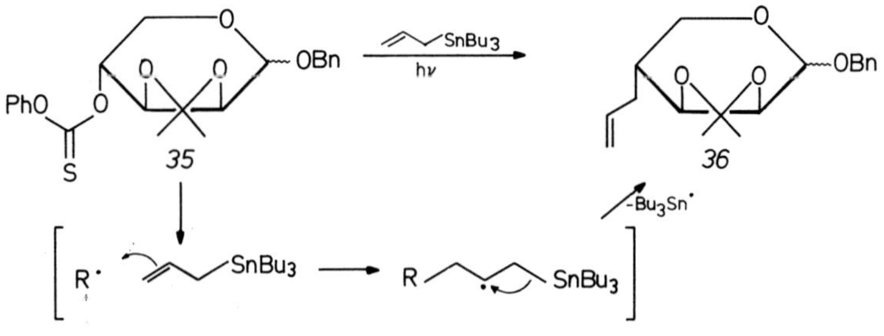

Scheme 18

All these syntheses of non-anomeric C-branched sugars show the influence of neighboring oxygenated groups as for the the preceding photosubstitutions at the anomeric positions. The importance of stereoelectronic effects seems predominant for all these radical reactions.

3 Photoadditions

Photoaddition reactions occur at the double bond of unsaturated sugars by:
— cycloaddition with formation of three or four-membered carbo or heterocyclic compounds.
— creation of a new σ bond without cyclization.

3.1 Photocycloadditions

3.1.1 Syntheses of Three-Membered Cyclic Compounds

3.1.1.1 Carbocyclic Compounds

Cycloaddition to α–β-unsaturated carbonyl systems represented by different carbohydrate enones yield "annulated" osides. These bicyclic molecules are used for synthesis of carbocyclic systems using the stereochemical information issued from sugar moieties.

Unsaturated sugar *37* is used as the starting material for cyclopropanation by non-photochemical means to form cyclopropanopyranoside *38* and, by further degradation and *cis-trans* isomerization, chrysanthemic derivatives *39* (Scheme 19) [41].

A stereoselective photochemical addition of diazomethane to unsaturated uronate *40* undergoes a pyrazoline derivative *41* which, after photolysis, leads to cyclopropanofuranoside *42* [42].

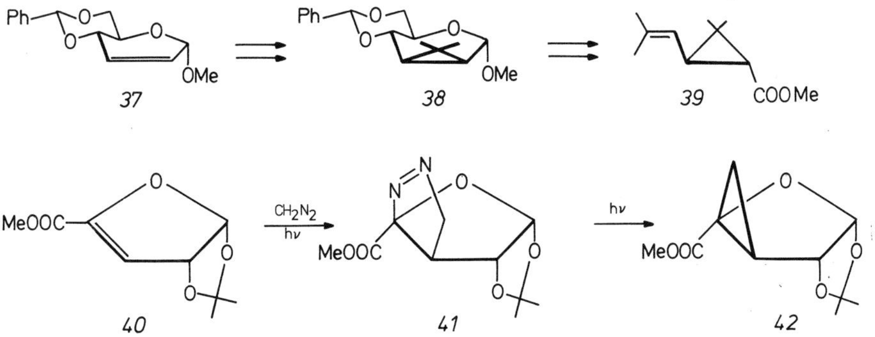

Scheme 19

These annulated bicyclic systems are of interest for further synthesis of complex macrolide antibiotics.

3.1.1.2 Heterocyclic Compounds

Photoannelation of glycals by nitrenes affords the 1–2 aziridine intermediates which are transformed into aminosides *43* in the presence of alcohols [43]. A similar approach with a photochemical α-addition of *N*-haloamides to glycals is also possible to yield mainly 1–2 *trans* aminosides (Scheme 20) [44a]. This similar preferential α-addition is in contrast with the reported β-addition on the β-face of the thioacetyl radicals [44b].

This limited methodology must be compared to a new approach for the synthesis of 2-amino 2-deoxycarbohydrates based on the cycloaddition of azodicarboxylates on glycals. This |4 + 2| cycloaddition is initiated by irradiation at 350 nm and seems highly stereoselective. After hydrolysis and reduction, compounds like *43* (Z = Ac) are obtained in good yields [45].

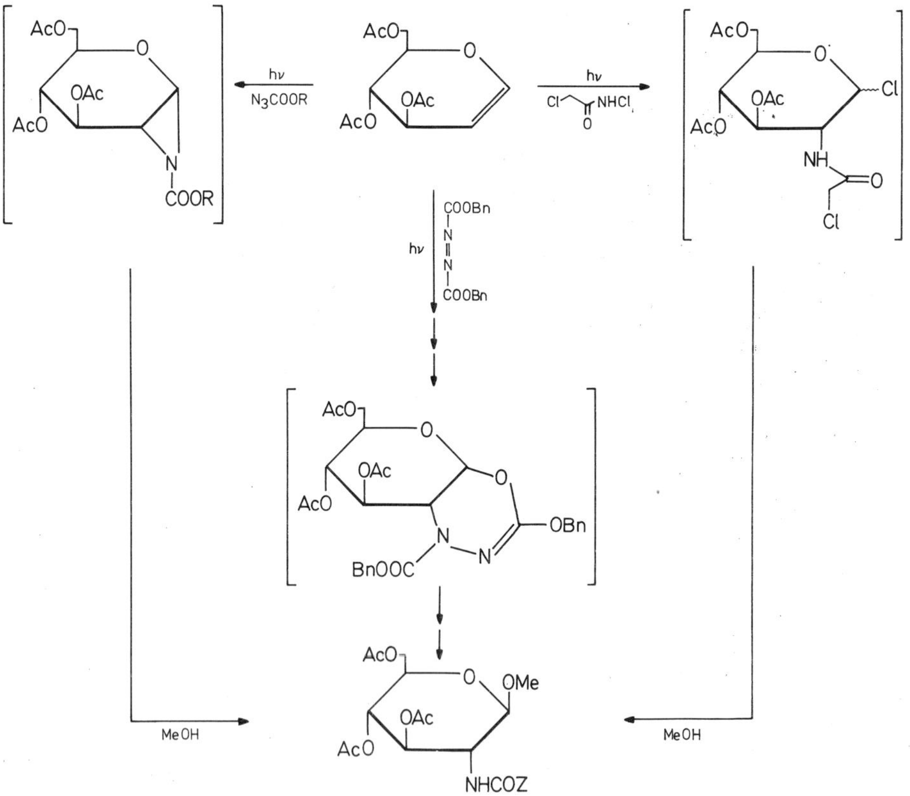

Scheme 20

3.1.2 Syntheses of Four-Membered Cyclic Compounds

3.1.2.1 Carbocyclic Compounds

A few examples of |2 + 2| photocycloadditions of olefins on carbohydrate enones have been recently reported. The main results are described in the preceding review [1] on photochemical reactions of sugars.

The use of these cyclobutano sugars has been developed for the synthesis of natural products such as grandisol [47]. Both enantiomers of this pheromone are available by |2 + 2| photoaddition of ethylene to methyl hex-2-enopyranosid-4 ulose *44*. After further manipulations, (+) and (−) grandisol (*45*, *46*) are synthesized following Scheme 21.

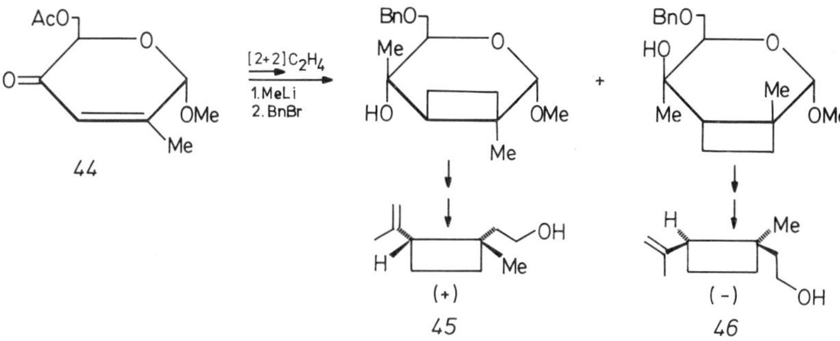

Scheme 21

More recently, acetylene is reported to give under irradiation a photoadduct *48* with enone *47* by α-attack [46], After deacetoxylation and ionic rearrangement, a simple approach to the synthesis of optically active trichothecene seems possible [46a] using the rearranged cyclopenteno bicyclic compound *48* (Scheme 22).

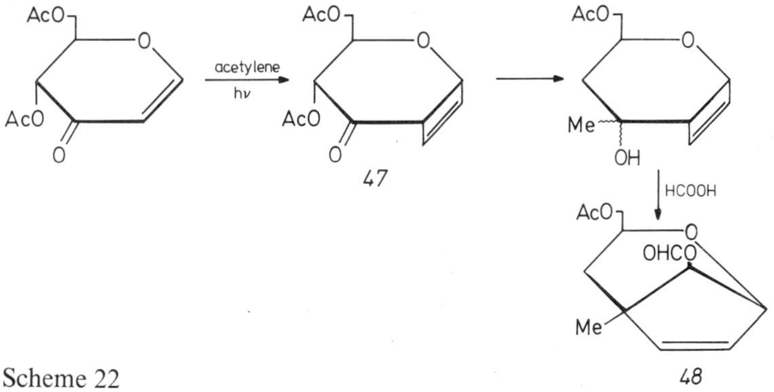

Scheme 22

3.1.2.2 Heterocyclic Compounds

The Paterno-Büchi photocycloaddition to glycals occurs with a total regioselectivity according to the stability of the supposed intermediate radicals. Acetone in presence of isopropanol adds to D-glucal to form oxetane *49* or *C*-glycoside *50*. These products are formed by competitive addition of alcohol and acetone depending on their concentrations [48]. (Scheme 23).

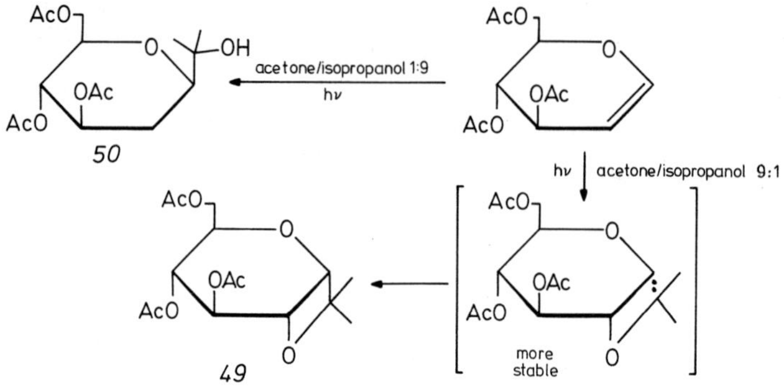

Scheme 23

The regio and α-stereoselectivity of this low-yield cycloaddition again prove the structural stability of the intermediate diradical with a preferential anomeric semi-occupied orbital in an axial orientation. Similar observations can be made in the furanoside series [48].

This type of |2 + 2| photocycloaddition [49] is possible with ketosugars on one of the double bond of furan. The less crowded adducts *51* and *52* are formed with equatorial C–C bonds as shown in Scheme 24.

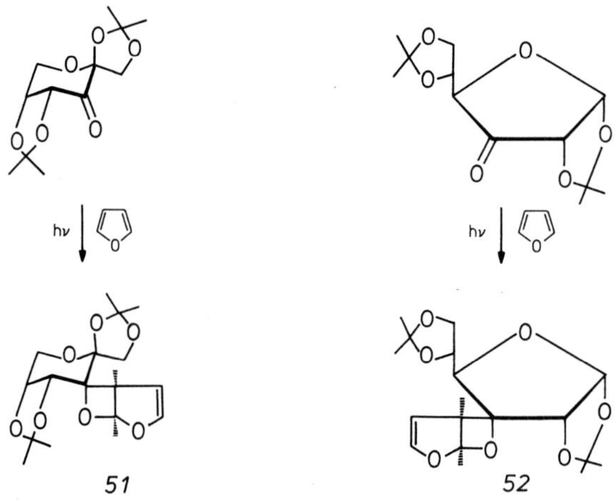

Scheme 24

3.2 Photoadditions without Cyclizations

The formation of a single bond by photoaddition on unsaturated carbohydrates is mainly performed on:
— enol ether double bonds
— enones and other unsaturated systems.

3.2.1 Photoaddition to Enol Ethers

3.2.1.1 Nitrogen Reagents

The addition of chloro azide ClN_3 on the double bond of glycals proceeds by either an ionic or a radical mechanism depending on experimental conditions. Under UV irradiation, in solvents of low polarity and in the absence of oxygen, radical addition is predominantly regio- and stereoselective [51, 52]. The double bond reactivity is affected by the substituent at C-3 position and its inductive effect. Therefore, the presence of acetates lowers the reactivity, but azidosides are formed following Scheme 25.

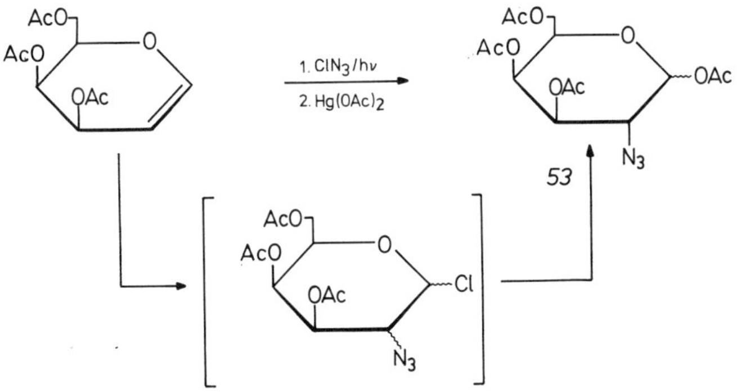

Scheme 25

Some other nitrogen compounds are also formed by carbamoylation of glycals using a photochemical regiospecific process which consists of the addition of the amide radical only to C-1 [53]. The addition mixture remains complex because of stereoisomers 54 (mainly α) and competitive addition products with acetone such as 55.

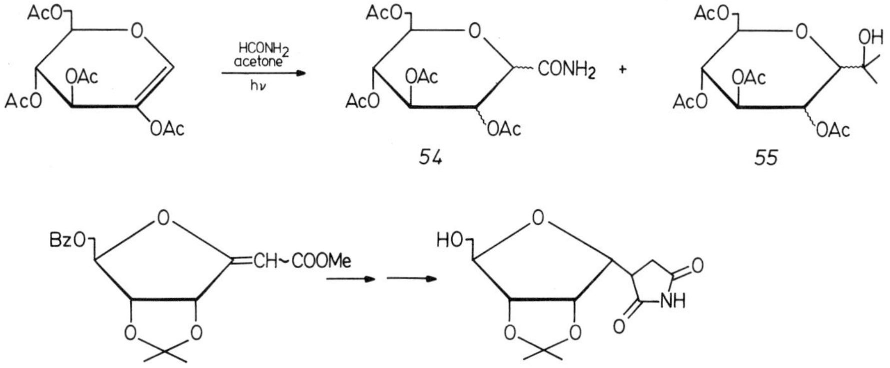

Scheme 26

The development of such addition reactions for synthetic purposes remains limited because of the complexity of the photochemical reaction mixtures. Nevertheless, a facile synthesis of dihydroshowdomycin [54] uses the same type of photoaddition of amides on unsaturated sugars following Scheme 26.

3.2.1.2 Oxycarbinyl Radicals

A general study of the photochemical addition of 2-propanol and 1,3-dioxolane to unsaturated sugars [48, 55, 56] shows the following order of reactivity (see Scheme 27) and yields mixtures of stereoisomers.

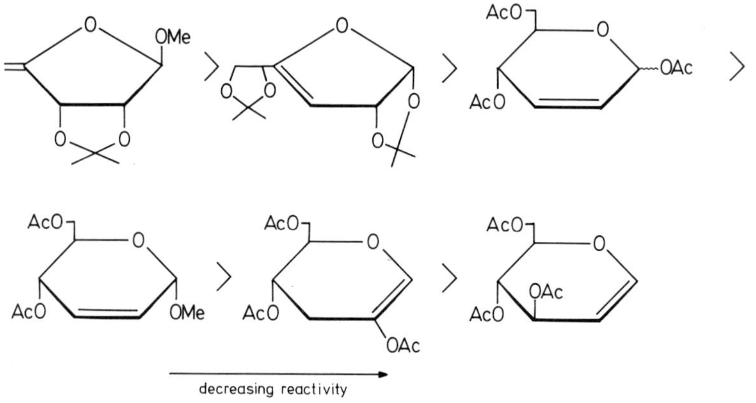

decreasing reactivity

Scheme 27

Recently, more efficient photochemical additions are described on the enol bonds at C-5, C-6 of unsaturated sugars. For example, chloracetonitrile reacts as a radical precursor on *56* and yields *57* in the presence of Bu_3SnH[5] (Scheme 28).

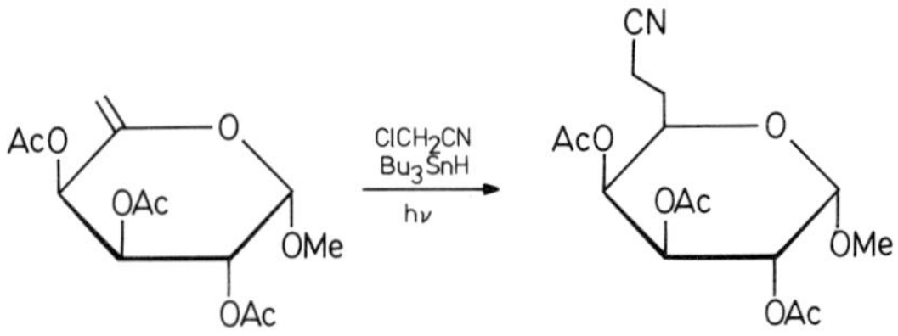

Scheme 28

3.2.2 Photoaddition to Enones and other Unsaturated Systems

3.2.2.1 Nitrenes

Photochemical α-addition of a nitrene to 1-isocyano sugars leads to the carbodiimide which adds water yielding glycosylurea or malonic acid to form potential glycosyl-barbiturate synthons [57] (Scheme 29).

58

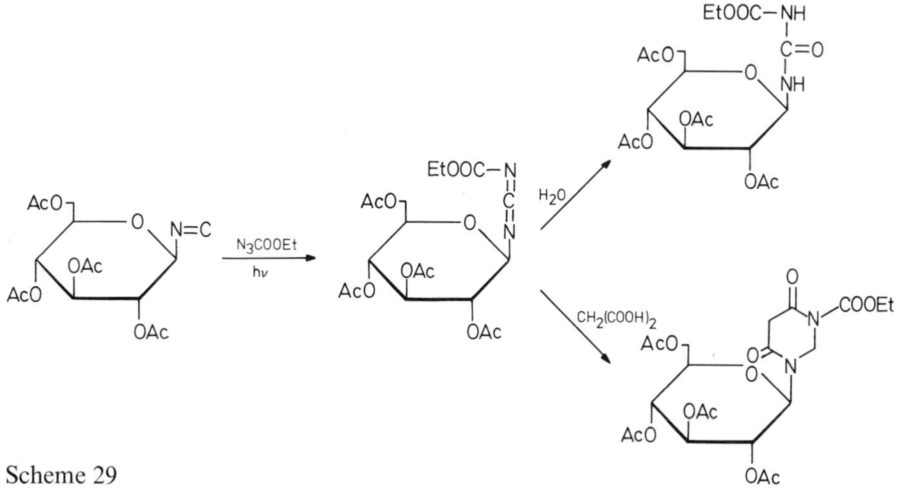

Scheme 29

3.2.2.2 Oxicarbinyl Species

The photosensitized 1–4 addition of alcohols to hexenopyranosuloses first reported by B. Fraser Reid and coworkers [58a] has been developed with other studies on photoadditions of oxycarbinyl species such as polyols, acetals, dioxolanes, aldehydes. A mechanistic study on this photoaddition has been recently detailed [58b] showing that the important photochemical event is hydrogen abstraction from methanol, for example, to form the hydroxymethyl radical.

The syntheses of these new C-branched sugars by regioselective and often stereoselective processes are obtained in better yields than with the precedingly mentioned glycals. The alkylation occurs at the less hindered side; however, with methanol, some mixtures of isomeric C-branched sugars are obtained (Scheme 30).

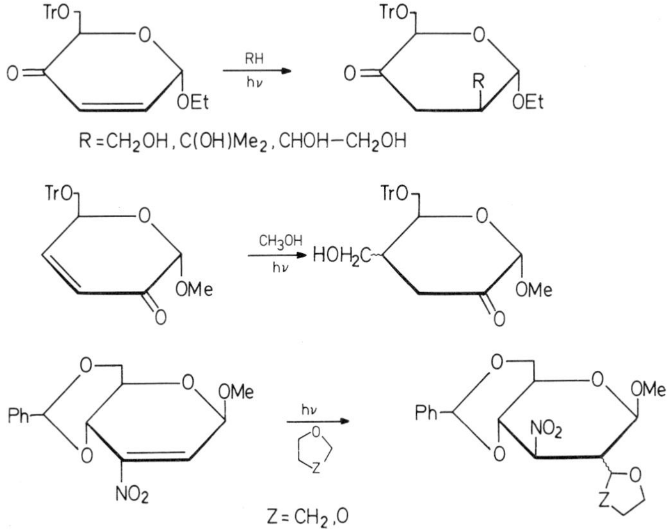

$R = CH_2OH, C(OH)Me_2, CHOH-CH_2OH$

$Z = CH_2, O$

Scheme 30

Similarly, irradiation of unsaturated nitro derivatives in 1,3 dioxolane affords the addition of the 1,3-dioxolan-2 yl group with a low stereoselectivity [58 c].

By photoaddition of other oxycarbinyl functionalized radicals, 1–4 ketols, 1–4 keto-ketals and 1–4 diketones are formed and a review of the main results has been published [59, 60]. Applications concerning the synthesis of natural C-branched sugars such as pillarose are given using this photoaddition methodology [61].

4 Intramolecular Phototransformations of Carbohydrates

Intramolecular photochemical reactions need the presence of light sensitive groups such as carbonyls to induce transformations by α-cleavage (Norrish I reaction) or by γ-H-abstraction (Norrish II reaction).

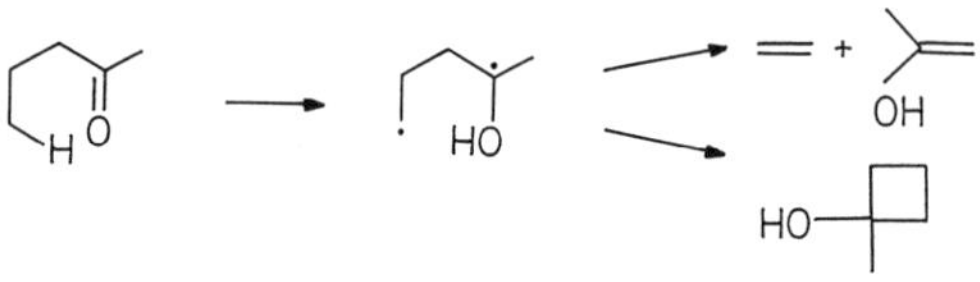

Scheme 31

The net effect is decarbonylation, cyclization and stereoisomerization which are linked to the structure of the carbohydrate moiety. The main results describe carbohydrates bearing a carbonyl group on the *aglycone* (Norrish I) or on the *ring* (Norrish II).

4.1 Photolysis of Ketosugars

The α-cleavage of ketosugars has been mainly studied by P. M. Collins and his group and their principal results have been reviewed [1].

The differentiation of ketosugars by the position of their carbonyl group on the ring allows a comparative study of their photolytic abilities.

4.1.1 2-Ketuloses

To avoid Norrish II photolysis of pyranosid-2-uloses derivatives containing aglycones derived from primary and secondary alcohols, t-butoxy derivatives *58* can be successfully photolyzed by a Norrish I reaction to yield, by C-1, C-2 cleavage and diradical formation, compounds *59* [30 %] and *60* [40 %] (Scheme 32) [62].

In a similar way 1,6 anhydro 2-ketopyranoses[63] of the *xylo, ribo* and *psico* series underwent stereoselective ring contraction by decarbonylation (Scheme 33).

Similar compounds, such as diluse *61* yield by photolytic decarbonylation reaction [64], pentulose derivatives *62* by a favoured cleavage at C-2, C-3.

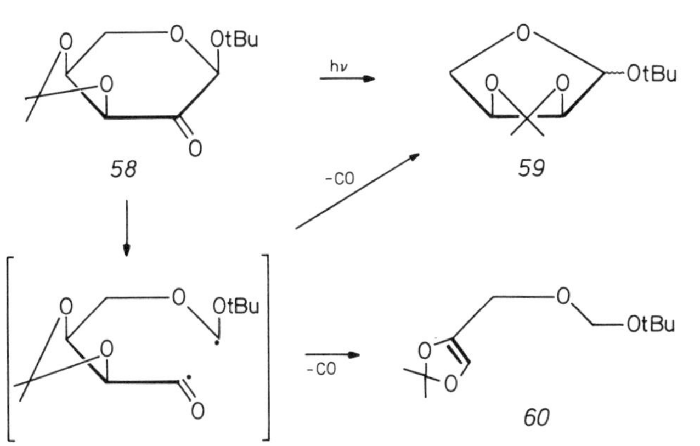

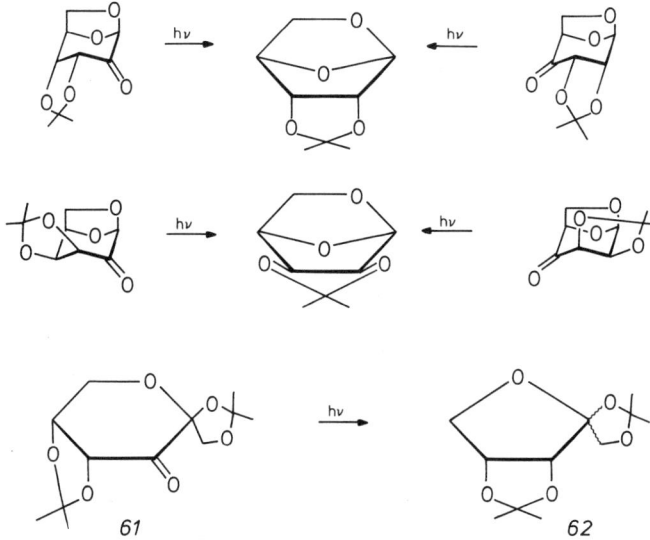

Scheme 32

Scheme 33

4.1.2 3-Ketuloses

For compounds such as *63*, a more complex rearrangement replaces the decarbonylation reaction [65]. This photoisomerization yields lactone *64* by α-cleavage at C-2,
C-3 position (Norrish I) and hydrogen transfer from C-1 to C-3 followed by a stereoselective nucleophilic attack at the carbonyl group by the terminal carbon of the electron-rich double bond and final ring closure (Scheme 34).

This mechanism, which does not clarify the stereoselectivity of these transforma-
tions, is in contest with an eventual intramolecular |2 + 2| cycloaddition of the alde-
hydic intermediate which can afford a |2,6| dioxa bicyclo |3.1.1.| heptane system which
could also be capable of giving *64*.

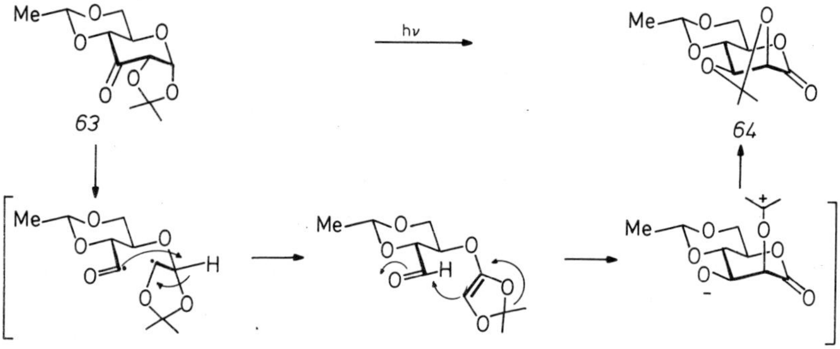

Gerard Descotes

Scheme 34

4.1.3 4-Ketuloses

Irradiation of pyranosid-4 ulose derivatives such as *65* undergo decarbonylation to yield furanoside *66* by preferential C-3, C-4 cleavage [66]. This is confirmed by the separation of its epimer *67* containing a 2-3 *trans*fused system derived from a diradical intermediate Scheme 35).

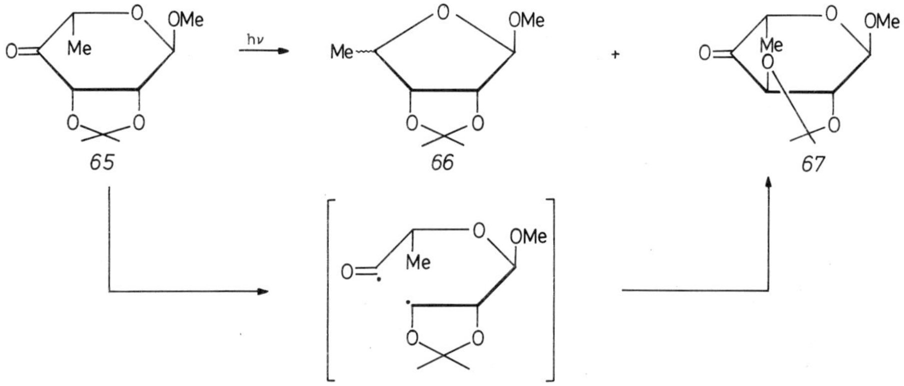

Scheme 35

In the 1,6-anhydro series [63], 4-ketuloses are photodecarbonylated to yield the same kind of compounds as the preceding 2-ketuloses (Scheme 33).

4.2 Photolysis of Oxoalkylglycosides

The photoabstraction of the anomeric hydrogen by any excited carbonyl group on the aglycone of *oxo*alkylglycosides *68* leads to anomeric diradicals which are transformed into lactones, spiro compounds, or *O*-vinyl glycosides depending on the length of the chain of the aglycone (Scheme 36).

If the irradiation of *68* ($n = 0$, $R = CH_3$) is closely connected to photoxidation processes which will be described later, the photocyclization reaction is the source

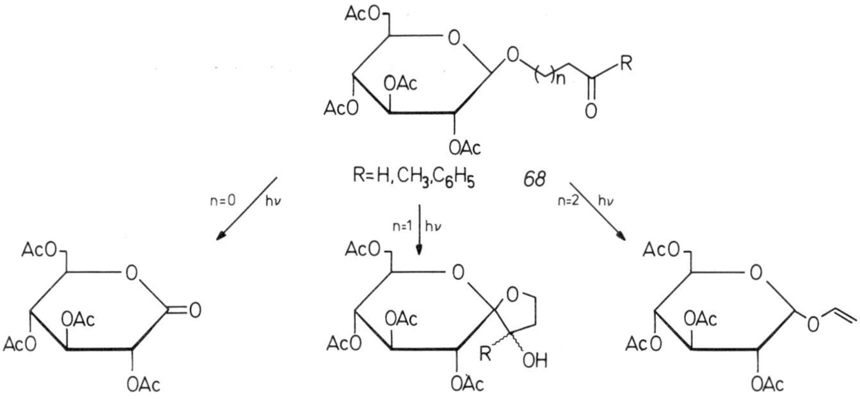

Scheme 36

of anomeric spiro compounds which are of interest for a synthetic approach to analogs of ionophore antibiotics.

The photocyclization reaction is more efficient with β-anomers, and the new C—C
bond created at the anomeric center is obtained by preferential axial anomeric hydrogen abstraction followed by carbocyclization. Norrish II reactions generally result
from hydrogen abstraction at the γ-position according to Scheme 31 to give cyclo-
butanols or degradation products. For 68, (n = 1) photoabstraction at the δ-position
of the anomeric hydrogen leads to cyclopentanols 69 (Scheme 37) [67].

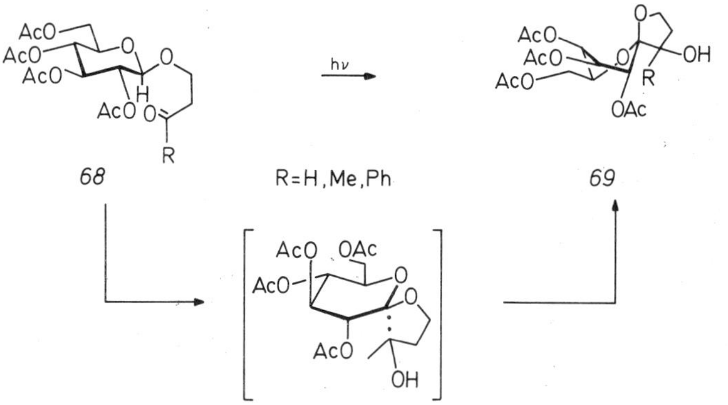

Scheme 37

This methodology is valuable for similar derivatives in the D-*manno* [68], L-*arabino* [69] and 2-deoxy or 2,3-deoxy [70] series. Parallel work [71] on the O-formyl and
O-acetyl phenyl β-D-glucopyranoside 70 and its α-anomer 71 yield compounds 72
and 73. The epimerization of 72 in acidic conditions leads to the more stable compound 73 by anomeric effect. Stereoelectronic effects can explain the twist-boat conformations of 69 and the chair conformation for the *manno* analog established by
x-ray cristallography [72–73].

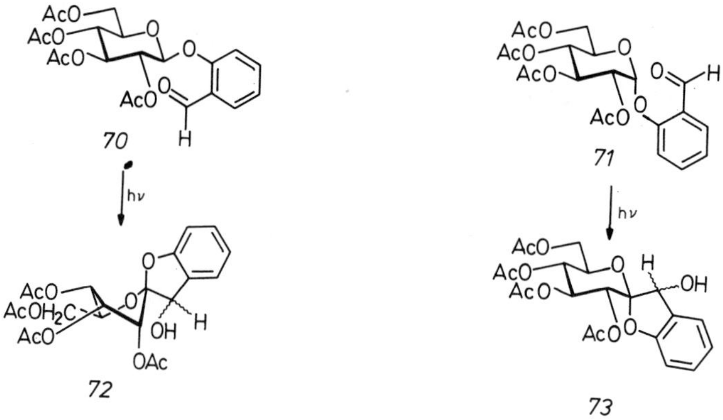

70

71

72

73

Scheme 38

The results clearly show that the same kinetic anomeric effects governs these intra-molecular cyclization reactions as well as the preceding anomeric hydrogen abstractions described in intermolecular photochemical substitutions.

The ε-anomeric hydrogen of compound 74 is too far away to permit abstraction and the Norrish II process results from the γ-hydrogen abstraction of the aglycone chain. In this way, O-vinylglycosides 75 can be formed [74] (Scheme 39).

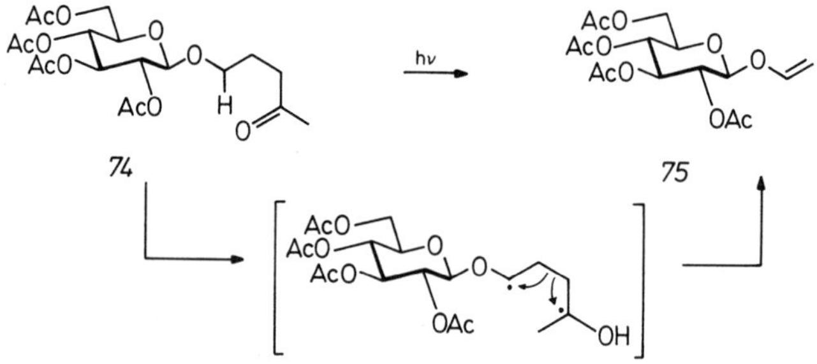

74

75

Scheme 39

Similarly, the photolysis of unsaturated α-glycoside 76 leads to the vinylglycoside 77 in low yields. This on thermolysis undergoes a |3.3| sigmatropic rearrangement to yield the C-branched sugar 78 (Scheme 40) [74].

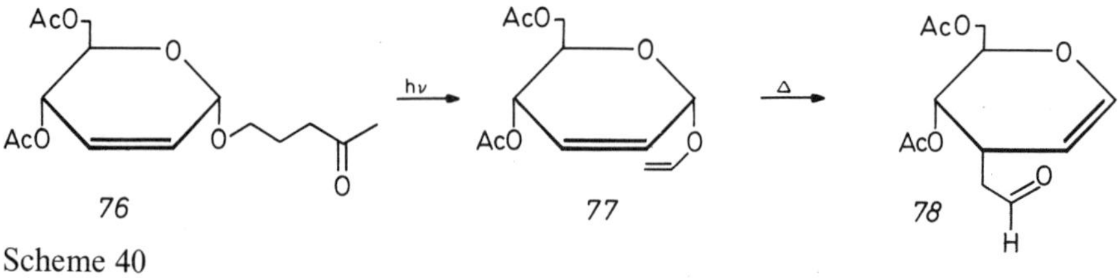

76

77

78

Scheme 40

5 Photoreduction and Photoxidation Reactions of Sugars

5.1 Photoreduction

5.1.1 Photodeoxygenation

Radical deoxygenation of sugars can be realized with suitable alcohol derivatives such as esters (acetates, pivaloates), sulfur compounds (xanthates, thiocarbamates, trifluoromethylsulfonates) according to the general Scheme 41.

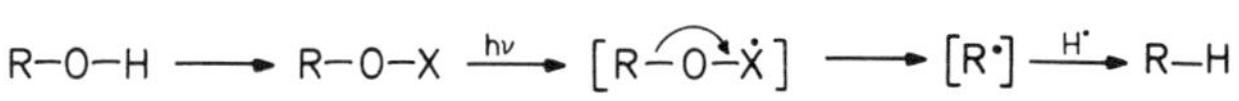

R–O–H $\longrightarrow$ R–O–X $\xrightarrow{h\nu}$ [R–O–X] $\longrightarrow$ [R$^{\cdot}$] $\xrightarrow{H^{\cdot}}$ R–H

Scheme 41

5.1.1.1 Acetates

Irradiation of non-absorbing monoacetates on sugars at any positions on the ring in a hexamethyl phosphotriamide solution (HMPA) with 5 % water leads in excellent yields to deoxy sugars [75–79]. This methodology is of particular interest for sterically hindered hydroxyl groups found in D-fructopyranose derivative 79 as well as in α-L-fucopyranoside 80 which are difficult to obtain by classical methods (Scheme 42).

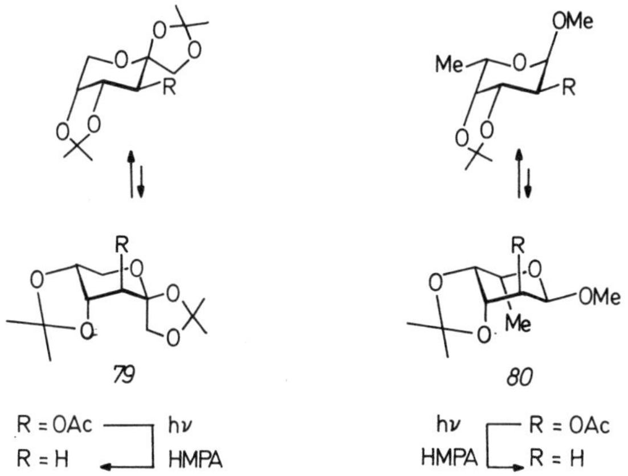

$$R = OAc \overset{h\nu}{\underset{HMPA}{\rightleftarrows}} R = H$$ (for 79)

$$\overset{h\nu}{\underset{HMPA}{\rightleftarrows}} \begin{array}{l} R = OAc \\ R = H \end{array}$$ (for 80)

Scheme 42

In the case of 2,3-diacetylglucosides, 2,3-dideoxyhexopyranosides are also obtained. This method shortens reaction paths and is used in photolytic trideoxygenation to give, for example, amicetose [80].

If this process seems particularly useful for small quantities of acetates; some difficulties are observed for the deoxygenation of larger quantities of substrates which necessitate longer times of photolysis and leads to lower yields of deoxygenation products [81].

5.1.1.2 Pivaloates

The pivaloyl group which can be regioselectively introduced in carbohydrates [82], is photolyzed under preceding conditions to deoxygenated compounds [83], in better yields than for acetates or aromatic esters.

Different protecting groups such as acetonides, *t*-butyl dimethyl silyl derivatives proved to be stable under these experimental conditions, but the benzylidene group is completely cleaved. The rate of photodeoxygenation is greater at the secondary positions [82], as indicated in Scheme 43, with the regioselective photolysis of dipivaloate to give a mixture of monodeoxy and dideoxy derivatives (*81*) in low yields.

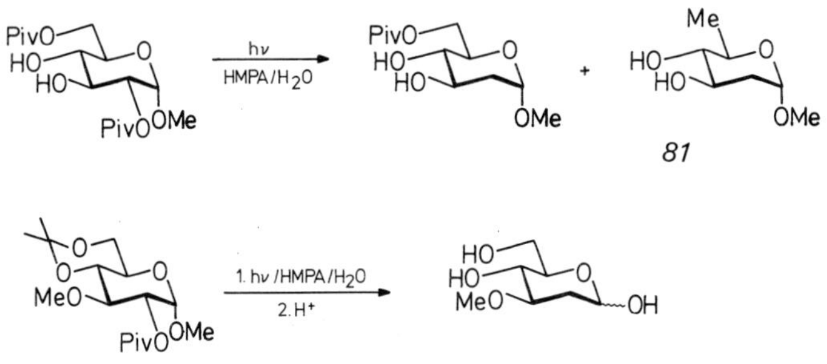

Scheme 43

This photochemical deoxygenation was applied to the synthesis of *81*, and *82* [84].

5.1.1.3 Benzoates

A recently reported regioselective photoreduction of benzoates by photosensitized electron transfer reaction was applied to nucleosides [85]. In presence of *N*-methyl-carbazole as the electron donor sensitizer and in an isopropanol, water solution, *m*-trifluoromethylbenzoates of adenosine *83* or benzoates of uridine *84* give deoxygenated products in good yields (73%).

This interesting methodology broadens the preceding electron transfer reduction of acetates and pivaloates, and seems very efficient for synthesizing protected 2'-deoxy uridine such as *85* in high yields (85%) (Scheme 44).

5.1.1.4 Trifluoromethane Sulfonates, Thiocarbamates and Xanthates

The deoxygenation reaction by similar single electron transfer process is also useful for the reduction of mesylates, while tosylates regenerate by hydrolysis the alcohol [86, 87] in the presence of base without epimerization. High yields are obtained in general, while in alcoholic solution, *N,N*-dimethyl thiocarbamates lead to reduction products in low yields with a predominant deprotection process [88] (Scheme 45).

In the case of xanthates, no photoreduction is observed but a more complex deprotection of acetonide groups at C-5, C-6 is obtained with subsequent formation of *ortho*trithiocarbonates in the presence of air [89].

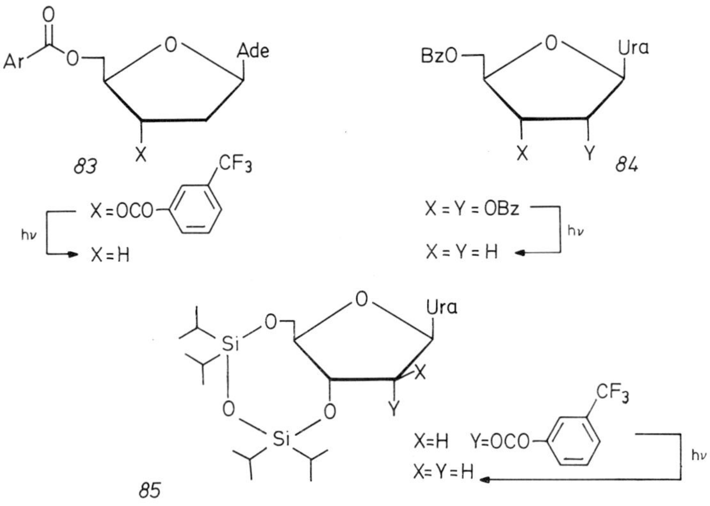

Scheme 44

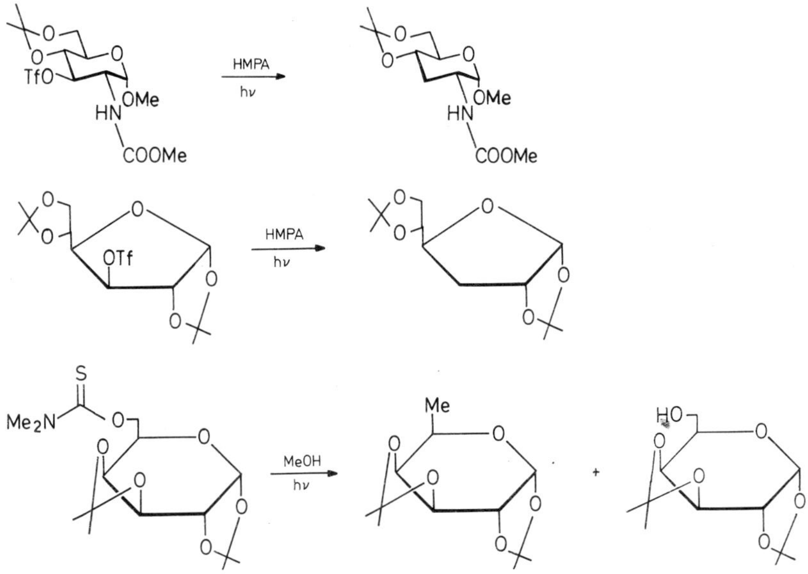

Scheme 45

5.2 Photoxidation

5.2.1 Photoxidation of Protected Carbohydrates

The previously described Norrish II reaction is used in the photolysis of pyruvate esters of carbohydrates and nucleosides [90–94]. The presence of different protecting groups (esters, tritylethers, benzylidene acetals) does not affect this useful oxidation process (Scheme 46).

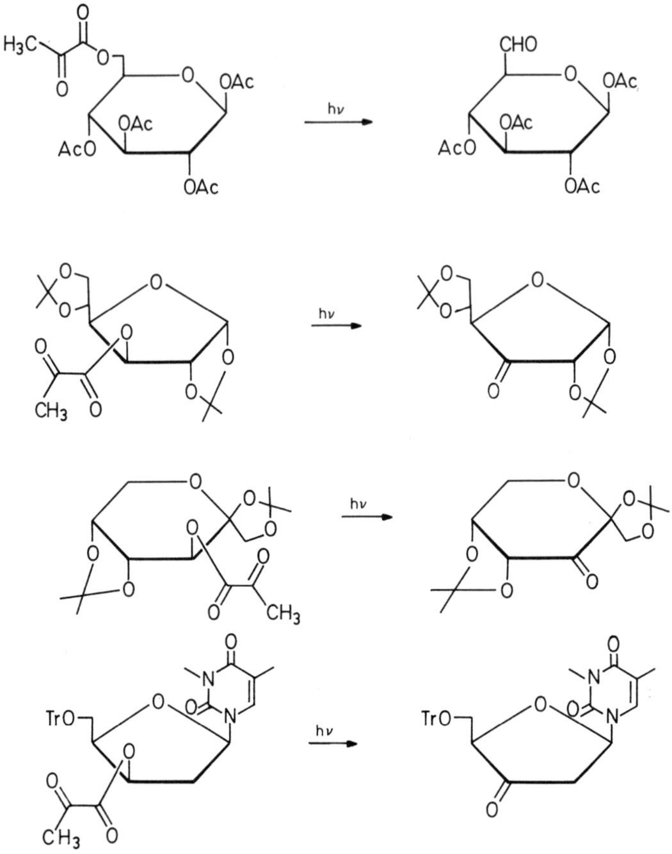

Scheme 46

This methodology was applied to the synthesis of L-streptose [91] and methyl α-D-mycaroside [95a]. The difficulty of this photochemical oxidation originates from the sensitivity of pyruvates to hydrolysis. This approach seems limited to small quantities of substrates but does not need any separation technique as in oxidation by pyridinium chlorochromate [95b]. Nethertheless, photolysis of pyruvates of partially protected derivatives of α-D-*gluco*furanose and β-D-fructofuranose in benzene yields the corresponding oxidized products in excellent yields [96].

The Norrish II type rearrangement can also explain the photochemical reactivity of glycosides 86 [97] leading to lactones 88 as for less stable pyruvates 87.

The high-yield photolytic oxidation of 86 does not depend on the configuration of the anomeric center and similar photolyses are possible with unsaturated sugars such as 89 to give unsaturated lactones 90 in lower yields. The hydrogenated derivative 91 is also photolyzed into the corresponding saturated lactone 92 [97] (Scheme 47).

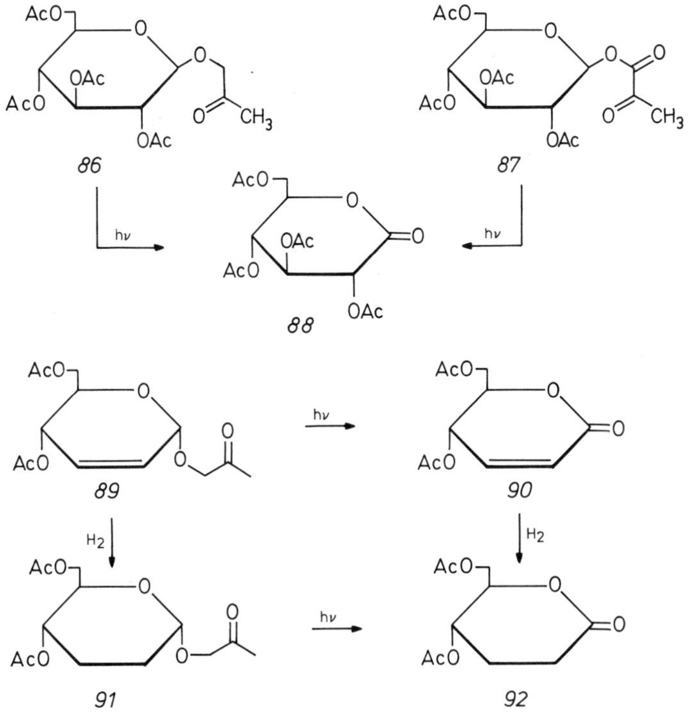

Scheme 47

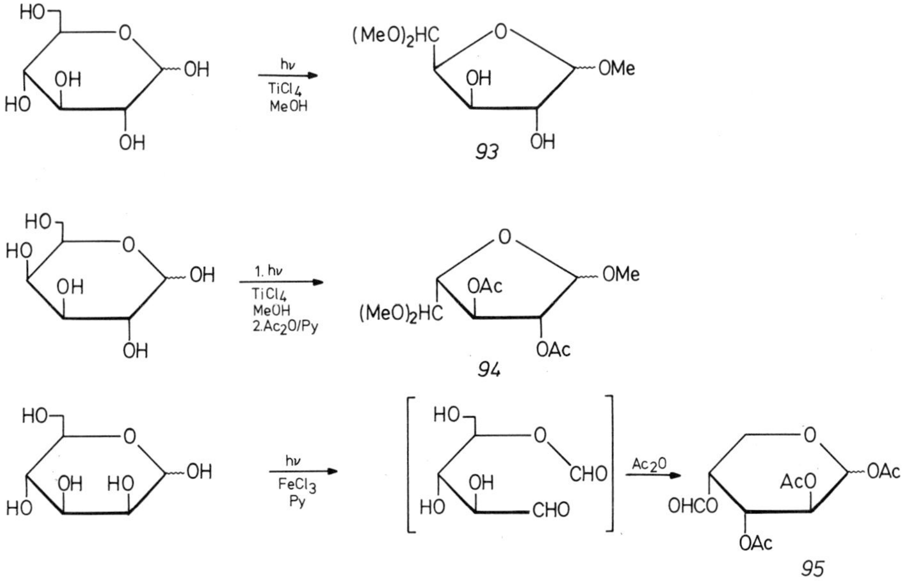

Scheme 48

5.2.2 Photodegradation of Free Carbohydrates

Photoreactions of *aldoses* such as D-glucose and D-galactose in methanol, in the presence of titanium (IV) chloride induced a regioselective bond cleavage, at C-5, C-6 position and gave pentodialdose derivatives *93* and *94* [98]. In contrast, under the same photolytic conditions but in the presence of iron (III) chloride. D-glucose, D-mannose and D-galactose in pyridine provided 4-*O*-formyl aldopentopyranoses in a clean reaction and in moderate yield, after a selective bond cleavage at the C-1, C-2 position [100]. This reaction is depicted for D-mannose to give the derivative *95*.

With *ketoses* such as D-fructose, photoreaction in the presence of iron (111) chloride yields D-erythrose as the single product [99]. This transformation provides high yield and results from a C-2, C-3 cleavage of the intermediate ferric complex.

All these photoreactions are of interest for purposes of mechanistic studies but have not yet been applied to synthetic targets.

6 Photoremovable Protecting Groups of Carbohydrates

The use of protecting groups, for which the regeneration of any protected function could be realized by a photochemical way, is rather attractive to avoid rigorous chemical treatment of sensitive substrates such as carbohydrates. This approach has been tested with different functional groups (acetals, carbonates, nitrates, dithiocarbamates, and carbamates) in complex carbohydrate syntheses and was reviewed in 1980 [101].

6.1 Benzyl Acetals

6.1.1 *O*-Benzylidene Acetals

Simple benzylidene acetals of carbohydrates are photochemically cleaved in a way similar to the well-used *N*-bromosuccinimide process regioselective ring opening process into bromodeoxysugars.

In contrast, the regioselectivity of this photochemical cleavage does not depend on the configuration of the acetalic benzylidene carbon. Thus, there is no need to separate diastereoisomeric pairs prior to ring opening [102] (scheme 49). The same product *97* is formed regardless of whether the phenyl group in *96* is *endo* or *exo* to the pyranose ring. The key intermediate results from the photobromination of the benzylidene group which is hydrolyzed to give an unstable *ortho*acid. This *ortho*acid is transformed into hydroxybenzoates *97*.

This light-initiated NBS reaction is similar to the photolysis of *o*-nitrobenzylidene acetals but is a one-step process instead of the following two-step procedure.

6.1.2 *o*-Nitrobenzaldehyde Acetals

o-Nitrobenzaldehyde acetals of carbohydrates are good temporary blocking groups which, on UV irradiation, followed by oxidation of the resulting nitroso groups, yield 2-nitrobenzoyl sugars with a free hydroxyl group.

The substitutions by *o*-nitrobenzyl groups are used for dioxolane and 1,3-dioxane derivatives, but the photochemical cleavage is only partially regioselective. However,

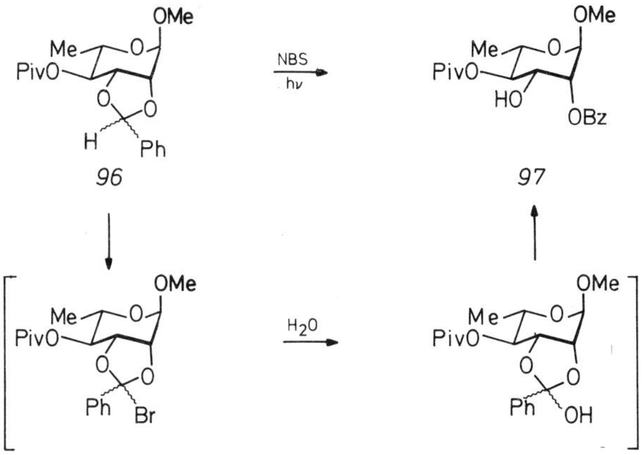

Scheme 49

the scope of this photoremoval was broadened by their application to anomeric 2-nitrobenzyl glycosides [103] in different mono and oligosaccharide series with the indicated major regioisomer formed (Scheme 50).

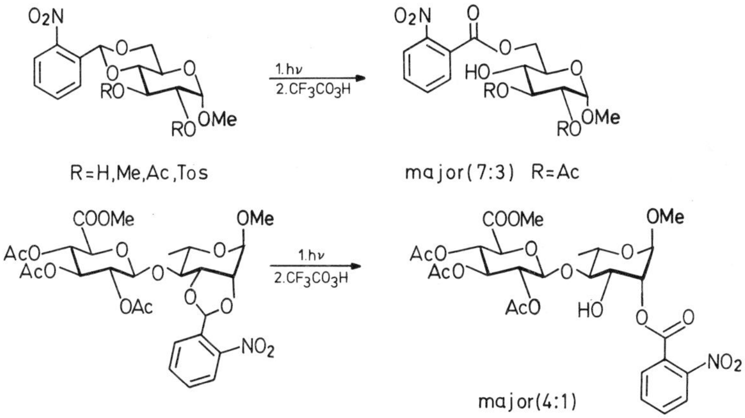

Scheme 50

The o-nitrobenzyl group was also used in nucleotide synthesis with the preparation of 2-O-(o-nitrobenzyl) ribonucleoside for oligoribonucleotide synthesis (Scheme 51) [104–106].

Resistant to acids and bases and without a tendency to migrate, this group is stable for all the reactions employed in the synthesis of oligonucleotides. It is photolytically cleaved under a controlled pH without affecting the pyrimidine or purine bases [107, 108]. It has also been used as a photoremovable protecting group for the phosphate function [109].

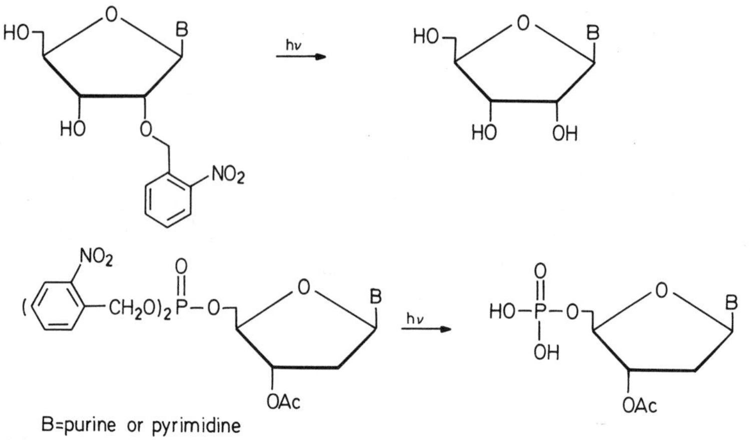

B=purine or pyrimidine

Scheme 51

6.2 Carbamates and Nitrates

6.2.1 N-Benzyloxycarbonyl Derivatives

The benzyloxycarbonyl has been used as a photoremovable protecting group for the amino function in aminosugars [110].

The photolysis of carbamates remains limited to mono derivatives but more complex molecules containing three or more N-benzyloxycarbonyl groups lead to mixtures.

Nevertheless, under the same conditions, methyl-2,6-dideoxy-2(benzyl-oxycarbonyl) amino 6-bromo-α-D-glucopyranoside *98* yields methyl 2,6-dideoxy-2-amino-6-bromo-α-D-glucopyranoside *99* which shows a greater stability of the C—Br bond in the photolytic treatment (Scheme 52).

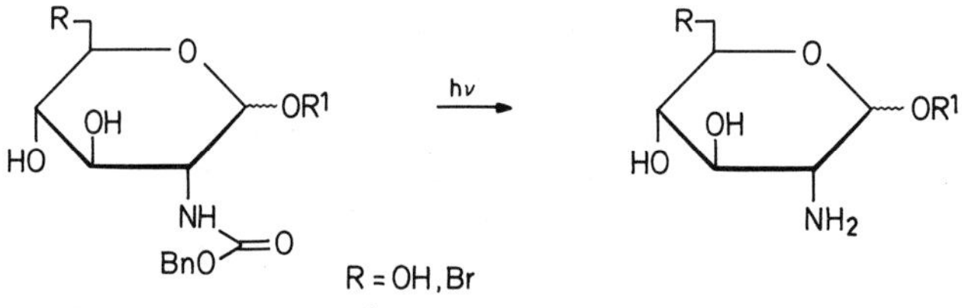

$R = OH, Br$
$R^1 = H, Me$

Scheme 52

6.2.2 Nitrates and other Nitrogen Protecting Groups

The interest of nitrates as protecting groups for carbohydrates is that they are stable in acidic conditions, but their stability in the presence of base is limited. Their deprotection is carried out by catalytic hydrogenolysis or by nucleophilic attack of hydrazine.

Photolysis of sugar nitrates in the presence of hydrogen donors represents an efficient technique for their cleavage with the intermediate formation of alkoxy radicals.

$$RO-NO_2 \xrightarrow{h\nu} [RO \cdot \ \cdot NO_2] \xrightarrow{alcohol} ROH$$

This alkoxy radical can explain the observed inversion of configuration at the C-3 position of 1,2–5,6 di-O-isopropylidene 3-O-nitro-α-D-*allo*furanose *100* during the deprotection. The cleavage of the C2, C-3 bond is a possible explanation for the epimerization [111] into *101* (Scheme 53).

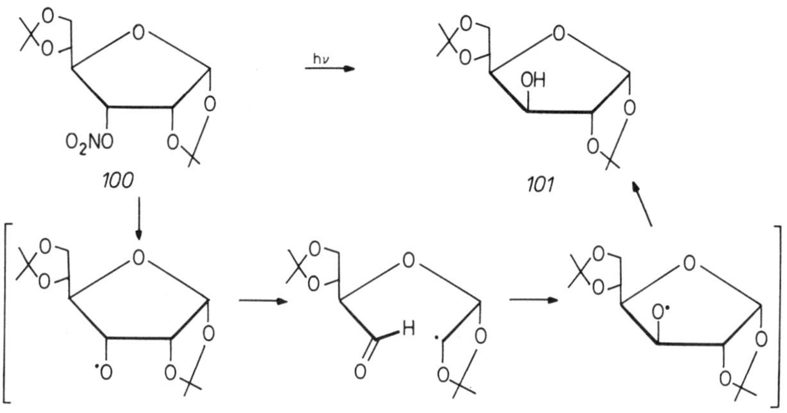

Scheme 53

Except for this particular example, quantitative deprotections were observed for other sugar nitrates without modification of the sugar moiety. Some other N-derivatives such as diphenyl hydrazino substituents introduced by triflate displacement at the C-6 position of galactose residues were photolyzed but in low yields [112]. This group remains of limited interest for any further use in synthesis.

7 Conclusions

The recent applications of photochemical reactions to carbohydrates are in accordance with the great development of radical reactions in organic synthesis this last decade.

Photochemistry provides mild reaction conditions for the formation of different new bonds especially C—C bonds. The study by ESR methods of glycosyl radicals and theoretical approach confirming the importance of the "*anomeric effect*" in stabilizing radicals explain the remarkable stereoselectivity of these reactions. The preferential α-attack of anomeric radicals is one of the most interesting results published in the recent literature. The development of new methodologies using sugar radicals or/and radical reagents should be an alternative way to many transformations of carbohydrates over the next years.

After a long period of studies using the electrophilic character of the anomeric center of sugars, some recent result show the possibility of "umpolung" of this carbon using strong electron-withdrawing groups. According to *"captodative effects"* which stabilize radicals and confirming the importance of *"stereoelectronic effects"* in carbohydrate chemistry, photochemical reactions with sugars should be regio- and stereocontrolled and should find a wide use in total syntheses of natural products using the *"chiron approach"*.

Photochemical reactions are usually run in homogeneous solutions but recent developments are found in the literature on photoreactions in solide state, on solid matrix, or in a micellar environment [113]. These new methodologies have not yet been applied to carbohydrates but should be of great interest in the near future.

The increasing use of spin-labeled carbohydrates [114] and the syntheses of enzyme-activated irreversible inhibitors (suicide substrates) such as diazoketo sugars [115] and azideoxy sugars [116] for glycosidases implicate that those compounds will serve as potential photoaffinity-labeling reagents fo arbohydrate-binding proteins.

With increasing theoretical and practical knowledge in photochemistry, synthetic applications of photochemical reactions to carbohydrates are a powerful tool for further research.

Acknowledgements: The author thanks Dr. L. Ludwikowska for her help in the preparation of the manuscript.

8 References

1. Binkley RW (1981) Adv. Carb. Chem. Biochem. 28: 105
2. Viehe HG, Janousek Z, Merenyi R (1985) Acc. Chem. Res. 18: 148
3. Mc Kelvey RD, Iwamura H (1985) J. Org. Chem. 50: 402
4. Deslongchamps P (1983) Stereoelectronic effects in organic chemistry, Pergamon, Oxford
5. Giese B (1986) Radicals in organic synthesis: formation of carbon-carbon bond, Pergamon, Oxford
6. Descotes G (1982) Bull. Soc. Chim. Belges, 91: 973
7. Box VGS (1984) Heterocycles, 22: 891
8. Somsak L, Batta G, Farkas I (1983) Carbohydr. Res., 124: 43
9. Praly JP, Descotes G (1987) Tetrahedron Lett., 28: 1405
10. Lichtenthaler FW, Frieder W, Jarglis P (1982) Angew. Chem., 948: 643
11. Lichtenthaler F, Jarglis P, Hempe W (1983) Liebigs Ann. Chem. 1959
12. Lichtenthaler FW, Kaji E, Weprek S (1985) J. Org. Chem., 50: 3505
13. Lichtenthaler FW, Kaji E (1985) Liebigs Ann. Chem. 1659
14. Collins PM, Manro A, Oppara Mottah C, Ali MH (1988) J. Chem. Soc. Chem. Commun. 272
15. a) Praly JP, Descotes G (1982) Tetrahedron Lett., 23: 849
    b) Praly JP, Descotes G, Grenier-Loustalot MF, Metras F (1984) Carbohydr. Res., 128:21
16. Wright DE (1979) Tetrahedron 35: 1207
17. a) Praly JP (1983) Tetrahedron Lett., 24: 3075
    b) Somsak L, Butta G, Farkas I (1986) Tetrahedron Lett., 27: 5877
18. a) Giese B, Dupuis J (1983) Angew. Chem. Int. Ed. Eng., 22: 622
    b) Giese B, Dupuis J, Nix M (1987) Org. Synth., 65. 236
19. Adlington RM, Baldwin JE, Basak A, Kozyrod RP J. Chem. Soc. Chem. Commun. 1983: 944
20. Korth HE, Sustmann R, Dupuis J, Giese B (1986) J. Chem. Soc. Perkin Trans 11: 1452
21. Giese B, Witzel T (1986) Angew. Chem. Int. Ed. Engl., 25: 450
22. Aebischer B, Meuwly R, Vasella A (1984) Helv. Chim. Acta., 67: 2236
23. Hashimoto S, Kurimoto I, Fujii Y, Noyori F (1985) J. Am. Chem. Soc., 107: 1427
24. Ferrier RJ, Furneaux RH J. Chem. Soc. Perkin 1. 1977: 1996
25. Ferrier RJ, Tyler PC J. Chem. Soc. Perkin 1, 1980: 2767
26. Blattner R, Ferrier RJ J. Chem. Soc. Perkin 1, 1980: 1523

27. Ferrier RJ, Furneaux RH (1980) Aust. J. Chem., 33: 1025
28. Chiba T, Sinay P (1984) Carbohydr. Res., 151: 379
29. Blattner R, Ferrier RJ (1986) Carbohydr. Res., 150: 151
30. a) Ferrier RJ, Haines SR J. Chem. Soc. Perkin I, 1984: 1675
    b) Ferrier RJ, Haines SR, Gainsford GJ, Gabe EJ: J. Chem. Soc. Perkin I, 1984: 1683
31. Ferrier RJ: J. Chem. Soc. Perkin I, 1985: 2413
32. Ferrier RJ: J. Chem. Soc. Perkin I, 1979: 1455
33. Blattner R, Ferrier RJ, Prasit P: J. Chem. Soc. Perkin I, 1980: 944
34. Ohrui H, Horiki H, Kishi H, Meguru H (1983) Agric. Biol. Chem., 47: 1101
35. Ohrui H, Yoshihiro N, Hiroshi M (1984) Agric. Biol. Chem., 48: 1049
36. Ohrui H, Tsutami M, Hiroshi M (1984) Agric. Biol. Chem., 48: 1825
37. Giese B, Gonzales-Gomez JA, Witzel T (1984) Angew. Chem., 96: 51
38. Giese B, Roninger KG (1984) Tetrahedron Lett., 25: 2743
39. Keck GE, Kachensky DF, Enholm EJ (1984) J. Org. Chem., 49: 1462
40. Keck GE, Yates JB (1982) J. Am. Chem. Soc., 104: 5829
41. Fitzsimmons BJ, Fraser-Reid B (1979) J. Am. Chem. Soc., 101: 6123
42. Bonjouklian R, Ganem B (1979) Carbohydr. Res., 76: 245
43. Kozlowska-Gramsz E, Descotes G (1982) Canad. J. Chem., 60: 558
44. a) Lessard J, Mondon M, Touchard D (1981) Canad. J. Chem., 59: 431
    b) Igarashi K, Honma T (1970) J. Org. Chem., 35: 606
45. Fitzsimmons BJ, Leblanc Y, Rokach J (1987) J. Am. Chem. Soc., 109: 285
46. a) Fetizon M, Ducdokhac Nguyen Dinh Tho (1986) Tetrahedron Lett., 26: 1777
    b) Matsui T, Morooka T, Akayama MN (1987) Bull. Chem. Soc. Japan., 60: 417
47. Fraser-Reid B, Anderson RC (1980) Fortschr. Chem. Org. Naturst., 39: 1
48. Araki Y, Senna K, Matsuura K, Ishido Y (1978) Carbohydr. Res., 60: 389
49. Jarosz S, Zamojski A (1983) Pol. J. Chem., 57: 57; (1982) Tetrahedron, 83: 1453
50. Araki Y, Nagasawa J, Ishido Y: J. Chem. Soc. Perkin I, 1981: 12
51. Bovin NV, Zurabyan SE, Khorlin AY (1981) Carbohydr. Res. 98: 25
52. Bovin NV, Zurabyan SE, Khorlin AY (1983) J. Carb. Chem. 2: 249
53. Chmielewski M, Bemiller JN, Cerretti D (1981) J. Org. Chem., 46: 3903
54. Rosenthal A, Chow J (1980) J. Carbohydr. Nucleosides, Nucleotides, 7: 77
55. Araki Y, Nishiyama K, Matsuura K (1978) Carbohydr. Res. 63: 288
56. Araki Y, Nishiyama K, Senna K, Matsuura K, Ishido Y (1978) Carbohydr. Res. 64: 119
57. Kozlowska-Gramsz E, Descotes G (1982) Tetrahedron Lett. 23: 1585
58. a) Fraser-Reid B, Holder NL, Hicks DR, Walker DL (1977) Can. J. Chem., 55: 3978
    b) Benko Z, Fraser-Reid B, Mariano PS, Beckwith ALJ (1988) J. Org. Chem. 53: 2066
    c) Sakakibara T, Nakagawa T (1987) Carbohydr. Res., 163: 239
59. Holder NL (1982) Chem. Rev. 82: 287
60. Hicks DR, Anderson RC, Fraser-Reid B Synth. Commun. 1976: 417
61. Fraser-Reid B, Walker DR (1980) Can. J. Chem., 58: 2694
62. Collins PM, Iyer R, Travis AS (1978) J. Chem. Res., 5: 446
63. Heyns K, Neste HP, Thiem J (1981) Chem. Ber., 114: 891
64. Collins PM, Gupta R, Travis AS J. Chem. Soc. Perkin I 1980: 277
65. Collins PM, Farnia F, Munasinghe VRN, Oparaech NN, Siavoshy F J. Chem. Soc. Chem. Commun. 1985: 1038
66. Collins PM, Farnia F, Travis AS J. Chem. Res., (S) 1979: 266
67. Remy G, Cottier L, Descotes G (1980) Can. J. Chem., 58: 2660
68. Remy G, Cottier L, Descotes G (1982) J. Carbohydr. Res. 1: 37
69. Remy G, Descotes G (1983) J. Carbohydr. Res. 2: 159
70. Remy G, Cottier L, Descotes G (1983) Can. J. Chem. 61: 434
71. Bernasconi C, Cottier L, Descotes G, Praly JP (1983) Carbohydr. Res. 115: 105
72. Remy G, Cottier L, Descotes G, Faure R. Loiseleur H (1980) Acta Cryst. B36: 873
73. Remy G, Cottier L, Descotes G, Faure R, Loiseleur H (1982) Cryst. Struct. Commun. 11: 235
74. Cottier L, Remy G, Descotes G Synthesis 1979: 711
75. Pete JP, Portella C, Monneret C, Florent JC, Khuong Huu Q Synthesis 1977: 774
76. Portella C, Deshayes H, Pete JP, Scholler D (1984) Tetrahedron. 40: 3635
77. Portella C, Pete JP (1985) Tetrahedron Lett. 26: 211

78. Pete JP, Portella C, Scholler D (1984) J. Photochem., 27: 128
79. Pete JP, Portella C: Bull. Soc. Chim. Fr. 1985: 195 and references herein
80. Collins PM, Ranjit V, Munasinghe Z: J. Chem. Soc., Chem. Commun., 1977: 1927
81. Hartwig W (1983) Tetrahedron 39: 2609
82. Klausener A, Muller E, Runsink J, Scharf HD (1985) Carbohydr. Res., 116: 295
83. Dornhagen J, Scharf HD J. Carbohydr. Chem. 5: 115 (1986)
84. Dornhagen J, Klausener A, Runsink J, Scharf HD Liebigs Ann. Chem. 1985: 1838
85. Saito I, Ikehira H, Kasatani R, Watanabe M, Matsuura T (1986) J. Am. Chem. Soc., 108: 3115
86. Kishi T, Tsuchiya T, Umezawa S (1979) Bull. Chem. Soc. Japan, 52: 3015
87. Tsuchiya T, Nakamura F, Umezawa S (1979) Tetrahedron Lett., 30: 2805
88. Bell RH, Horton D, Williams DM, Winter-Mihaly E (1977) Carbohydr. Res. 58: 109
89. Descotes G, Faure A, Kryczka B, Bouchu MN (1979) Bull. Soc. Sci. Pol., 27: 173
90. Binkley RW, Bankaitis D (1982) J. Carbohydr. Chem., 1: 1
91. Roth RC, Binkley RW (1985) J. Org. Chem. 50: 690
92. Binkley RW, Fan JC (1982) J. Carbohydr. Chem. 1: 213
93. Binkley RW (1977) J. Org. Chem. 42: 1216
94. Binkley RW, Hehemann DG, Binkley WB (1978) J. Org. Chem., 43: 2573
95. a) Binkley RW (1985) J. Carbohydr. Chem. 4: 227
    b) Binkley RW (1980) Methods Carbohydr. Chem. 8: 321
96. Dendruver L, Holzapfel CW, Vandick MS, Kruger GJ (1987) Carbohydr. Res., 161: 65
97. Bernasconi C, Cottier L, Descotes G, Remy G (1979) Bull. Soc. Chim. Fr. 1979: 332
98. Sato T, Takahashi K, Ichikawa S Chemistry Lett. 1983: 1589
99. Araki K, Sakuma M, Shiraishi S Chemistry Lett. 1983: 665
100. Ichikawa S, Sato T (1986) Tetrahedron Lett., 27: 45
101. Pillai VNR: Synthesis 1980: 1
102. Binkley RV, Goewey GS, Johnston JC (1984) J. Org. Chem. 49: 992
103. Collins PM, Munasinghe V J. Chem. Soc. Perkin I 1983: 921
104. Collins PM, Eder H: J. Chem. Soc. Perkin I 1983: 927
105. Collins PM, Munasinghe V J. Chem. Soc. Perkin I, 1983: 1879
106. Baines FC, Collins PM, Farnia F (1985) Carbohydr. Res. 136: 27
107. Ohtsuka E, Tanaka S, Ikehara M: Synthesis 1977: 453
108. Hayes JA, Brunder MJ, Gilham PT, Gough GR (1985) Tetrahedron Lett., 26: 2407
109. Rubinstein M, Amit B, Patchornik A: Tetrahedron Lett. 1975: 1445
110. Hanessian S, Masse R (1977) Carbohydr. Res., 54: 142
111. a) Binkley RW, Koholic DJ (1984) J. Carbohydr. Chem. 3: 85
     b) Masnovi J, Koholic DJ, Berki RJ, Binkley RW (1987) J. Am. Chem. Soc. 109: 2851
112. Flechtner FW, Pohlid H (1984) J. Carbohydr. Chem. 3: 107
113. Ramamurthy V (1986) Tetrahedron 42: 5753
114. Gnewuch T, Sosnovsky G (1986) Chem. Rev. 86: 203
115. Myers RW, Chuan Lee Y (1986) Carbohydr. Res. 152: 143
116. Lehmann J, Thieme R Liebigs Ann. Chem. 1986: 525

Synthesis of Glycolipids

Jill Gigg and Roy Gigg

Laboratory of Lipid and General Chemistry, National Institute for Medical Research, Mill Hill, London NW7 1AA/U.K.

Table of Contents

Topics in Current Chemistry, Vol. 154
© Springer-Verlag Berlin Heidelberg 1990

The chemical syntheses of the carbohydrate portions of the naturally occurring glycolipids derived from the sphingosine bases (as well as the syntheses of the complete glycosphingolipids) and of myo-inositol-containing lipids derived from esters of glycerol, which were reported between 1977 and 1987, are reviewed. This article therefore brings up to date a previous review published in Chemistry and Physics of Lipids (1980) 26: 287. Since this is a very active field of research, most of the new synthetic techniques introduced into carbohydrate chemistry in the last decade are discussed. The syntheses of sphingosine bases and glycerol derivatives from carbohydrate precursors acting as 'chiral templates' are also reviewed with a more historical perspective.

1 Introduction

This chapter will cover work published in this field in the last ten years and will thus be an update of a previous article [1] which reviewed the field up to the end of 1977. Because of the large amount of work published in this area, this review is selective rather than complete. The synthesis of glycoglycerolipids (see Ref. [2] for an excellent review of the structures of these lipids) has been reviewed recently [3] as has the synthesis of the serologically active glycolipids from *Mycobacterium leprae* [4]. The extensive early work on the synthesis of lipid A, based on an incorrect structure, has been reviewed [5, 6] (see Refs. [7–11] for more recent work). This review will therefore deal mainly with the work on glycosphingolipids (see Refs. [12–20] for excellent reviews of the structures and biochemistry of these lipids) and also covers the synthetic work on *myo*-inositol-containing lipids. Significant advances have been made during the last ten years in oligosaccharide synthesis and this is reflected in the more sophisticated approach to glycolipid synthesis. The use of 2-azido-2-deoxy sugar derivatives by Paulsen and his co-workers has allowed the synthesis of 1,2-*cis*-glycosides of 2-amino-2-deoxy sugars which are present in some immunologically interesting glycolipids and the continued introduction of new protecting groups and manipulative methods has also made an impact. The synthesis of glycosides of *N*-acetyl-neuraminic acid is being studied in greater detail and this will ultimately lead to the synthesis of the more complex gangliosides which are of general scientific significance. Some excellent review articles [22–32] on oligosaccharide synthesis have been written by researchers active in the area of glycolipid and glycoprotein synthesis.

Since the synthesis of glycolipids involves the synthesis of both the lipid and the oligosaccharide portions, we shall also discuss that part of the recorded lipid synthetic work which has involved the use of carbohydrates. In fact, some of the first-recorded applications of carbohydrate molecules as "chiral templates" are to be found in the lipid field particularly with the use of 1,2:5,6-di-*O*-isopropylidene-D-mannitol [33] as a precursor of chiral glycerol derivatives for the synthesis of phospholipids and glycolipids based on glycerol and with the use of glucosamine derivatives for the synthesis of phytosphingosines [34] and since this area has not previously been reviewed, it will be treated with a more historical perspective.

As well as the synthesis of natural glycolipids, we shall discuss the synthesis of portions of the oligosaccharide of the natural lipid which may be the epitope of an immunologically interesting glycolipid and/or glycoprotein since the linking of the oligosaccharide portion to the lipid is an extension of this work which is not always reported.

2 Use of Carbohydrates as Chiral Templates for the Synthesis of the Lipid Portion of Glycolipids

2.1 Sphingosine Bases

The first recorded use of carbohydrates as templates for the synthesis of the sphingosine bases was for the preparation of *N*-benzoylphytosphingosine (6) [34, 35]. The oxazoline derivative (1) which is readily prepared from D-glucosamine [36] was converted by very mild methanolysis and subsequent mesylation and reformation of an oxazoline into the phenyloxazoline (2). This, on methanolysis and subsequent periodate oxidation, gave the aldehyde (3) which was condensed with a long-chain Wittig

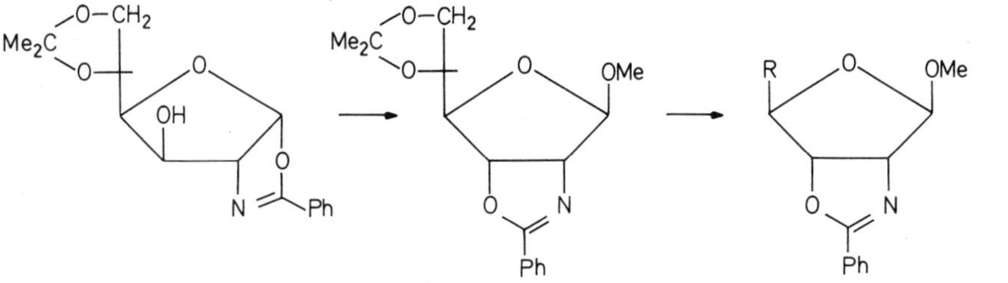

1*   2   3   R=CHO
                                         4   R=$(CH_2)_{13}$Me

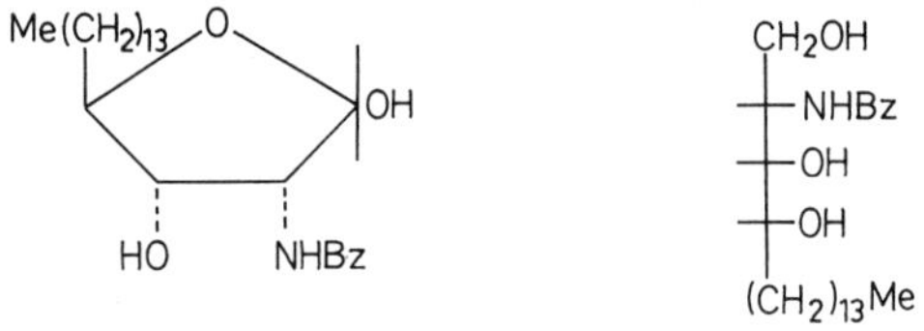

5   6

*   Ac = CH_3CO- ; Bz = PhCO− ; Bn = $PhCH_2-$;
All = $CH_2=CH-CH_2-$; Phth = phthaloyl ;
Ms = SO_2Me

reagent and the product hydrogenated to give (4). Opening of the oxazoline ring by acid gave an *O*-benzoyl derivative which was converted into a benzamido derivative under basic conditions; subsequent hydrolysis of the glycosidic bond gave (5) which was reduced with sodium borohydride to *N*-benzoylphytosphingosine (D-*ribo*-2-benzamido1,3,4-trihydroxyoctadecane) (6).

A further synthesis of phytosphingosine (11) was achieved [37] from D-galactose via 3,4,6-tri-*O*-benzyl-D-galactose. Reduction of the latter and subsequent acetonation gave the galactitol derivative (7), and this on mesylation and reaction with potassium

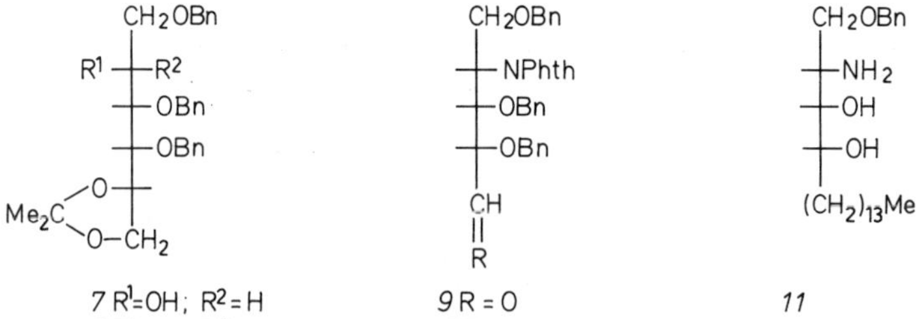

7 R^1=OH; R^2=H   9 R = O   11
8 R^1= H ; R^2= NPhth   10 R = $CH(CH_2)_{11}$Me

phthalimide gave the phthalimido derivative (8). Hydrolysis of the isopropylidene derivative and periodate oxidation of the product gave the aldehyde (9) which was condensed with a long-chain Wittig reagent to give (10). Hydrogenation to remove benzyl groups and reduce the olefinic group and subsequent removal of the phthalimido group with hydrazine gave phytosphingosine (11). Mulzer and Brand [38] have prepared D- and L-*ribo*- and *arabino*-C_{18}-phytosphingosine from D-mannitol via 2,3-*O*-isopropylidene-D-glyceraldehyde. Condensation of the latter with a long-chain Wittig reagent and eventual replacement of the secondary hydroxyl group, with inversion, by a phthalimido group (using triphenylphosphine, diethylazodicarboxylate and phthalimide — Mitsunobo reaction) and hydroxylation of the double bond with osmium tetroxide and *N*-methylmorpholine *N*-oxide gave a mixture of isomers which were separated by chromatography.

Dihydrosphingosine (20) was prepared by Reist and Christie [39] from 1,2:5,6-di-*O*-isopropylidene-D-glucofuranose which was converted into the 3-amino-3-deoxyallose derivative (12). Hydrolysis of (12) and subsequent periodate oxidation gave the aldehyde (14) which was condensed with the Wittig reagent and the product hydrolysed to give a mixture of *cis*- and *trans*-isomers (16) in which the *cis*-isomer predominated. Periodate oxidation and subsequent reduction with sodium borohydride gave (18) and this on hydrogenolysis gave the dihydrosphingosine (20).

12 R=Bn
13 R=Et

14 R=Bn
15 R=Et

16 R=Bn
17 R=Et

18 R=Bn
19 R=Et
20 R=$(CH_2)_{14}Me$
21 R= $HC{=}CH(CH_2)_{12}Me$

20
21

18
19

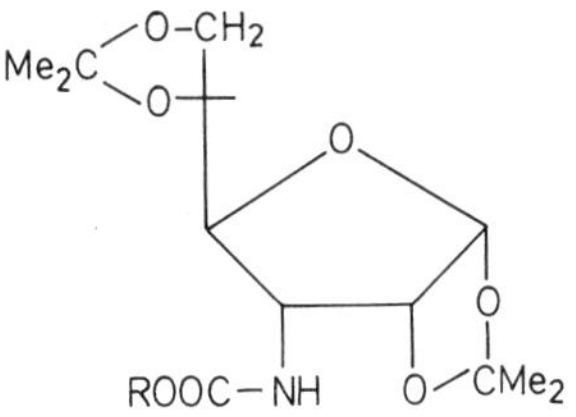

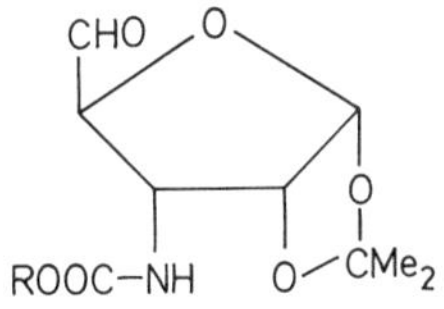

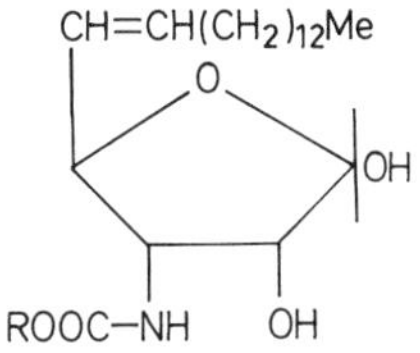

For the preparation of sphingosine (21), which has a *trans*-olefinic linkage, Reist and Christie [40] started from the ethoxycarbonyl derivatives (13) and (15) and modified the conditions of the Wittig condensation so that the product (17) was predominantly the *trans*-olefin. Periodate oxidation and subsequent borohydride reduction of (17) gave (19) and the ethoxycarbonyl group was removed with barium hydroxide to give sphingosine (21).

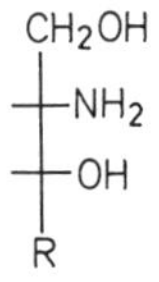

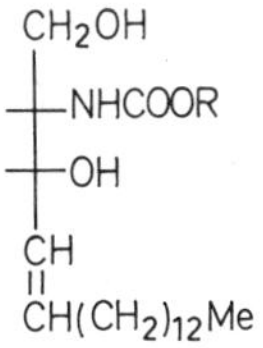

Ogawa and co-workers [41, 42] have also described a synthesis of the *N*-acyl sphingosine (ceramide) (27) from the aldehyde (22) derived from 1,2:5,6-di-*O*-isopropylidene-D-glucofuranose. This was condensed with a Wittig reagent to give an approximately equal mixture of the *cis*- and *trans*-olefins (23) and irradiation of the mixture gave a product containing 94% of the *trans*-olefin. The *trans*-olefin (23) was converted into the mesylate and the isopropylidene group hydrolysed to give (24) which was oxidised with periodate and the product reduced with borohydride.

Subsequent treatment with ethyl vinyl ether gave the ethoxyethyl ether (25). The mesyl group was replaced by sodium azide to give (26) and the azide group reduced to an amine which was acylated with lignoceric acid (tetracosanoic acid); removal of the protecting groups with acid gave the ceramide (D-*erythro*-2-acylamido-1,3-dihydroxy-octadec-*trans*-4-ene) (27).

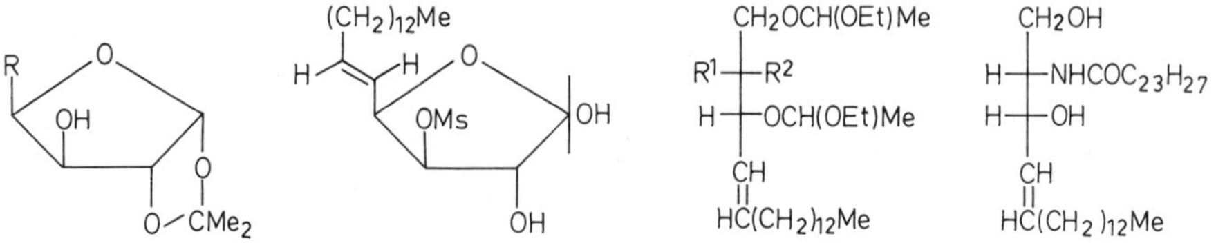

22   R=CHO
23   R=CH=CH(CH₂)₁₂Me

24

25   R¹= OMs; R²=H
26   R¹= H; R²= N₃

27

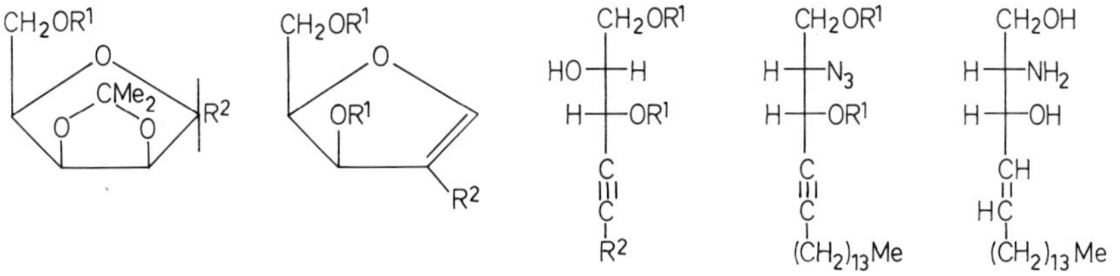

28   R²= Bn
29   R²=Cl

30   R²=H
31   R²=Br

32   R²=H
33   R²=(CH₂)₁₃Me

34

35

R¹ = CH₂OMe

Obayashi and Schlosser [43] have briefly described syntheses of *erythro*-sphingosine and *threo*-sphingosine from D-mannose and D-ribono-1,4-lactone, respectively. For the synthesis of *erythro*-sphingosine (35) D-mannose was converted into benzyl 2,3:5,6-di-*O*-isopropylidene-manno-furanosides and hydrolysis of the 5,6-*O*-isopropylidene group followed by periodate oxidation and borohydride reduction and protection gave the methoxymethyl ether (28). This was converted into the chloride (29) which gave the 2,3-dihydrofuran (30) in 56% yield on treatment with lithium — ammonia and subsequent protection of the free hydroxyl group. Bromination of (30) and treatment of the product with base gave the bromo-2,3-dihydrofuran (31) which was treated with butyl lithium followed by water to give the acetylene (32). Alkylation of (32) gave the alkyl acetylene (33) in 64% yield which was converted via the mesylate into the azide (34) which gave the *erythro*-sphingosine (35) in ca. 15% yield from (33).

For the synthesis of *threo*-sphingosine (40), D-ribono-1,4-lactone was converted into the protected derivative (36) which after reduction was converted into the chloride (37). This was converted via the 2,3-dihydrofurans (38) and (39), as described above for the synthesis of the *erythro*-sphingosine, into the *threo*-sphingosine (40).

Schmidt and Zimmerman [44] have prepared C₂₀-*erythro*-sphingosine (44) from 4,6-*O*-benzylidene-D-galactose. This was cleaved with periodate to give the D-threose derivative (41) which, with a Wittig reagent in the presence of lithium bromide, gave predominantly the *trans*-olefin (42). This was converted via the trifluoromethane sulfonate into the azide (43) which was deprotected and reduced with hydrogen sulfide to give the C20-sphingosine (44) which was converted into the known triacetate.

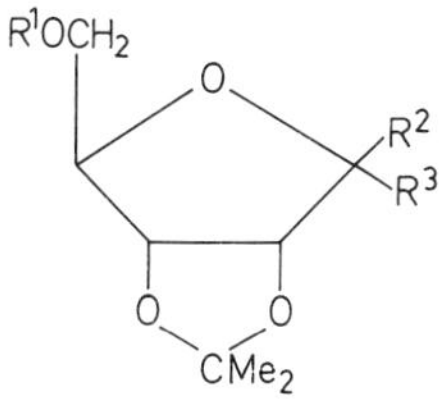

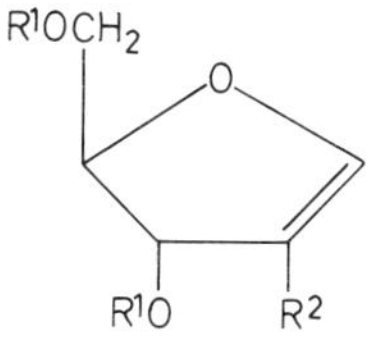

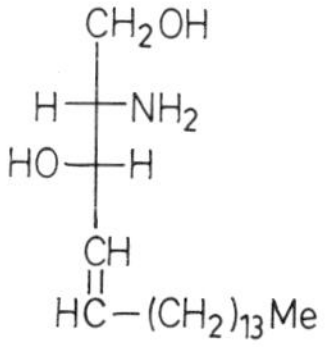

CH2OH
H—NH2
HO—H
CH
HC—(CH2)13Me

36 R2,R3 = O
37 R2,R3 = H,Cl

38 R2=H
39 R2=Br

40

R1 = CH2OMe

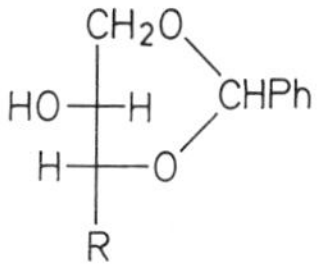

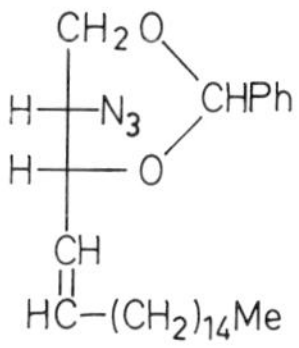

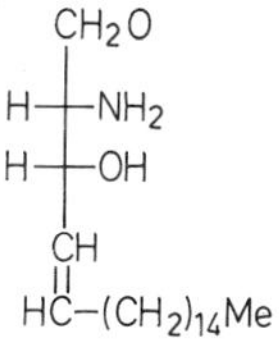

41 R=CHO
42 R=C=C(CH2)14Me

43

44

Kiso and his co-workers [45, 46] have also prepared *erythro*-sphingosine and the *N*-acyl derivatives ("ceramides", 48) from galactose or xylose using similar procedures. D-Xylose was converted into 3,5-*O*-isopropylidene-D-xylofuranose which was oxidised with periodate to the aldehyde (45) (which was also obtained from 4,6-*O*-isopropylidene-D-galactopyranose on treatment with periodate). The Wittig reaction with (45) gave a mixture of the *cis*- and *trans*-isomers (46) which were separated in crystalline form by chromatography. Each was converted via the mesylates into the azides (47) which were reduced to the amines and acylated to give the ceramides (48) on acid hydrolysis. Alternatively, the azides (47) were deacetonated and reduced to give the free sphingosines.

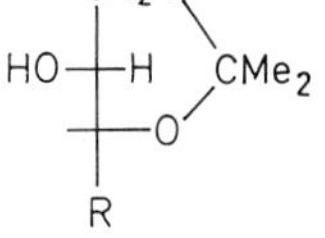

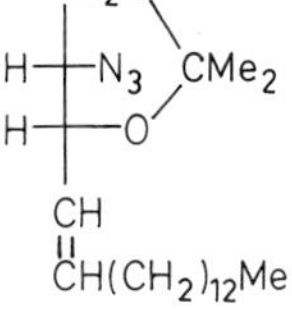

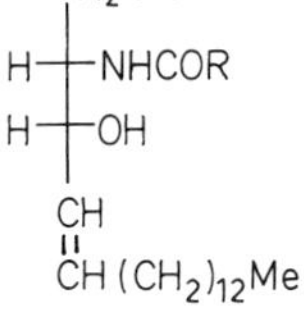

45 R=CHO
46 R=CH=CH(CH2)12Me

47

48

Gigg and Conant [47] have described the synthesis of the oxirane (49) as an intermediate for the synthesis of the protected sphingosine derivative (52). The oxazoline (1) was converted into the allyl furanoside (cf. 2) which gave the oxirane (49). The copper-catalysed Grignard reaction on (49) should give (50) which can be converted into (51) and this should give the *trans*-sphingosine derivative (52) on ene formation using the Corey-Winter procedure [48, 49]. Phenyloxazoline derivatives of sphingosine (52) have been used by Shapiro et al. [50, 51] for the synthesis of sphingomyelin.

Several elegant syntheses of sphingosine bases — not involving carbohydrate derivatives — have also been reported recently [52–62, 313].

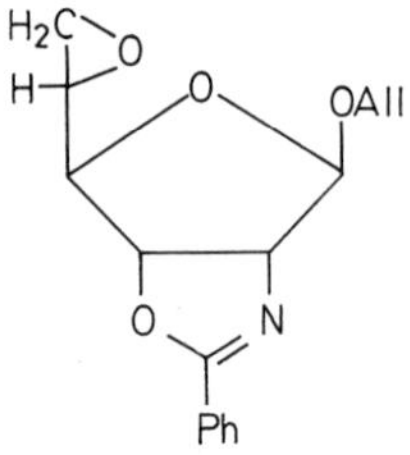

49 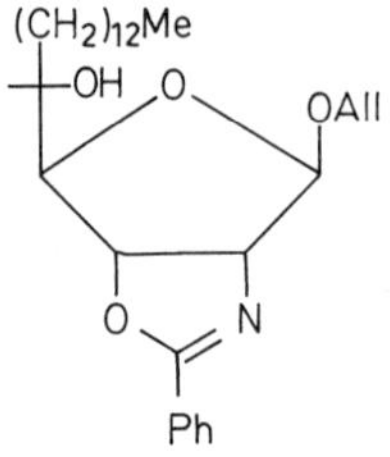   50 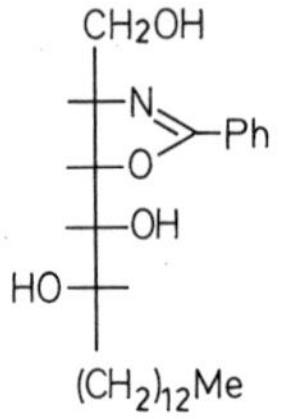   51 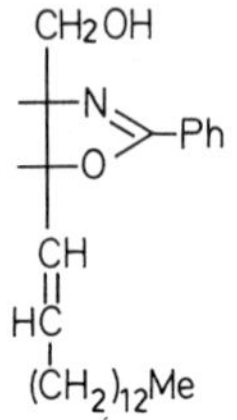   52

2.2 Glycerol Derivatives

Baer and Fischer [33] recognised early the value of D-mannitol for the preparation of chiral derivatives of glycerol in lipid synthesis. 1,2:5,6-Di-*O*-isopropylidene-D-mannitol (53) was cleaved with lead(IV) acetate to give the glyceraldehyde derivative (54) and this was reduced to give the 1,2-*O*-isopropylidene L-glycerol (55) which was used as a precursor for the synthesis of glycerol-containing lipids. This series of reactions for the synthesis of (54) and (55) has been extensively investigated in order to improve the conditions and yields (for reviews see Refs [63–65]). The method has also been used for making the enantiomer (56) starting from L-mannitol, but the latter is not readily available and ascorbic acid (57) is a better starting material. The acetonide

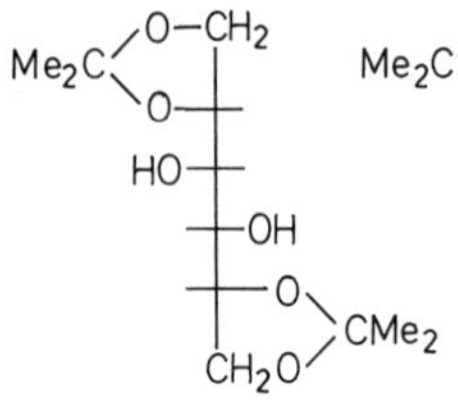

53 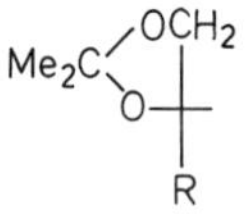   54 R=CHO   56   57 R¹=R²=H
       55 R=CH₂OH   58 R¹, R²=>CMe₂

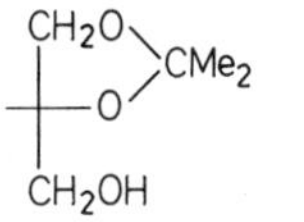

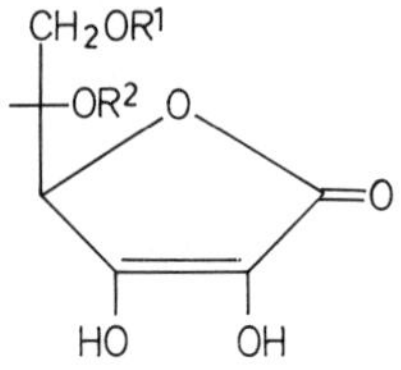

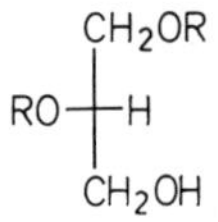

59 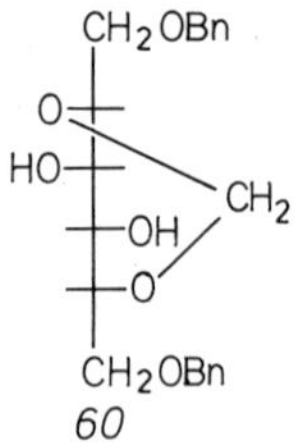   60 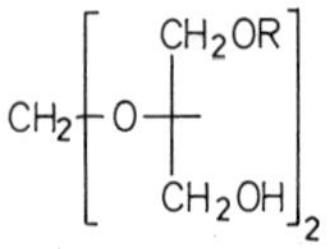   61 R=Bn
             62 R=COR¹

(58) of ascorbic acid was converted [66–68] by borohydride reduction, alkali treatment and subsequent lead (IV) acetate oxidation into 1,2-O-isopropylidene-D-glycerol (56). The latter has also been prepared from L-arabinose [69]. The conversion of the enantiomeric glycerol acetonides into lipids based on glycerol via their allyl or benzyl ethers has been used extensively [70, 71].

3,4-O-Isopropylidene-D-mannitol [72] is readily prepared by hydrolysis of 1,2:3,4: 5,6-tri-O-isopropylidene-D-mannitol and after alkylation gives the 1,2,5,6-tetra-O-alkylmannitol on acidic hydrolysis and this is readily converted into 1,2-di-O-alkyl-L-glycerol (59). The benzyl (59, R = Bn) [73, 74], allyl (59, R = All) [75] and but-2-enyl (59, R = CH_2—CH=CH—Me) [75, 76] ethers have been prepared in this way as intermediates for the synthesis of lipids.

The readily available [77, 78] 1,6-di-O-benzoyl-3,4-O-benzylidene-2,5-O-methylene-D-mannitol has been converted [79] into the benzyl ether (60), which on periodate oxidation and borohydride reduction gives the methylene acetal (61) which readily gives 1-O-benzyl-L-glycerol for use in lipid synthesis. The acyl derivative (62) has been used [80] in glycolipid synthesis by glycosidation of the hydroxyl group.

3 Monoglycosyl Ceramides

Schmidt and Klager [81] have synthesized the glucosylceramide (67) by condensation of the synthetic racemic erythro-3-O-benzoyl-ceramide (63) with O-(2,3,4,6-tetra-O-acetyl-α-D-glucopyranosyl) trichloroacetimidate (64) in dichloromethane with boron trifluoride etherate as a catalyst. The diastereoisomeric products (66) were separated by column chromatography and hydrolysed by base to give the cerebroside (67, with natural configuration) as well as the isomer with the opposite configuration in the sphingosine portion.

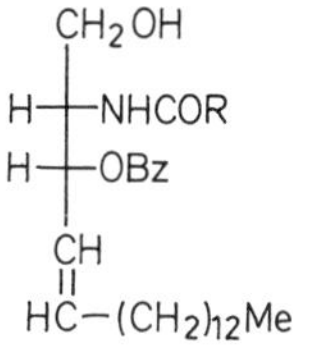

63

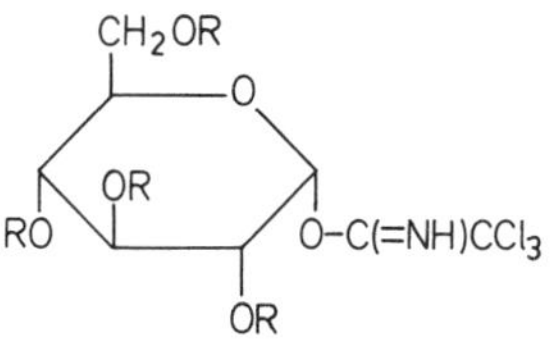

64 R= Ac
65 R = Me₃CCO

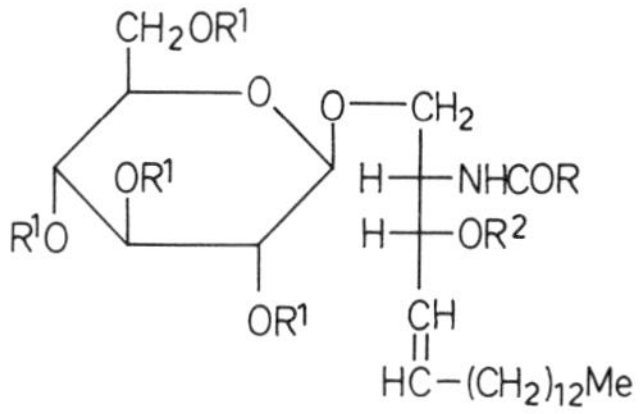

66 R¹=Ac ; R²=Bz
67 R¹=R² =H

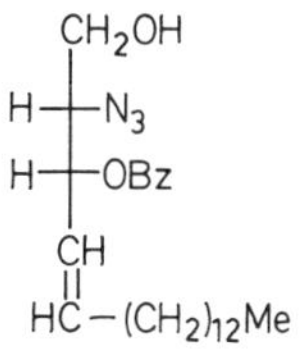

68

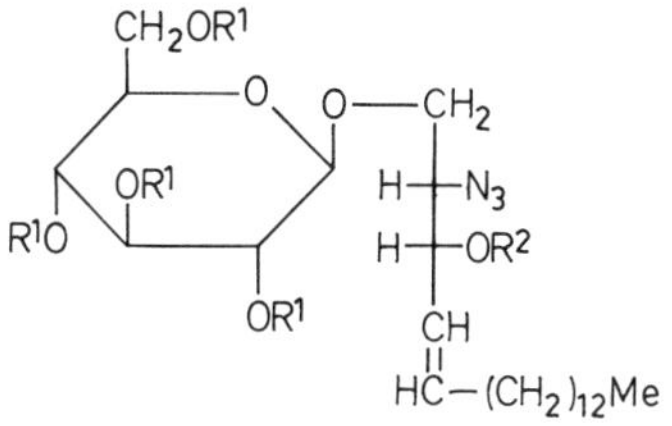

69 R¹ =Me₃CCO ; R²=Bz
70 R¹ =R² =H

Schmidt and Zimmerman [82] have also condensed the imidate (65) with the synthetic azido derivative (68) to give (69) in high yield, which on basic hydrolysis gave (70). The azido group of (70) was reduced to the free amine with hydrogen sulfide in aqueous pyridine to give the glucosyl sphingosine ("glucopsychosine") which was N-acetylated to give glucosylceramides. (For a review of the synthetic work in this area see Ref. [28]).

Ogawa and co-workers [83] have used the galacto-analog of the imidate (64) to prepare galactosylceramides (73) from ceramides (72) containing either the (R)- or (S)-isomers of the t-butyldimethylsilyl ether of 2-hydroxytetracosanoic acid which were prepared from the protected sphingosine derivative (71). Condensation of the ceramide (72) containing the (R)-acid derivative with the galactosyl imidate gave, after deprotection, a galactosyl ceramide with an NMR spectrum identical to that of the natural lipid.

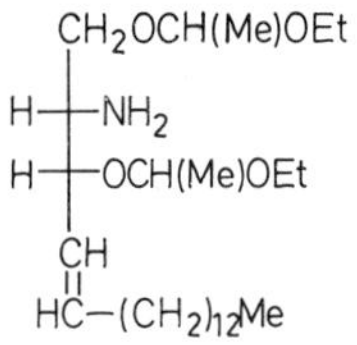

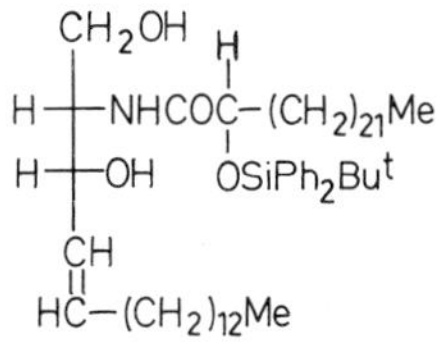

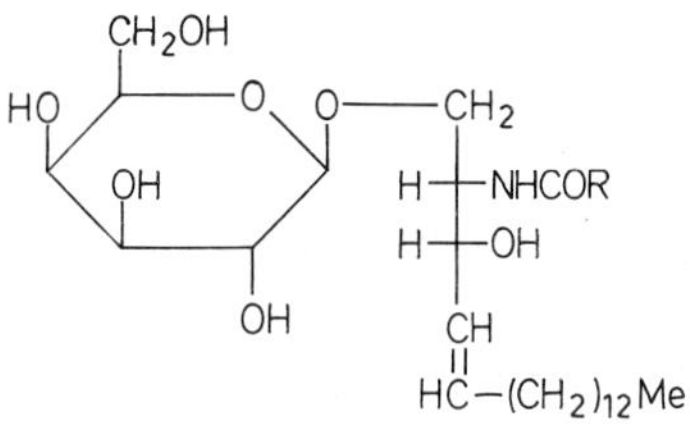

71   72   73

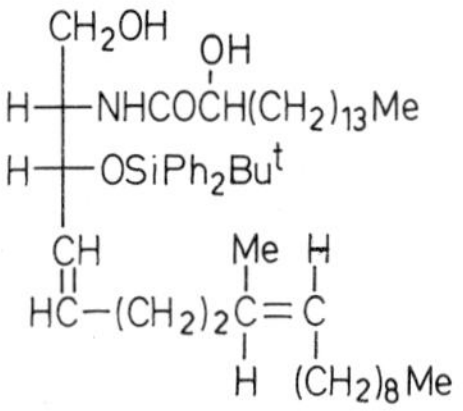

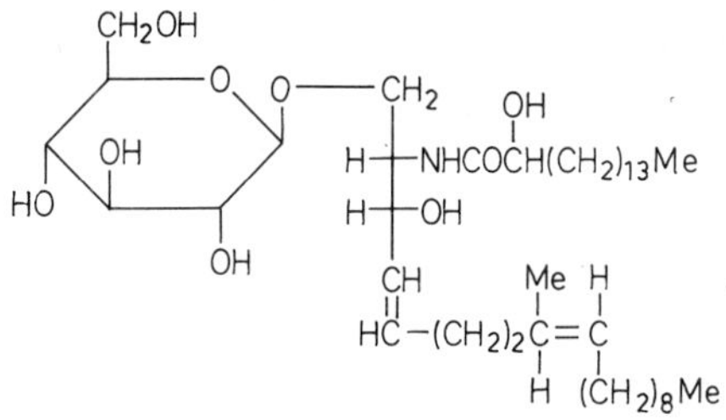

74   75

Mori and Funaki [84] have prepared a glucosyl ceramide from the protected ceramide (74) containing the (R)-2-hydroxyhexadecanoic acid derivative of sphingadiene. Condensation of (74) with acetobromoglucose in the presence of mercury(II) cyanide and subsequent deprotection gave the glucosyl ceramide (75) which is biologically active and induces fruiting in the fungus *Schizophyllum commune* [85]. Ceramides protected in this way were first used by Tkaczuk and Thornton [55] for the synthesis of cerebrosides.

A glucosyl ceramide containing L-glucose (and also containing ^{14}C labelling in the palmitoyl group of the ceramide) has been synthesised [86] from acetobromo-L-glucose and a derivative of racemic N-palmitoyl-*erythro*-sphingosine in the presence of mercury(II) cyanide. The product was not affected by a glucocerebrosidase which hydrolyses the natural cerebroside, containing D-glucose, and thus it should accumulate in vivo and mimic the effects of the accumulation of the normal D-glucosyl ceramide in Gaucher disease.

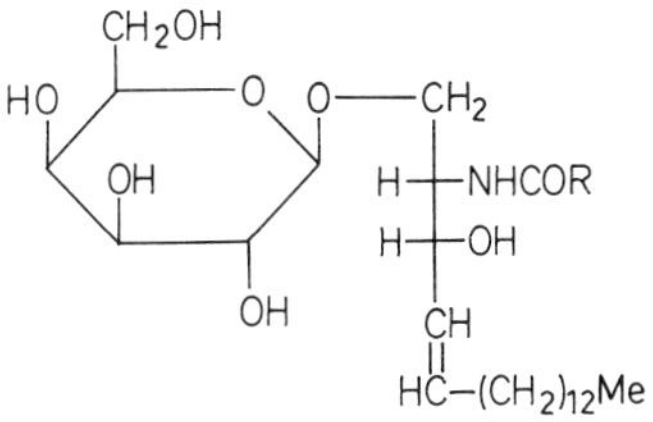

CH₂OH structures...

76 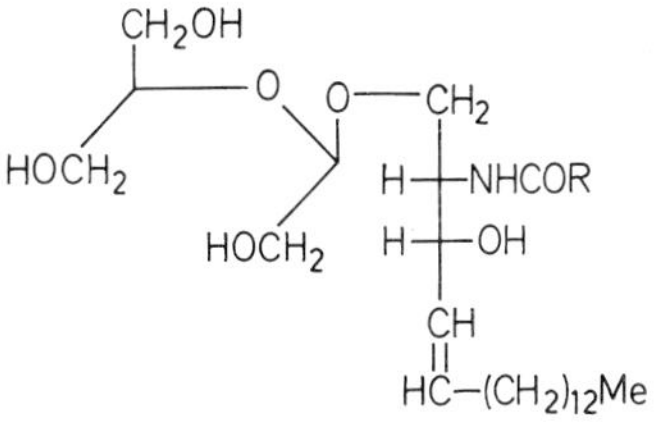 *77* 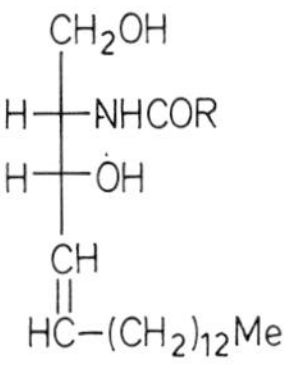 *79*

Ceramides have often been prepared [87], for partial syntheses of other glycolipids, from the readily available natural galactosyl ceramides (76) of bovine brain using the Smith degradation. Periodate oxidation of (76) and subsequent reduction gave the acetal (77) of glycolaldehyde which was very readily hydrolysed by dilute acid to give the ceramide (79). Direct acid hydrolysis of the cerebroside (76) required much more vigorous acidic conditions resulting in the cleavage of the *N*-acyl group and allylic rearrangements in the sphingosine molecule. These ceramides contain the same mixture of sphingosine bases and fatty acids found in the natural lipid.

Ceramides prepared in this way have been used for the partial synthesis of the α-L-fucopyranosyl ceramide (84) (a compound previously isolated from metastatic human carcinoma). Condensation [88] of 2,3,4-tri-*O*-benzyl-α-L-fucopyranosyl bromide (80) with an unprotected ceramide in the presence of tetraethylammonium bromide in dichloromethane and subsequent chromatographic purification gave (81) which on catalytic hydrogenation gave the saturated fucosyl ceramide (84). The dichloroacetamido derivative (82) was prepared similarly and converted into the free sphingosine derivative (83) by the action of barium hydroxide.

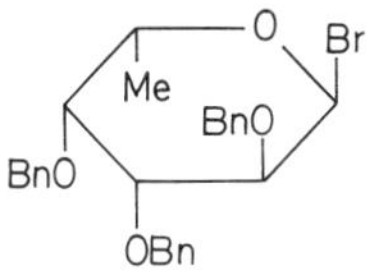

80 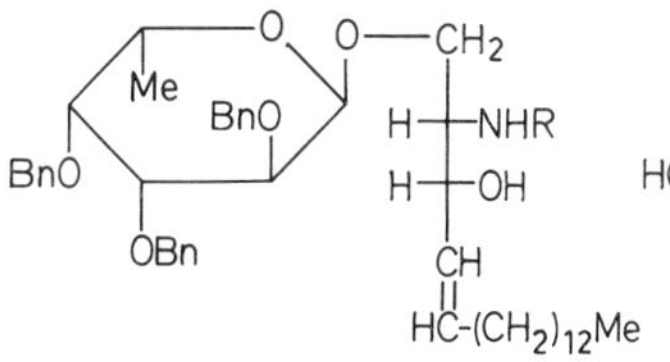   *81* R=CO(CH₂)ₙMe   *84* R = (CH₂)ₙMe
        82 R=COCHCl₂
        83 R=H

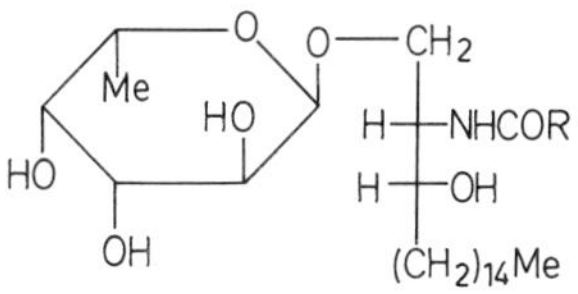

Shapiro and co-workers have described [89] (in a paper submitted just before the death of this pioneer in the field of glycosphingolipid synthesis — for a review of his extensive work in this area see Ref. [1]) a synthesis of the thio-analogue (88) of glucosyl ceramide (and the saturated derivative) from 2,3,4,6-tetra-*O*-acetyl-1-mercapto-β-D-glucopyranose (86) and the 1-deoxy-1-iodoceramide derivative (85) in the presence of 1,8-diazabicyclo[5.4.0]undec-7-ene. The product (87) was deacylated with methanolic barium methoxide to give (88).

A partial synthesis of cerebroside sulfate ['sulfatide', the glycoside of ceramide with galactose 3-sulfate, (90)] was achieved [90] by acylating the sphingosine galactosyl 3-sulfate (89) (obtained by basic hydrolysis of natural sulfatide) with palmitoyl chloride or D-2-acetoxypalmitoyl chloride (and subsequent basic hydrolysis of the

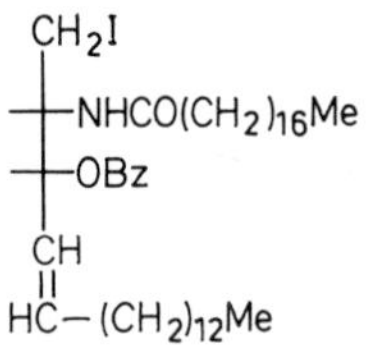

85 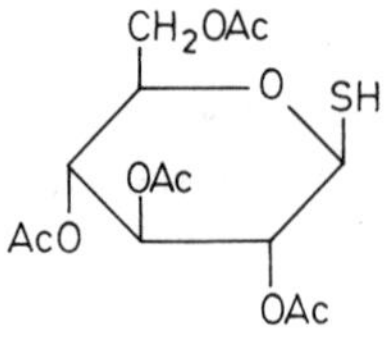   86 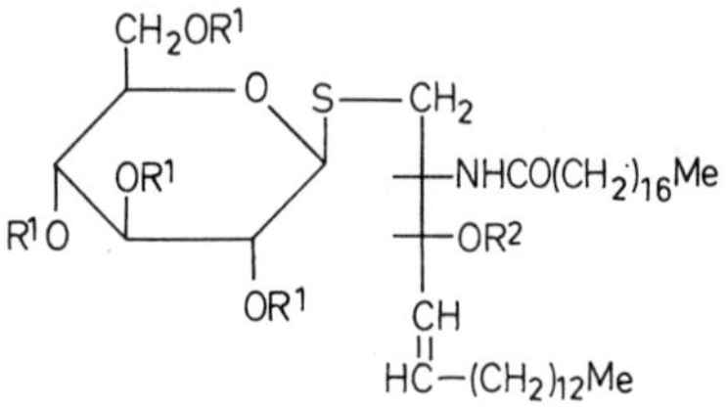   87 $R^1 = Ac$; $R^2 = Bz$
88 $R^1 = R^2 = H$

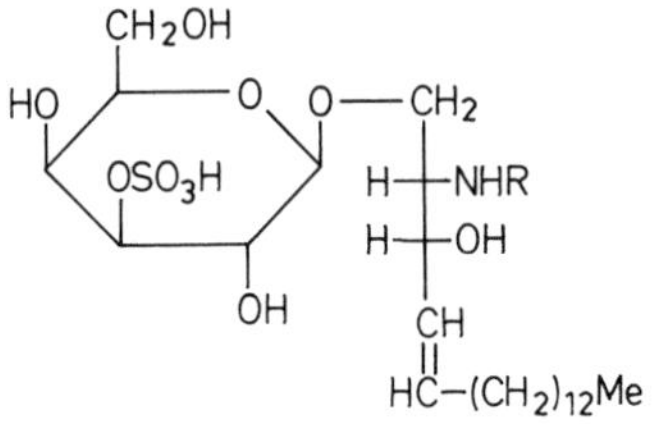

89 R = H   90 R = COR^1

acetyl group). Partial syntheses of glucosyl ceramide by reacylation of glucosyl sphingosine (obtained by basic hydrolysis of natural glucosyl ceramide) with palmitic acid in the presence of dicyclohexylcarbodiimide have also been described [91].

Deuteration of galactosyl ceramide by sonication in deuterium oxide — tetrahydrofuran in the presence of Raney nickel at 40 °C has been investigated [92]. Most of the incorporation of deuterium occurred at C-3 and C-4 of the galactose residue and into the allylic double bond. Tritium labelling of the 6-positions of the sugars in glucosyl and galactosyl ceramides was achieved by oxidation of the hydroxymethyl groups with chromium trioxide-graphite and subsequent reduction of the aldehydes produced with tritiated potassium borohydride [93].

4 Diglycosyl Ceramides

4.1 The gal-β-(1 → 4)-glc (Lactose) Linkage

Schmidt and Zimmermann [82] condensed the glycosyl imidate (91) with the azidoderivative (68) in the presence of boron trifluoride — etherate to give the β-lactoside in high yield. Further processing as described above for the glucosyl sphingosine derivative gave the lactosyl ceramide (93). Ogawa and his co-workers [95] have also prepared the lactosyl ceramide (93) in good yield by condensation of the glycosyl fluoride (92) with the sphingosine derivative (63) in the presence of tin(II) chloride, silver perchlorate and molecular sieves in chloroform and subsequent deprotection. Kanemitsu and Sweeley [96] have prepared lactosyl ceramide with two different radioactive labels by first preparing a 1-^{14}C fatty acid-labelled ceramide using the Ogawa method [as described for the preparation of (27)]. This unprotected ceramide was converted into a lactosyl ceramide by reaction with acetobromolactose and mercury(II) cyanide in benzene and subsequent deprotection. The product was oxidised to the 6′-aldehydo derivative using galactose oxidase and subsequent reduction with tritiated sodium borohydride gave the doubly-labelled lactosyl ceramide.

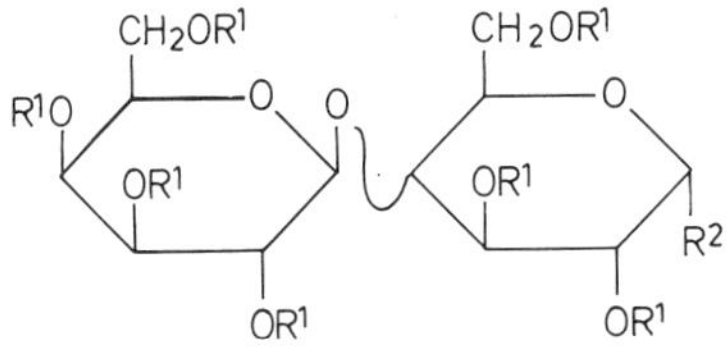

91 R^1=Me$_3$CCO ; R^2=OC(=NH)CCl$_3$
92 R^1=Ac ; R^2=F

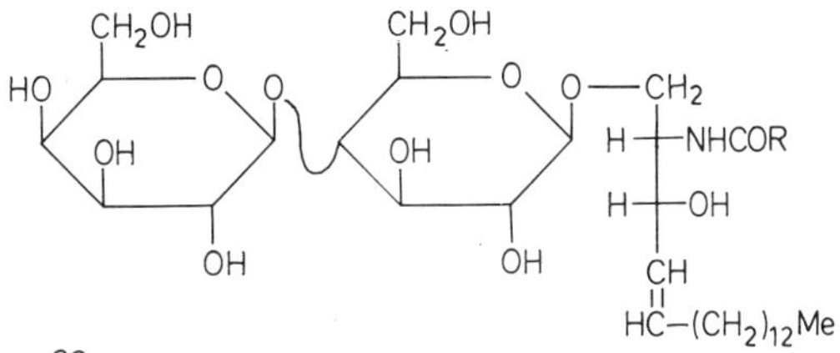

93

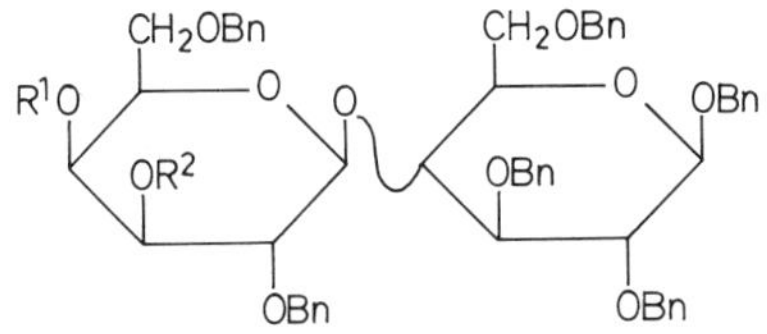

94 R^1,R^2 = $>$CHPh
95 R^1=Bn ; R^2=H
96 R^1=R^2=H

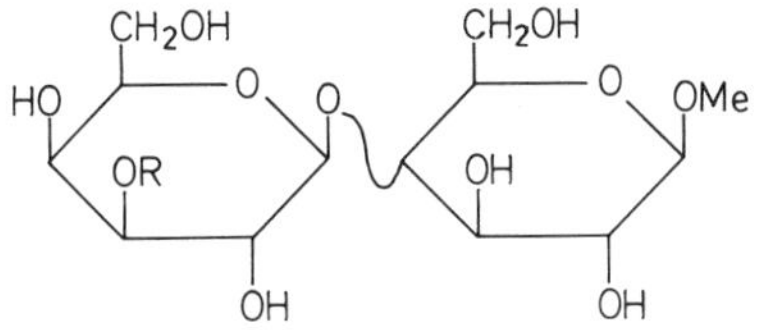

97 R= All

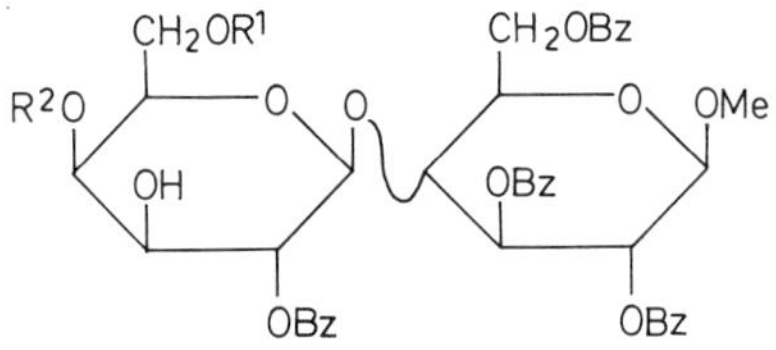

98 R^1,R^2 = $>$CHPh
99 R^1=R^2=H

Because lactose is the basic unit of all mammalian glycosphingolipids, a considerable amount of research has been devoted to providing suitably protected derivatives for further elaboration into oligosaccharides related to those occurring in glycolipids.

The benzylidene derivative (94) has been converted into the alcohol (95) [97, 98] using the diborane — trimethylamine — aluminium chloride reagent [99] and into the diol (96) [95, 100, 101]. Veyrières has converted methyl β-lactoside (and the corresponding allyl lactoside) into the 3-O-allyl ether (97) in good yield [102] by alkylation of the dibutylstannylene derivative in the presence of tetrabutylammonium iodide [103, 104], and this was converted into the alcohol (98) and the triol (99) [105]. Veyrières [106] has also converted (97) into the per-p-bromobenzyl derivative and deallylated the product to give a derivative with a free 3'-hydroxyl group. The diol (100) [107] has been converted by the stannylation procedure [108, 109] into the alcohol (101) [110, 111]. The partially acetylated benzyl β-lactoside (103) [101, 112, 113] has been converted into the alcohol (104) via the orthoacetate [113].

The perbenzoyl derivative (105) [114, 115] has been prepared via the 4',6'-O-isopropylidene derivative of lactose and the methyl lactoside (106) via the 4',6'-O-benzylidene derivative [116]. Benzoylation of lactose has given a perbenzoyl derivative with a free 3-hydroxyl group [117, 118].

The 6'-O-benzyl-methylthioglycoside (107) was prepared [119] by opening the 4',6'-O-benzylidene derivative with the sodium cyanoborohydride — hydrogen chloride reagent [120, 121]. Bundle [122] has prepared a mixture of the 3′,4′- and 4′,6′-*O*-cyclohexylidene derivatives of methyl β-lactoside which on benzylation and subsequent hydrolysis of the cyclohexylidene groups gave a mixture of the corresponding diols which were converted into the perbenzylated lactose derivative (102) with a free 4′-hydroxyl group by benzylation of the tin derivatives.

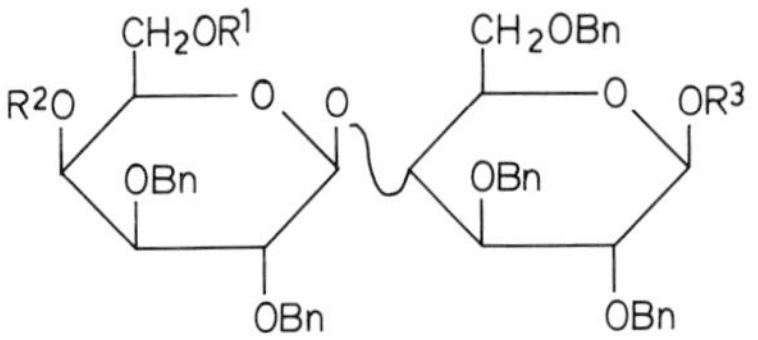

100 $R^1=R^2=H$; $R^3=Bn$
101 $R^1=R^3=Bn$; $R^2=H$
102 $R^1=Bn$; $R^2=H$; $R^3=Me$

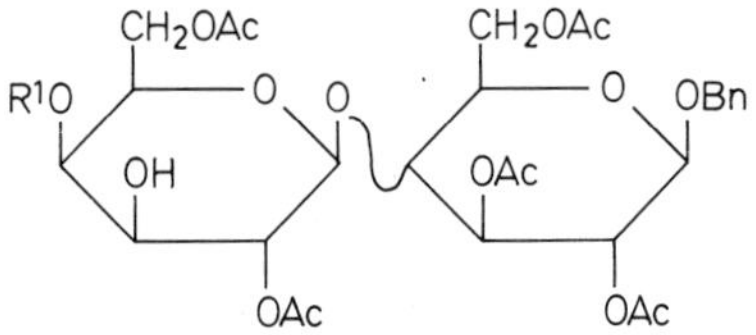

103 $R^1=H$
104 $R^1=Ac$

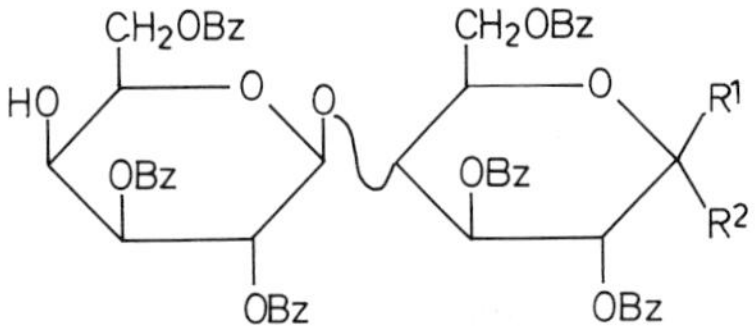

105 $R^1=H$; $R^2=OBz$
106 $R^1=OMe$; $R^2=H$

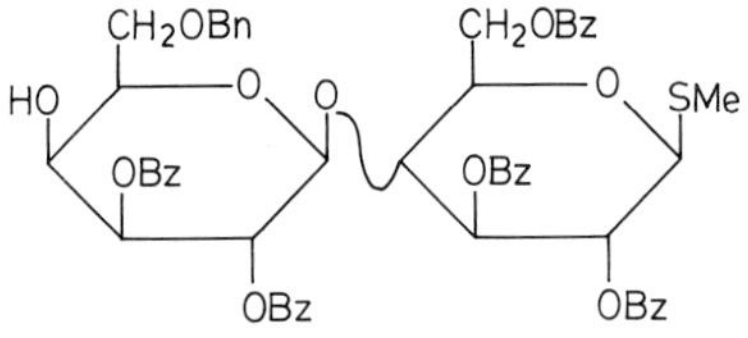

107

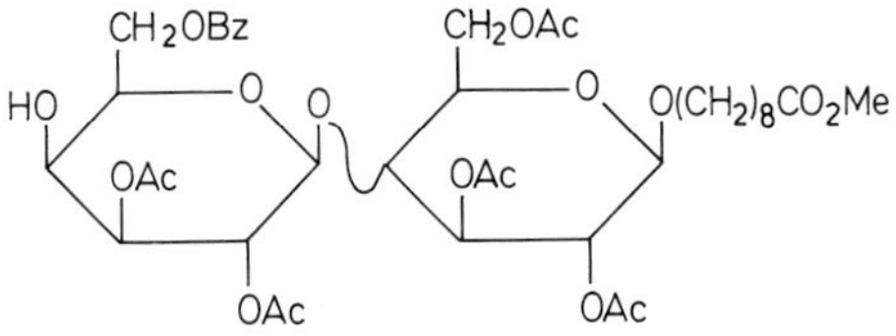

108

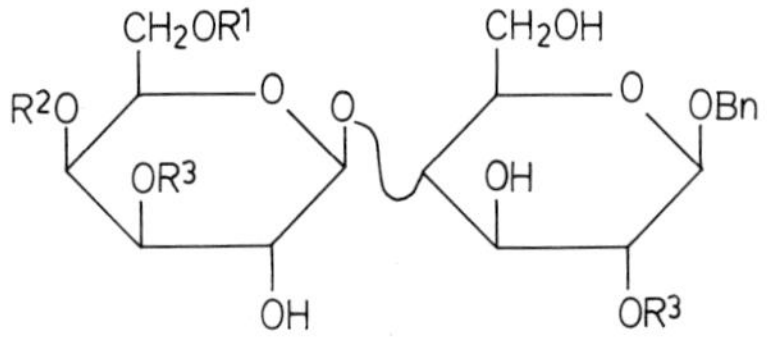

109 $R^1,R^2=$>$CHPh$; $R^3=H$
110 $R^1,R^2=$>$CHPh$; $R^3=Bn$
111 $R^1=Bz$; $R^2=H$; $R^3=Bn$

The 8-methoxycarbonyloctyl derivative of lactose has been prepared from lactose octaacetate in the presence of tin(1V) chloride or from acetobromolactose in the presence of silver triflate [123, 124] and converted into the derivative (108) with a free 4′-hydroxyl group [125] via the 4′,6′-*O*-benzylidene derivative. The benzyl lactoside (109) was converted into the benzyl ether (110) by the action of benzyl bromide and tetrabutylammonium bromide on the dibutylstannylene derivative [126] and this was converted into the benzoate (111). Phase transfer benzylation of the benzyl lactoside (112) has given the lactose derivative (113) with a free 3-hydroxyl group in 38 % yield [118]. 1,6-Anhydrolactose has also been used for the preparation of the protected lactose derivative (114) [127].

The open-chain derivative (115) of lactose prepared by the action of dimethoxypropane [128–130] has also been used to prepare the lactose derivative (116) containing a free 2′-hydroxyl group [131].

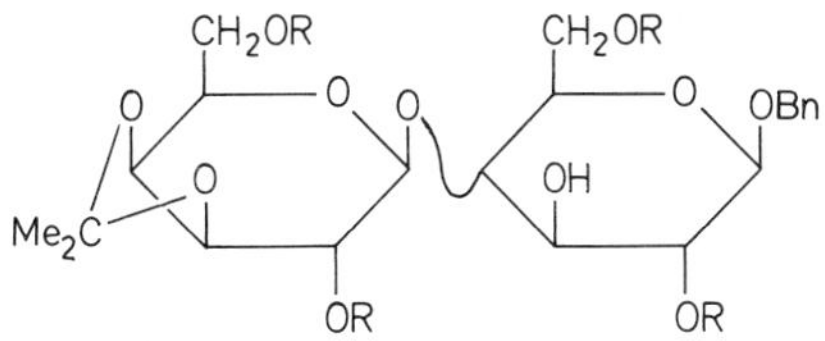

CH₂OR ... (chemical structures)

112 R = H
113 R = Bn

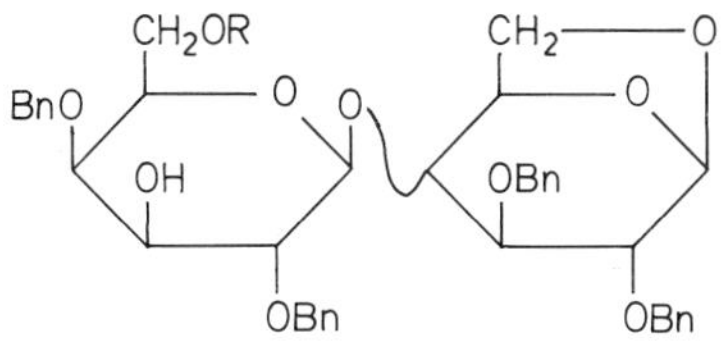

114 R = Bn
114a R = H

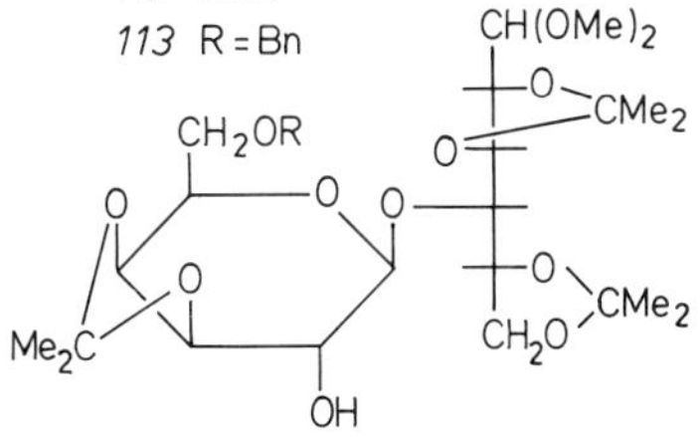

115 R = H
116 R = Bz

4.2 Glycosyl Ceramides Containing the gal-α-(1 → 4)-gal Linkage

The normal mammalian triglycosyl ceramide which accumulates in Fabry's disease (due to the genetic deficiency of an α-galactosidase) has the structure:

$$\text{gal-}\alpha\text{-}(1 \rightarrow 4)\text{-gal-}\beta\text{-}(1 \rightarrow 4)\text{-glc-}\beta\text{-}(1 \rightarrow 1)\text{-CER} .$$

This trisaccharide sequence is also known as the p^K antigenic determinant and it is also believed that glycolipids containing the gal-α-(1 → 4)-gal-β-(1 → x) linkage may act as receptor sites in the epithelium of the urinary tract for adhesion by pathogenic fimbriated *Escherichia coli* [132, 133]. There has thus been considerable incentive to investigate the synthesis of molecules containing the gal-α-(1 → 4)-gal linkage and several methods have been developed since the previous review [1].

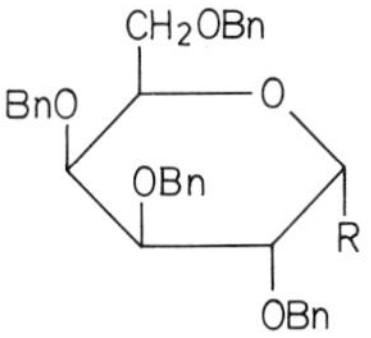

117 R = Br
118 R = Cl

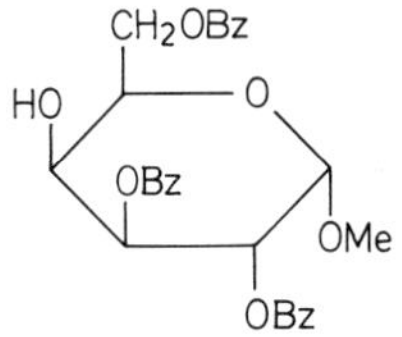

119

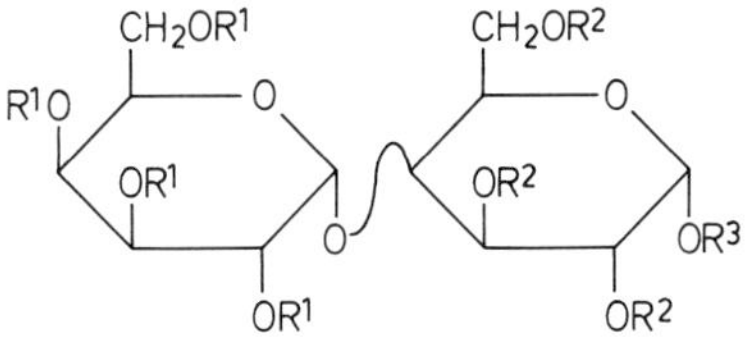

120 R¹ = Bn; R² = Bz ; R³ = Me
121 R¹ = R² = R³ = Ac

Reist and co-workers [134] condensed the galactosyl bromide [117] with the protected galactose derivative (119) under "halide ion-catalysed" conditions [25] to give the disaccharide (120) and this was deprotected, acetylated and acetolysed to give the acetylated gal-α-(1 → 4)-gal disaccharide (121). Under similar conditions, condensation of (117) with the protected lactose derivative (106) gave the trisaccharide (122) [116] and this was converted into the acetylated methyl glycoside (123) containing the complete trisaccharide sequence of the Fabry glycolipid. A ^{13}C-NMR study of the derivatives (121) and (123) was also described [135].

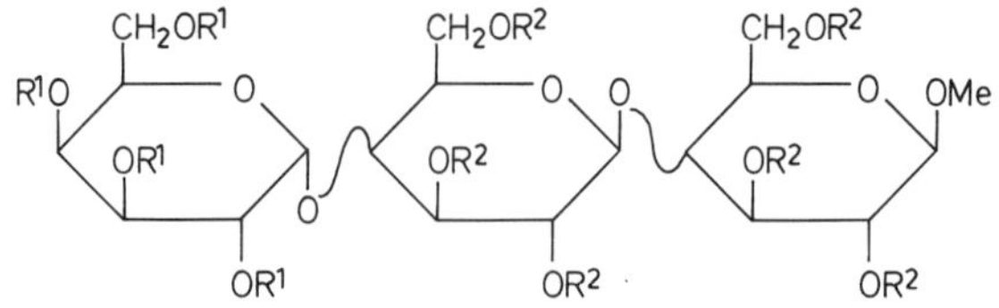

122 R^1=Bn ; R^2=Bz
123 R^1=R^2=Ac

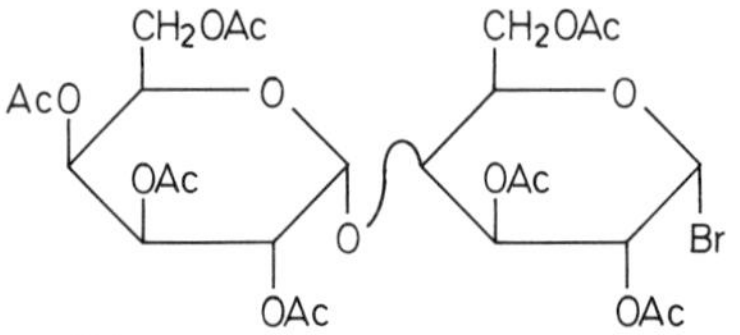

124

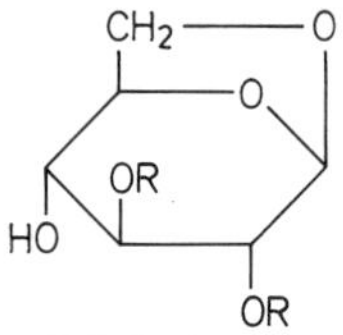

125 R=Ac
126 R=Bz

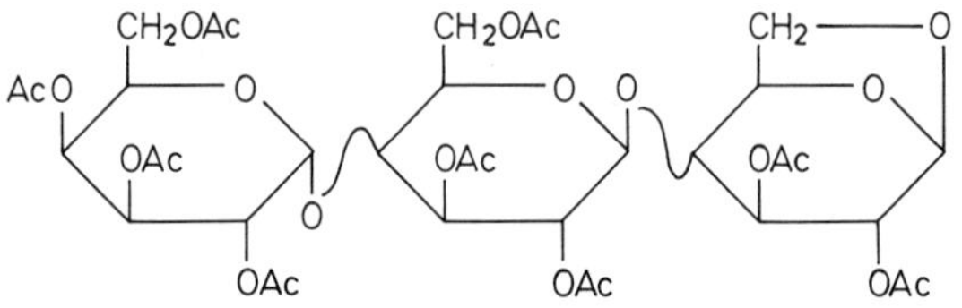

127

Shapiro and Acher [136] condensed the gal-α-(1 → 4)-gal derivative (124) [prepared by the action of hydrogen bromide in acetic acid on the previously described peracetyl derivative [137] with the 1,6-anhydroglucose derivative (125) [or (126) which was superior since acyl migration was reduced] to give the trisaccharide (127). This was acetolysed and converted into the glycosyl chloride and condensation of the latter with the benzoyl ceramide (63) and subsequent deprotection gave the complete Fabry glycolipid.

Garegg and Hultberg [114] condensed the galactosyl chloride (118) with the benzoate (128) in the presence of silver triflate to give (129) in high yield. Hydrogenolysis of (129) and subsequent benzoylation gave the perbenzoate (130) which was converted into the bromide (131). Condensation of (131) with p-nitrophenol in the presence of silver imidazolate and zinc chloride and subsequent deprotection gave the β-linked p-nitrophenylglycoside of the gal-α-(1 → 4)-gal disaccharide. Condensation of the galactosyl chloride (118) with the benzoylated lactose derivative (105) gave [114] the trisaccharide (132) which on hydrogenolysis and benzoylation gave the perbenzoyl derivative (133) and this was converted into the bromide (134). Condensation of (134) with methanol or p-nitrophenol in the presence of silver imidazolate and zinc chloride followed by basic hydrolysis gave the corresponding β-glycosides of the trisaccharide. Jacquinet and Sinaÿ [115] have described a similar route to the free trisaccharide (135) via the derivatives (132) and (133),

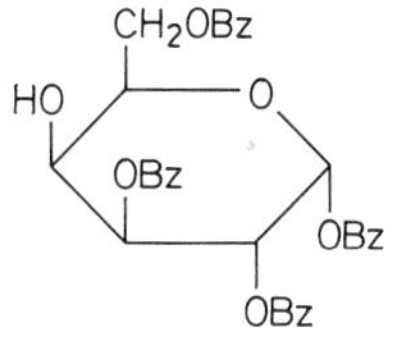

CH₂OBz ... the chemical structures are shown.

128

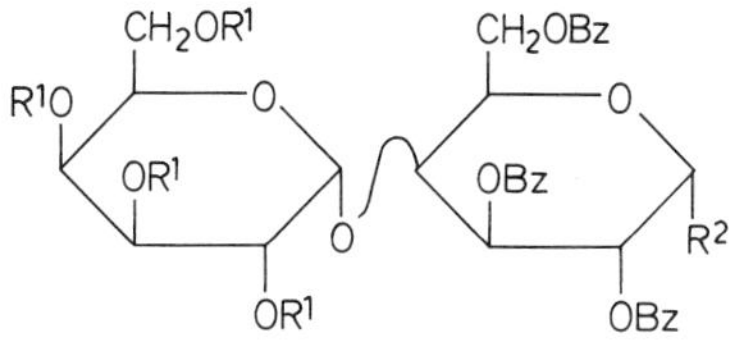

129 R¹=Bn ; R²=OBz
130 R¹=Bz ; R²=OBz
131 R¹=Bz ; R²=Br

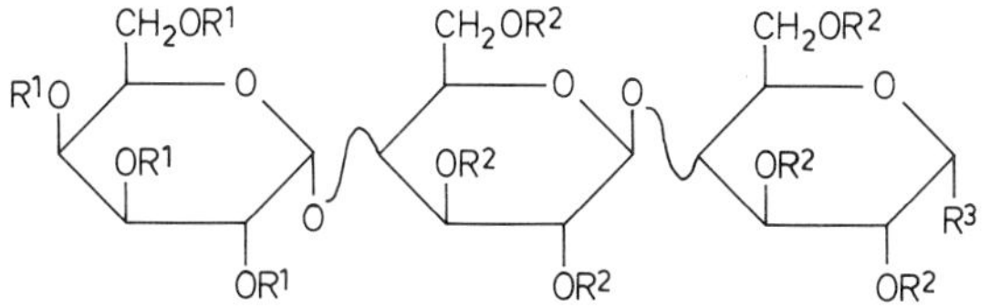

132 R¹=Bn; R²=Bz ; R³=OBz
133 R¹=R²=Bz ; R³=OBz
134 R¹=R²=Bz ; R³=Br
135 R¹=R²=H ; R³=OH

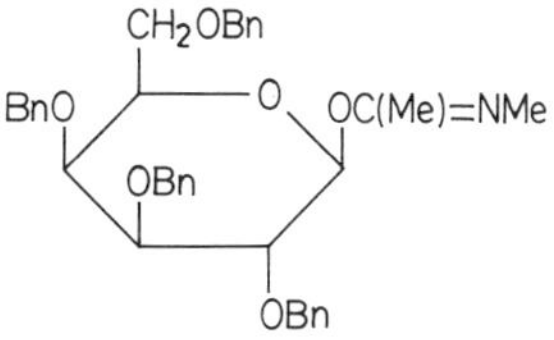

136

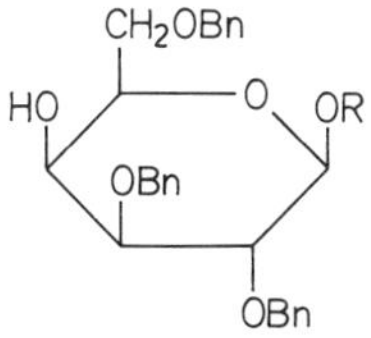

137 R= Bn or Me
138 R= CH₂CH₂Br

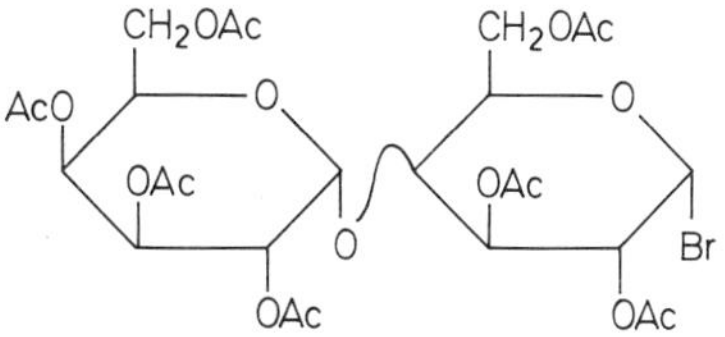

139

Sinaÿ and co-workers [138] have also described the preparation of the free disaccharide gal-α-(1 → 4)-gal in good yield by condensation of the imidate (136) with the galactose derivatives (137) and subsequent deprotection (hydrogenolysis or acetolysis). Magnusson and co-workers [139] have prepared this disaccharide by a combination of enzymatic and chemical degradations of polygalacturonić acid. Magnusson and co-workers [140] also converted the free disaccharide via the peracetate into the glycosyl bromide (139) which in the presence of silver triflate and base reacted with 2-bromoethanol to give the 2-bromoethyl glycoside which on deacetylation gave (140). This bromo-compound (140) could be coupled with methyl 3-mercaptopropionate to give the disaccharide (141) with a spacer arm suitable for coupling to proteins [141, 142]. Magnusson and co-workers [143] also coupled the glycosyl bromide (139) with the gluco analog of (138) in the presence of silver triflate to give a β-linked trisaccharide which on catalytic hydrogenation in glacial acetic acid (which did not affect the bromine atom) and acetylation gave (142). Reaction of (142) with methyl 3-mercaptopropionate and cesium carbonate and deacetylation gave (143) suitable for coupling to protein or reaction of (142) with octadecanethiol and deacetylation gave the "neoglycolipid" (144), with the Fabry-p^K oligosaccharide sequence.

In order to investigate in more detail the combining site on the β-methyl glycoside of the gal-α-(1 → 4)-gal disaccharide ("methyl urobioside") recognised by the *p*-fimbriae of *E. coli*, Garegg and Oscarson [144] prepared the 6-deoxy, 6'-deoxy and 6,6'-dideoxy derivatives of the β-methyl glycoside of the disaccharide. Condensation of the chloride (118) with the methyl galactoside (145) in the presence of silver triflate and subsequent hydrogenolysis gave the disaccharide (146). Compound (146) was converted into the 4',6'-*O*-benzylidene derivative (147) and the product acetylated and treated with *N*-bromosuccinimide to give the bromide (149a). This on deacylation and catalytic hydrogenation gave the 6'-deoxy derivative [D-fuc-α-(1 → 4)-gal-OMe].

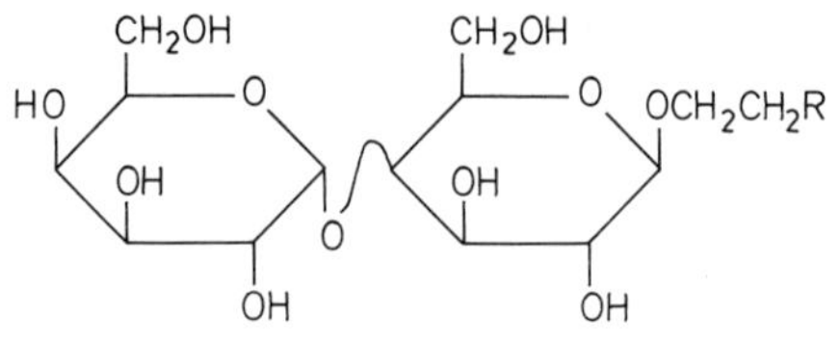

140 R=Br
141 R=S(CH₂)₂CO₂Me

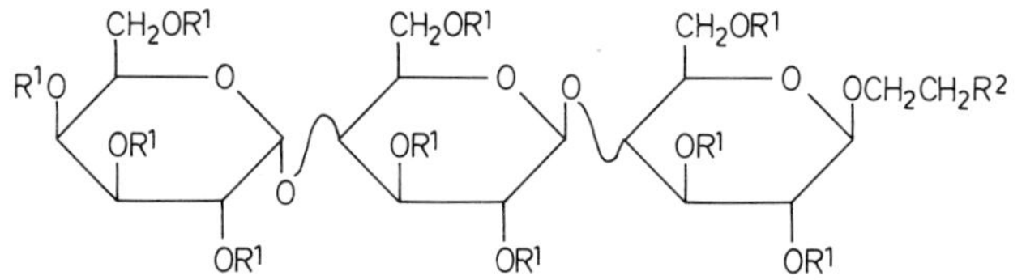

142 R¹=Ac ; R²=Br
143 R¹=H ; R²=S(CH₂)₂CO₂Me
144 R¹=H ; R²=S(CH₂)₁₇Me

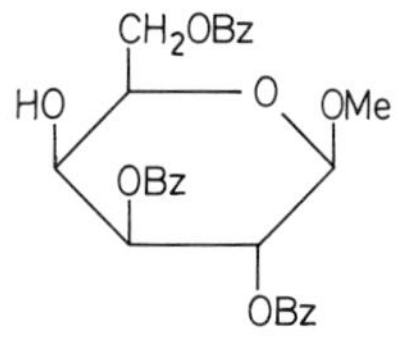

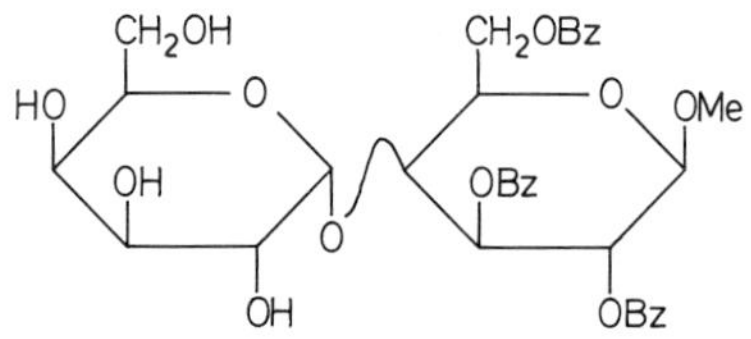

145   146

The 4',6'-*O*-benzylidene derivative (147) was selectively debenzoylated to give (148) and this was converted into the 6-deoxy-6-iodo derivative (149) by treatment with the triphenylphosphine — imidazole — iodine reagent [145]. Compound (149) was hydrogenolysed and deacetylated to give the 6-deoxy derivative [gal-α-(1 → 4)-D-fuc-OMe]. Acetylation of compound (149) and subsequent treatment with *N*-bromosuccinimide gave the 6'-bromo-6-iodo-dideoxy derivative (149b) which was converted by hydrogenolysis and deacylation into the 6,6'-dideoxy derivative [D-fuc-α-(1 → 4)-D-fuc-OMe].

For similar reasons, that is to investigate the inhibition of adhesion of pathogens to glycolipid receptors in host cells, Magnusson and co-workers [146] have prepared the 3-*O*-methyl-, 3-deoxy-3-*C*-methyl and the 3-deoxy derivatives of the β-methyl glycoside of the disaccharide gal-α-(1 → 4)-gal. The β-methyl galactoside (150) was converted into the benzyl ether (151) using benzyl trichloracetimidate [147] and then

94

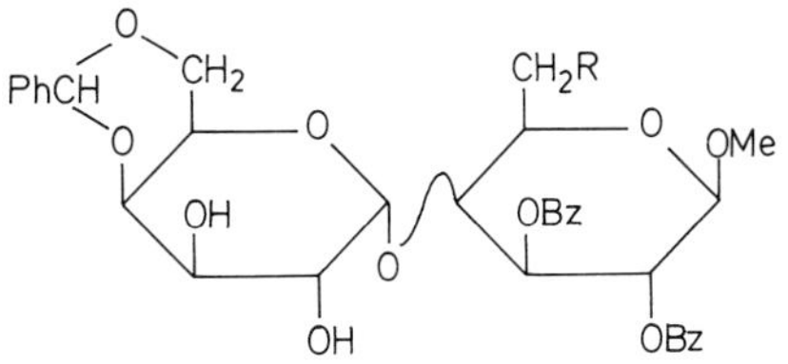

O
PhCH   CH₂
       O   CH₂R
   O   O   OMe
   OH   OBz
            O   OBz
       OH   OBz

147 R = OBz
148 R = OH
149 R = I

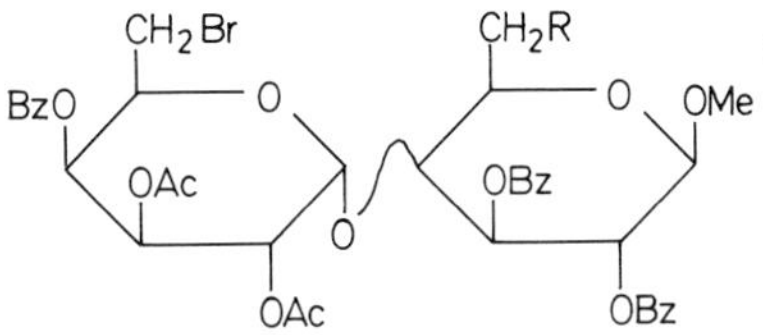

CH₂Br   CH₂R
BzO   O   O   OMe
   OAc   OBz
        O
       OAc   OBz

149a R = OBz
149b R = I

O
PhCH   CH₂
       O   O   OMe
   O
      R¹
           OR²

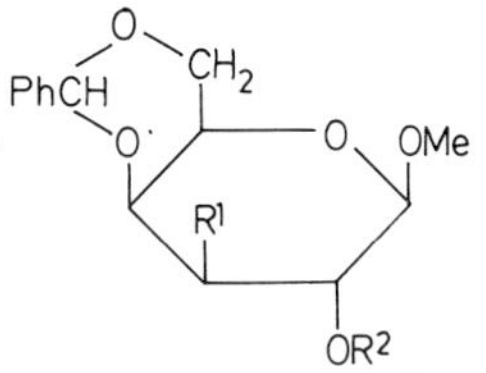

150 R¹ = OBz; R² = H
151 R¹ = OBz; R² = Bn
152 R¹ = OH; R² = Bn
153 R¹ = OMe; R² = Bn
154 R¹ = Me; R² = Bn
155 R¹ = OH; R² = Ac
156 R¹ = H; R² = Bn

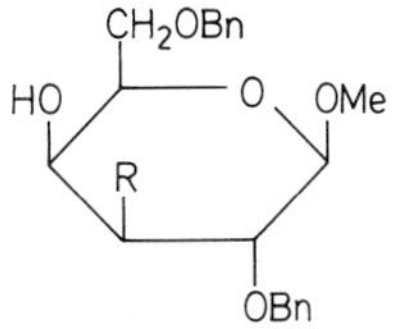

CH₂OBn
HO   O   OMe
   R
       OBn

157 R = OMe
158 R = Me
159 R = H

this was hydrolysed to (152) and methylated to give (153). Oxidation of (152) with oxalyl chloride in dimethylsulfoxide gave the 3-keto derivative which was treated with the Wittig reagent (derived from triphenylmethylphosphonium bromide) and the product reduced to give the 3-deoxy-3-C-methyl derivative (154). Deoxygenation of the S-methyl dithiocarbonate [148] of compound (155) with subsequent deacetylation and benzylation gave the 3-deoxy-derivative (156). The benzylidene group in compounds (153), (154) and (156) was opened by sodium cyanoborohydride — hydrogen chloride [120, 121] to give the derivatives (157), (158), and (159), respectively. These were condensed with the galactosyl bromide (117) in the presence of tetraethylammonium bromide to give the α-linked disaccharide and subsequent deprotection gave the required derivatives of the β-methyl glycoside of gal-α-(1 → 4)-gal.

Norberg et al. [149] have also synthesised the disaccharides gal-α-(1 → 2)-L-rha and gal-α-(1 → 2)-D-man as analogs of "methyl urobioside" to investigate the binding by the fimbriae of virulent *E. coli* strains.

The human blood-group P1-antigenic determinant has the terminal trisaccharide sequence:

gal-α-(1 → 4)-gal-β-(1 → 4)-glcNAc

and this trisaccharide has been synthesised by two groups. Sinaÿ and co-workers [150] condensed acetobromogalactose with the glucosamine derivative (160) [151] in the presence of silver triflate and 2,4,6-trimethylpyridine to give the β-linked disaccharide and subsequent deacetylation gave (161) and this was converted into the 3',4'-O-isopropylidene derivative which was benzylated and the product hydrolysed to give (162).

95

Benzylation of the dibutylstannylene derivative of (162) gave the derivative (163) in which only the 4′-hydroxyl group was free. Compound (163) was condensed [152] with the chloride (118) in the presence of silver triflate, 2,4,6-trimethylpyridine and molecular sieves to give the α-linked trisaccharide (164) in high yield and this on hydrogenolysis gave the trisaccharide (165) which was active as a selective inhibitor of the P1-*anti*-P1 system. A similar sequence of reactions gave the trisaccharide derivative (166) which was also converted into (165).

160

161 $R^1 = R^2 = H$
162 $R^1 = Bn$; $R^2 = H$
163 $R^1 = R^2 = Bn$

164 R=Bn
165 R=H

166 $R^1 = Bn$; $R^2 = Bz$

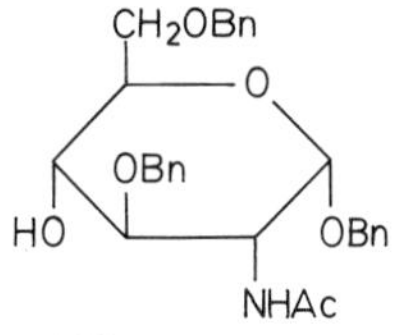

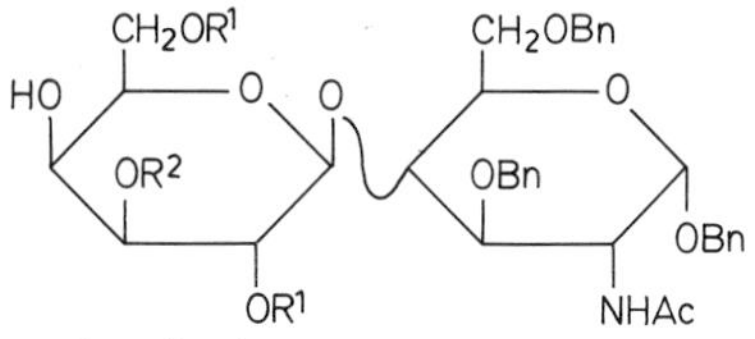

Magnusson and co-workers [153] prepared the bromoethyl glycoside of the same trisaccharide by condensing the acetobromodisaccharide (139) with the glucosamine derivative (167) [154] in the presence of silver triflate and tetramethylurea to give the derivative (168). Hydrogenolysis in acetic acid followed by acetylation gave (169) which was condensed with methyl 3-mercaptopropionate to give (170). Depththaloylation, *N*-acetylation and de-*O*-acetylation then gave the trisaccharide with a spacer-arm suitable for coupling to protein.

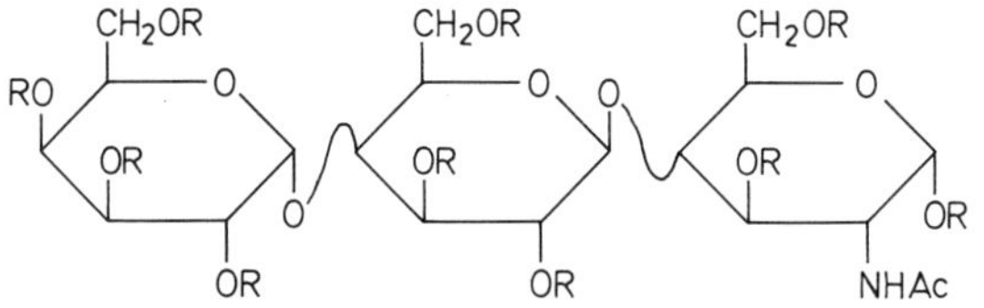

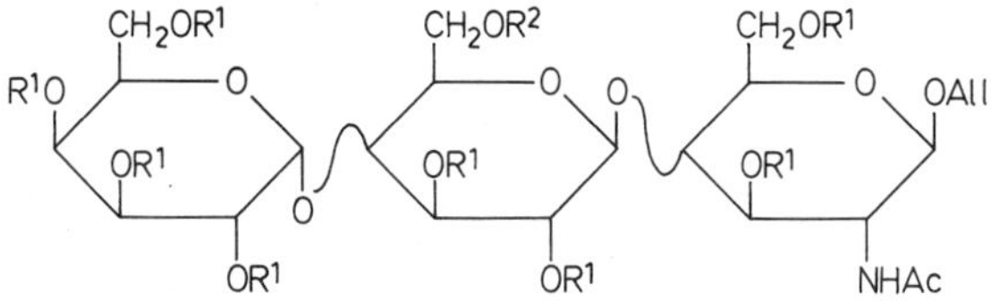

The same trisaccharide sequence has been prepared by Anderson and co-workers [155] in the following way. Condensation of the galactosyl chloride (172) with the glucosamine derivative (171) in the presence of silver triflate, tetramethylurea and 2,6-dimethylpyridine (to preserve the tetrahydropyranyl group) gave the *N*-acetyllactosamine derivative (173) which was converted into (174) by acid hydrolysis. Condensation of (174) with the bromide (117) in the presence of silver carbonate and silver triflate [156] gave the protected trisaccharide (175). In Anderson's study this was converted by basic hydrolysis into the alcohol (176) which was condensed with tribenzylfucopyranosyl bromide to give an α-fucosyl derivative. Deprotection gave a

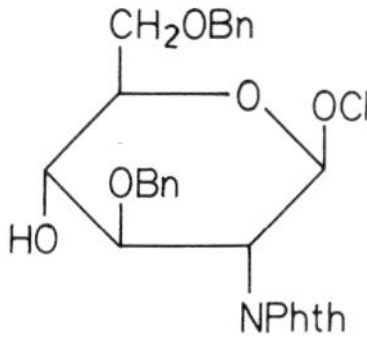

CH$_2$OBn OCH$_2$CH$_2$Br
OBn
HO
NPhth
167

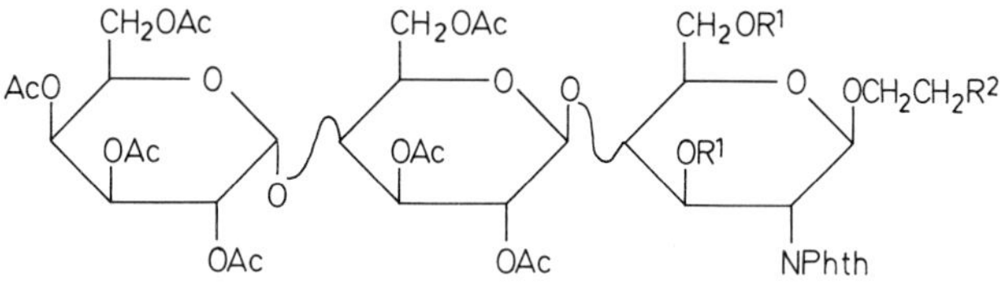

CH$_2$OAc
AcO
OAc
OAc
168 R^1=Bn ; R^2=Br

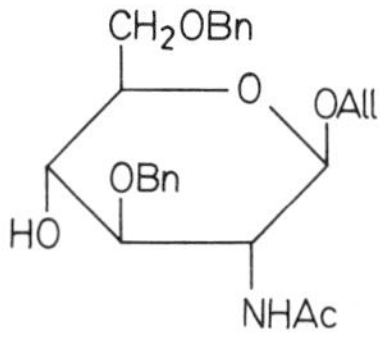

CH$_2$OAc
OAc
OAc
169 R^1=Ac ; R^2=Br

CH$_2$OR^1 OCH$_2$CH$_2$R^2
OR^1
NPhth
170 R^1=Ac ; R^2=S(CH$_2$)$_2$CO$_2$Me

CH$_2$OBn OAll
OBn
HO
NHAc
171

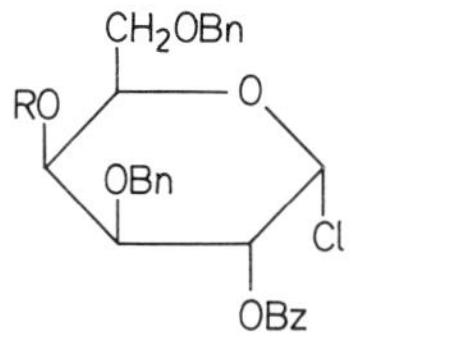

CH$_2$OBn
RO
OBn
Cl
OBz
172 R=tetrahydropyranyl

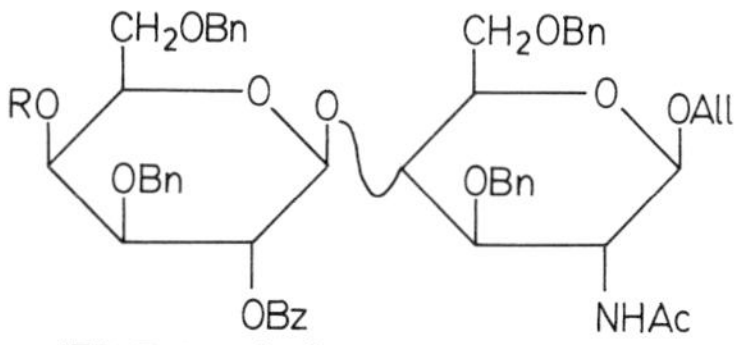

CH$_2$OBn
RO
OBn
OBz
CH$_2$OBn OAll
OBn
NHAc
173 R=tetrahydropyranyl
174 R=H

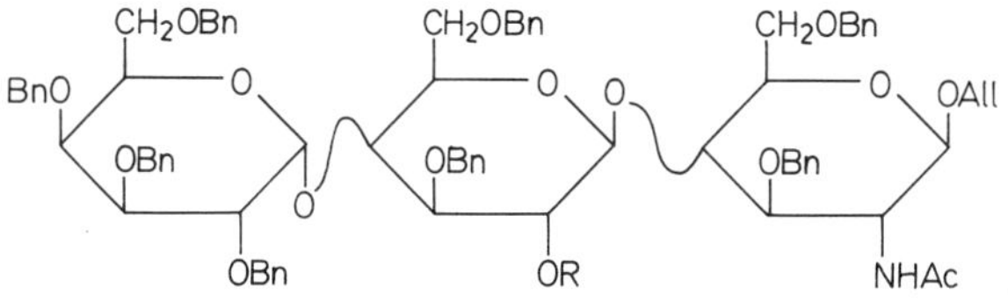

CH$_2$OBn
BnO
OBn
OBn
CH$_2$OBn
OBn
OR
CH$_2$OBn OAll
OBn
NHAc
175 R=Bz
176 R=H

tetrasaccharide which is an isomer of the blood-group B (type 2) antigenic determinant.

The major glycolipid of human erythrocytes known as globoside (P-antigen) has the structure:

$$\text{NAcgal-}\beta\text{-}(1 \rightarrow 3)\text{-gal-}\alpha\text{-}(1 \rightarrow 4)\text{-gal-}\beta\text{-}(1 \rightarrow 4)\text{-glc-}\beta\text{-}(1 \rightarrow 1)\text{-CER}$$

being derived biosynthetically from the Fabry glycolipid (p^K antigen) by addition of a β-linked N-acetyl galactosamine residue. The carbohydrate sequence was first synthesised by Paulsen and co-workers [111]. Condensation of the azido derivative (177) [157] with the 1,6-anhydrogalactose derivative (178) in the presence of silver silicate gave the β-linked disaccharide (179) which was converted into the acetate (180). Acetolysis of (180) using trifluoroacetic acid — acetic anhydride and subsequent treatment of the product with titanium(IV) bromide gave the glycosyl bromide (181). This was condensed with the lactose derivative (101) in the presence of silver carbonate and silver perchlorate to give the α-linked tetrasaccharide (182) in 32% yield. Further processing involving conversion of the azido group to the acetamido group (reduction with nickel(II) chloride — sodium borohydride and acetylation), de-O-acetylation and hydrogenolysis of the benzyl groups gave the tetrasaccharide sequence of the globoside. Swedish workers [119] have prepared the thiomethyl glycoside of the same tetrasaccharide. They prepared the protected thiomethyl lactoside (107) and condensed this with the unstable glycosyl bromide (184) [containing a free 2-hydroxyl

97

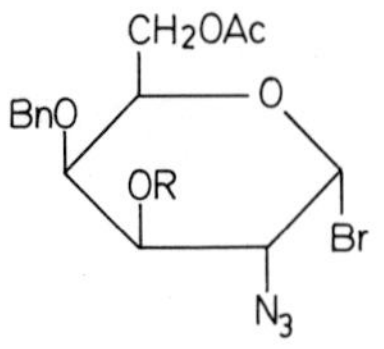

177 R=COCCl₃

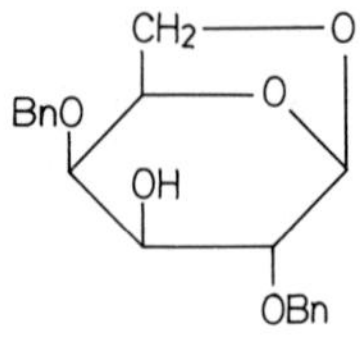

178

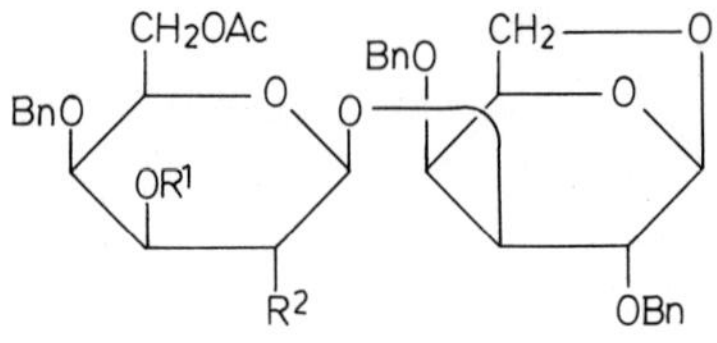

179 R¹=COCCl₃ ; R²=N₃
180 R¹=Ac ; R²=N₃

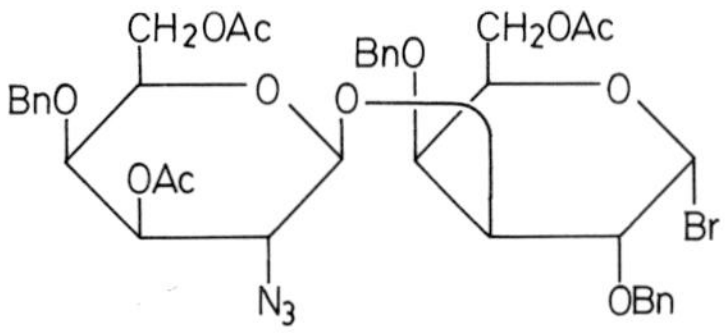

181

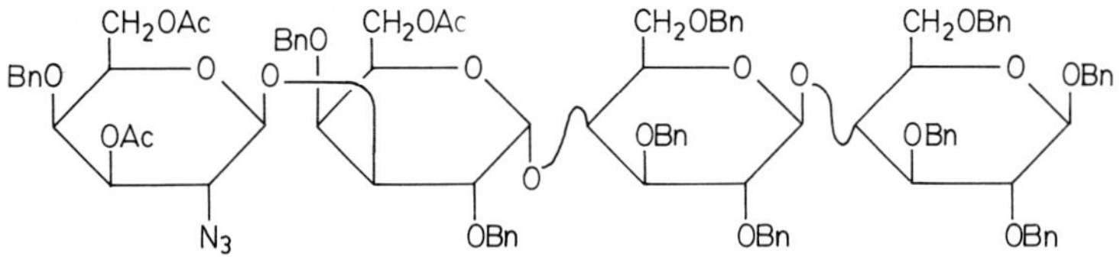

182

group and prepared by the action of bromine on the thiomethyl glycoside (183)] in the presence of silver triflate to give the α-linked tetrasaccharide (185) in 66% yield. This could be converted into the corresponding glycosides (186) by the action of methyl triflate and an alcohol (R′OH) and subsequent deprotection gave glycosides of the free tetrasaccharide. They recommend the thioglycosides since they are stable

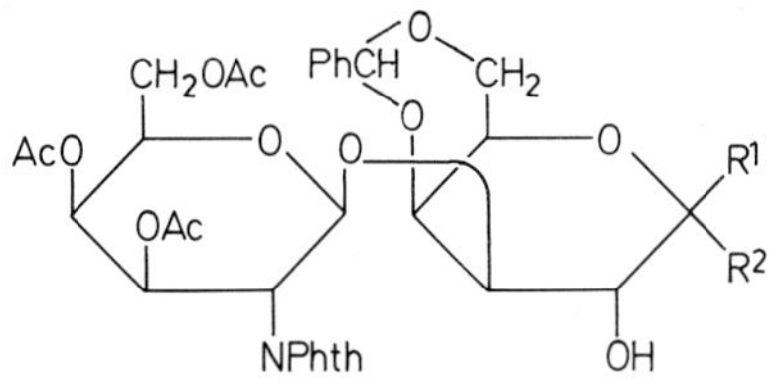

183 R¹=SMe; R²=H
184 R¹=H; R²=Br

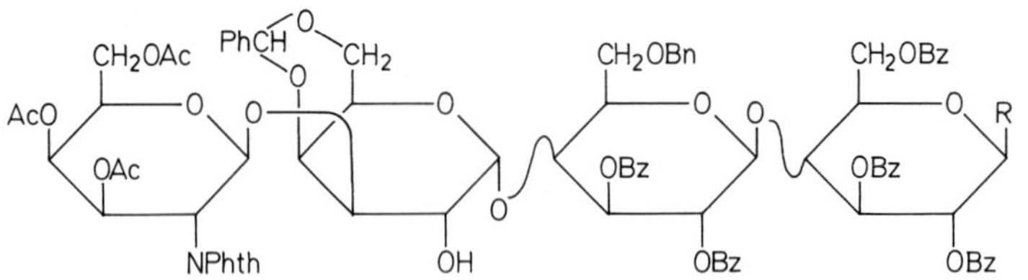

185 R=SMe
186 R=OR¹

under conditions for glycosidation and manipulation of protecting groups, can be converted into glycosyl-donating halides by the action of halogens and act as glycosyl donors in the presence of methyl triflate [158].

5 Glycosphingolipids Containing *N*-Acetylgalactosamine

5.1 β-Linked *N*-Acetylgalactosamine

5.1.1 Asialoganglioside GM2

Tay Sachs globoside *N*Acgal-β-(1 → 4)-gal-β-(1 → 4)-glc-β-(1 → 1)-CER is the asialo derivative of the ganglioside (Tay-Sachs ganglioside, GM2) which accumulates in Tay-Sachs disease (a genetic disorder due to the deficiency of a β-*N*-acetyl-galactos-aminadase) in which the *N*-acetylneuraminic acid is attached in α-linkage to the 3-position of the galactose residue.

It was first synthesised by Shapiro in 1973 (for a review see Ref. [1]) and recently several syntheses of the oligosaccharide portion have been described using modern methods of glycoside synthesis.

Bundle and co-workers [122] prepared a glycosyl bromide (192) of a protected galactosamine derivative from the glucosamine derivative (187) [159] as follows. Condensation of the bromide (187) with *t*-butanol in the presence of silver salicylate gave the glycoside (188) which was deacetylated and partially benzoylated to give (189) and this was converted into the mesylate (190). The mesylate was treated with sodium acetate and a crown ether in *N,N*-dimethylformamide to give the galactosamine derivative (191) in good yield and this was converted into the bromide (192) with hydrogen bromide in acetic acid or methyl dibromomethyl ether. The bromide (192) was condensed with the protected β-methyl lactoside (102) in the presence of silver perchlorate — silver carbonate (1:20) to give the trisaccharide (194) in 49% yield. This was deacylated and the phthalimido group removed with hydrazine hydrate [174] and the product *N*-acetylated and hydrogenolysed to give the β-methyl glycoside of Tay Sachs globoside (GM2).

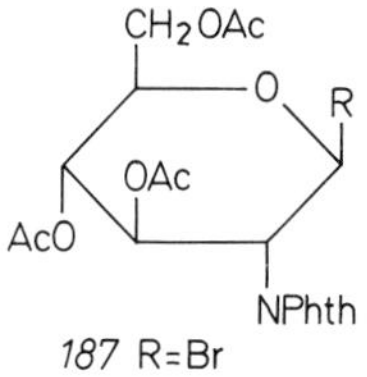

187 R=Br
188 R=OCMe$_3$

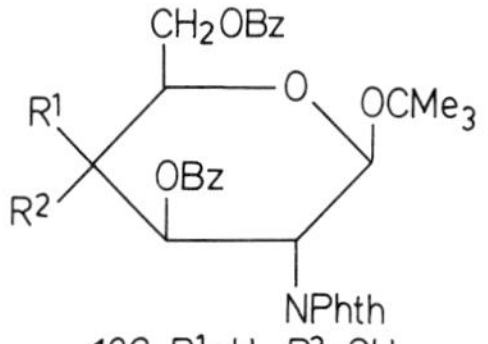

189 R^1=H; R^2=OH
190 R^1=H; R^2=OSO$_2$Me
191 R^1=OAc; R^2=H

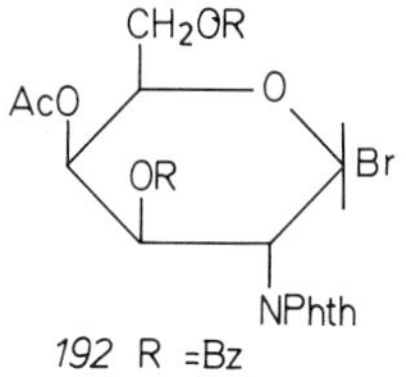

192 R =Bz
193 R = Ac

The lack of availability of galactosamine has stimulated other workers to reinvestigate the inversion of the 4-hydroxyl group of glucosamine for the preparation of this sugar for use in glycoside synthesis. Thus, Nashed [160] has treated the *p*-bromo-

194

195 R^1 = H; R^2 = OH
196 R^1 = OBz; R^2 = H

197

198 R = H
199 R = Bn

200

201

202

203

benzene sulfonate of the glucosamine derivative (195) with sodium benzoate in hexamethylphosphorotriamide to give the galactosamine derivative (196) in high yield. The allyl group was isomerised to the prop-1-enyl group by the action of palladium-charcoal and this was converted into the methyl oxazoline (197) by the action of mercury(II) chloride — mercury(II) oxide [161]. The oxazoline (197) is a potential glycosidating agent for β-*N*-acetylgalactosamine synthesis. More recently, Nashed [162] has used cesium benzoate in *N,N*-dimethylformamide for the conversion of the *p*-bromobenzenesulfonate of (195) into (196) in high yield. Debenzoylation of (196) and conversion of the product into the 4,6-*O*-benzylidene derivative followed by benzylation and acidic hydrolysis gave the benzyl ether (198) which was converted into the dibutylstannylene derivative and reaction of this with benzyl bromide and tetrabutylammonium iodide in toluene [103] gave the dibenzyl ether (199) in high yield. The cesium benzoate-*N,N*-dimethylformamide method was first used by Anderson and Nashed [163] for the conversion of the *p*-bromobenzenesulfonate of the glucosamine derivative (200) into the benzoate of (199).

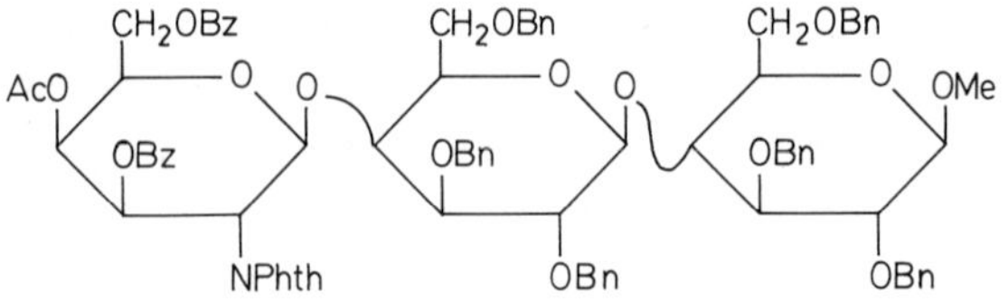

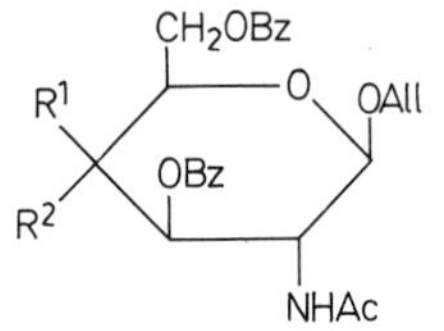

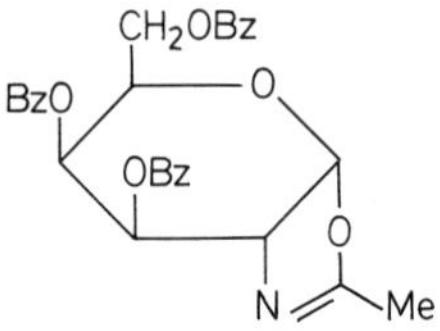

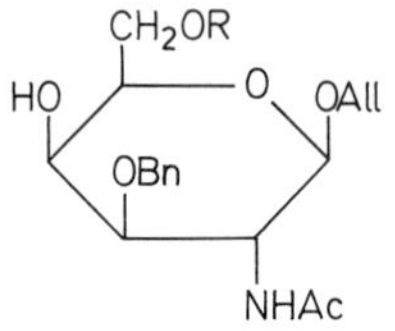

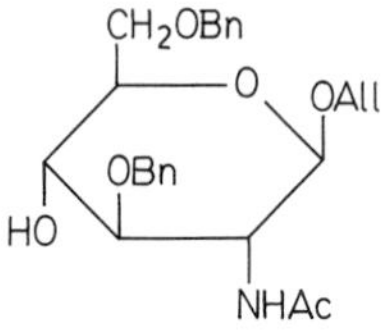

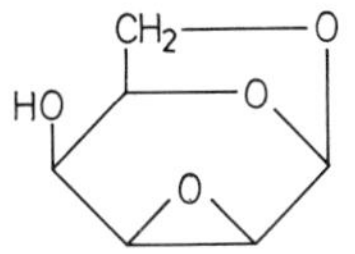

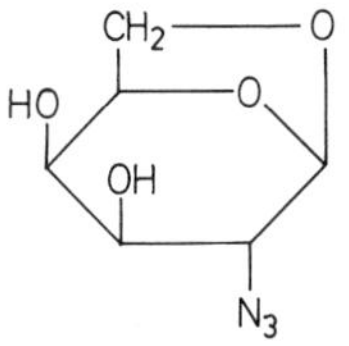

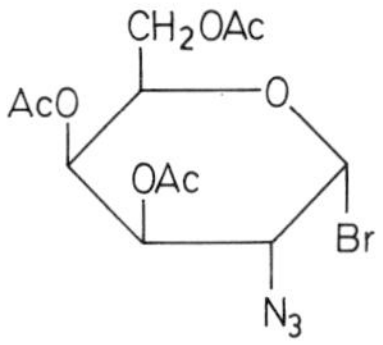

The azido analogs of galactosamine derivatives were developed by Paulsen primarily for *cis*-1,2-glycoside synthesis in the amino-sugar series (see Sect. 5.2) but the preparation of these derivatives also provides a route to galactosamine derivatives from more readily available sugars and they have more recently been used for 1,2-*trans*-glycoside synthesis in the amino-sugar series. Paulsen initially used the epoxide (201) derived from 1,6-anhydrogalactose which could be readily converted into the azide (202) and its O-protected derivative and hence into the bromide (203) (and other derivatives) by acetolysis and subsequent reaction with titanium(IV) bromide [164] and these α-bromides could be converted into β-glycosides using silver carbonate as a catalyst [165]. Lemieux and Ratcliffe [166, 167] developed a route to the azidogalactose derivative (207) by the addition of sodium azide in the presence of ceric ammonium nitrate to the galactal (204) to give a mixture in which the anomeric azidonitrates (206) were the major products (75%). Treatment of the crude product with lithium iodide in acetonitrile converted these to the α-azido iodide (207) and brief treatment of this with tetraethyl ammonium chloride in acetonitrile gave the crystalline β-chloride (208) which was isolated in 31% yield from triacetyl D-galactal (204). Lemieux and co-workers had earlier developed a successful route to glucosamine derivatives by the reduction of the 2-oximino derivatives but this was not so successful in the galactosamine series since the reduction of the 2-oximino derivatives gave a large proportion of the 2-amino-2-deoxy-D-talose derivatives together with the 2-amino-2-deoxy-D-galactose derivatives [168]. Khorlin and co-workers [169, 170] prepared the azido galactose derivatives by the addition of halogen azides to glycals. Thus, treatment of (204) with chlorine azide and subsequent treatment of the product with mercury(II) acetate gave the anomeric acetates (209) in 66% yield. Addition of the halogen azide to the benzyl glycal (205) was much more rapid and the corresponding 2-azido-2-deoxy-D-galactose derivative was isolated in high yield (82%). The effect of a mixture of alkyl or acyl groups on the rate of reaction of chlorine azide with substituted D-galactal was studied in detail [171] and it was found that an acetyl group on the 3-position was important in slowing the reaction.

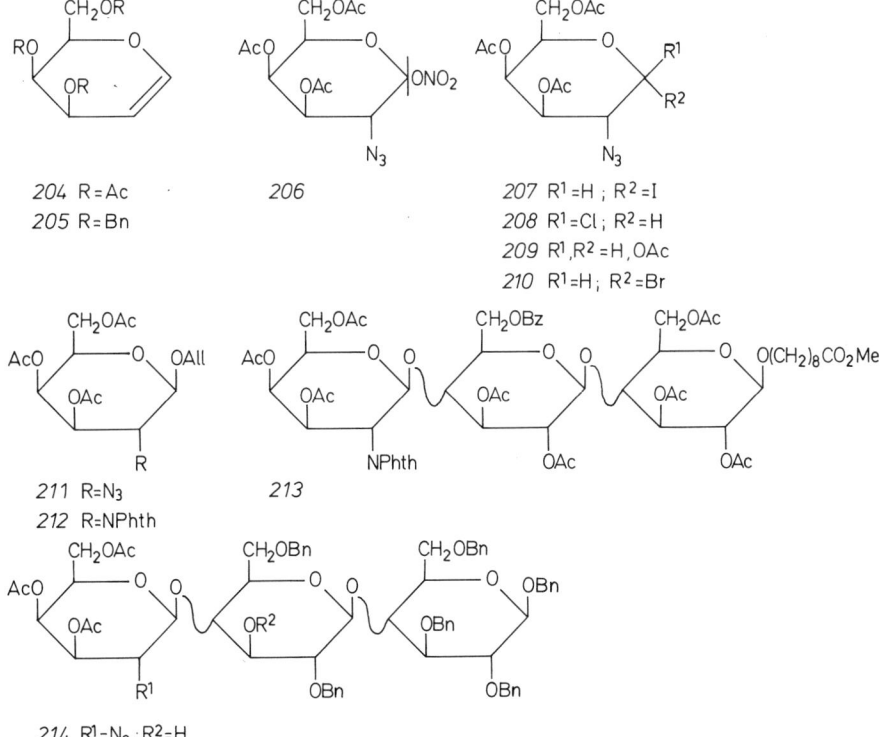

204 R=Ac
205 R=Bn

206

207 R^1=H ; R^2=I
208 R^1=Cl; R^2=H
209 R^1,R^2=H,OAc
210 R^1=H; R^2=Br

211 R=N$_3$
212 R=NPhth

213

214 R^1=N$_3$;R^2=H
215 R^1=NPhth;R^2=Bn

Lemieux and co-workers [125] have used the azidobromide (210) prepared as described above by the "azido-nitration" route to prepare the phthalimido bromide (193). The glycosyl bromide (210) was converted into the allyl glycoside (211) and this was deacetylated and the azido group reduced with hydrogen sulfide [172] and the amino group converted into the phthalimido group with phthalic anhydride. Acetylation gave (212) which was deallylated using the Wilkinson catalyst [173] and the free sugar was converted into the glycosyl bromide (193) by using the Vilsmeier reagent [172]. Condensation of the bromide (193) with the 8-carboxymethyloctyl lactose derivative (108) using silver triflate and 2,4,6-trimethylpyridine in nitromethane gave the trisaccharide derivative (213) and this on deprotection [by deacylation with sodium methoxide in methanol and dephthaloylation with hydrazine hydrate [174] followed by N-acylation] gave the β-(8-carboxymethyloctyl) glycoside of the asialo GM2 trisaccharide suitable for coupling to protein.

Paulsen and co-workers [175, 101, 100] studied the glycosidation of the benzyl β-lactoside (96) with different glycosylating agents derived from galactosamine. The bromide (193) in the presence of silver nitrate gave β-glycosides at both the 3- and 4-hydroxyl groups in equal proportions whereas the bromide (203) in the presence of silver silicate or silver carbonate gave the β-glycoside (214), linked at the 4-position in 76% yield. By reduction and deprotection this was converted into the trisaccharide of asialo GM2.

Ogawa and co-workers [107] have condensed the bromide (193) with the lactose derivative (101) in the presence of silver triflate and molecular sieves in dichloroethane to give the trisaccharide (215) in high yield. This was deacetylated and then dephthaloylated with butylamine in boiling methanol and the product N-acetylated to give (216). On hydrogenolysis and acetylation this gave the peracetate (217) which was deacetylated at the anomeric position by the action of hydrazine hydrate [176] to give the free sugar (218). This was treated with sodium hydride and trichloroacetonitrile [28, 177] to give the imidate (219). Condensation of the imidate with the benzoyl ceramide (63)

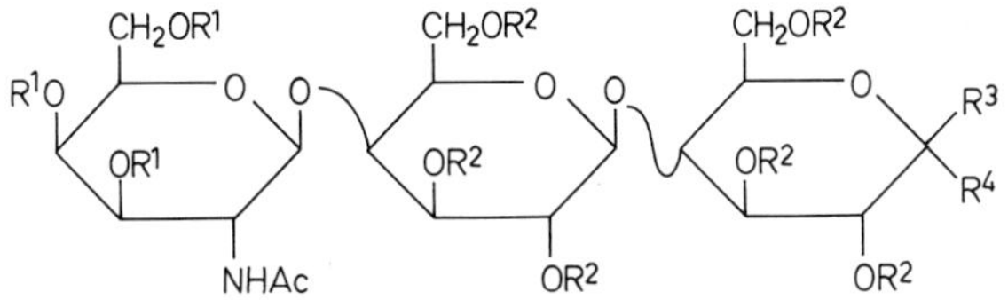

216   R^1=H; R^2=Bn; R^3=OBn; R^4=H
217   R^1=R^2=Ac; R^3=H; R^4=OAc
218   R^1=R^2=Ac; R^3=H; R^4=OH
219   R^1=R^2=Ac; R^3=H; R^4=OC(=NH)CCl$_3$

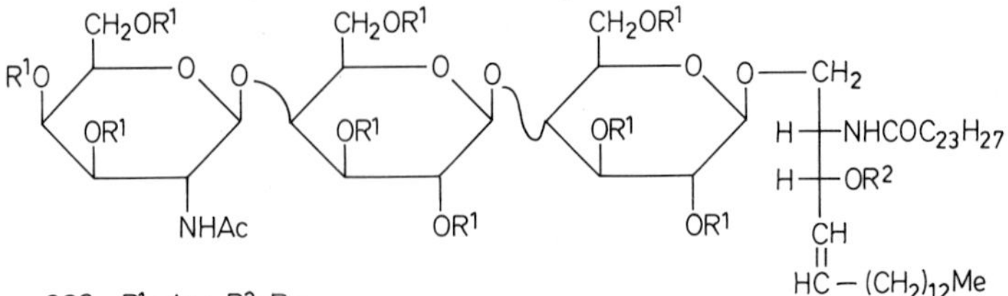

220   R^1=Ac; R^2=Bz
221   R^1=R^2=H

in the presence of boron trifluoride etherate and molecular sieves gave the protected asialo GM2 (220) in 38% yield and this on de-O-acylation gave asialo GM2 (221).

A related glycolipid (222) has been synthesised by Ogawa and co-workers [95]. Condensation of the glucosyl-phthalimido bromide (187) with the lactose derivative (96) in the presence of silver triflate gave glycosidation preferentially at the 3'-hydroxyl group (72%, with 23% on the 4'-hydroxyl group) and the product was then glycosylated on the 4'-hydroxyl group with the galactosyl-phthalimido bromide (193) in the presence of silver silicate to give the tetrasaccharide (223) in 76% yield. This was dephthaloylated by reduction with sodium borohydride [178] and converted into the peracetate by further deprotection and acetylation. The 1-O-acetyl group was removed with hydrazine acetate [176] and the product converted into the glycosyl fluoride [179, 180] with diethylaminosulfur trifluoride (DAST). Condensation of the fluoride with 3-O-benzoyl ceramide (63) in the presence of silver perchlorate — tin(II) chloride and subsequent treatment of the orthoester produced with trimethylsilyl triflate gave the β-glycoside in 22% yield. Deprotection then gave the glycolipid (222).

$$NAcgal-\beta-(1\rightarrow4)-$$
$$NAcglc-\beta-(1\rightarrow3)-$$ $$gal-\beta-(1\rightarrow4)-glc-\beta-(1\rightarrow1)CER$$

222

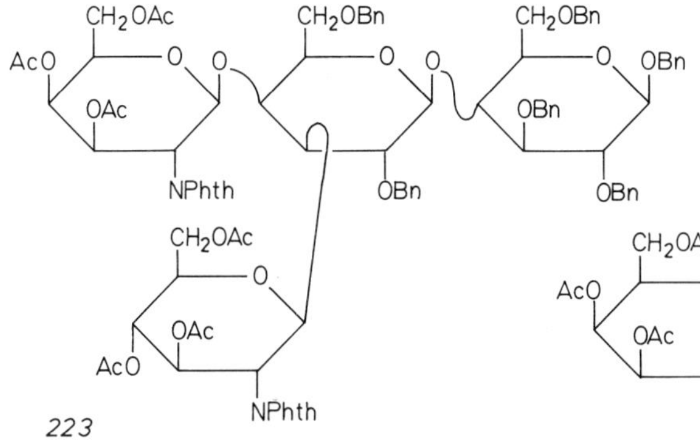

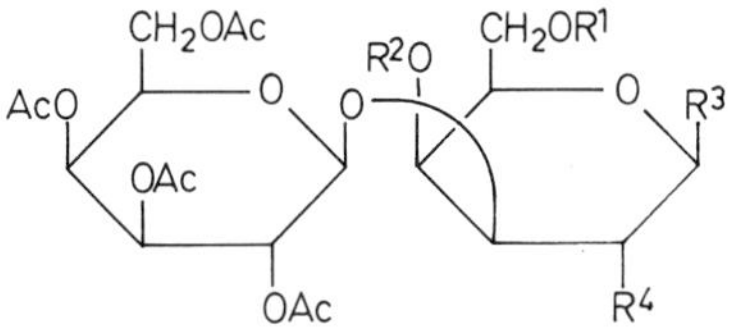

224 $R^1, R^2 = >CHPh$; $R^3 = OAll$; $R^4 = N_3$
225 $R^1, R^2 = Ac$; $R^3 = Br$; $R^4 = NPhth$

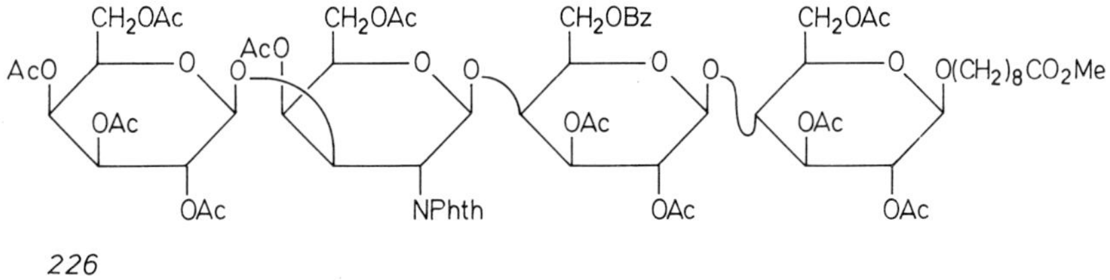

223

226

5.1.2 Asialoganglioside GM1

The ganglioside GM1 contains another galactose residue joined in β-linkage to the 3-position of N-acetylgalactosamine of Tay Sachs ganglioside and asialo GM1 thus has the structure:

$$gal-\beta-(1\rightarrow3)-NAcgal-\beta-(1\rightarrow4)-gal-\beta-(1\rightarrow4)-glc-\beta-(1\rightarrow1)-CER$$

103

The β-8-methoxycarbonyloctyl glycoside of the oligosaccharide portion of asialo GM1 was first prepared by Sabesan and Lemieux [125] as an extension of their work on the synthesis of the corresponding derivative of asialo GM2 described above. The azido compound (211) was deacetylated and converted into the 4,6-O-benzylidene derivative and this was condensed with acetobromogalactose in the presence of silver triflate to give the disaccharide (224) which was converted into the phthalimido bromide (225) using a similar route to that described above for the preparation of the bromide (210). Condensation of the bromide (225) with the 8-methoxycarbonyloctyl lactoside (108) using silver triflate gave the tetrasaccharide derivative (226) which was deprotected as described above for the deprotection of (194) to give the β-8-methoxycarbonyloctyl glycoside of the oligosaccharide portion of asialo GM1.

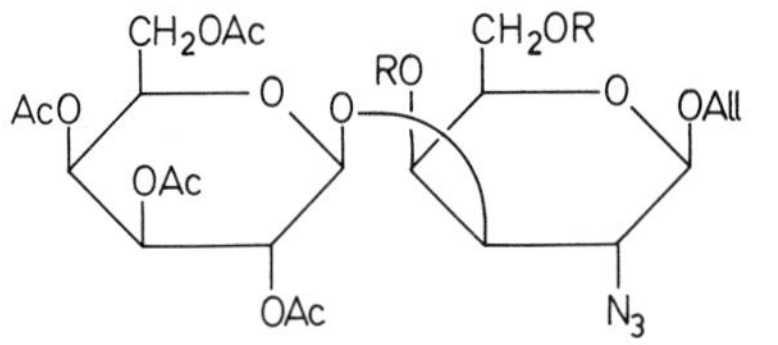

227 R = Bz
228 R = Ac

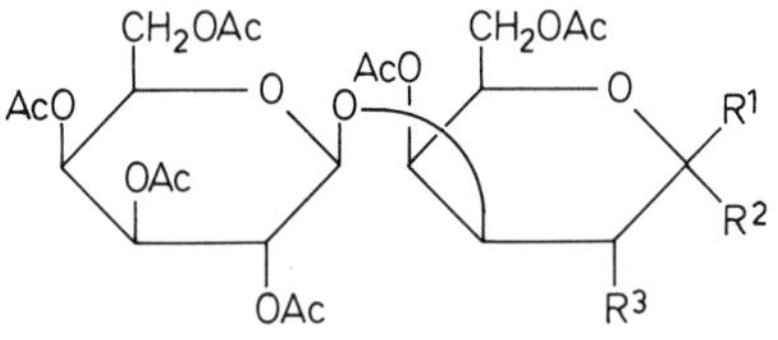

229 R¹ =OAll; R²=H; R³=NPhth
230 R¹ =Br; R²=H; R³=NPhth
231 R¹ =H; R²=Br; R³=N₃

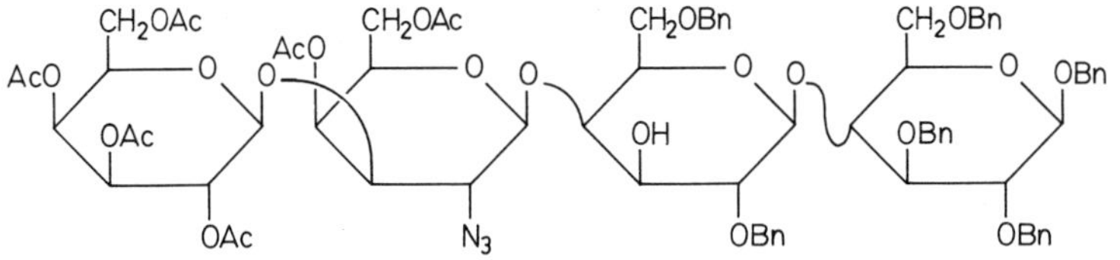

232

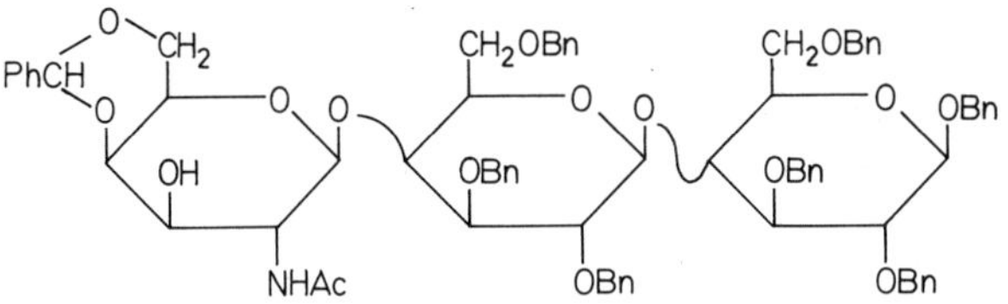

233

Paulsen and Paal [181] prepared the disaccharide (227) by condensation of 1,2,3,4,6-penta-O-acetyl-β-D-galactopyranose with allyl 2-azido-4,6-di-O-benzoyl-2-deoxy-β-D-galactopyranoside in the presence of trimethylsilyltriflate and converted this into the acetate (228) and then into the N-phthaloyl derivative (229). This was deallylated and converted into the 1-O-acetate which gave the glycosyl bromide (230) on reaction with titanium(IV) bromide. Condensation of the bromide (230) with the lactose derivative (96) in the presence of silver silicate or mercury salts gave only the β-(1 → 3) linked tetrasaccharide. However, the azido-bromide (231), in the presence of silver carbonate or silver silicate gave the β-(1→4) linked tetrasaccharide derivative (232) which on processing as described above for compound (214) gave the free tetrasaccharide corresponding to the oligosaccharide portion of asialo GM1.

In the synthesis of the glycolipid asialo GM1 described by Ogawa and co-workers [107] the trisaccharide derivative (215) (used in the synthesis of asialo GM2) was converted into the 4,6-*O*-benzylidene derivative (233) and this was condensed with acetobromogalactose in the presence of mercury(II) cyanide and molecular sieves to give the β-linked acetylated galactose derivative in 97% yield. The product was deprotected and acetylated to give the peracetyl derivative (234) of the tetrasaccharide which was converted into the free sugar (235) with hydrazine hydrate. Compound (235) was converted into asialo GM1 via the imidate as described above for the synthesis of asialo GM2 from the imidate (219).

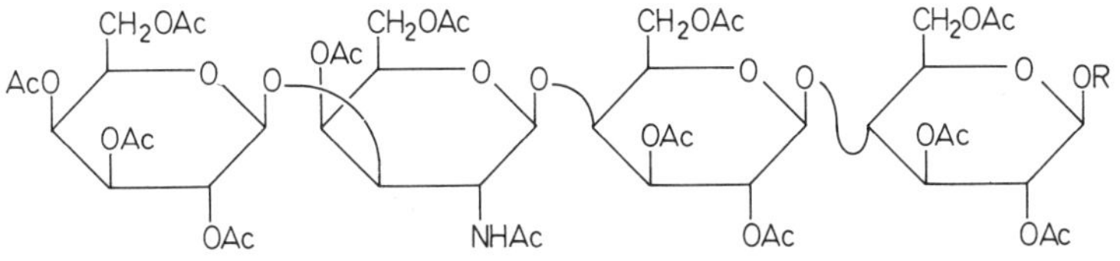

234 R = Ac
235 R = H

5.2 Glycosphingolipids Containing α-linked *N*-Acetylgalactosamine Residues

The α-linked *N*-acetylgalactosamine residue is present in the immunologically interesting glycolipid known as the Forssman antigen and related glycolipids, as well as being part of the determinant of the blood-group A glycolipids and glycoproteins. The efficient preparation of α-linked derivatives of glucosamine and galactosamine had been a major obstacle in oligosaccharide syntheses until Paulsen and co-workers in 1975 introduced the use of the 2-azido-2-deoxy derivative. Since the previous review [1] considerable progress has been made in this field including efficient syntheses of the pentasaccharide of the Forssman antigen and oligosaccharide portions of the blood-group A substances. The routes used by various workers for the synthesis of 2-azido-2-deoxygalactosyl halides, since the pioneering work of Paulsen, have been reviewed in Sect. 5.1

5.2.1 The Pentasaccharide of the Forssman Antigen

In early work [165] on the synthesis of the pentasaccharide (236), the azide (237) was condensed with (238) [an intermediate in the preparation of (237)] in the presence of silver perchlorate and polyvinylpyridine to give the α-linked disaccharide (239) in 60% yield and this on acetolysis gave the disaccharide (240) which contains the potential terminal disaccharide unit of the Forssman antigen. Compound (240) was converted into the glycosyl bromide with titanium(IV) bromide under carefully controlled conditions [182] and condensed with 1,6-anhydro-2,4-di-*O*-benzyl-D-galactopyranose in the presence of silver carbonate to give the potential terminal trisaccharide (241) of the Forssman antigen.

NAcgal-α-(1→3)-NAcgal-β-(1→3)-gal-α-(1→4)-gal-β-(1→4)-glc
236

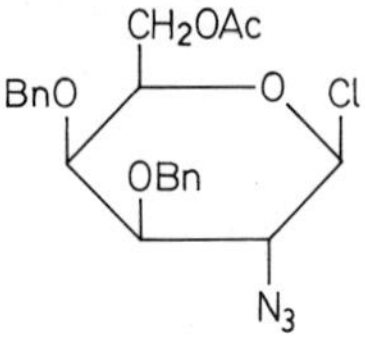
CH2OAc
BnO
O
Cl
OBn
N3

237

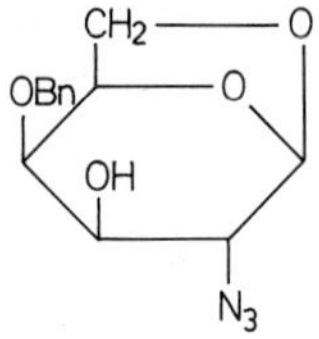
CH2
O
OBn
O
OH
N3

238

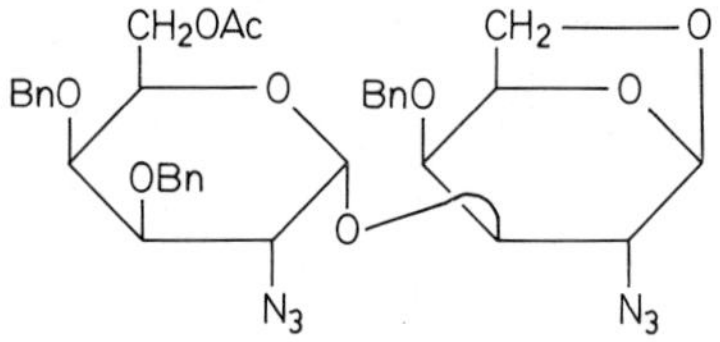
CH2OAc
BnO
O
OBn
O
N3
CH2
O
BnO
O
N3

239

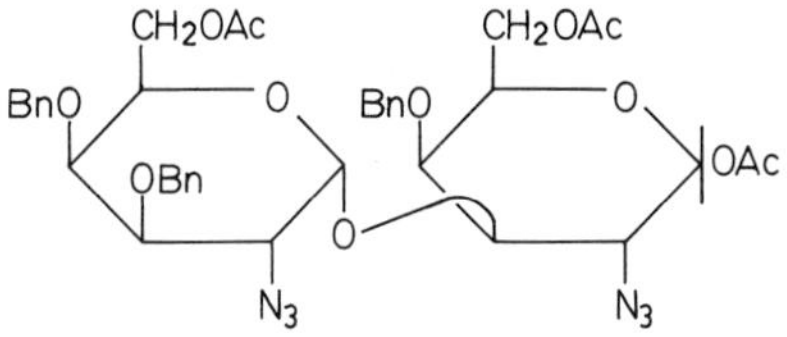
CH2OAc
BnO
O
OBn
O
N3
CH2OAc
BnO
O
OAc
N3

240

After extensive investigations by Paulsen and co-workers on the reaction parameters for the synthesis of the glycosidic linkage [23, 156, 164, 165, 183], e.g., the effects of substituents on the reactivity of glycosyl halides (ether protection giving glycosyl halides with higher reactivity than those with acyl protection), the effects of substituents on the reactivity of the acceptor alcohol and the activities of various catalysts (silver carbonate — silver perchlorate being a highly active catalyst and tetra-

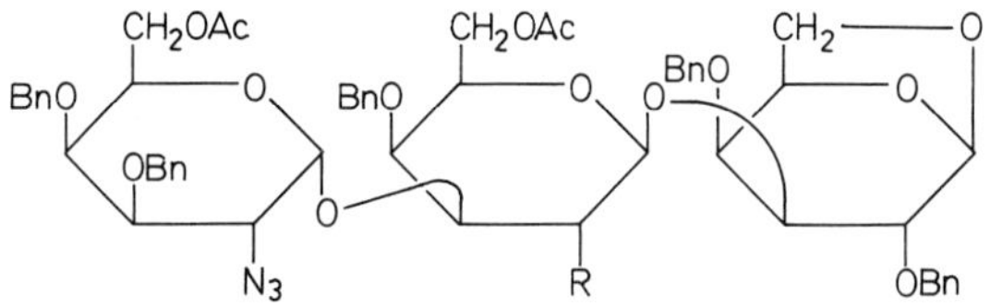
CH2OAc
BnO
O
OBn
O
N3
CH2OAc
BnO
O
O
R
CH2
O
BnO
O
OBn

241 R = N_3
242 R = NHAc

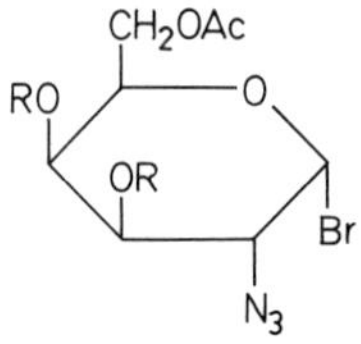
CH2OAc
RO
O
OR
Br
N3

243 R = Bn
244 R = Bz

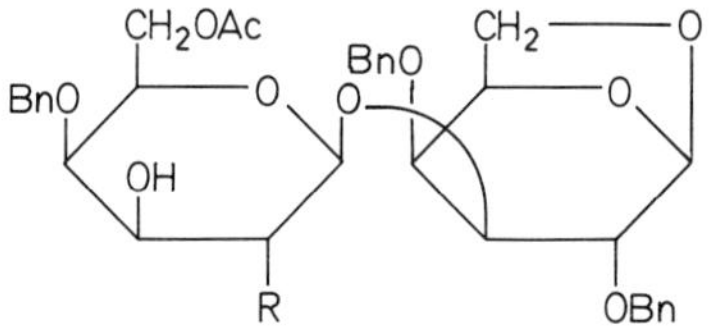
CH2OAc
BnO
O
OH
R
BnO
O
CH2
O
O
OBn

245 R = N_3
246 R = NHAc

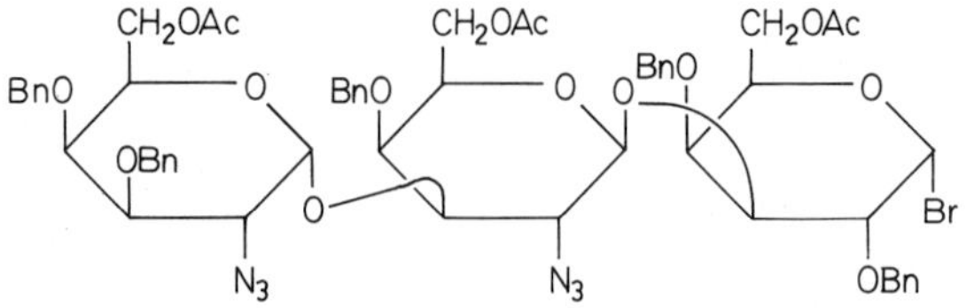
CH2OAc
BnO
O
OBn
O
N3
CH2OAc
BnO
O
N3
O
BnO
CH2OAc
O
Br
OBn

247

ethylammonium bromide being considered weakly active) they adopted a different approach [184] for the synthesis of the trisaccharide (241) from that described above.

In this study they condensed the α-glycosyl bromide (243) with the disaccharide (245) in the presence of silver carbonate — silver perchlorate to give the α-linked trisaccharide (241) in 58% yield or the bromide (244) with the disaccharide (246) in the presence of the mixed silver catalysts to give the α-linked trisaccharide (242) in 63% yield. In the earlier approach, the preparation of the β-chloride (237) required a previous treatment of the α-bromide with tetraethylammonium chloride under carefully controlled conditions.

Acetolysis of (241) and subsequent treatment of the 1,6-di-O-acetate produced with titanium(IV) bromide gave the α-glycosyl bromide (247). This was condensed with the lactose derivative (248) (substitution with deuterated benzyl groups made it possible to interpret the NMR spectra [185]) in the presence of silver carbonate — silver perchlorate to give the α-linked pentasaccharide (249) in 38% yield, and this was deprotected to give (236).

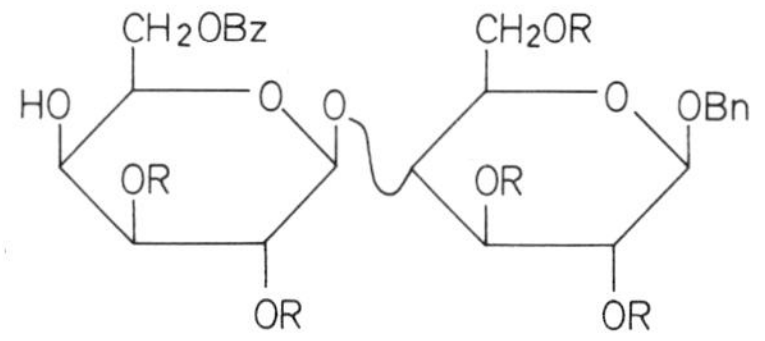

248 R = C^{2}H$_2$Ph

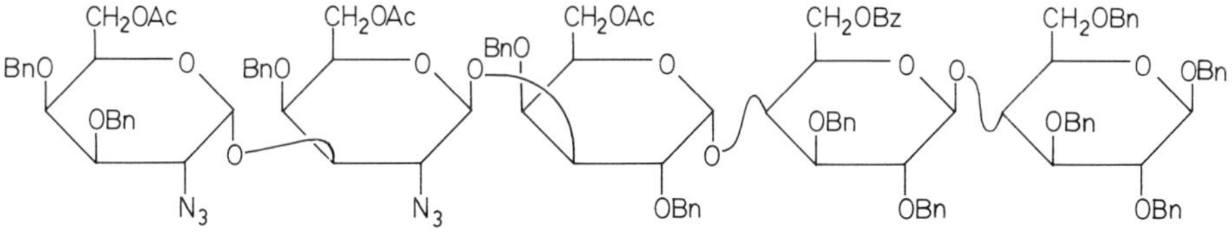

249

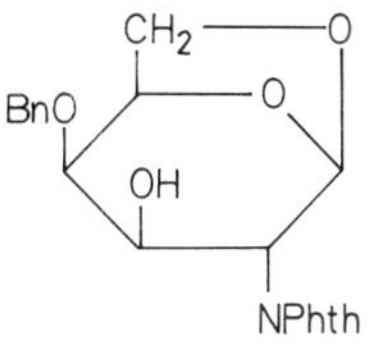

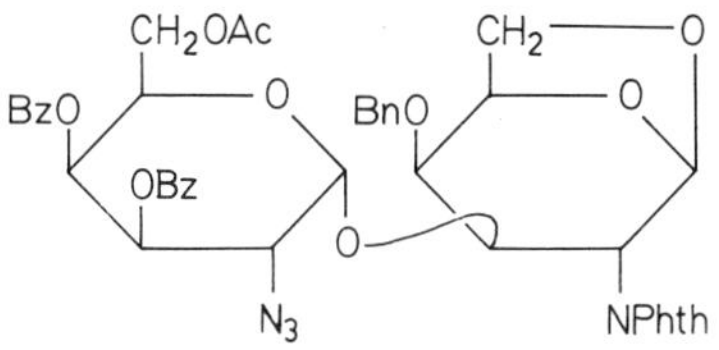

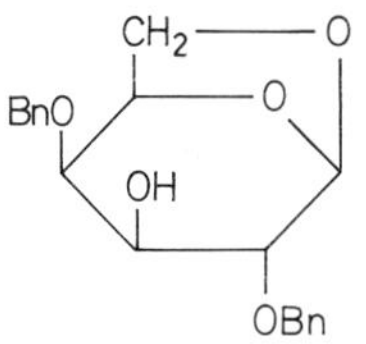

250   251   252

In a different approach to the pentasaccharide chain of the Forssman antigen, Paulsen and Bünsch [182, 186] condensed the bromide (244) with the phthalimido-derivative (250) in the presence of silver carbonate — silver perchlorate to give the α-linked disaccharide (251) which was acetolysed and the product converted into the glycosyl bromide and condensed with the galactose derivative (252) in the presence of silver triflate to give the β-linked trisaccharide (253). This was acetolysed and the product treated with titanium(IV) bromide and the glycosyl bromide produced condensed with the lactose derivative (248) in the presence of silver carbonate — silver perchlorate to give the α-linked pentasaccharide (254).

Jill Gigg and Roy Gigg

The azido group was reduced with sodium borohydride — nickel(II) chloride and the resulting amine acetylated. The O-acyl groups were removed with sodium methoxide and the phthalimido group then cleaved with hydrazine and the resulting amino group acetylated. Finally, the benzyl groups were removed by hydrogenolysis to give the pentasaccharide (236).

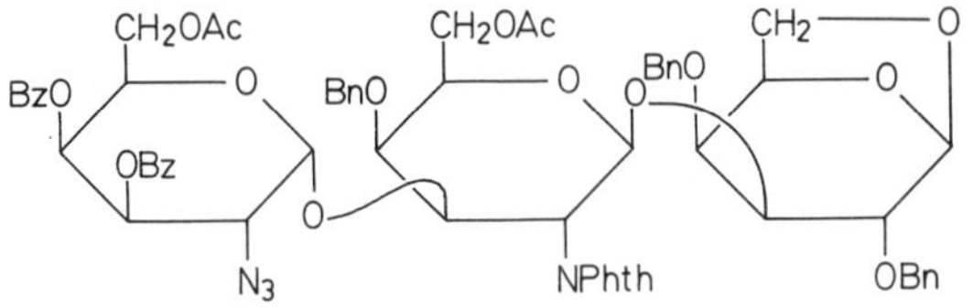

253

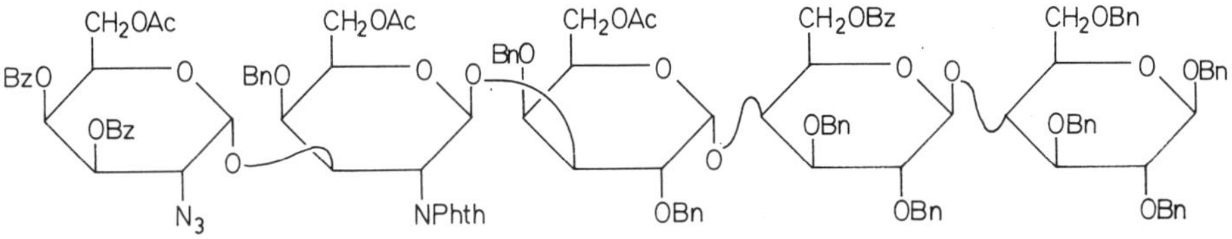

254

The synthetic pentasaccharide showed similar antigenic specificity to the Forssman glycolipid but because of the absence of the lipid portion did not form a stable complex with the Forssman antibody [187].

The crystal structure and solution conformation of the fully acetylated NAcgal-α-(1 → 3)-NAcgal disaccharide have been studied [188].

5.2.2 The Determinant of the Blood-group A Glycolipid

The first-recorded synthesis of the terminal trisaccharide (255) of the blood-group A substance was recorded by David and his coworkers [189, 190] using an entirely novel approach to glycoside synthesis involving a cycloaddition reaction between the 3-O-(but-1,3-dienyl) ether of galactose (256) and the (—)-menthyl ester of glyoxylic acid. After isomerisation of the product with boron trifluoride etherate the α-linked disaccharide (257) was isolated in 43% yield. This was deallylated and the product condensed with 2,3,4-tri-O-benzyl-α-L-fucopyranosyl bromide to give (258). After a series of reactions involving lithium aluminium hydride reduction, epoxide formation, Sharpless isomerisation and tritylation, the product (259) was oxidised to the

L-fuc-α-(1→2)-
NAcgal-α-(1→3)- } gal

255

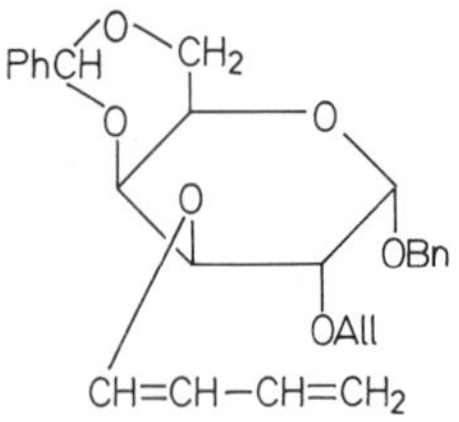

CH=CH−CH=CH₂
256

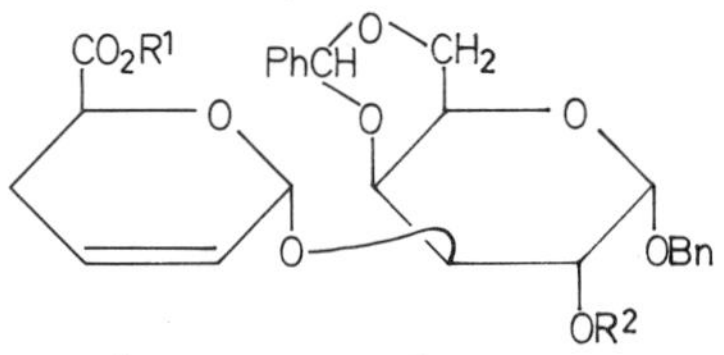

257 R¹=(−)-menthyl; R²=All
258 R¹=(−)-menthyl; R²=α-L-perbenzylfucos

ketone, the double bond hydroxylated and the ketone converted to the oxime which was acetylated to give (260) in 85% yield from (259). Reduction of the oxime (260) with lithium aluminium hydride gave the α-linked galactosamine derivative [(261) (54%) together with the talosamine derivative (262) (11%)]. Deprotection of (261) gave the trisaccharide (255) of the blood-group A substance. Reduction of the intermediate ketone to the galactose derivative also gave the B blood-group trisaccharide and various other analogs could be made from the intermediates.

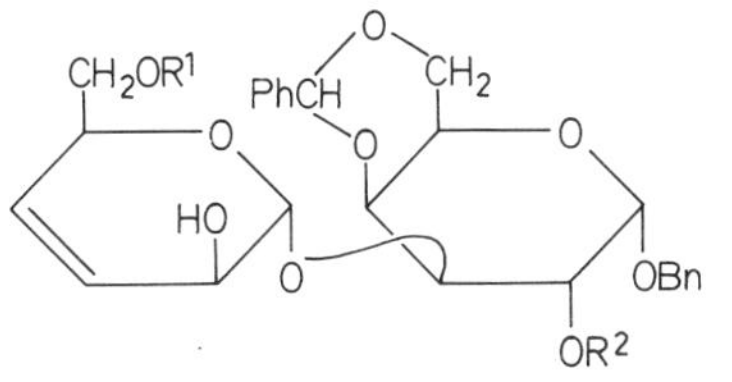

259 R¹ = CPh₃ ; R² = α-L-perbenzylfucose

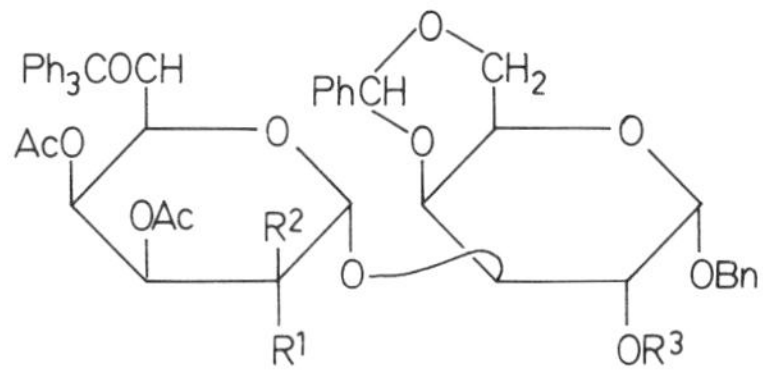

260 R¹, R² = N-OAc ; R³ = α-L-perbenzylfucose
261 R¹ = NHAc ; R² = H ; R³ = α-L-perbenzylfucose
262 R¹ = H ; R² = NHAc ; R³ = α-L-perbenzylfucose

Paulsen and co-workers [191] synthesised the trisaccharide (263) representing the afucosyl end group of the blood-group A (type 1) determinant from the disaccharide (264). This was acetylated and deacetonated and the product converted into the acetate (265) via the 3′,4′-orthoacetate. Condensation of (265) with the azido-chloride (237) in the presence of silver perchlorate gave the α-linked trisaccharide (266) which was deprotected to give (263).

For the synthesis of the tetrasaccharide unit (267) of the blood-group substance A (type 1), Paulsen and Kolář [192, 193] prepared the benzoate (269) by partial benzoylation of the diol (268) and condensed this with 2,3,4-tri-O-benzyl-α-L-fucosyl bromide in the presence of tetraethylammonium bromide to give (270) in 82% yield. The benzoyl group was removed to give (271) and this was condensed with the chloride (274) in the presence of silver carbonate — silver perchlorate to give the tetrasaccha-

NAcgal-α-(1→3)-gal-β-(1→3)-NAcglc
263

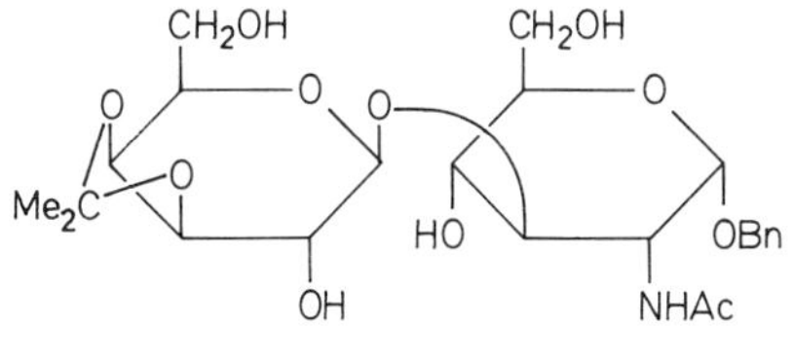

264 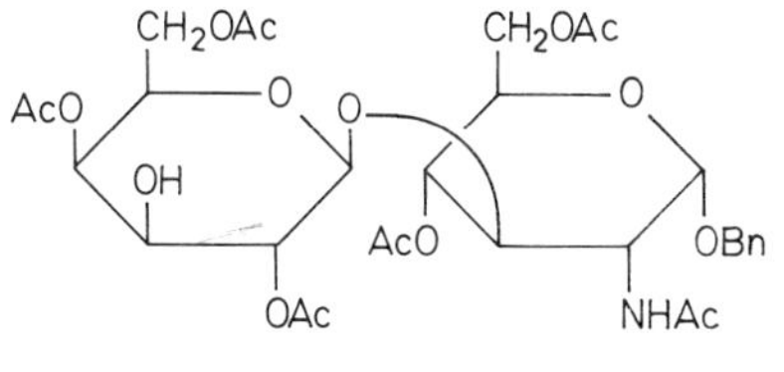   265

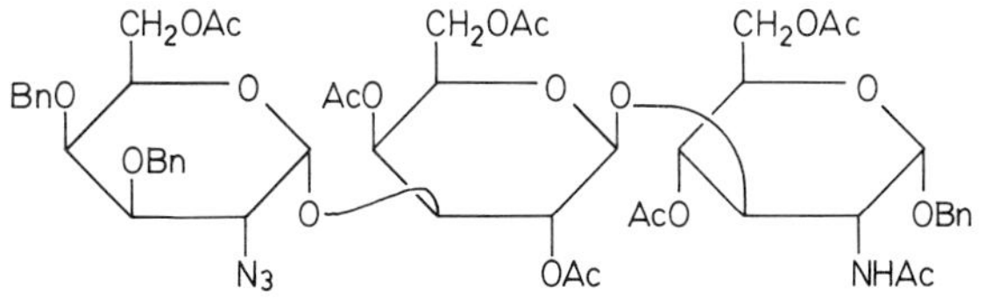

266

ride (276) in 46% yield or with the bromide (275) in the presence of mercury(II) cyanide — mercury(II) bromide to give (276) in 63% yield. Deprotection of (276) gave the tetrasaccharide (267).

The intermediate trisaccharide (270) on deprotection gave the blood-group H (type 1) trisaccharide (278), and (271) was also condensed with tetra-*O*-benzyl-galactopyranosyl bromide in the presence of silver carbonate — silver perchlorate or mercury(II) cyanide to give a tetrasaccharide derivative which was converted into the blood-group B (type 1) tetrasaccharide (279). The blood-group A (type 2) tetra-saccharide (280) was prepared from the α-anomer of the bromide (275) by Paulsen and co-workers [156] via an intermediate trisaccharide used for the preparation of the blood-group B (type 2) tetrasaccharide (see Sect. 6.1).

Khorlin and co-workers [194, 195] prepared the blood-group A (type 1) tetra-saccharide (267) from the chloroacetate (272). This was condensed with 2-*O*-benzyl-3,4-di-*O*-*p*-nitrobenzoyl-α-L-fucopyranosyl bromide to give the 2-*O*-α-L-fucosyl derivative and after removal of the chloroacetyl group with thiourea the product was

NAcgal-α-(1→3)-
L-fuc-α-(1→2)- } gal-β-(1→3) NAcglc
267

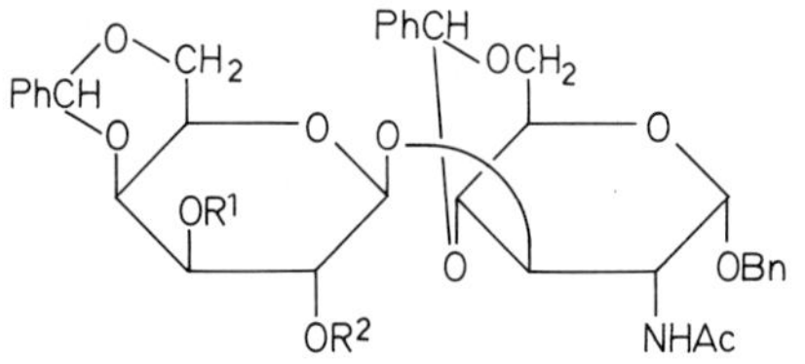

268 R¹=R²=H
269 R¹=Bz ; R²=H
270 R¹=Bz ; R²= α-L-perbenzylfucose
271 R¹=H ; R²= α-L-perbenzylfucose
272 R¹=COCH₂Cl ; R²=H
273 R¹=H ; R²=Ac

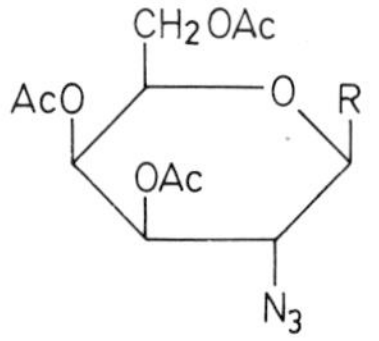

274 R=Cl
275 R=Br

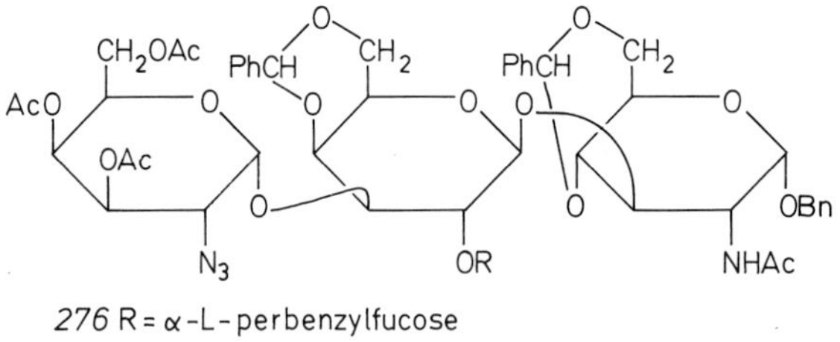

276 R= α-L-perbenzylfucose
277 R= 2-O-benzyl-3,4-di-O-p-nitrobenzoyl-α-L-fucose

L-fuc-α-(1→2)-gal-β-(1→3)-NAcglc
278

gal-α-(1→3)-
L-fuc-α-(1→2)- } gal-β-(1→3)-NAcglc
279

NAcgal-α-(1→3)-
L-fuc-α-(1→2)- } gal-β-(1→4)-NAcglc
280

110

condensed with the chloride (274) in the presence of silver carbonate — silver perchlorate to give the tetrasaccharide (277) in 82% yield. They also condensed the disaccharide (271) with the chloride (274) under similar conditions to give (276) in 60% yield. Deprotection of (276) or (277) gave the tetrasaccharide (267). The H (type 1) trisaccharide (278) was also prepared in this work by deprotection of the intermediate trisaccharides and the afucosyl A (type 1) trisaccharide (263) was also prepared by reaction of the chloride (274) with the disaccharide (273), prepared from (272) by acetylation and dechloracetylation with thiourea.

6 Glycosphingolipids Containing the Gal-β-(1 → 4)-NAcglc (N-Acetyllactosamine) and the Gal-β-(1 → 3)-NAcglc Residues

6.1 The N-Acetyllactosamine Residue

Many mammalian glycolipids including the blood-group substances have the N-acetyllactosamine residue as part of their oligosaccharide sequence (see Refs. [12–18]). This disaccharide unit defines the blood-group substances of "type 2" to distinguish them from those containing the gal-β-(1→3)-NAcglc unit which are described as "type 1". The occurrence of this linkage in the glycolipids has stimulated many synthetic approaches to derivatives of this disaccharide which can be elaborated into more complex oligosaccharides.

Ponpipom et al. [113] converted peracetyl lactal by the nitrosyl chloride route [25] into the lactosamine derivative (281) which was converted into the glycosyl chloride (282). Lemieux and co-workers [172] studied the preparation of the chloride

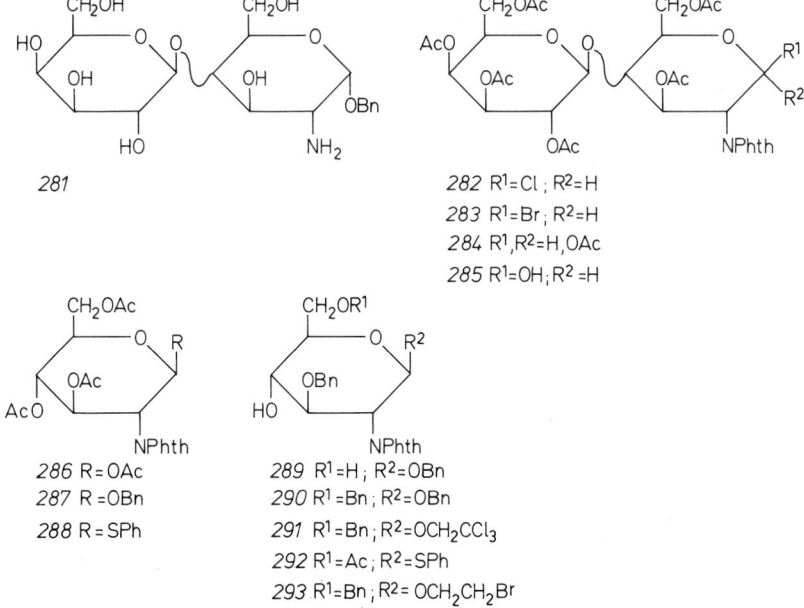

(282) from the peracetyl lactal in detail using both the nitrosyl chloride route and the azido nitrate route [166] and condensed [196] the chloride (282) with 8-methoxy-carbonyloctyl 2-*O*-benzoyl-4,6-*O*-benzylidene-β-D-galactopyranose. The corresponding bromide (283) has also been prepared by Arnap and Lönngren [197] from peracetyl lactal using the azido nitration route. Ogawa and co-workers [198] have also prepared the chloride (282) as a "lactosaminyl donor" by a different route. Treatment of the acetate (286) with the tributyltin derivative of benzyl alcohol and tin(IV) chloride gave the benzyl glycoside (287) in high yield and this was converted via the 4,6-*O*-benzylidene derivative into the diol (289). Compound (289) was converted by benzylation of the tributyltin derivative [108, 199] into (290). This was condensed with acetobromogalactose in the presence of mercury(II) bromide and molecular sieves and the product processed to give the chloride (282). Using similar techniques Ogawa and co-workers [200] have also prepared the lactosamine derivative from the glucosamine derivative (291) and also prepared [201] the thioglycoside (292) as an acceptor molecule for syntheses of this type. The β-acetate (286) was converted into the β-thioglycoside (288) in 65% yield by the action of the tributyltin derivative of thiophenol and tin(IV) chloride. Magnusson and co-workers [154] have also prepared a lactosamine derivative, with a spacer-arm suitable for coupling to protein, from the bromoethyl glycoside (293).

Ponpipom et al. [113] condensed the chloride (282) with the lactose derivative (104) in the presence of silver triflate to give the β-linked tetrasaccharide in high yield and this on deprotection gave the β-benzyl glycoside of gal-β-(1→4)-*N*Acglc-β-(1→3)-gal-β-(1→4)-glc which has the oligosaccharide sequence of the glycolipid known as paragloboside.

The oxazoline derivative (294) of lactosamine has been used extensively by Khorlin and co-workers [112], by Veyrières and co-workers [202] and by Tejima and co-workers [203] as a glycosyl donor. Veyrières [202] prepared a series of derivatives of galactose using this glycosylating agent and it was found that the trisaccharide gal-β-(1→4)-*N*Acglc-β-(1→6)-gal was recognised by antibodies directed against the I blood-group glycolipid, whereas the gal-β-(1→4)-*N*Acglc-β-(1→3)-gal trisaccharide

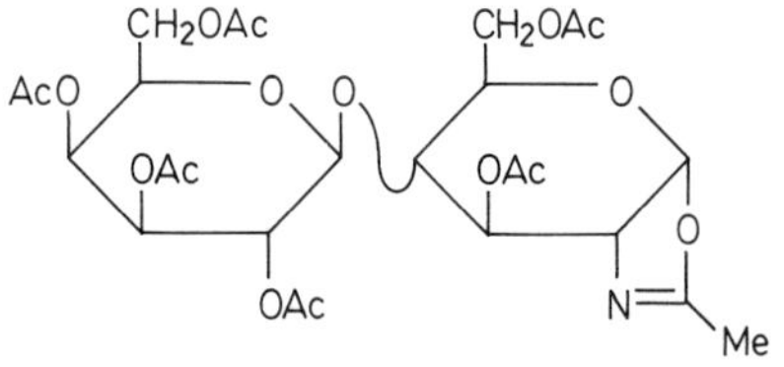

294

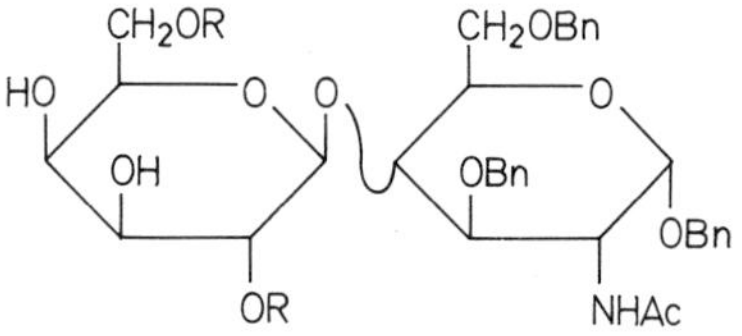

295 R = Bn
296 R = H

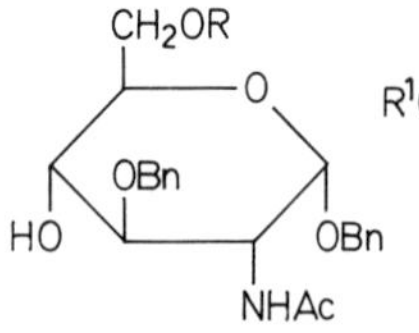

297 R = H
298 R = Bn

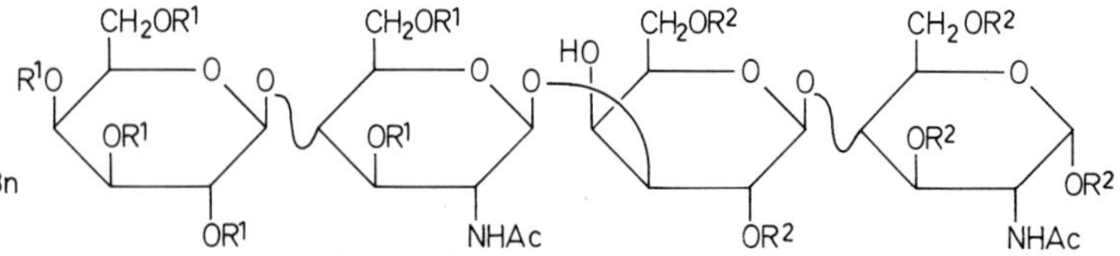

299 R¹ = Ac ; R² = Bn
300 R¹ = R² = H

was more active with antibodies directed against the *i*-glycolipid [204]. Veyrières et al. [110] prepared a dimer of *N*-acetyllactosamine by condensing the oxazoline (294) with the lactosamine derivative (295). Compound (295) was prepared from the glucosamine derivative (297) as follows. The dibutylstannyl ether of (297) was benzylated in the presence of tetraethylammonium bromide to give the 6-*O*-benzyl derivative (298) in high yield and this was converted into (296) and hence (295) by conventional methods. Condensation of the oxazoline (294) with the diol (295) in the presence of toluene *p*-sulfonic acid gave the tetrasaccharide derivative (299) in 52% yield with preferential reaction at the 3-hydroxyl group. Deprotection of (299) gave the dimer of *N*-acetyllactosamine (300) which was used for studies of the Ii blood-group activity. Lactosaminyl derivatives of mannose were also prepared by Veyrières and co-workers [205, 206] from the oxazoline (294) for investigations of the Ii blood-group activity and other lactosaminyl derivatives of mannose were prepared by Arnarp and co-workers using glycosyl bromide (283) [197, 207–209].

Veyrières and co-workers [210] improved a previously reported [211] method for the preparation of *N*-acetyllactosamine by the reduction of 2-benzylamino-2-deoxy-4-*O*-(β-D-galactopyranosyl)-D-gluconitrile. The 2-amino-2-deoxy-4-*O*-(β-D-galactopyranosyl)-D-glucose formed in this reaction was used by Veyrières and co-workers [212] as an intermediate for a new synthesis of the chloride (282). The lactosamine derivative (284) was treated with hydrazine — acetic acid [176] or with piperidine in tetrahydrofuran to de-*O*-acetylate specifically the anomeric position and give the free sugar (285) which was converted by the Vilsmeier reagent [172] into the chloride (282). From the chloride (282) they prepared [212] the protected methyl glycoside (301) and this was condensed with the chloride (282) in the presence of silver triflate and 2,4,6-trimethylpyridine [159] to give the tetrasaccharide (303) in good yield and this on acetolysis gave the acetate (304). The acetate (304) was converted into the

301 R = Ac
302 R = H

chloride (305) and this was condensed with the diol (301) in the presence of silver triflate to give the β-(1→3)-linked hexasaccharide in 40% yield. On deprotection and *N*-acetylation this gave the β-methyl glycoside of gal-β-(1→4)-*N*Acglc-β-(1→3)-gal-β-(1→4)-*N*Acglc-β-(1→3)-gal-β-(1→4)-*N*Acglc which was found in biological tests to be the most potent synthetic inhibitor of anti-i antibodies so far prepared.

The methyl 3'-*O*-allyl-β-lactoside (97) prepared by Veyrières by the allylation of the dibutylstannylene derivative in the presence of tetrabutylammonium bromide [102] was converted into the per-*p*-bromobenzyl ether and deallylated to give (306). This was condensed [106] with the chloride (282) in the presence of silver triflate to give a tetrasaccharide in 68% yield. This was far superior to the condensation of the oxazoline (294) with (306) which gave a very low yield (9%) of tetrasaccharide. Depro-

113

tection of the tetrasaccharide gave the β-methyl glycoside of gal-β-(1→4)-NAcglc-β-(1→3)-gal-β-(1→4)-glc which was also used for studies with the anti-Ii system.

Veyrières et al. [102] also showed that allylation of the dibutylstannylene derivative of the lactosamine derivative (302) in the presence of tetrabutylammonium bromide gave the 3′-O-allyl ether in 56% yield.

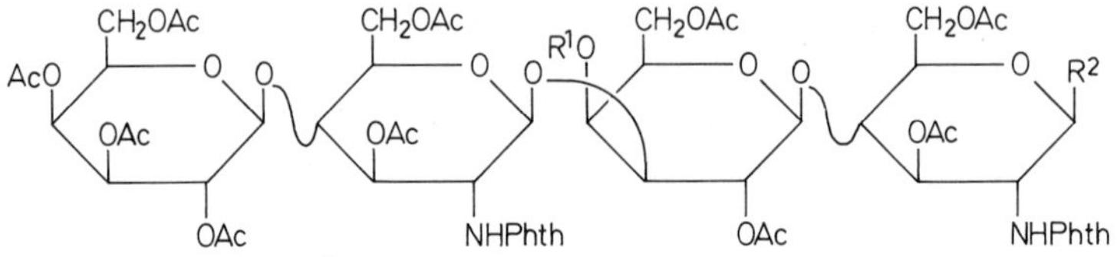

CH₂OAc · · · CH₂OAc · R¹O · CH₂OAc · · · CH₂OAc
AcO · · · OAc · · · · OAc
OAc · · · NHPhth · · · OAc · · · NHPhth

303 R¹=H; R²=OMe
304 R¹=Ac; R²=OAc
305 R¹=Ac; R²=Cl

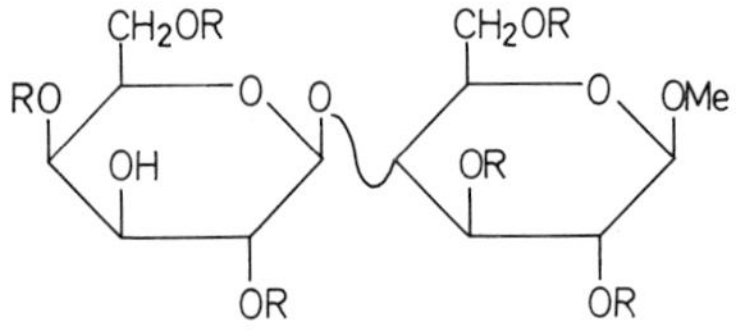

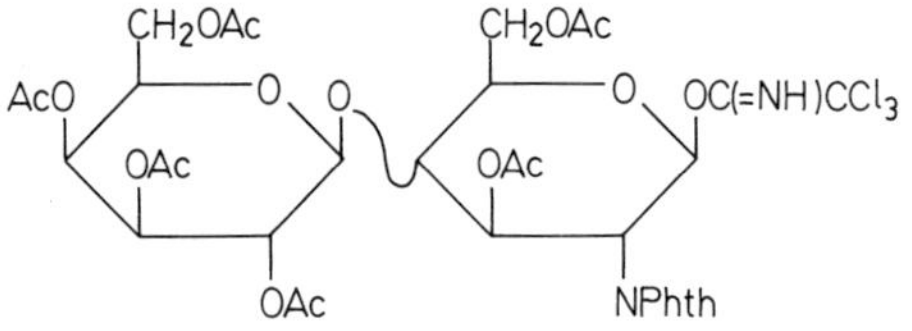

CH₂OR · CH₂OR · · · · CH₂OAc · CH₂OAc
RO · · · OMe · · · AcO · · · OC(=NH)CCl₃
OH · · · OR · · · · OAc · · · OAc
OR · · · OR · · · · OAc · · · NPhth

306 R = p-BrPhCH₂ · · · 307

CH₂OAc · CH₂OAc · R²O · CH₂OR¹ · CH₂OBz
AcO · · · · OMe
OAc · · · OAc · · · · OBz · · · OBz
OAc · · · NPhth · · · OBz · · · OBz

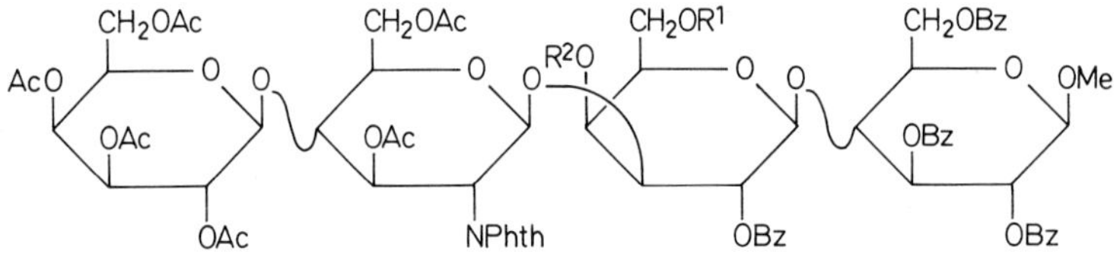

308 R¹,R² = >CHPh
309 R¹=R² = H

In continuation of their studies of the blood-group Ii-active glycolipids, Veyrières et al. [105] converted methyl 3′-O-allyl lactoside (97) into the benzylidene derivative (98) and condensed this with the lactosamine imidate derivative (307) [213] in the presence of boron trifluoride — etherate at −20 °C to give the tetrasaccharide (308) in 75% yield. The benzylidene group was hydrolysed to give the diol (309) and this was condensed with the imidate (307) in the presence of trimethylsilyl triflate with reaction specifically at the primary alcohol to give a hexasaccharide in 80% yield. Deprotection and N-acetylation gave the hexasaccharide with two β-N-acetyllactosaminyl groups attached to methyl β-lactoside at the 3′ and 6′ positions. The same hexasaccharide has been prepared [214] using an immobilised cyclic multienzyme system capable of generating UDP-D-gal and transferring this to the 4-hydroxyl group of glucosamine in synthetic methyl 3′,6′-di-O-(2-acetamido-2-deoxy-β-D-glucopyranosyl)-β-lactoside, i.e. generating lactosaminyl residues. Similar enzymatic systems had been described previously by Whitesides and co-workers [215, 216] and by David [217]. Tejima and co-workers [203] prepared the same hexasaccharide by the condensation of the oxazoline (294) with the lactose derivative (114a) and separation of the 6'-lactosaminyl derivative (24.5 %) from the 3',6'-lactosaminyl derivative (53.5 %) by chromatography. Deprotection gave the 3',6'-*bis*-lactosaminyl lactose.

Tejima et al. [218] found that partial tosylation of the 1,6-anhydrolactose derivative (310) gave the 2-*O*-tosyl derivative (311) in 25 % yield and this was converted into the epoxide (312) which could be converted into the anomeric azides (313) and hence into *N*-acetyllactosamine. Tejima et al. [219] have also prepared peracetyl lactosamine from the 1,6-anhydro derivative (314).

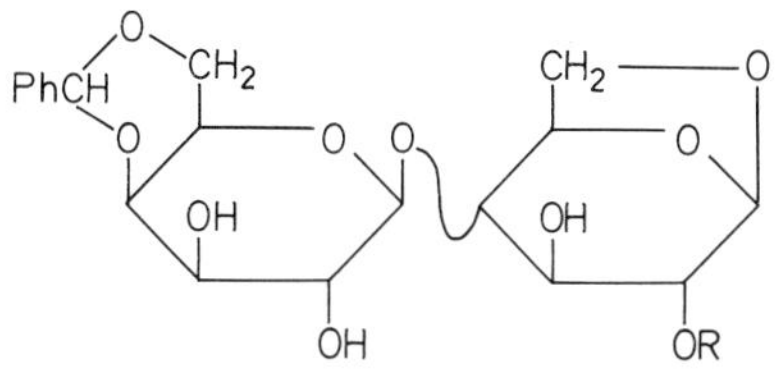

310 R=H
311 R=SO$_2$PhpMe

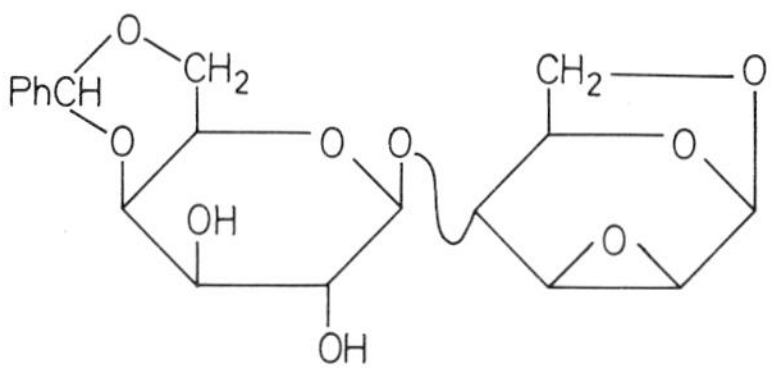

312

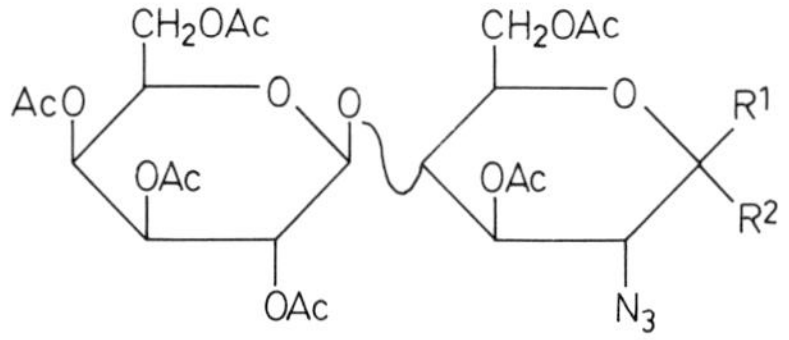

313 R^1,R^2 =OAc,H

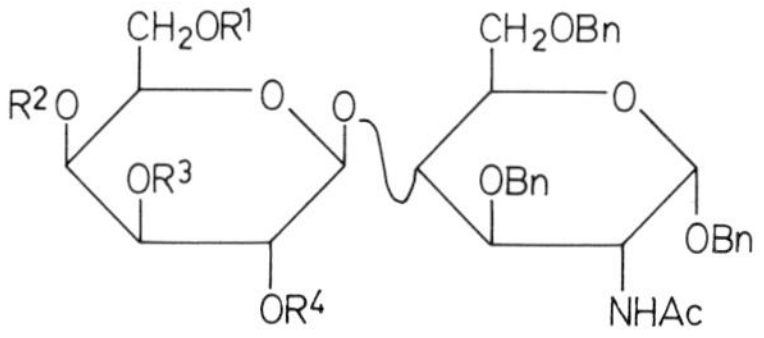

315 R^1=R^2=R^3=R^4 =Ac
316 R^1=R^4=Bn ; R^2=R^3=H
317 R^1=R^3=R^4=Bn; R^2=H
318 R^1,R^2= >CHPh; R^3=Bz ; R^4 =H
319 R^1,R^2= >CHPh; R^3=Bz ; R^4= α-L-perbenzylfucose
320 R^1=Bz ; R^2=H; R^3=R^4 =Bn

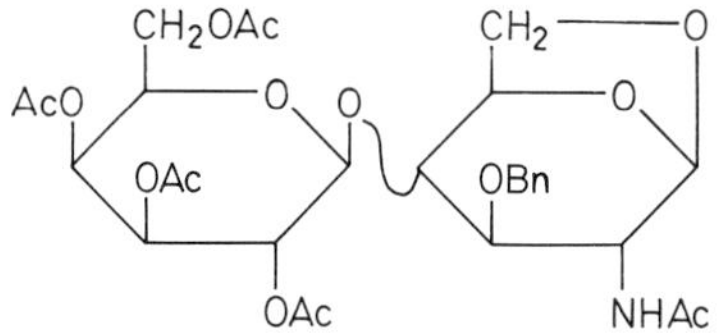

314

Sinaÿ and his co-workers [150] prepared the *N*-acetyllactosamine derivative (315) by the condensation of acetobromogalactose with the glucosamine derivative (298) which was prepared [151] by reaction of the 6-*O*-tosylate of (297) with the sodium salt of benzyl alcohol. Compound (315) was converted via the 3',4'-*O*-isopropylidene derivative into the diol (316) which was condensed with the imidate (136) [138, 220] in the presence of toluene *p*-sulfonic acid to give the α-galactosyl derivative at the 3'-position of the diol (316) in 74 % yield. Deprotection gave the trisaccharide gal-α-(1→3)-gal-β-(1→4)-*N*Acglc which had previously been isolated as a hydrolysis product of the blood-group B (type 2) substance. Attempted galactosidation [150] of (317) with the imidate (136) gave no reaction under conditions where the corresponding galactose derivatives did react and Sinaÿ therefore concluded that the axial 4'-hydroxyl group of lactosamine is much less reactive than the same position in galactose derivatives.

In continuation of this work, Sinaÿ and co-workers [221] prepared the terminal tetrasaccharide of the blood-group B (type 2) glycolipid which has *N*-acetyllactosamine substituted at the 2'-position with an α-L-fucosyl residue and at the 3'-position with an α-D-galactosyl residue. The lactosamine derivative (318) was condensed with the L-fucosyl imidate (321) [222] to give the 2'-O-α-L-fucosyl derivative (319) in 89% yield. The benzoyl group was hydrolysed and the product condensed with the D-galactosyl imidate (136) to give the 3'-O-α-D-galactosyl derivative (323) in 90% yield. Deprotection of (323) gave the B (type 2) blood-group tetrasaccharide (324). The terminal trisaccharide (325) of the B blood-group first prepared by Lemieux and co-workers [25] has also been prepared by Jacquinet and Sinaÿ [220] using the imidate procedure. Augé and Veyrières [223] also prepared (325) by condensation of 2,3,4,6-tetra-O-benzyl-α-D-galactosyl bromide with the stannylene derivative (326) in the presence of lithium iodide in hexamethylphosphorotriamide. The equatorial 3-hydroxyl group was preferentially glycosidated to give predominantly the α-linked disaccharide (327) (74%). Deallylation of (327) and subsequent α-fucosylation at the 2-hydroxyl group (32%) and deprotection gave (325). Martin-Lomas and co-workers [224] have also prepared (325) via the disaccharide (328).

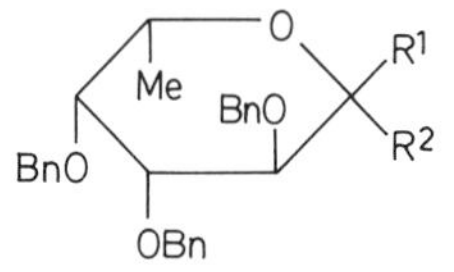

321 R¹=H; R² =OC(Me)=N—Me
322 R¹=Br; R²=H

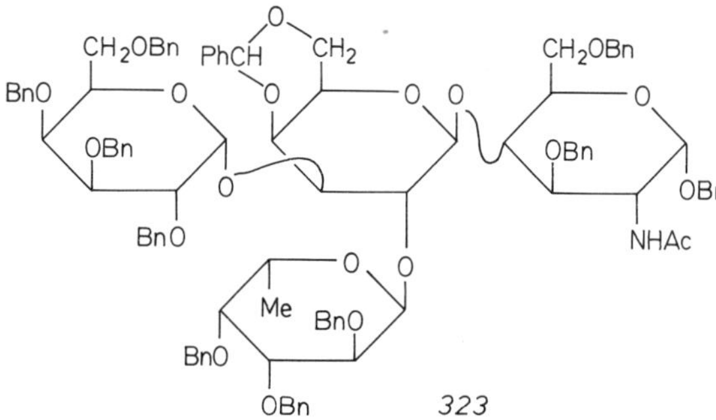

$$gal-\alpha-(1\rightarrow3)-$$
$$L-fuc-\alpha-(1\rightarrow2)-$$ gal$-\beta(1\rightarrow4)$ NAcglc

324

$$gal-\alpha-(1\rightarrow3)-$$
$$L-fuc-\alpha-(1\rightarrow2)-$$ gal

325

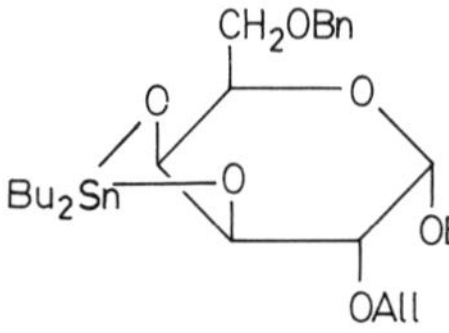

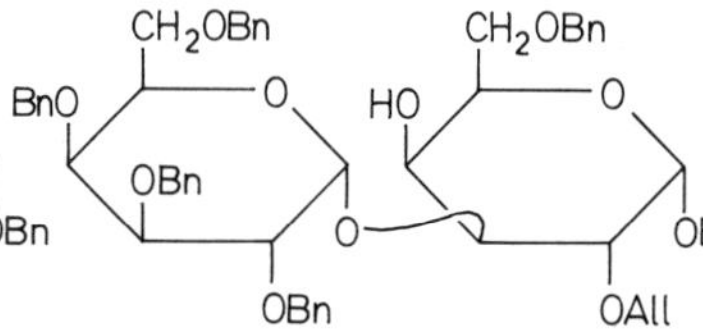

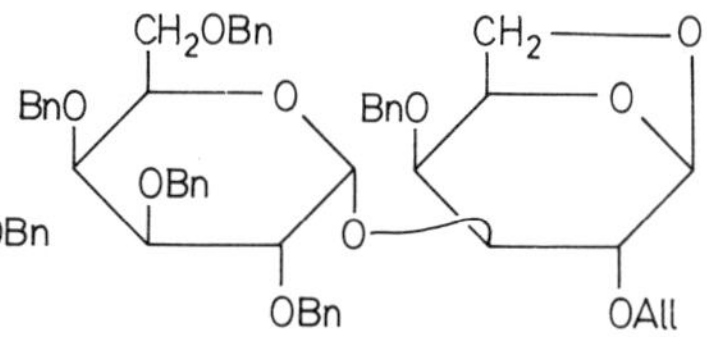

326   327   328

Paulsen and co-workers [156] have also prepared the blood group B (type 2) tetrasaccharide (324) from the lactosamine derivative (318). They attached the 2'-O-L-fucosyl residue using the fucosyl bromide (322) in the presence of mercury(II) bromide and molecular sieves and obtained (319) in 80% yield. Removal of the benzoyl group and glycosidation with 2,3,4,6-tetra-O-benzyl-α-galactopyranosyl bromide in the presence of mercury(II) bromide and molecular sieves at 20 °C or with silver triflate — silver carbonate at −25 °C gave the 3'-O-α-D-galactosyl derivative (323) in 80% yield. In this paper Paulsen discusses in detail the conditions (reactivity of alcohol, halide and catalysts) affecting the outcome of α-glycoside synthesis. The intermediate protected trisaccharide (319) was also used [156] for the synthesis of the related tetra-saccharide of the blood-group A (type 2) substance (see Sect. 5.2).

Paulsen and Schnell [225] have also converted the lactosamine derivative (315) into the derivative (320) and condensed this with the azidoglucosyl bromide (329) in the presence of silver perchlorate — silver carbonate to give an α-linked trisaccha-ride which was deprotected to give the trisaccharide NAcglc-α-(1→4)-gal-β-(1→4)-NAcglc which has been shown to be present in blood-group substances.

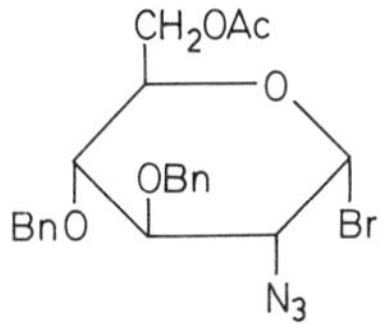

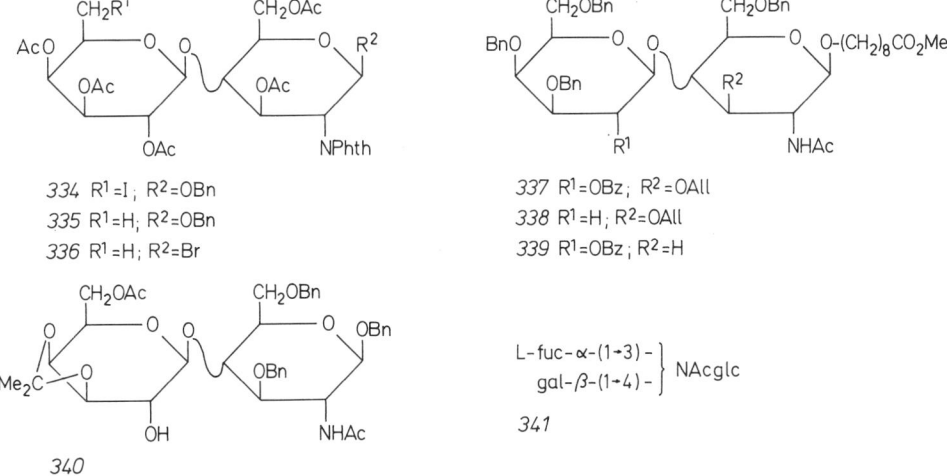

329

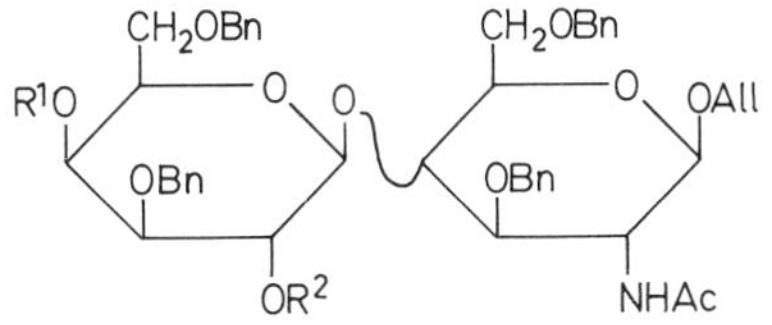

330 R¹=tetrahydropyranyl; R²=Bz
331 R¹=H; R²=Bz
332 R¹=tetrahydropyranyl; R²=H
333 R¹=H; R²=α-L-perbenzylfucose

Nashed and Anderson [155] prepared an isomer of the blood-group B (type 2) tetrasaccharide with lactosamine substituted at the 4'-position with an α-D-galactosyl residue and at the 2'-position with an α-L-fucosyl residue. For this purpose, the lactos-amine derivative (331), prepared via (330), was α-glycosylated with tetrabenzyl-galactosyl bromide in the presence of silver carbonate and silver triflate and the benzoyl group removed from the product so that the 2'-position could be fucosylated with the bromide (302) in the presence of bromide ion. If the fucose residue was added first to the lactosamine derivative (332), then the product (333) [which was converted into the trisaccharide of the blood group H (type 2)] would not react with the tetrabenzyl-galactosyl bromide.

334 R¹=I; R²=OBn
335 R¹=H; R²=OBn
336 R¹=H; R²=Br

337 R¹=OBz; R²=OAll
338 R¹=H; R²=OAll
339 R¹=OBz; R²=H

340

$$L-fuc-\alpha-(1{\to}3)-\atop gal-\beta-(1{\to}4)-\Big\}NAcglc$$

341

In an extensive investigation Lemieux and co-workers [226] prepared several fucosyl derivatives of 8-methoxycarbonyloctyl β-lactosaminide. The 2'-O-α-L-fucosyl derivative [blood-group H (type 2)] trisaccharide and the related 6-deoxy-2'-O-α-L-fucosyl derivative, the 3-O-α-L-fucosyl derivative ("X-hapten") and 3,2'-di-O-α-L-fucosyl derivative ("Y-Hapten") were prepared for binding studies with the lectin I of *Ulex europaeus*. Lemieux and co-workers [227] have also condensed the chloride (282) with 1,2:3,4-di-O-isopropylidene-6-C-methyl-α-D-galactopyranose and after deprotection used the trisaccharide for studies of the binding site of anti-I antibodies [228]. For similar reasons the 6'-deoxy derivative (335) of lactosamine was prepared by reduction of the 6'-iodo derivative (334) with tributyltin hydride and this was converted into the glycosyl bromide (336) which was condensed with 1,2:3,4-di-O-isopropylidene-D-galacto-pyranose and the product deprotected to give the 6-deoxy-D-gal-β-(1→4)-NAcglc-β-(1→6)-D-gal.

Further research by the same group [229] involved conversion of the lactosamine derivative (337) into the 2'-deoxy- (338) or 3-deoxy- (339) derivatives by conversion of the corresponding alcohols into the phenylthiocarbonate derivatives followed by reduction with tributylstannane.

The blood-group H (type 2) trisaccharide L-fuc-α-(1→2)-gal-β-(1→4)-NAcglc was also prepared by Matta and co-workers [230] by glycosidation of the lactosamine derivative (340) with the fucosyl bromide (322) in the presence of bromide ion and subsequent deprotection.

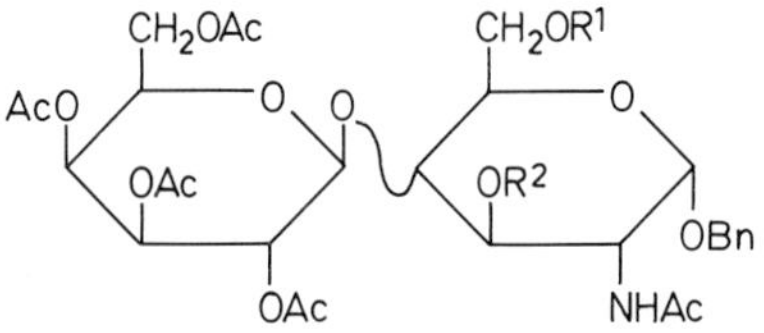

342 R² =All; R¹ =Ac or Bn
343 R² =H; R¹ =Ac or Bn
344 R¹ =Ac or Bn; R² = α-L-perbenzylfucose

345

Jacquinet and Sinaÿ [231] synthesised the trisaccharide (341) considered to be the possible determinant of the Lewis c (Le^c) antigen. Condensation of (322) with compounds (343), both prepared via the allyl ethers (342), in the presence of bromide ion gave the trisaccharides (344) in 85% yield in both cases. Deprotection of these gave (341).

6.2 The Gal-β-(1 → 3)-NAcglc Residue

6.2.1 The Lewis *a* Determinant

The Lewis *a* (Le^a) blood-group trisaccharide determinant (345) first synthesised by Lemieux and co-workers [25] was also prepared by Sinaÿ and co-workers [222] using the imidate procedure [232]. Condensation of the fucosyl imidate (321) with the disaccharide (346), prepared via the allyl ether (347), in the presence of toluene *p*-

sulfonic acid in nitromethane at 20 °C gave the trisaccharide (348) (85%) which was deprotected to give (345). Attempts to condense acetobromogalactose with the di-saccharide (351), prepared from (350), were unsuccessful.

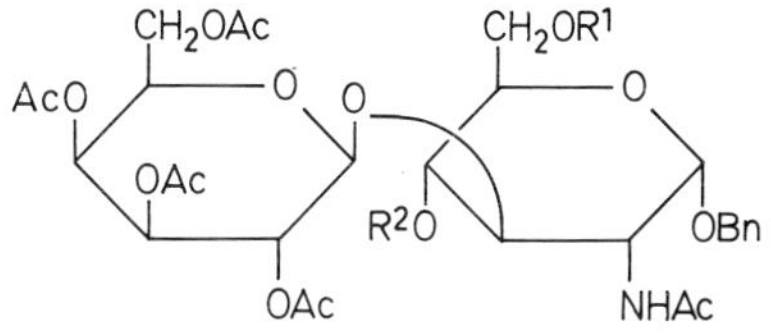

346 R^1=Bn ; R^2=H
347 R^1=Bn; R^2=All
348 R^1=Bn; R^2=α-L-perbenzylfucose
349 R^1 = Ac; R^2=H

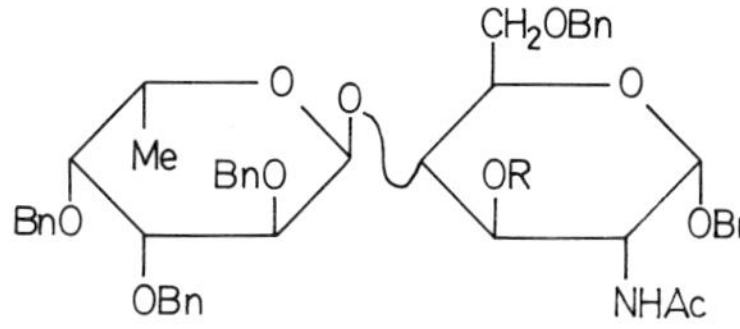

350 R=All
351 R=H

Khorlin and co-workers [233] also prepared (345) using a combination of chloro-acetyl and 2-tetrahydrofuranyl protecting groups to prepare the disaccharide (349) which was converted into the diphenylcyclopropenyl ether and fucosylated in the presence of silver perchlorate in 34% yield. They also observed that a glucosamine derivative glycosylated at the 4-hydroxyl group could not be glycosylated subsequently at the 3-hydroxyl group.

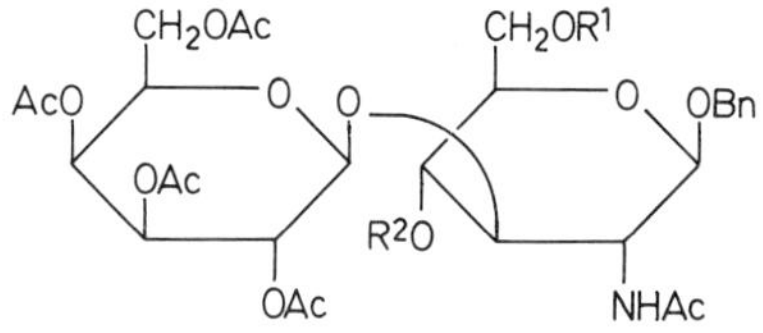

352 R^1,R^2= $>$CHPh
353 R^1=Bn; R^2=H
354 R^1=Bn; R^2= α-L-perbenzylfucose
355 R^1,R^2= $>$CMe$_2$

$$\left.\begin{array}{l} \text{L - fuc - }\alpha\text{-(1}\rightarrow\text{4)-} \\ \text{L-fuc-}\alpha\text{-(1}\rightarrow\text{2)-gal -}\beta\text{-(1}\rightarrow\text{3)-} \end{array}\right\} \text{NAcglc}$$

356

In a further synthesis of (345) by Rana and Matta [234], the disaccharide (353), prepared by the action of sodium cyanoborohydride — hydrogen chloride [120, 121] on the benzylidene derivative (352), was condensed with the fucosyl bromide (322) to give the trisaccharide (354) in 82% yield.

6.2.2 The Lewis b Determinant

The Lewis b (Le^b) blood-group tetrasaccharide determinant (356) was synthesised by Matta and co-workers [235] via the disaccharide (355) which was deacetonated and deacetylated and the product converted into the pertrimethylsilyl ether (357). Treat-ment of (357) with pyridine — acetic anhydride — acetic acid [236] caused acetylation of the primary hydroxyl groups and subsequent treatment with acidic methanol gave (358). This was converted into the 3',4'-O-isopropylidene derivative (359) which was condensed with the bromide (322) in the presence of bromide ion to give the tetra-saccharide (360) which gave (356) on deprotection.

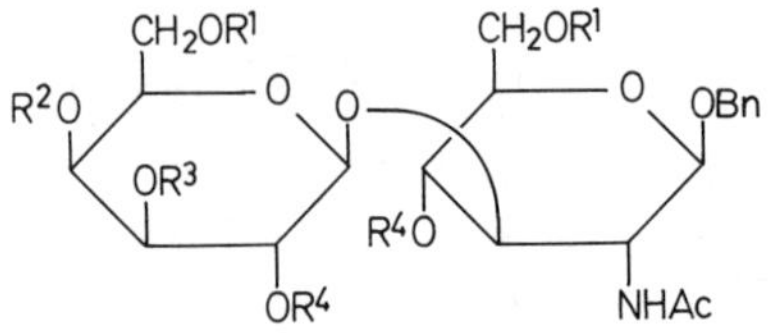

357 R¹=R²=R³=R⁴=SiMe₃

357 $R^1=R^2=R^3=R^4=SiMe_3$
358 $R^1=Ac$; $R^2=R^3=R^4=H$
359 $R^1=Ac$; $R^2,R^3=$ CMe_2 ; $R^4=H$
360 $R^1=Ac$; $R^2,R^3=$ CMe_2 ; $R^4=\alpha$-L-perbenzylfucose

$$\left. \begin{array}{l} L\text{-fuc-}\alpha\text{-}(1\rightarrow2)\text{-gal-}\beta\text{-}(1\rightarrow3)\text{-} \\ L\text{-fuc-}\alpha\text{-}(1\rightarrow4)\text{-} \end{array} \right\} NAcglc\text{-}\beta\text{-}(1\rightarrow3)\text{-gal-}\beta\text{-}(1\rightarrow4)\text{-glc-}\beta\text{-}(1\rightarrow1)\,CER$$

361

Ogawa and co-workers [97] have synthesised the complete glycosphingolipid (361) containing the terminal tetrasaccharide (356) of the Le^b antigen attached to lactosyl ceramide. The lactose derivative (95), prepared by the action of trimethylamine — borane and aluminium chloride [99] on the benzylidene derivative (94), was condensed with the thioglucoside (362) in the presence of methyl triflate and molecular sieve to give the trisaccharide (363). This was deacetylated and the product (364) condensed

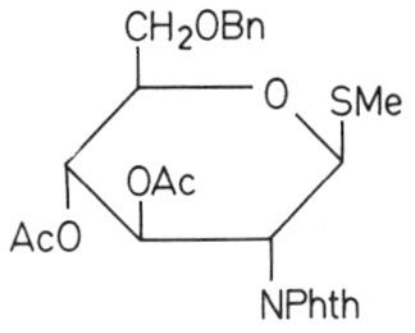

362

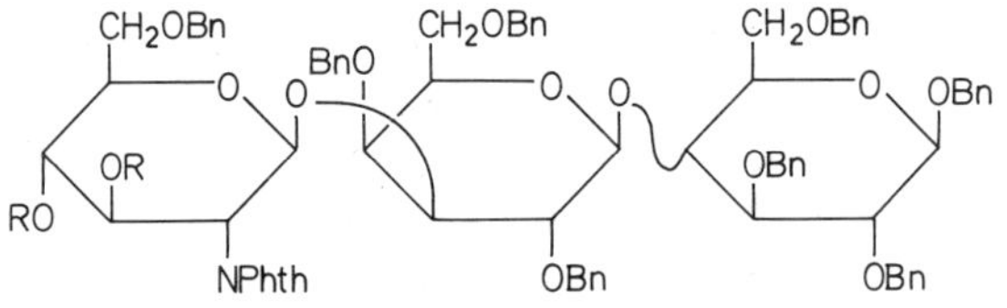

363 R=Ac
364 R=H

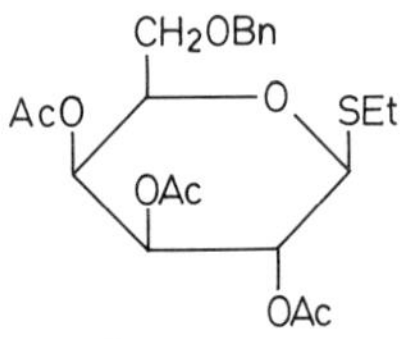

365

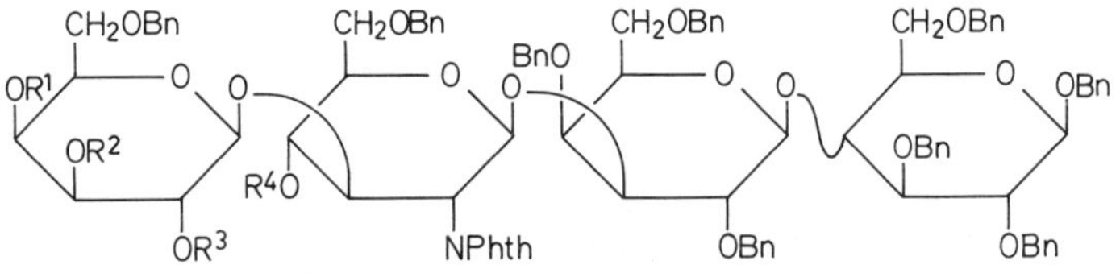

366 $R^1=R^2=R^3=Ac$; $R^4=H$
367 $R^1,R^2=$ CMe_2; $R^3=R^4=H$
368 $R^1,R^2=$ CMe_2; $R^3=R^4=\alpha$-L-perbenzylfucose

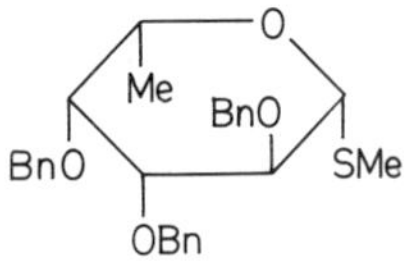

369

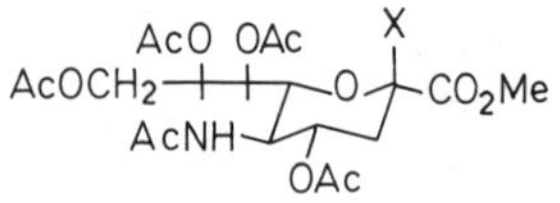

370 X=Cl
371 X=Br
372 X=F

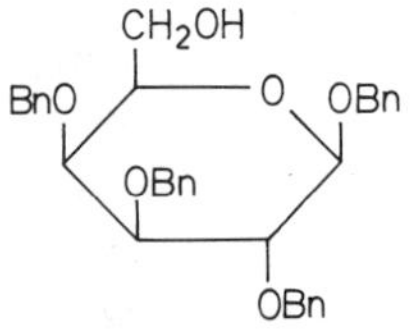

373

with the thiogalactoside (365) in the presence of copper(II)bromide, tetrabutyl-ammonium bromide, mercury(II) bromide and molecular sieves [98] to give the tetrasaccharide (366) with a β-galactosyl linkage at the 3-hydroxyl group. Deacetylation of (366) and acetonation gave (367) and this was condensed with the β-thiofucoside (369) in the presence of the same catalysts to give the *bis*-fucosylated product (368) in 68% yield. This was deprotected and converted into the peracetylhexasaccharide and the 1-*O*-acetyl group removed with hydrazine — acetic acid [176]. The free hydroxyl group was treated with diethylaminosulfur trifluoride (DAST) to give the glycosyl fluoride. This was condensed with a 3-*O*-benzoyl ceramide (63) in the presence of silver triflate — tin(II) chloride and molecular sieve in low yield (9%) to give the protected glycolipid which on de-*O*-acylation gave (361).

7 Sialic Acid-containing Glycosphingolipids — Gangliosides

Synthetic work in this area has been hindered by several factors; despite some recent excellent chemical and biochemical syntheses of *N*-acetylneuraminic acid and derivatives [237–247, 283] and the development of improved methods for its isolation from natural sources [248–250], this compound is still a very expensive starting material for the synthetic chemist. Moreover, the problem of obtaining α-glycosides (such as occur in the gangliosides) with high specificity and in good yield remains a challenge and has stimulated some interesting research.

During the period under review many methods have been developed for the synthesis of simple alkyl glycosides of *N*-acetylneuraminic acid with the object of getting high yields and anomeric specificity [94, 251–254] and these have been extended to the synthesis of oligosaccharides. The synthesis of alkyl thioglycosides has also been studied [255]. We shall first discuss the work developed on the condensation of *N*-acetylneuraminic acid derivatives with the primary hydroxyl group of other sugars and then the more recently developed investigations with the secondary hydroxyl groups since both of these are relevant to the syntheses of gangliosides (see Refs. [12, 13, 15, 18] and [20] for the structures of the natural gangliosides).

7.1 Condensation with Primary Hydroxyl Groups of Sugars

In 1982, a Dutch group [256] condensed the chloride (370) with the galactose derivative (373) in the presence of silver salicylate in benzene at 20 °C to give the α-linked disaccharide (374) in 65% yield together with 3% of the β-linked disaccharide. However, with the glucose derivative (375) no disaccharide was obtained with silver salicylate but with silver triflate [257] and 2,4,6-trimethylpyridine in a mixture of ether and nitromethane the α-linked disaccharide was obtained in 11.5% yield but the major product (46%) was the β-linked disaccharide together with a common by-product in these reactions, the unsaturated derivative (376) (40%). Brandstetter and Zbiral [258] found that the condensation of (370) with the α- or β-methyl glycosides of 2,3,4-tri-*O*-benzyl-D-glucopyranoside in the presence of silver triflate and 2,4,6-trimethylpyridine in dichloromethane gave 40% of the β- and 10% of the α-linked disaccharide.

In 1982, Paulsen and Tietz [259, 260] condensed the chloride (370) with the disaccharide derivative (379) (with preferential reaction at the primary hydroxyl group) using mercury(II) cyanide — mercury(II) bromide (3:1) in dichloromethane at 20 °C for 3 days to give 22% of the α- and 23% of the β-linked disaccharides together with the unsaturated derivative (376). Under these conditions the galactose derivative (373) and the chloride (370) gave 36% α- and 48% β-linked disaccharide. Using silver carbonate with (370) and (373) the major product was the unsaturated derivative (376) together with 20% of α-linked disaccharide whereas with the chloride (370) and 1,2:3,4-di-O-isopropylidene-α-D-galactopyranose in the presence of silver carbonate, 67% of the α-linked disaccharide was formed.

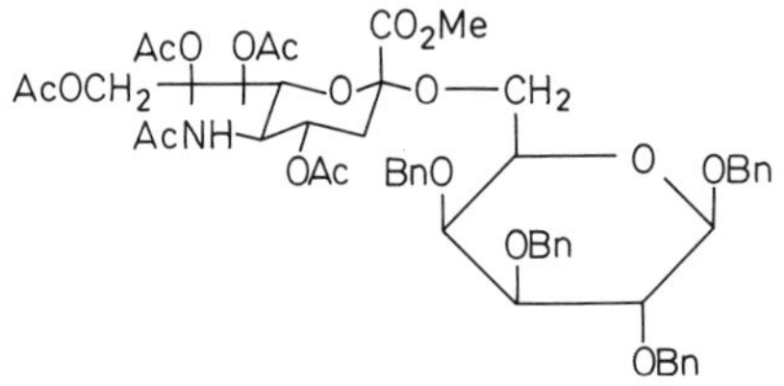

374

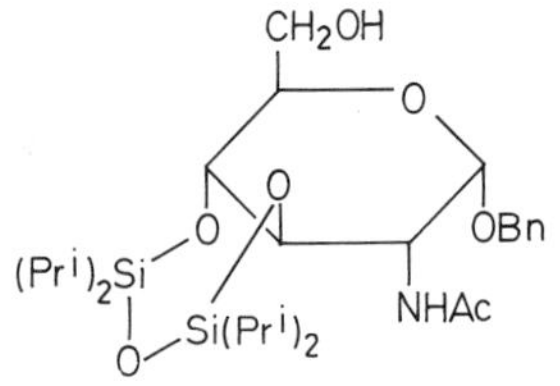

375

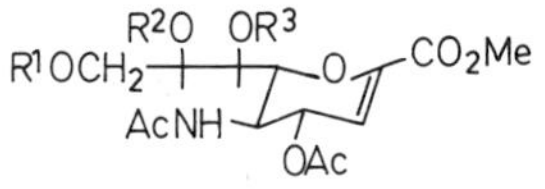

376 $R^1=R^2=R^3=Ac$
377 $R^2=H$; $R^1=R^3=Ac$
378 $R^1=H$; $R^2=R^3=Ac$

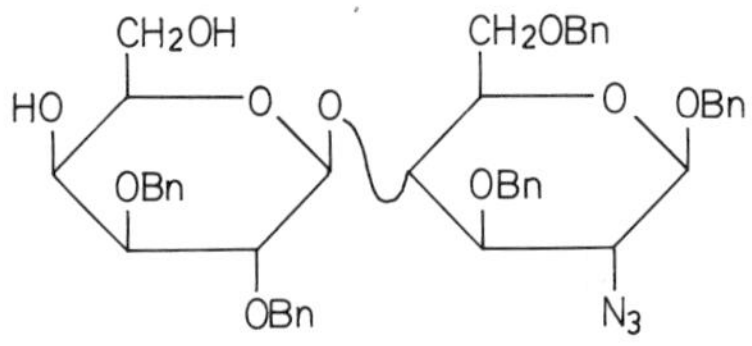

379

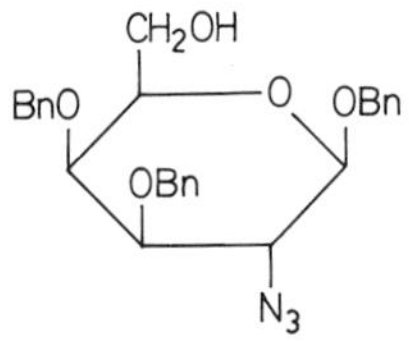

380

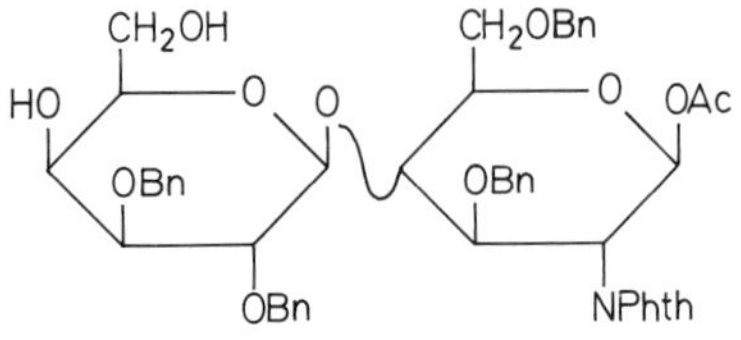

381

Paulsen and his co-workers [261] condensed an excess of the chloride (370) with the galactose derivative (380) using mercury(II) cyanide — mercury(II) bromide (3:1) in dichloromethane during 5 days to give 42% of α- and 36% of the β-linked disaccharides [yields based on (380)]. Silver silicate, triflate or salicylate gave only ca. 10% of disaccharides. Using the same mixture of mercury salts together with molecular sieves Paulsen and Tietz [262, 263] condensed an excess of the chloride (370) with the disaccharide derivative (381) to give approximately equal amounts (25% each) of the α- and β-linked disaccharides [on the primary hydroxyl group of (381)] together with the elimination product (376). The products were acetylated and used as trisaccharide donors and condensed with a mannose derivative containing a free 2-hydroxyl group in the presence of trimethylsilyl triflate to give tetrasaccharides.

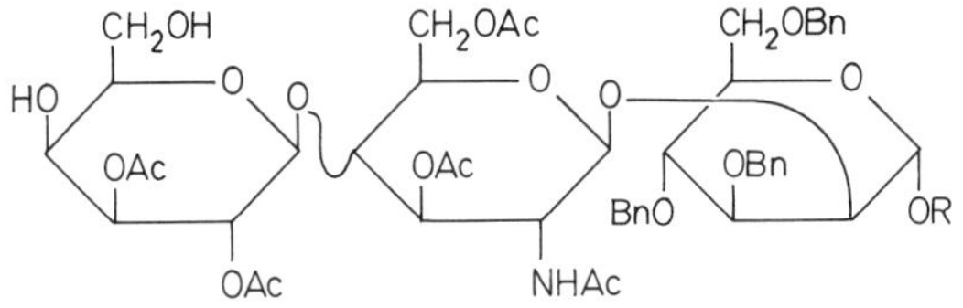

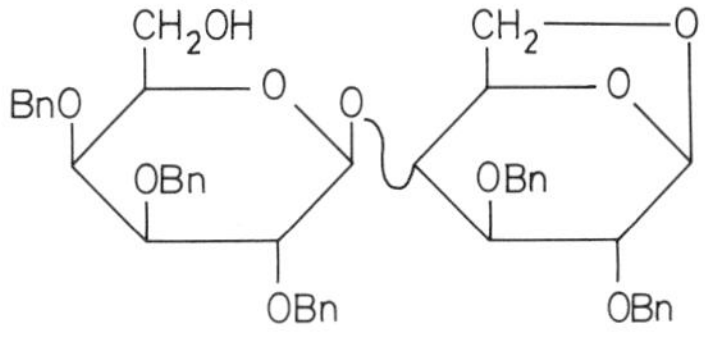

CH₂OH ... O O (382 trisaccharide structure)

382 R=Bn
383 R=All

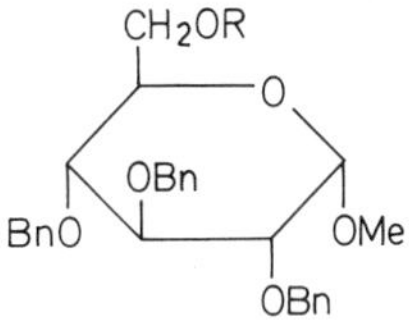

CH₂OR
385 R=H
386 R=CH(CO₂Me)P(OMe)₂ (O)

384

CH=R
OBn
AcNH
O=
O—CMe₂
O—CMe₂
CH₂O

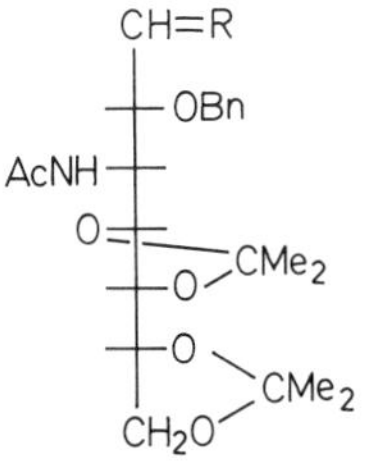

387 R=O
388 R=CH₂—OC(CO₂Me)=

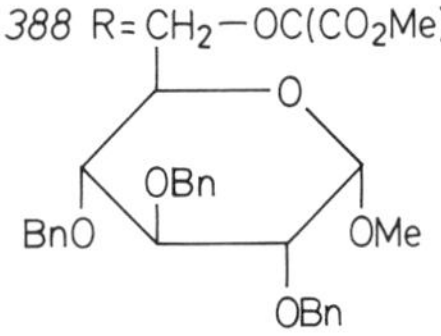

OBn / BnO / OMe / OBn

Kitajima et al. [264] used the mixture of mercury salts and molecular sieves in 1,2-dichloromethane to condense (370) with the trisaccharide derivative (382) to give similar tetrasaccharides [on the primary hydroxyl group of (382)] with again equal amounts (30% each) of α- and β-linked derivatives. With the allyl glycoside (383) under the same conditions 48% of α- and 33% of β-linked glycosides were obtained [265]. The α-isomer was used to prepare [265] the undecasaccharide characteristic of the complex type of glycan of glycoproteins.

Furuhata et al. [266] condensed the chloride (370) with the disaccharide derivative (384) using the mixture of mercury salts in dichloromethane at 20 °C for 60 hours to give 21% of α- and 8% of β-linked disaccharide together with 60% of elimination product (376).

The chloride (370) has been converted [250] into the fluoride (372) which is so much more stable than the chloride that both the acetyl and carbomethoxy groups can be hydrolysed with base without affecting the fluoro group. The corresponding allyl ester of the fluoride (372) has been condensed [267] with 1,2:3,4-di-O-isopropylidene-α-D-galactopyranose in the presence of boron trifluoride — etherate to give the disaccharides with an α:β ratio of 1:5. The corresponding allyl ester of the chloride (370) with silver carbonate and Drierite gave [267] the disaccharide with an α:β ratio of 6:1.

A novel approach to glycosidation with N-acetylneuraminic acid was adopted by Sinaÿ [268]. Reaction of (385) with diazotrimethylphosphonoacetate in the presence of rhodium acetate gave the phosphonate (386) as a mixture of diastereoisomers which were allowed to react with the aldehyde (387) in the presence of sodium hydride in tetrahdrofuran to give a mixture of E- and Z-isomers (388) which were separated. The isopropylidene groups were hydrolysed and the products cyclised with mercury(II) trifluoroacetate each isomer cyclising stereospecifically to give the α-anomer (389) or the β-anomer. Demercuration of (389) with triphenyltin hydride gave the α-anomeric disaccharide derivative (390).

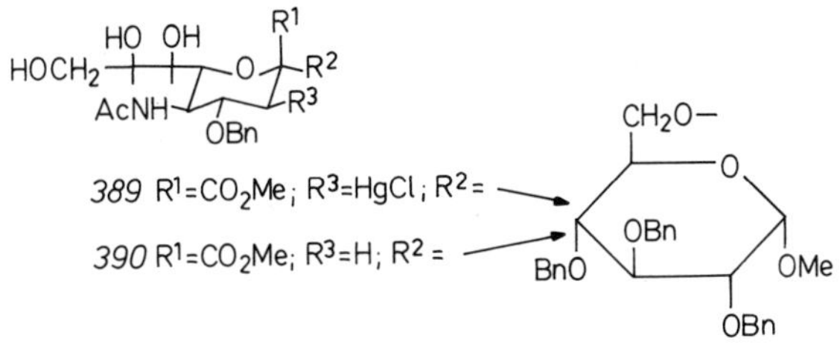

389 R¹=CO₂Me; R³=HgCl; R²=

390 R¹=CO₂Me; R³=H; R² =

7.2 Condensation with Secondary Hydroxyl Groups of Sugars

The work described in the previous section has set the scene for the more difficult problem of the glycosidation of secondary hydroxyl groups of carbohydrate derivatives where elimination to give the unsaturated derivative (376) is a more severe problem.

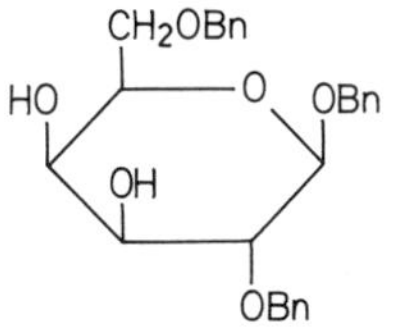

391

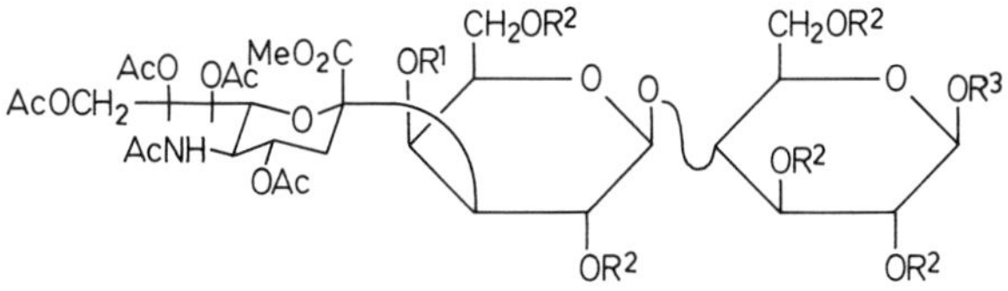

392 R¹=H; R²=R³=Bn
393 R¹=R²=R³=Ac
394 R¹=R²=Ac; R³=H
395 R¹=R²=Ac; R³=C(=NH)CCl₃
396 R²=R³=Bn; R¹= CH₂OAc

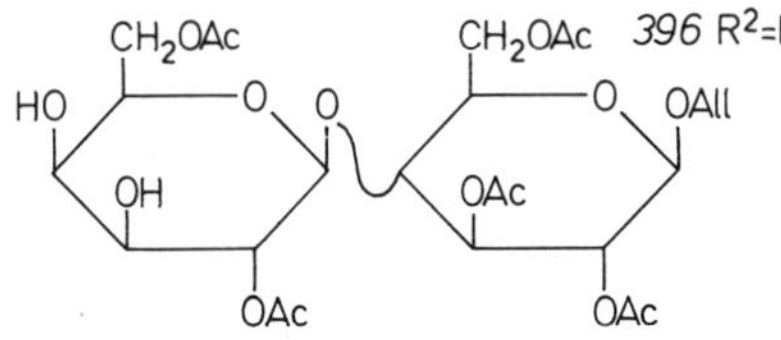

397

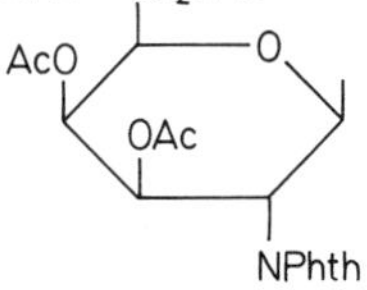

Ogawa and Sugimoto [269] condensed the chloride (370) with the galactose derivative (391) (the equatorial 3-hydroxyl group reacting preferentially) with mercury(II) cyanide — mercury(II) bromide and molecular sieves to give a 15% yield of disaccharide [based on chloride (370)] with an α:β ratio of 2:3. Condensation [269] of (370) with the lactose derivative (96) under the same conditions gave an 18% yield of disaccharide with an α:β ratio of 1:2. The isomers were separated by chromatography and deprotection of the α-linked derivative (392) gave a sialosyl lactose identical with the natural trisaccharide. Acetylation of (392) [270] and subsequent hydrogenolysis and acetylation gave (393) which was treated with hydrazine acetate to give (394). This was converted into the trichloroacetimidate (395), by the action of sodium hydride and trichloroacetonitrile [28], and this was condensed with a 3-O-benzoylceramide (63) in the presence of boron trifluoride-etherate and molecular sieves to give a β-linked glycoside in 37% yield. Deprotection gave hematoside (ganglioside GM3) completely identical with the natural glycolipid, this being the first recorded synthesis of a ganglioside.

Condensation [270] of the chloride (370) with the allyl lactoside (397) as described above for the preparation of (392) gave a 6% yield of the β-linked N-acetylneuraminic acid derivative (on the 3'-hydroxyl group) and none of the α-linked material. The product was acetylated and deallylated to give (398) and this was converted into the trichloroacetimidate (399) which was condensed with the 3-O-benzoylceramide to give the β-N-acetyl-neuraminyl derivative of hematoside ("epi-hematoside").

The synthetic hematoside was immunologically identical [271] with the natural material when compared using a mouse monoclonal antibody (M2590) directed against syngenic B-16 melanoma cells but the "epi-hematoside" did not react with the monoclonal antibody.

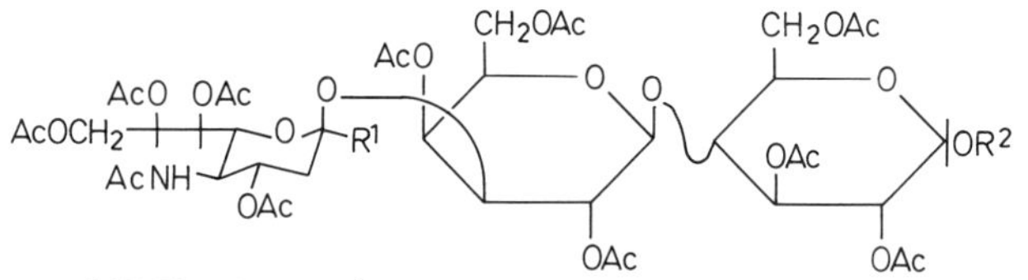

398 R¹=CO₂Me ; R²=H
399 R¹ =CO₂Me ; R² =C(=NH)CCl₃

NAcgal-β-(1→4)-⎫
NANA - α- (2→3)-⎬ gal-β-(1→4) - glc-β-(1→1)-CER
GM2 400

gal-β-(1→3)-NAcgal -β-(1→4)-⎫
         NANA - α- (2→3)- ⎬ gal-β-(1→4)-glc-β-(1→1)-CER
GM1 401

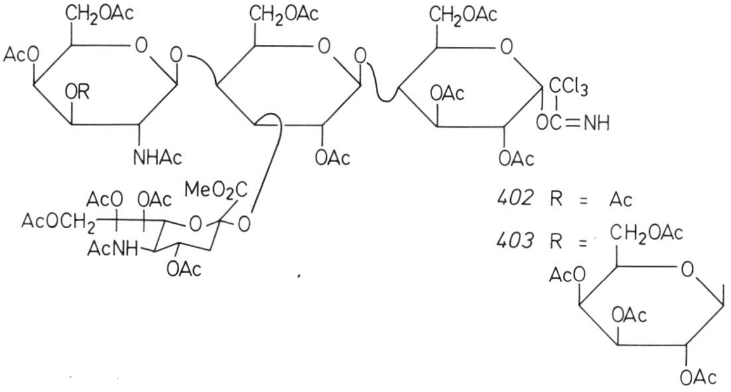

402 R = Ac

403 R = CH₂OAc

Paulsen and von Deessen [272] improved the yield using the bromide (371) instead of the chloride (370) in condensations with the lactose derivative (96) in the presence of silver carbonate — silver perchlorate (30:1) and molecular sieve in toluene at

0 °C during 2 days. A 36% yield of a 1:1, α:β mixture was obtained but again the unsaturated derivative (376) was the major product due to elimination.

Ogawa et al. [273] have described the total synthesis of the gangliosides GMI (401) and GM2 (400) from (392). Condensation of the trisaccharide (392) with the phthalimido galactosyl bromide (193) in the presence of silver triflate and molecular sieves gave a 60% yield of the β-galactosyl derivative (396). The carbomethoxy group was first de-esterified with lithium iodide in pyridine and then, as a result of a series of deprotection and reprotection manipulations (as described in previous sections), (396) was converted into imidate (402) which was condensed with a 3-O-benzoyl-ceramide and the product (11%) deprotected to give the ganglioside GM2 (400).

Similarly, condensation of the disaccharide derivative (404) [125] via the imidate (405) with the trisaccharide (392) in the presence of boron trifluoride — etherate gave a pentasaccharide in 40% yield and this was converted into the imidate (403), using the same techniques as described for the preparation of (402). Condensation (33% yield) of the imidate (403) with a 3-O-benzoylceramide (63) and deprotection of the product gave the ganglioside GM1 (401).

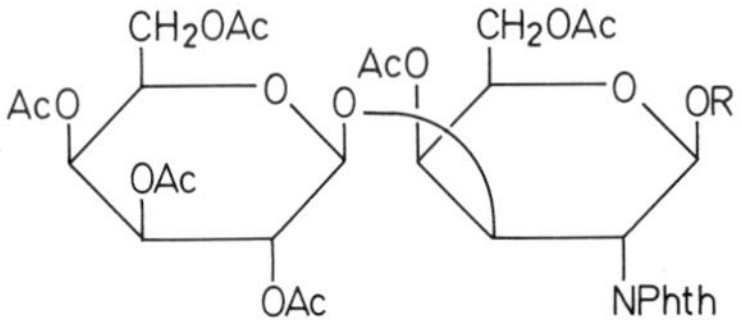

404 R = H
405 R = C(=NH)CCl₃

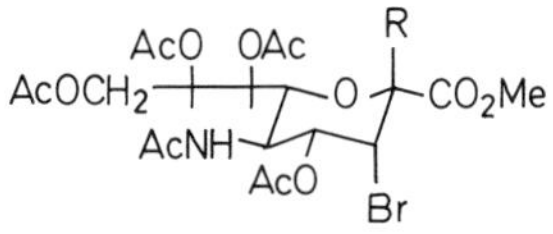

406 R = Br
407 R = OR¹

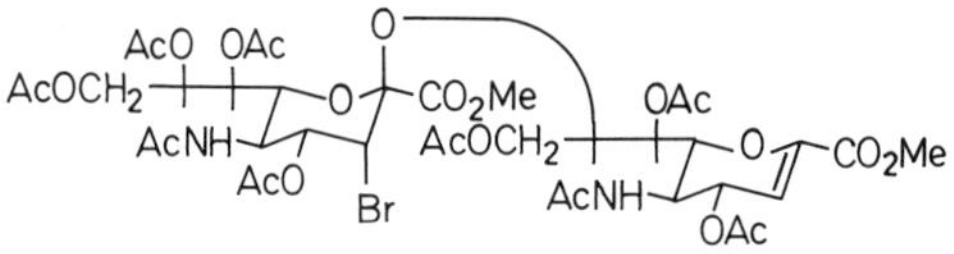

408

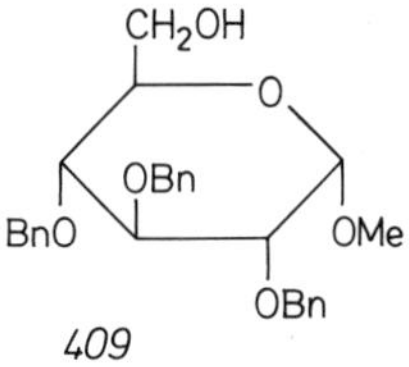

409

In order to overcome the major problem of elimination in the glycosidation of secondary hydroxyl groups of sugars by N-acetylneuraminic acid derivatives Oka-moto and co-workers [274–276] have developed some new derivatives of N-acetyl-neuraminic acid for glycosidation reactions. Bromination of the unsaturated derivative (376) gave the crystalline dibromide (406) [276] and condensation of this with model alcohols in the presence of silver triflate gave good yields [cholesterol (88%), methyl 2,3,4-tri-O-benzyl-α-D-glucopyranoside (70%) and methyl 2,4,6-tri-O-benzyl-β-D-galactopyranoside (50%)] of disaccharides. Due to steric hindrance these were un-fortunately entirely β-glycosides (407) but no elimination occurred. These glycosides were readily debrominated with tri-n-butylstannane to give β-sialosyl derivatives.

Condensation [276] of the bromide (406) with the unsaturated derivative (377) gave the β-linked disaccharide (408) (58%) which was brominated at the double bond and the glycosyl bromide produced condensed with the glucoside (409) to give, in

42% yield, a trisaccharide containing two β-linked neuraminic acid derivatives (substituted with bromine at the 3-position) which was readily debrominated.

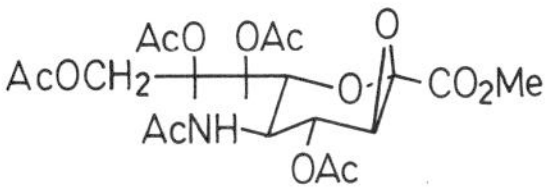

Okamoto et al. [275] converted the unsaturated derivative (376) into the epoxide (410) which with boron trifluoride-etherate and the corresponding titanium(IV) halide gave the glycosyl halides (411) and (412). Condensation of the chloride (411) with (409) in the presence of silver triflate gave α-linked (21%) and β-linked (18%) disaccharides containing a 3-hydroxy-N-acetylneuraminic acid derivative. The hydroxyl group was removed by first converting it into the phenoxythiocarbonate [R-OC(S)OPh] followed by reduction with tri-n-butylstannane and azobisisobutyronitrile in toluene.

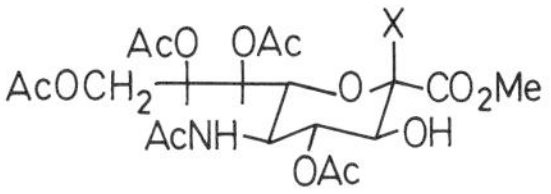

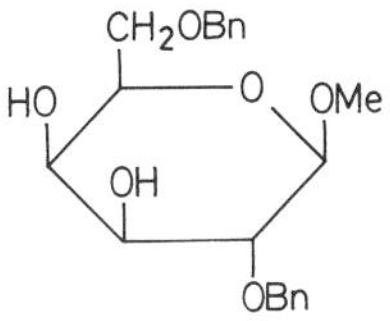

With the bromide (412) and the glucoside (409) in the presence of silver triflate in benzene at 20 °C both the α-linked (28%) and the β-linked (53%) disaccharides were obtained. In toluene at −10 °C more of the α- (64%) and less of the β- (15%) linked disaccharides were formed. The major product was the β-linked disaccharide (32%) using mercury(II) cyanide and mercury(II) bromide as catalysts.

Reaction of the bromide (412) [275] with the galactose derivative (413) (with preferential reaction at the 3-hydroxyl group) in the presence of silver triflate in benzene at 20 °C gave an α:β yield ratio of 23:48% whereas in toluene at −15 °C it was 37:15% for the two disaccharides.

Condensation [274] of the bromide (412) with the unsaturated derivative (378) using silver triflate gave an α:β yield ratio of 42:21% and condensation of (412) with (377) gave 26:8%. From these results the authors concluded that the 3-β-hydroxyl group on the neuraminic acid halide prevents dehydrohalogenation and assists glycosidation.

7.3 Enzymatic Syntheses of Sialyloligosaccharides

In order to try to overcome some of the problems associated with chemical synthesis of oligosaccharides containing N-acetylneuraminic acid, Sabesan and Paulson [277] have used a combination of chemical and enzymatic methods using purified sialyltransferases in the presence of CMP-N-acetylneuraminic acid and synthetic acceptor molecules to give sialyl derivatives of oligosaccharides which were characterised by NMR. Thus, methyl β-D-galactopyranoside, methyl β-D-lactoside and N-acetyllactosamine were converted into the 6-sialosyl derivatives using the enzyme β-galacto-side-α-2,6-sialyltransferase. In all, ten sialyloligosaccharides were prepared on the 10–20 μmol scale using these techniques.

Thiem and Treder [278] have used immobilised enzymes [216] for the separate syntheses of *N*-acetyllactosamine and CMP-*N*-acetylneuraminic acid. These were allowed to react in solution in the presence of the sialyltransferase to give the 6'-*N*-acetylneuraminyl glycoside on the 50 μmol scale.

7.4 Partial Syntheses of Gangliosides

Alkaline hydrolysis of isolated natural gangliosides [279, 280] has given derivatives in which the *N*-acyl group on sphingosine and the *N*-acetyl group of neuraminic acid were hydrolysed preferentially with only a small amount of hydrolysis of the *N*-acetyl group of galactosamine. These diamino-derivatives were reacylated preferentially at the sphingosine amino-group in a two-phase system using fluorescent fatty acids or with a temporary protecting group and the neuraminic acid amino group subsequently reacetylated. Removal [279] of the protecting group on the sphingosine amino-group gave "lyso-gangliosides" which were reacylated with a labelled fatty acid. Double labelling [279] was also achieved with labelled acetate (for the neuraminic acid) and labelled fatty acid (for the sphingosine).

Gangliosides have also been radioactively labelled [281] by oxidation of the galactose or *N*-acetylgalactosamine residue with galactose oxidase and reduction of the aldehydo-group produced with tritiated sodium borohydride or by oxidation of the allylic hydroxyl group of sphingosine with dichlorodicyanoquinone and reduction of the ketone produced in the same way. The 3-axial-hydrogen of free neuraminic acid exchanges specifically with deuterium oxide at pH 9 during 6 hours [282].

8 *Myo*-Inositol-Containing Glycolipids

The large amount of work in this area reported by Russian workers and covered in our previous review [1] has also been reviewed by the Russian workers [284]. These glycolipids have recently acquired a vastly increased biological interest since the discovery that various agonists at the cell surface stimulate a phospholipase which releases Ď-*myo*-inositol 1,4,5-trisphosphate from the membrane-bound phosphatidyl-inositol-4,5-bisphosphate. The released inositol trisphosphate acts as a "second messenger" by mobilising intra-cellular calcium ions [285–290].

Mannosides of phosphatidylinositol are important serologically active components of *Mycobacterium tuberculosis*. For the synthesis of the 2-*O*-mannosyl derivative (421), Stepanov et al. [291] treated the chiral prop-1-enyl ether (414) [and the corresponding racemic prop-1-enyl ether and racemic benzoate (415)] with the ortho-esters (416) or (417) to give the disaccharide (418) [from chiral (414) and acetate (416)] in moderate yield. The disaccharide obtained from racemic (414) contained a high proportion of (418) as a result of asymmetric synthesis. Acidic hydrolysis of (418) gave in low yield (30%) the alcohol (419).

Condensation [292] of (419) with 1,2-di-*O*-palmitoyl-L-glycerol 3-phosphate in the presence of 2,4,6-triisopropylbenzene sulfonyl chloride in pyridine gave (420) in 88% yield. Deprotection by hydrogenolysis of the benzyl groups and specific removal of the acetyl groups with hydrazine hydrate gave the naturally occurring glycolipid (421) in 65% yield.

Since the benzyl ether protecting groups preclude the synthesis of the phospholipid with unsaturated fatty acids, Shvets et al. [293] also studied the corresponding derivatives of 2-*O*-mannopyranosyl-*myo*-inositol protected by acetyl groups on the inositol residue. For this purpose the racemic acetate (422) was condensed with β-benzoylpropionic acid in the presence of dicyclohexylcarbodiimide to give the derivative (423) which was glycosylated with the orthoester (425) and the β-benzoylpropionate group removed with hydrazine [294] to give the mixture of diastereoisomers (424).

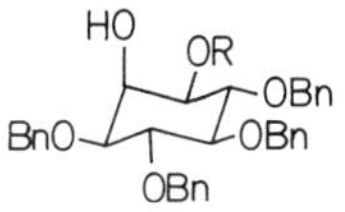

414 R=CH=CH—Me
415 R=Bz

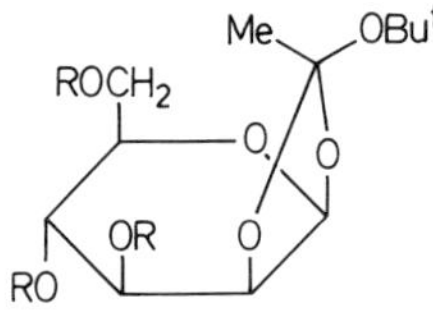

416 R=Ac
417 R=Bn

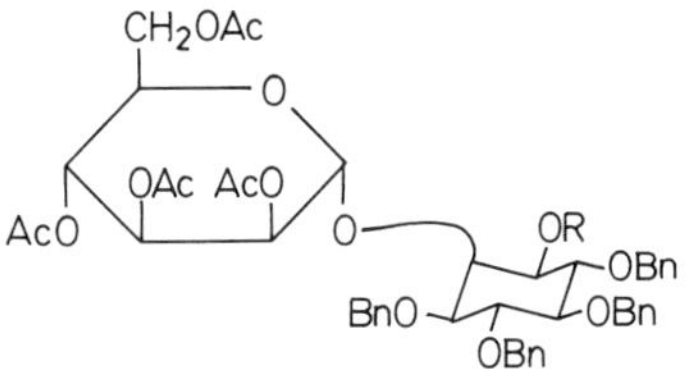

418 R = CH=CH—Me
419 R = H

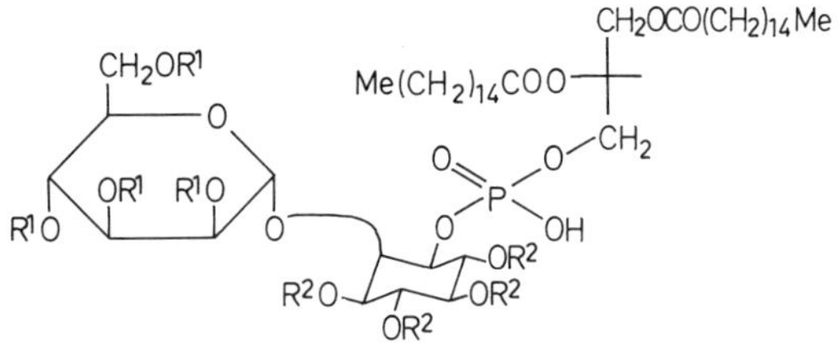

420 R¹=Ac; R²=Bn
421 R¹=R²=H

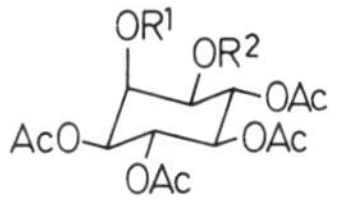

422 R¹=R²=H
423 R¹=H; R²=COCH₂CH₂COPh
424 R¹=α-D-peracetylmannose; R²=H

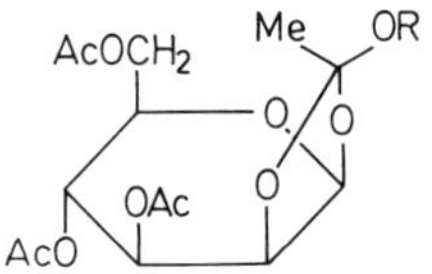

425 R=Et
426 R= 1,2:3,4-di-o-cyclohexylidene-myo-inositol
427 R= chiral 1,2:5,6-di-o-cyclohexylidene-myo-inositol
428 R= chiral 2,3:4,5-di-o-cyclohexylidene-myo-inositol

Because of problems encountered in the removal of the prop-1-enyl group in the above synthesis of the mannosyl-phosphatidylinositol, Shvets et al. [295] investigated the formation of the mono-tetrahydropyranyl ethers of the racemic inositol derivatives (429) and (430) and separated the isomeric mono-tetrahydropyranyl ethers from both by chromatography. Although the 1-*O*-tetrahydropyranyl ether preponderated in the product from (430) the selectivity was not good and approximately equal quantities of the mono-ethers were obtained from (429).

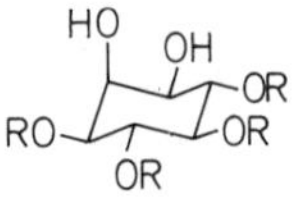

429 R=Ac
430 R=Bn

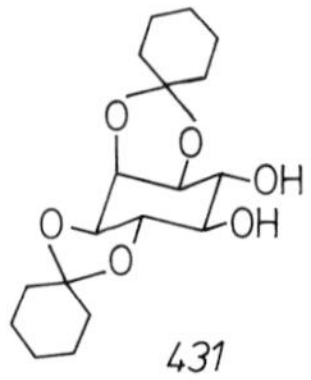

431

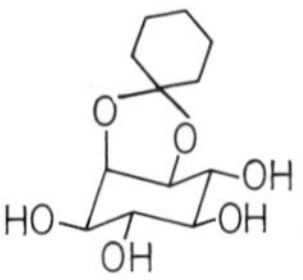

432

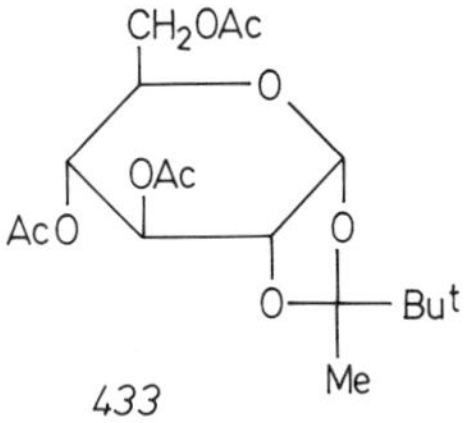

433

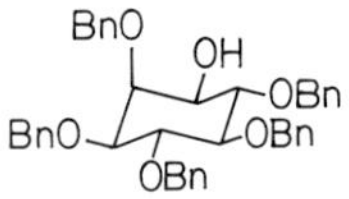

434

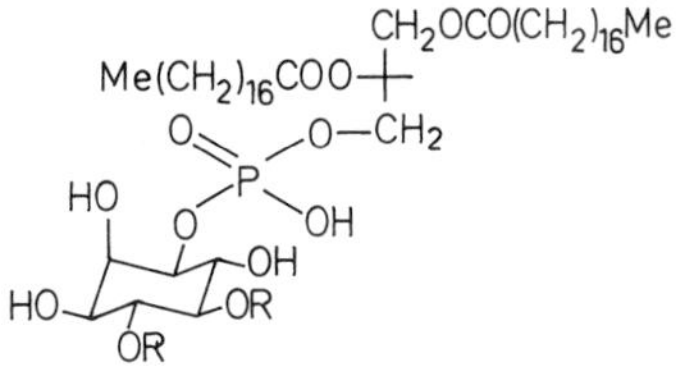

435 R=H
436 R=P(O)(OH)$_2$

In continuation of their studies of the resolution of *myo*-inositol derivatives via their orthoesters with sugar derivatives, Evstigneeva et al. [296] converted the racemic 1,2:3,4-di-*O*-cyclohexylidene-*myo*-inositol (431) by transesterification with the mannose orthoester (425) into a mixture of diastereoisomers (426) formed by esterification of the 5- and 6-positions of (431). One of the four possible isomers was separated by crystallisation and the other three were obtained by preparative TLC. Partial hydrolysis of the resolved isomers gave both enantiomers of 1,2-*O*-cyclohexylidene-*myo*-inositol (432).

The same group of workers [297] has used the mannose orthoester (425) and the glucose orthoacetate (433) for glycosylations of racemic 1,2,4,5,6-penta-*O*-benzyl-*myo*-inositol [434]. The diastereoisomeric acetylated α-mannosides of (434) (16% and 20%) and acetylated β-glucosides of 434 (16% and 13%) were separated in low yield by preparative TLC and converted into the chiral benzyl ethers (434). The isolation and separation of the diastereoisomeric glucosides was the most practical, but the yields in both cases indicate a low degree of asymmetry in the syntheses. Similar glycosidations of racemic 1,4,5,6-tetra-*O*-benzyl-*myo*-inositol (430) and isolation of the products by preparative-TLC and subsequent benzylation and hydrolysis of the glycosides also gave the chiral benzyl ethers (434).

Shvets and co-workers (298) have also prepared chiral 1-*O*-benzyl-*myo*-inositols by benzylation of the mannose orthoesters (427) and (428) with subsequent acidic hydrolysis. They were also prepared by partial benzylation of chiral 1,2:5,6-di-*O*- and 2,3:4,5-di-*O*-cyclohexylidene-*myo*-inositols. These chiral monobenzyl ethers are potential intermediates for the synthesis of chiral inositol pentakisphosphates. Reaction of the orthoester (427) [299] with benzoyl chloride or diphenylphosphochloridate in pyridine at 20 °C gave little reaction, whereas at 80 °C the orthoester group was replaced and both hydroxyl groups of the inositol derivative were acylated (or phosphorylated).

130

Racemic 4-O-benzyl-1,6:2,3-di-O-cyclohexylidene-*myo*-inositol has been resolved via the L(+)-O-acetylmandelates to give both chiral 4-O-benzyl-*myo*-inositols [300]. Benzoylation of these and removal of the benzyl groups by hydrogenolysis followed by condensation with the fucosyl imidate (321) and deprotection gave 4-O-α-L-fucopyranosyl-D-*myo*-inositol identical with a compound isolated from human urine.

Shvets and co-workers [301] have described a new synthesis of the phosphatidyl-inositol (435) by phosphorylation of the orthoester (437), derived from 2,3:4,5-di-O-cyclohexylidene-L-*myo*-inositol, with 1,2-di-O-stearoyl-L-glycerol 3-phosphate in

437 R¹=H
R²=CH₂OAc

438 R=H
439 R=P(NHPh)₂
        ‖
        O

(PhNH)₂PCl
        ‖
        O

440

the presence of 2,4,6-triisopropylbenzene sulfonyl chloride in pyridine and subsequent removal of cyclohexylidene and orthoester groups with dilute sulfuric acid. By this method phosphatidylinositols with unsaturated fatty acids should also be available.

Phosphatidylinositols containing spin-labelled fatty acids have been prepared by the same group [302] by acylating the primary hydroxyl group of the phosphodiester formed from 2,3,4,5,6-penta-O-acetyl-DL-*myo*-inositol and DL-glycerol 1-phosphate in the presence of the potassium salt and anhydride of the spin-labelled fatty acid and then acylating the secondary hydroxyl group with palmitic acid in the presence of carbodiimidazole. Hydrazinolysis was used to specifically remove the acetyl groups from the inositol.

Phosphatidylinositol containing fluorescent fatty acids has also been prepared [303] by synthesising a phosphatidylcholine containing fluorescent fatty acids and then using a phospholipase D, in the presence of inositol, to effect an ester inter-change.

Shvets et al. [304] prepared six chiral inositol bisphosphates (L-*myo*-inositol 4,5-, 1,6-, 1,4-, 5,6-, 3,4-, and 3,6-bisphosphates) by phosphorylation of the corresponding chiral di-O-cyclohexylidene-*myo*-inositols with diphenylphosphochloridate in pyridine followed by removal of the phenyl-protecting groups by hydrogenation with Adams catalyst and acidic hydrolysis of the cyclohexylidene groups. However, phosphorylation of the diol (438) [305] with diphenylphosphochloridate was not successful due to cyclic phosphate formation, but (438) was phosphorylated with the reagent (440). Hydrolysis of the product (439) with acetic acid gave the diol (441) in 50% yield.

Condensation of (441) [306, 307] with 1,2-di-O-stearoyl-DL-glycerol 3-phosphate in the presence of 2,4,6-triisopropylbenzene sulfonyl chloride in pyridine gave a mixture of the required products (442) (52%) and the isomer (443) (26%) which were separated by chromatography. The anilino groups were removed from (442) by the action of isopentyl nitrite in pyridine — acetic acid and the benzyl groups removed by hydrogenolysis to give the mixture of isomers of phosphatidylinositol bisphosphate (436).

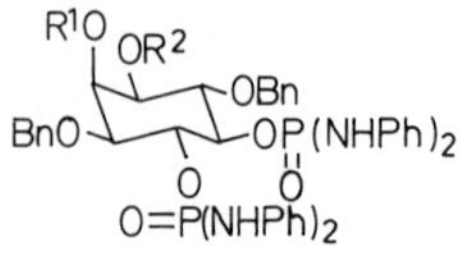

R^1O OR^2
  OBn
BnO   OP(NHPh)$_2$
   O   O
   O=P(NHPh)$_2$

441 R^1=R^2=H
442 R^1=H; R^2= distearoylglycerol phosphate
443 R^1= distearoylglycerol phosphate; R^2=H

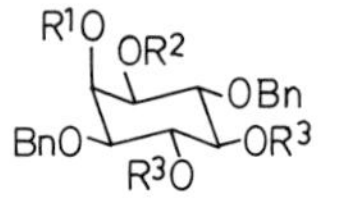

R^1O OR^2
  OBn
BnO   OR^3
  R^{3}O

HO O=P(OH)$_2$ ...
HO O–P(OH)$_2$
   OH
R^1O OR^2
   O
   O=P(OH)$_2$

447 R^1=Bn; R^2=R^3=H
448 R^1=R^2=H; R^3=All
449 R^1=H; R^2= l–menthoxyacetyl
450 R^1=H; R^2=R^3=All
451 R^1=Bn; R^2=R^3=All
452 R^1=R^2=Bn; R^3=H
453 R^1=R^2=Bn; R^3=P(OCH$_2$CH$_2$CN)$_2$
   O

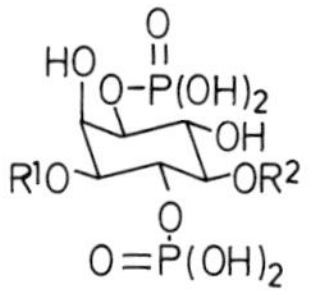

444 R^1=H; R^2=P(O)(OH)$_2$
445 R^1=R^2=P(O)(OH)$_2$
446 R^1=P(O)(OH)$_2$; R^2=H

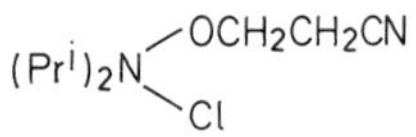

(Pr^i)$_2$N   OCH$_2$CH$_2$CN
   Cl

454

Because of the current biological interest [312] in the inositol trisphosphate (444) [derived from the glycolipid (436)] and its metabolites (445) and (446), synthesis in this field is currently very active although little has been published so far. Racemic 1,2:4,5-di-*O*-isopropylidene-*myo*-inositol was synthesised [308] and converted into racemic 1,2,4-tri-*O*-benzyl-*myo*-inositol (447) [309] as an intermediate for the synthesis of the trisphosphate. Ozaki and co-workers [310] have described the resolution of racemic (448) via the 1-menthoxyacetate (449). The enantiomer (448) was converted via the allyl ether (450) into (451) and this on deallylation gave chiral (447) which was phosphorylated with the reagent (440) and the product deprotected as described by Shvets et al. in the synthesis of the lipid (436) to give D-*myo*-inositol 1,4,5-trisphosphate (444). The racemic tetra-*O*-benzyl ether (452) has been phosphorylated using the reagent (454) and the product condensed with 2-cyanoethanol and oxidised to give (453) which was deprotected with sodium in liquid ammonia to give racemic inositol 4,5-bisphosphate [311].

9 References*

1. Gigg R (1980) Chem Phys Lipids 26: 287
2. Ishizuka I, Yamakawa T (1985) Glycoglycerolipids, in: Wiegandt H (ed) Glycolipids Elsevier, Amsterdam, p 101
3. Gigg J, Gigg R (in press) Synthesis of glycoglycerolipids, in: Handbook of lipid research, Glycoglycerolipids (ed) Kates M, Plenum, New York
4. Gigg J, Gigg R, Payne S, Conant R (1986) Synthetic studies on the major serologically active glycolipid from *Mycobacterium leprae*, in: Klein RA, Schmitz B (eds) Topics in lipid research — from structural elucidation to biological function, Royal Society of Chemistry p. 119
5. Anderson L, Unger FM (1983) Bacterial lipopolysaccharides, ACS Symp Ser 231
6. Morrison DC, Alving CR (1984) Molecular concepts of lipid A, Rev Infect Diseases 6: 427
7. Imoto M, Kusumoto S, Shiba T, Rietschel ET, Galanos C, Lüderitz O (1985) Tetrahedron Lett 26: 907
8. Imoto M, Yoshimura H, Sakaguchi N, Kusumoto S, Shiba T (1985) Tetrahedron Lett 26: 1545
9. Takahashi T, Nakamoto S, Ikeda K, Achiwa K (1986) Tetrahedron Lett 27: 1819
10. Kiso M, Tanaka S, Tanahashi M, Fujishima Y, Ogawa Y, Hasegawa A (1986) Carbohydr Res 148: 221
11. Paulsen H, Stiem M, Unger FM (1986) Tetrahedron Lett 27: 1135
12. Weigandt H (1985) Glycolipids Elsevier Amsterdam
13. Kanfer JN, Hakomori S (eds) (1983) Handbook of lipid research, vol 3, Sphingolipid biochemistry, Plenum, New York
14. Gigg R (1983) Phospholipids and Glycolipids, in: Ansell MF (ed) Supplement to 2nd ed of Rodd's Chemistry of Carbon Compounds vol 1E, Elsevier, Amsterdam, p 425
15. Hakomori S (1984) Ann Rev Immunol 2: 103
16. Hakomori S (1986) Chem Phys Lipids 42: 209
17. Makita A, Handa S, Taketomi T, Nagai Y (eds) (1982) New vistas in glycolipid research, Plenum, New York.
18. Hakomori S (1981) Ann Rev Biochem 50: 733
19. Hakomori S, Kannagi R (1986) Carbohydrate antigens in higher animals, in: Weir, DM (ed) Handbook of experimental immunology vol 1, Chap 9, Blackwell, Oxford
20. Weigandt H (1982) Adv Neurochem 4: 149
21. Walborg EF (ed) Glycoproteins and glycolipids in disease processes, ACS Symp Ser 80 1979
22. Paulsen H (1977) Pure and Appl Chem 49: 1169
23. Paulsen H (1982) Angew Chem, Int Ed Engl 21: 155
24. Paulsen H (1984) Chem Soc Rev 13: 15
25. Lemieux RU (1978) Chem Soc Rev 7: 423
26. Sinaÿ PA (1978) Pure and Appl Chem 50: 1437
27. Ogawa T, Yamamoto H, Nukada T, Kitajima T, Sugimoto M, (1984) Pure and Appl Chem 56: 779
28. Schmidt RR (1986) Angew Chem, Int Ed Engl 25: 212
29. van Boeckel CAA, Beetz T (1985) Recl Trav Chim Pays-Bas 104: 171, 174
30. van Boeckel CAA, Beetz T, van Aelst SF (1984) Tetrahedron 40: 4097
31. Garegg PJ (1984) Pure and Appl Chem 56: 845
32. van Boeckel CAA (1986) Recl Trav Chim 105: 35
33. Baer E, Fischer HOL (1939) J Biol Chem 128: 463
34. Gigg R, Warren CD, Cunningham J (1965) Tetrahedron Lett 1303
35. Gigg J, Gigg R, Warren CD (1966) J Chem Soc C 1872
36. Konstas S, Photaki I, Zervas L (1959) Chem Ber 92: 1288
37. Gigg J, Gigg R (1966) J Chem Soc C 1876
38. Mulzer J, Brand C (1986) Tetrahedron 42: 5961
39. Reist EJ, Christie PH (1970) J Org Chem 35: 3521
40. Reist EJ, Christie PH (1970) ibid 35 4127
41. Koike K, Nakahara Y, Ogawa T (1984) Glycoconjugate J 1: 107

* The numbers in brackets after the page numbers of Russian journals refer to the page numbers in the cover-to-cover English translations of these journals.

42. Koike K, Numata M, Nakahara Y, Ogawa T (1986) Carbohydr Res 158: 113
43. Obayashi M, Schlosser M (1985) Chem Lett 1715
44. Schmidt RR, Zimmermann P (1986) Tetrahedron Lett 27: 481
45. Kiso M, Nakamura A, Nakamura J, Tomita Y, Hasegawa A (1986) J Carbohydr Chem 5: 335
46. Kiso M, Nakamura A, Tomita Y, Hasegawa A (1986) Carbohydr Res 158: 101
47. Gigg R, Conant R (1977) J Chem Soc, Perkin 1: 2006
48. Corey EJ, Winter RAE (1963) J Am Chem Soc 85: 2677
49. Corey EJ, Hopkins PB (1982) Tetrahedron Lett 23, 1979
50. Shapiro D, Flowers HM, Spector-Shefer S (1959) J Am Chem Soc 81: 4360
51. Shapiro D, Flowers HM (1962) ibid 84: 1047
52. Funaki Y, Kawai G, Mori K (1986) Agric Biol Chem 50: 615
53. Mori K, Funaki Y (1985) Tetrahedron 41: 2369
54. Garigipati RS, Weinreb SM (1983) J Am Chem Soc 105: 4499
55. Tkaczuk P, Thornton ER (1981) J Org Chem 46: 4393
56. Mori K, Umemura T (1982) Tetrahedron Lett 23: 3391
57. Roush WR, Adam MA (1985) J Org Chem 50: 3752
58. Julina R, Herzig T, Bernet B, Vasella A (1986) Helv Chim Acta 69: 368
59. Cardillo G, Orena M, Sandri S, Tomasini C (1986) Tetrahedron 42: 917
60. Hino T, Nakayama K, Taniguchi M, Nakagawa M (1986) J Chem Soc, Perkin Trans 1: 1687
61. Kulmacz RJ, Kisic A, Schroepfer GJ (1979) Chem Phys Lipids 23: 291
62. Shoyama Y, Okabe H, Kishimoto Y, Costello C (1978) J Lipid Res 19: 250
63. Hirth G, Walter W (1985) Helv Chim Acta 68: 1863
64. Jurczak J, Pikul S, Baner T (1986) Tetrahedron 42: 447
65. Takano S, Kurotaki A, Takahashi M, Ogasawara K (1986) Synthesis 403
66. Jung ME, Shaw TJ (1980) J Am Chem Soc 102: 6304
67. Takano S, Numata H, Ogasawara K (1982) Heterocycles 19: 327
68. Hubschwerlen C (1986) Synthesis 962
69. Kanda P, Wells MA (1980) J Lipid Res 21: 257
70. Gigg R (1979) J Chem Soc, Perkin Trans 1: 712
71. Eibl H, Woolley P (1986) Chem Phys Lipids 41: 53
72. Wiggins LF (1946) J Chem Soc 13
73. Bering HFG, Boren HB, Garegg PJ (1967) Acta Chem Scand 21: 2083
74. van Boeckel CAA, Visser GM, van Boom JH (1985) Tetrahedron 41: 4557
75. Gent PA, Gigg R (1976) Chem Phys Lipids 17: 111
76. van Boeckel CAA, van Boom JH (1985) Tetrahedron 41: 4545
77. Ness AT, Hann RM, Hudson CS (1943) J Am Chem Soc 65: 2215
78. Wickberg B (1958) Acta Chem Scand 12: 1187
79. Gigg J, Gigg R (1967) J Chem Soc C: 1865
80. Wehrli HP, Pomeranz Y (1969) Chem Phys Lipids 3: 357
81. Schmidt RR, Kläger R (1985) Angew Chem, Int Ed Engl 24: 65
82. Schmidt RR, Zimmerman P (1986) ibid 25: 725
83. Koike K, Sugimoto M, Nakahara Y, Ogawa T (1985) Glycoconjugate J 2: 105
84. Mori K, Funaki Y (1985) Tetrahedron 41: 2379
85. Kawai G, Ikeda Y (1985) J Lipid Res 26: 338
86. Gal AE, Pentchev PG, Massey JM, Brady RO (1979) Proc Nat Acad Sci 76: 3083
87. Carter HE, Rothfus JA, Gigg R (1961) J Lipid Res 2: 228
88. Yoshino T, Watanabe K, Hakomori S (1982) Biochemistry 21: 928
89. Weis AL, Brady RO, Shapiro D (1985) Chem Phys Lipids 38: 391
90. Koshy KM, Boggs JM (1983) ibid 34: 41
91. Ariga T, Murata T, Oshima M, Maezawa M, Miyatake T (1980) J Lipid Res 21: 879
92. Cioffi EA, Prestegard JH (1986) Tetrahedron Lett 27: 415
93. Usuki S, Nagai Y (1986) Analytical Biochem 152: 172
94. Ogura H, Furuhata K, Itoh M, Shitori Y (1986) Carbohydr Res 158: 37
95. Ito Y, Sugimoto M, Sato S, Ogawa T (1986) Tetrahedron Lett 27: 4753
96. Kanemitsu K, Sweeley CC (1986) Glycoconjugate J 3: 143
97. Sato S, Ito Y, Ogawa T (1986) Carbohydr Res 155: C1
98. Sato S, Mori M, Ito Y, Ogawa T (1986) ibid 155: C6

99. Ek M, Garegg PJ, Oscarson S (1983) J Carbohydr Chem 2: 305
100. Paulsen H, Hadamczyk D, Kutschker W, Bünsch A (1985) Annalen 129
101. Paulsen H, Paal M (1985) Carbohydr Res 137: 39
102. Alais J, Maranduba A, Veyrières A (1983) Tetrahedron Lett 24: 2383
103. David S, Thieffry A, Veyrières A (1981) J Chem Soc, Perkin Trans 1: 1796
104. David S, Hanessian S (1985) Tetrahedron 41: 643
105. Maranduba A, Veyrières A (1986) Carbohydr Res 151: 105
106. Maranduba A, Veyrières A (1985) ibid: 135: 330
107. Sugimoto M, Horisaki T, Ogawa T (1985) Glycoconjugate J 2: 11
108. Ogawa T, Matsui M (1978) Carbohydr Res 62: C1
109. Ogawa T, Matsui M (1981) Tetrahedron 37: 2363
110. Veyrières A (1981) J Chem Soc, Perkin Trans 1: 1626
111. Paulsen H, Bünsch A (1982) Carbohydr Res 101: 21
112. Zurabyan SE, Markin VA, Pimenova VV, Rozynov BV, Sadovskaya VL, Khorlin AY (1978) Bioorg Khim 4: 928 (679)
113. Ponpipom MM, Bugianesi RL, Shen TY (1978) Tetrahedron Lett 1717
114. Garegg PJ, Hultberg H (1982) Carbohydr Res 110: 261
115. Jacquinet JC, Sinaÿ P (1985) ibid 143: 143
116. Cox DD, Metzner EK, Reist EJ (1978) ibid 63: 139
117. Martin-Lomas M, Bernabe M, Garcia-Montes P (1981) An Quim Ser C, 77: 230
118. Fernandez-Mayoralas A, Martin-Lomas M, Villanueva D (1985) Carbohydr Res 140: 81
119. Leontein K, Nilsson M, Norberg T (1985) ibid 144: 231
120. Garegg PJ, Hultberg H (1981) ibid 93: C10
121. Garegg PJ, Hultberg H, Wallin S (1982) ibid 108: 97
122. Wessel HP, Iversen T, Bundle DR (1984) ibid 130: 5
123. Banoub J, Bundle DR (1979) Can J Chem 57: 2085
124. Banoub J, Bundle DR (1979) ibid 57: 2091
125. Sabesan S, Lemieux RU (1984) ibid 62: 644
126. Takeo K, Tei S (1985) Carbohydr Res 141: 159
127. Takamura T, Chiba T, Ishihara H, Tejima S (1979) Chem Pharm Bull 27: 1497
128. Baer HH, Abbas SA (1979) Carbohydr Res 77: 117
129. Baer HH, Abbas SA (1980) ibid 84: 53
130. Hough L, Richardson AC, Thelwall LAW (1979) ibid 75: C11
131. Abbas SA, Barlow JJ, Matta KL (1981) ibid 88: 51
132. Karlsson KA (1986) Chem Phys Lipids 42: 153
133. de Graaf FK, Mooi FR (1986) Adv Microbial Physiol 26: 65
134. Cox DD, Metzner EK, Reist EJ (1978) Carbohydr Res 62: 245
135. Cox DD, Metzner EK, Cary LW, Reist EJ (1978) ibid 67: 23
136. Shapiro D, Acher AJ (1978) Chem Phys Lipids 22: 197
137. Chacon-Fuertes ME, Martin-Lomas M (1975) Carbohydr Res 43: 51
138. Milat ML, Zollo PA, Sinaÿ P (1982) ibid 100: 263
139. Dahmén J, Frejd T, Lave T, Lindh F, Magnusson G, Noori G, Pålsson K (1983) ibid 113: 219
140. Dahmén J, Frejd T, Grönberg G, Lave T, Magnusson G, Noori G (1983) ibid 116: 303
141. Dahmén J, Frejd T, Magnusson G, Noori G, (1982) ibid 111: C1
142. Dahmén J, Frejd T, Grönberg G, Lave T, Magnusson G, Noori G (1983) ibid 118: 292
143. Dahmén J, Frejd T, Magnusson G, Noori G, Carlström AS (1984) ibid 127: 15
144. Garegg PJ, Oscarson S (1985) ibid 137: 270
145. Garegg PJ, Samuelsson B (1979) J Chem Soc, Chem Commun 978
146. Kihlberg J, Frejd T, Jansson K, Magnusson G (1986) Carbohydr Res 152: 113
147. Wessel HP, Iversen T, Bundle DR (1985) J Chem Soc, Perkin Trans 1: 2247
148. Barton DHR, McCombie SW (1975) ibid 1574
149. Norberg T, Oscarson S, Szönyi M (1986) Carbohydr Res 152: 301
150. Jacquinet JC, Duchet D, Milat ML, Sinaÿ P (1981) J Chem Soc, Perkin Trans 1: 326
151. Petit JM, Jacquinet JC, Sinaÿ P (1980) Carbohydr Res 82: 130
152. Zollo PHA, Jacquinet JC, Sinaÿ (1983) ibid 122: 201
153. Dahmén J, Frejd T, Magnusson G, Noori G, Carlström AS (1984) ibid 129: 63
154. Dahmén J, Frejd T, Magnusson G, Noori G, Carlström AS (1984) ibid 125: 237

155. Nashed MA, Anderson L (1983) ibid 114: 43
156. Paulsen H, Kolář C (1981) Chem Ber 114: 306
157. Paulsen H, Bünsch H (1981) ibid 114: 3126
158. Lönn H (1985) Carbohydr Res 139: 105
159. Lemieux RU, Takeda T, Chung BY (1976) ACS Symp Ser 39: 90
160. Nashed MA (1979) Carbohydr Res 71: 299
161. Nashed MA, Slife CW, Kiso M, Anderson L (1980) ibid 82: 237
162. Nashed MA, El-Sokkary RI, Rateb L (1984) ibid 131: 47
163. Nashed MA, Anderson L (1983) ibid 114: 53
164. Paulsen H, Richter A, Sinnwell V, Stenzel W (1978) ibid 64: 339
165. Paulsen H, Kolář C, Stenzel W (1978) Chem Ber 111: 2358
166. Lemieux RU, Ratcliffe RM (1979) Can J Chem 57: 1244
167. Lemieux RU, Ratcliffe RM (1978) Ger Offen 2, 816, 340/1978; (1979) Chem Abs 90: 87846k
168. Lemieux RU, Stick RV (1978) Austral J Chem 31: 901
169. Bovin NV, Zurabyan SE, Khorlin AY (1979) Bioorg Khim 5: 1257 (945)
170. Bovin NV, Zurabyan SE, Khorlin AY (1981) Carbohydr Res 98: 25
171. Bovin NV, Zurabyan SE, Khorlin AY (1983) J Carbohydr Chem 2: 249
172. Lemieux RU, Abbas SZ, Burzynska MH, Ratcliffe RM (1982) Can J Chem 60: 63
173. Corey EJ, Suggs JW (1973) J Org Chem 38: 3224
174. Bundle DR, Josephson S (1979) Can J Chem 57: 662
175. Paulsen H, Paal M, Hadamczyk D, Steiger KM (1984) Carbohydr Res 131: C1
176. Excoffier G, Gagnaire D, Utille JP (1975) ibid 39: 368
177. Schmidt RR, Michel J (1980) Angew Chem, Int Ed Engl 19: 731
178. Osby JO, Martin MG, Ganen B (1984) Tetrahedron Lett 25: 2093
179. Rosenbrook W, Riley DA, Lartey PA (1985) ibid 26: 3
180. Posner GH, Haines SR (1985) ibid 26: 5
181. Paulsen H, Paal M (1984) Carbohydr Res 135: 53
182. Paulsen H, Bünsch A (1980) Angew Chem, Int Ed Engl 19: 902
183. Paulsen H, Lockhoff O (1981) Chem Ber 114: 3079
184. Paulsen H, Bünsch A (1981) Annalen 2204
185. Paulsen H, Röben W, Heiker FR (1980) Tetrahedron Lett 21: 3679
186. Paulsen H, Bünsch A (1982) Carbohydr Res 100: 143
187. Kijimoto-Ochiai S, Makita A, Bünsch A, Paulsen H (1983) Biochim Biophys Acta 756: 247
188. Luger P, Vangehr K, Bock K, Paulsen H (1983) Carbohydr Res 117: 23
189. David S, Lubineau A, Vatèle JM (1978) J Chem Soc Chem Commun 535
190. David S, Lubineau A, Vatèle JM (1980) Nouveau J Chim 4: 547
191. Paulsen H, Kolář C, Stenzel W (1978) Chem Ber 111: 2370
192. Paulsen H, Kolář C (1978) Angew. Chem, Int Ed Engl 17: 771
193. Paulsen H, Kolář C (1979) Chem Ber 112: 3190
194. Bovin NV, Zurabyan SE, Khorlin AY (1983) Carbohydr Res 112: 23
195. Bovin NV, Zurabyan SE, Khorlin AY (1982) Izvest Akad Nauk SSSR, Ser Khim 1148 (31D, 1023)
196. Lemieux RU, Abbas SZ, Chung BY (1982) Can J Chem 60: 68
197. Arnarp J, Lönngren J (1981) J Chem Soc, Perkin Trans 1: 2070
198. Ogawa T, Nakabayashi S (1981) Carbohydr Res 97: 81
199. Ogawa T, Nukada T, Matsui M (1982) ibid 101: 263
200. Ogawa T, Nakabayashi S, Sasajima K (1981) ibid 96: 29
201. Ogawa T, Nakabayashi S, Sasajima K (1981) ibid 95: 308
202. Augé C, David S, Veyrières A (1979) Nouveau J Chim 3: 491
203. Takamura T, Chiba T, Tejima S (1981) Chem Pharm Bull 29: 2270
204. Wood E, Feizi T (1979) FEBS Lett 104: 135
205. Alais J, Veyrières A (1981) Carbohydr Res 92: 310
206. Alais J, Veyrières A (1981) J Chem Soc, Perkin Trans 1: 377
207. Arnarp J, Lönngren J, Ottosson H (1981) Carbohydr Res 98: 154
208. Arnarp J, Haraldsson M, Lönngren J (1981) ibid 97: 307
209. Arnarp J, Lönngren J (1980) J Chem Soc, Chem Commun 1000
210. Alais J, Veyrières A (1981) Carbohydr Res 93: 164

211. Lee RT, Lee YC (1979) ibid 77: 270
212. Alais J, Veyrières A (1983) Tetrahedron Lett 24: 5223
213. Grundler G, Schmidt RR (1985) Carbohydr Res 135: 203
214. Augé C, Mathieu C, Mérienne C (1986) ibid 151: 147
215. Wong CH, Haynie SL, Whitesides GM (1982) J Org Chem 47: 5416
216. Whitesides GM, Wong CH (1985) Angew Chem, Int Ed Engl 24: 617
217. Augé C, David S, Mathieu C, Gautheron C (1984) Tetrahedron Lett 25: 1467
218. Takamura T, Chiba T, Tejima S (1979) Chem Pharm Bull 27: 721
219. Itoh Y, Tejima S (1983) ibid 31: 727
220. Jacquinet JC, Sinaÿ P (1979) Tetrahedron 35: 365
221. Milat ML, Sinaÿ P (1981) Carbohydr Res 92: 183
222. Jacquinet JC, Sinaÿ P (1979) J Chem Soc, Perkin Trans 1: 319
223. Augé C, Veyrières A (1979) ibid 1825
224. Subero C, Jimeno ML, Alemany A, Martin-Lomas M (1984) Carbohydr Res 126: 326
225. Paulsen H, Schnell D (1981) Chem Ber 114: 333
226. Hindsgaul O, Norberg T, Le Pendu J, Lemieux RU (1982) Carbohydr Res 109: 109
227. Lemieux RU, Wong TC, Thørgersen H (1982) Can J Chem 60: 81
228. Lemieux RU, Wong TC, Liao J, Kabat EA (1984) Molecular Immunol 21: 751
229. Khare DP, Hindsgaul O, Lemieux RU (1985) Carbohydr Res 136: 285
230. Rana SS, Vig R, Matta KL (1982) J Carbohydr Chem 1: 261
231. Jacquinet JC, Sinaÿ P (1979) J Chem Soc, Perkin Trans 1: 314
232. Pougny JR, Nassr MAM, Naulet N, Sinaÿ P (1978) Nouveau J Chim 2: 389
233. Bovin NV, Zurabayan SE, Khorlin AY (1980) Bioorg Khim 6: 242 (121)
234. Rana SS, Matta KL (1983) Carbohydr Res 117: 101
235. Rana SS, Barlow JJ, Matta KL (1981) ibid 96: 231
236. Horton D, Lehmann J (1978) ibid 61: 553
237. Benzing-Nguyen L, Perry MB (1978) J Org Chem 43: 551
238. Beau JM, Sinaÿ P (1978) Carbohydr Res 65: 1
239. Beau JM, Sinaÿ P, Kamerling JP, Vliegenthart FG (1978) ibid 67: 65
240. Augé C, David S, Gautheron C (1984) Tetrahedron Lett 25: 4663
241. Augé C, David S, Gautheron C, Veyrières A (1985) ibid 26: 2439
242. Danishefsky SJ, DeNinno MP (1986) J Org Chem 51: 2615
243. Baumberger F, Vasella A (1986) Helv Chim Acta 69: 1205
244. Baumberger F, Vasella A (1986) ibid 69: 1535
245. Hagedorn HW, Brossmer R (1986) ibid 69: 2127
246. Mack H, Brossmer R (1987) Tetrahedron Lett 28: 191
247. Vliegenthart JFG, Kamerling JP (1982) Synthesis of sialic acids and sialic acid derivatives, in: Schauer R (ed) Sialic acids, chemistry, metabolism and function Springer-Verlag, New York, p 59
248. Martin JE, Tanenbaum SW, Flashner M (1977) Carbohydr Res 56: 423
249. Czarniecki MF, Thornton ER (1977) J Am Chem Soc 99: 8273
250. Sharma MN, Eby R (1984) Carbohydr Res 127: 201
251. Eschenfelder V, Brossmer R (1979) Z physiol Chem 360: 1253
252. Eschenfelder V, Brossmer R (1980) Carbohydr Res 78: 190
253. Beau JM, Schauer R, Haverkamp J, Dorland L, Vliegenthart JFG, Sinaÿ P (1980) ibid 82: 125
254. van der Vleugel DJM, van Heeswijk WAR, Vliegenthart JFG (1982) ibid 102: 121
255. Hasegawa A, Nakamura J, Kiso M (1986) J Carbohydr Chem 5: 11
256. van der Vleugel DJM, Wassenburg FR, Zwikker JW, Vliegenthart JFG (1982) Carbohydr Res 104: 221
257. van der Vleugel DJM, Zwikker JW, Vliegenthart JFG, van Boeckel CAA, van Boom JH (1982) ibid 105: 19
258. Brandstetter HH, Zbiral E (1983) Monatshefte 114: 1247
259. Paulsen H, Tietz H (1982) Angew Chem, Int Ed Engl 21: 927
260. Paulsen H, Tietz H (1984) Carbohydr Res 125: 47

261. Paulsen H, von Deessen U, Tietz H (1985) ibid 137: 63
262. Paulsen H, Tietz H (1985) Angew Chem, Int Engl Ed 24: 128
263. Paulsen H, Tietz H (1985) Carbohydr Res 144: 205
264. Kitajima T, Sugimoto M, Nukada T, Ogawa T (1984) ibid 127: C1
265. Ogawa T, Sugimoto M, Kitajima T, Sadozai KK, Nukada T (1986) Tetrahedron Lett 27: 5739
266. Furuhata K, Anazawa K, Itoh M, Shitori Y, Ogura H (1986) Chem Pharm Bull 34: 2725
267. Kunz H, Waldmann H (1985) J Chem Soc, Chem Commun 638
268. Paquet F, Sinaÿ P (1984) Tetrahedron Lett 25: 3071
269. Ogawa T, Sugimoto M (1985) Carbohydr Res 135: C5
269. Ogawa T, Sugimoto M (1985) Carbohydr Res 135: C5
270. Sugimoto M, Ogawa T (1985) Glycoconjugate J 2: 5
271. Hirabayashi Y, Sugimoto M, Ogawa T, Matsumoto M, Tagawa M, Taniguchi M (1986) Biochim Biophys Acta 875: 126
272. Paulsen H, von Deessen U (1986) Carbohydr Res 146: 147
273. Sugimoto M, Numata M, Koike K, Nakahara Y, Ogawa T (1986) ibid 156: C1
274. Okamoto K, Kondo T, Goto T (1986) Tetrahedron Lett 27: 5229
275. Okamoto K, Kondo T, Goto T (1986) ibid 27: 5233
276. Okamoto K, Kondo T, Goto T (1986) Chem Lett 1449
277. Sabesan S, Paulson JC (1986) J Am Chem Soc 108: 2068
278. Thiem J, Treder W (1986) Angew Chem, Int Engl Ed 25: 1096
279. Neuenhofer S, Schwarzmann G, Egge H, Sandhoff K (1985) Biochemistry 24: 525
280. Acquotti D, Sonnino S, Masserini M, Casella L, Fronza G, Tettamanti G (1986) Chem Phys Lipids 40: 71
281. Gazzotti G, Sonnino S, Ghidoni R, Orlando P, Tettamanti G (1984) Glycoconjugate J 1: 111
282. Dorland L, Haverkamp J, Schauer R, Veldink GA, Vliegenthart JFG (1982) Biochem Biophys Res Commun 104: 1114
283. Brossmer R, Rose U, Kasper D, Smith TL, Grasmunk H, Unger FM (1980) ibid 96: 1282
284. Stepanov AE, Shvets VI (1979) Chem Phys Lipids 25: 247
285. Berridge MJ (1986) Biochem Soc Symposium 52: 153
286. Majerus PJ, Connolly TM, Deckmyn H, Ross TS, Bross TE, Ishii, H, Bansal VS, Wilson DB (1986) Science 234: 1519
287. Abdel-Latif AA (1986) Pharmacol Rev 38: 227
288. Parthasarathy R, Eisenberg F (1986) Biochem J 235: 313
289 Troyer DA, Schwertz DW, Kreisberg JI, Venkatachalam MA (1986) Annu Rev Physiol 48: 51
290. Irvine RF, Moor RM (1986) Biochem J 240: 917
291. Stepanov AE, Shvets VI, Evstigneeva RP (1977) Zhur Org Khim. 13: 1410 (1295)
292. Stepanov AE, Shvets VI, Evstigneeva RP (1977) Zhur obshchei Khim 47: 1653 (1515)
293. Sibrikov YI, Stepanov AE, Shvets VI (1981) Zhur Org Khim 17: 2015 (1799)
294. Letsinger RL, Miller PS (1969) J Am Chem Soc 91: 3356
295. Stepanov AE, Sibrikov YI, Shvets VI (1980) Zhur Org Khim 16: 2284 (1946)
296. Krylova VN, Lyutik AI, Kobelkova NN, Shvets VI, Evstigneeva RP (1978) ibid. 14: 1858 (1725)
297. Sibrikov YI, Stepanov AE, Shvets VI (1984) ibid 20: 979 (891)
298. Kaplun AP, Krylova VN, Soloveichik GY, Klyashchitskii BA, Shvets VI (1978) ibid 14: 1863 (1730)
299. Kaplun AP, Krylova VN, Soloveichik GY, Shvets VI (1979) ibid 15: 307 (266)
300. Garegg PJ, Lindberg B, Kvarnström I, Svensson SCT (1985) Carbohydr Res 139: 209
301. Krylova VN, Lyutvik AI, Kosaeva AE, Shvets VI (1979) Zhur. Org Khim 15: 2323 (2104)
302. Shvets VI, Sukhanov VA, Okhanov VV, Zhdanov RI (1979) Chem Phys Lipids 23: 163
303. Shragin AS, Kuzmina YV, Borin ML, Kaplun AP, Shvets VI (1985) Bioorg Khim 11: 1669
304. Krylova VN, Kobelkova NI, Oleinik GF, Shvets VI (1980) Zhur Org Khim 16: 62 (59)
305. Krylova VN, Gornaeva NP, Oleinik GF, Shvets VI (1980) ibid 16: 315 (277)
306. Krylova VN, Gornaeva NP, Shvets VI, Evstigneeva RP (1979) Dokl Akad Nauk SSSR 246: 339
307. Krylova VN, Lyutvik AI, Gornaeva NP, Shvets VI (1979) Zhur obshchei Khim 51: 210 (183)
308. Gigg J, Gigg R, Payne S, Conant R (1985) Carbohydr Res 142: 132
309. Gigg J, Gigg R, Payne S, Conant R (1985) ibid 140: C1

310. Ozaki S, Watanabe Y, Ogasawara T, Kondo Y, Shiotani N, Nishii H, Matsuki T (1986) Tetrahedron Lett 27: 3157
311. Hamblin MR, Potter BVL, Gigg R (1987) J Chem Soc, Chem Commun 626
312. Houslay MD (1987) Trends Biochem Sci 12: 1
313. Boutin RH, Rapoport H (1986) J Org Chem 51: 5320

Conformational Aspects of Oligosaccharides

Bernd Meyer

Complex Carbohydrate Research Center, University of Georgia, 220 Riverbend Road, Athens, Ga. 30602, USA

Table of Contents

Topics in Current Chemistry, Vol. 154
© Springer-Verlag Berlin Heidelberg 1990

The three dimensional structure of oligosaccharides determines their interaction with receptors and hence is important for their biological activity. Conformational analysis of oligosaccharides makes the three dimensional structure available. The analysis of the conformation of oligosaccharides is usually determined by a combination of computational methods and experimental techniques. NMR spectroscopy is the most important experimental tool. The calculational techniques cover a wide range with most emphasis put into force field calculations. Conformational flexibility plays an important role in many though not in all oligosaccharide structures. Glycosidic linkages to a side chain of a pyranose ring are more flexible than are linkages to the pyranose ring. The major attempts are described to determine the three dimensional structure of oligosaccharides with the exception of homooligomers. This review covers conformational analyses of blood group antigens of N-linked and of O-linked oligosaccharide chains, of glycolipids, of oligosaccharides related to O-specific polysaccharides of bacteria, and of oligosaccharides related to proteoglycans.

1 Introduction

This review presents a brief account on the conformation of biologically active oligosaccharides. Emphasis will be put on the conformational analysis of heterooligomers. The conformational preferences of some disaccharides will be included for reference purposes. Conformational aspects of homo-oligomers and homopolymers will not be discussed.

Biological or chemical or physical properties of an oligosaccharide are largely determined by what is exposed to the outer surface. Most interactions with other molecules will occur at the surface of the molecules. Therefore, stereo plots of CPK models are used to illustrate the conformations of the three dimensional structures discussed.

The importance of the conformation of oligosaccharides towards an understanding of the biological function is widely accepted [1, 2, 3, 4, 5, 6]. In crystals most molecules adopt usually one single conformation. However, in solution there are generally some dynamic variations of the preferred conformation. Furthermore, in solution there may be more than one preferred conformation. The resulting mixture of conformations usually undergoes a fast interconversion and most experimental techniques, like NMR, show only results which are time averaged over all conformations. Because of the importance of the conformation of the oligosaccharides in solution, this review will present mainly results from experimental studies in solution and data from calculations.

The experimental determination of the conformation of oligosaccharides has been made possible in the recent years by the use of 2-D-NMR and especially NOE-NMR measurements. The 2-D-NMR experiments allow the assignment of all protons and carbons of even very complex structures at high magnetic field [7]. Starting from this point a precise determination of the NOE values across the glycosidic bond gives a good estimate of the preferred conformation of the glycosidic [8]. This, in turn, will then be confirmed by further interpretation of the chemical shifts in comparing them to reference compounds. All NMR techniques supply time averaged values, i.e. NOEs, chemical shifts, coupling constants. Usually NMR spectroscopy alone does not give an unambiguous assignment of a specific conformation but rather supplies limits for the conformations. Calculations of the preferred conformations of oligosaccharides are very important as they allow an assignment and an interpretation of the experimental parameters. Therefore, usually only a combination of experimental and theoretical approaches is sufficient to solve the solution conformation of an oligosaccharide.

2 Methods for the Determination of Oligosaccharide Conformations

Very many properties of oligosaccharides are determined by a unique group — the acetal fragment at C1 in aldoses or C2 in ketoses. This fragment shows unique effects: the anomeric effect and exo anomeric effect which represent two important facts ruling the conformations of oligosaccharides [9, 10, 11, 12, 13, 14, 15, 16, 17, 18, 19].

143

Bernd Meyer

2.1 Relevance of Conformations of Glycosidic Linkages

It is widely accepted that the major factor ruling the overall shape of an oligosaccharide is the conformation of the glycosidic linkage. The description of the dihedral angles is made with reference to the hydrogen atoms at both sides of the glycosidic link in the case where two secondary alcohols are encountered, i.e. 1–1, 1–2, 1–3, and 1–4 linkages in hexopyranoses. In an A-B-C-D fragment, the definition of the dihedral angle of the B-C bond and its sign is obtained by: i) oriented the molecule that you view parallel to the B-C bond with either B or C pointing to you, ii) rotating the front side bond vector, A-B or C-D, until the atoms A and D are in an eclipsed position (one is behind the other), and iii) if the rotation was clockwise the angle has a positive sign, if counterclockwise the angle has a negative sign. The two dihedral angles are named φ for the H1-C1-O1-$C_{aglycon}$ fragment and ψ for the C1-O1-C_{aglcon}-$H_{aglycon}$ fragment. In case of a 1–6 bond in hexopyranoses, or more generally a primary alco-

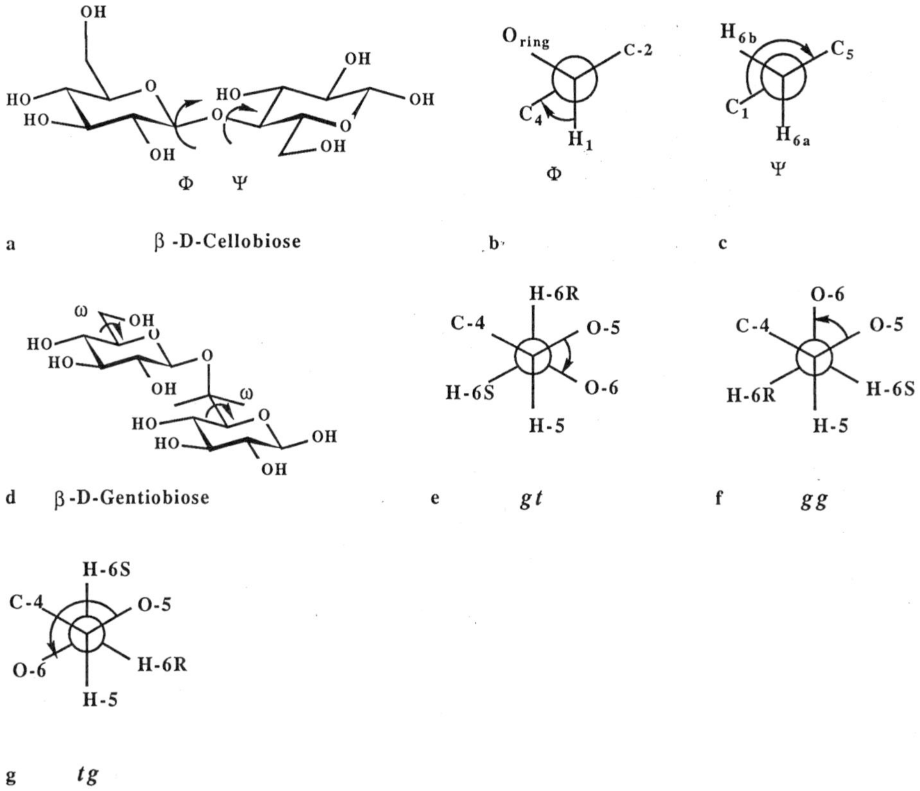

a   β-D-Cellobiose   b   c

d   β-D-Gentiobiose   e   *gt*   f   *gg*

g   *tg*

Fig. 1a–f. Definition of the dihedral angles φ, ψ and ω. **a** Glycosidic bond defining φ and ψ, **b** Newman projection along the O1'-C1' bond with definition of the sign of the φ angle, **c** Newman projection along the C4-O1' bond with the definition of the ψ angle, **d** 1,6 glycosidic bond with the definition of the ω angle, **e, f** Newman projections along the C5-C6 bond with two conformers representing the *gt* conformer, O6 *gauche* to O5 and *trans* to C4, and the *gg* conformation, O6 *gauche* to O5 and C4, of the ω angle

144

hol, the ψ angle is measured with reference to the carbon atom next to the primary alcoholic group, i.e. to C5 in hexoses. The additional degree of freedom in a 1–6 linkage is defined by the free rotation of the primary hydroxymethyl group around the C5–C6 bond. This conformation is described by the ω angle which is defined by the atoms O6–C6–C5–O5. In cases where an ambiguity may arise which atom is used as a reference atom, a superscript to the Greek letter designating the angle specifies the atom that is used as a reference. The three mainly populated conformers of a C5–C6 fragment are the staggered conformations which are termed *gg*, *gt*, and *tg* respectively (cf. Fig. 1). However, in rare occasions eclipsed conformations may also be found as in the X-ray structure of panose where a hydrogen bond causes one hydroxymethyl group to adopt an eclipsed conformation [20].

2.2 NMR-Determination of Oligosaccharide Conformations

There is a variety of methods for the determination of the structure of oligosaccharides. Different techniques are available for the determination of solution conformations and crystal structures respectively. The following section will deal with the most frequently used techniques.

NMR spectroscopy has developed during the last few years into a very powerful method to establish both the conformation of an oligosaccharide in solution and in the crystalline state. The solution conformation of a saccharide can be determined generally by the combination of ^{13}C-NMR and ^{1}H-NMR techniques. 2-D methods became available in the last few years and are prerequisites for the elucidation of the three dimensional structure of an oligosaccharide [7]. Both NOE measurements [21] as well as "traditional" determination of coupling constants and chemical shifts are important tools for the determination of preferred conformations. 2-dimensional methods have greatly improved the accessibility of these parameters from complex molecules.

Determination of NOEs across the glycosidic bond (Fig. 2) represent the most important tool for the conformational analysis of the glycosidic bond [8, 21]. As a result of the NOE measurement one obtains an averaged distance between the glycosidic and the aglyconic hydrogen atoms [22]. Subsequently, this distance can be converted into a range of dihedral angles ϕ and ψ which fit the observed NOE [23].

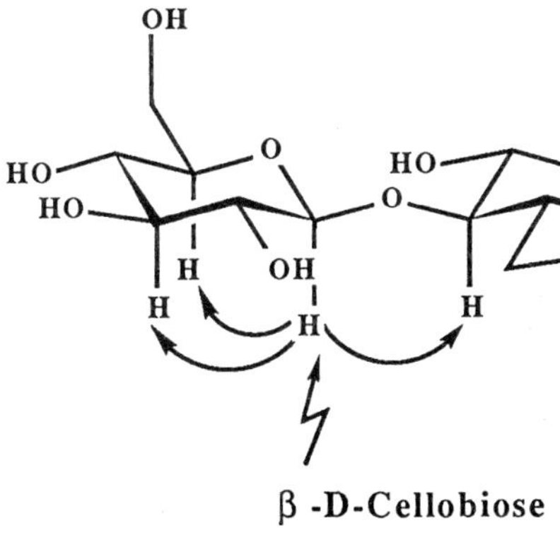

β -D-Cellobiose

Fig. 2. Determination of the NOE across the glycosidic bond to estimate the distance between the two protons neighboring the glycosidic bond. The distances of the irradiated proton to others within the same ring serve as calibrations for the interglycosidic NOE

145

Because of the dependence of the NOE with the r^{-6} the transcription of the observed NOE to the conformation of the glycosidic bond is not unequivocally possible. Conformations with a low statistical weight and eventually a large NOE are highly overrepresented in the time averaged interpretation of the NOEs. An increase in the distance of two protons by just 12% causes the NOE between these two protons to decrease by a factor of two. Thus the deduction of a single conformer from NOE experiments usually leaves some uncertainty because the determination of the distance between the two protons involved overestimates the conformations with short contact between them.

However, if there is not only one NOE across the glycosidic bond but two or more the determination of the conformation based solely on NMR spectroscopy becomes feasible. The occurrence of several interglycosidic NOEs gives much more precise values of the dihedral angles φ and ψ because the two parameter system of the two angles is then determined from two independent NOEs [24].

To overcome the problem of the precision of the determination of the conformation of the dihedral angle from the NOE measurements it is advisable to support the interpretation by the fine analysis of changes in chemical shifts of protons close to the glycosidic bond. They will often show marked changes in their values due to a change in the preferred conformation [23]. These changes might arise from anisotropies and/or from the relatively close proximity of a lone pair of an oxygen or nitrogen atom. These changes are dependent on r^{-3} and hence will give in accordance with the NOE data a higher reliance of the interpreted conformation.

Additionally, the determination of the $^3J(^{13}C, {}^1H)$ coupling constants across the glycosidic bond can improve the value of the conformational analysis [25]. These coupling constants are directly dependent on just one dihedral angle and would thus allow its direct determination. There is a restriction because the $^3J(^{13}C, {}^1H)$ coupling constants vary only slightly with the variation of the dihedral angles of the glycosidic bonds [26, 27]. For this reason the two parameters $^3J_{C, Hx}$ and $^3J_{Cx, H1}$ are not very precise measures of the preferred conformation but support the data obtained by other methods.

In special cases the measurement of the $^4J(^1H, {}^1H)$ coupling constant across the glycosidic bond gives useful information for the determination of the conformation of the glycosidic bond. This parameter is — as well as the NOE — dependent on two dihedral angles which will let the interpretation be ambiguous [28]. Thus, this parameter may just complement other data.

The determination of the conformation of 1,6-linkages requires that the conformation of the C5, C6 bond is established, in addition to conformation of the glycosidic bond. Besides the standard techniques for obtaining the coupling constants interpretation of the cross peak patterns in phase sensitive COSY spectra proved to be very valuable [29, 30].

Measuring deuterium isotopic multiplet patterns in ^{13}C-NMR spectra of partially OH — OD exchanged mono- or oligosaccharides gives information on the acceptor or donor function of hydroxygroups in hydrogen bonds [31]. The partially O-deuterated compounds were prepared by dissolving them in a 1:1 mixture of H_2O and D_2O dried and subsequently dissolved in dimethylsulfoxide for NMR spectroscopy. An analogous way was described using ^1H-NMR spectroscopy [32, 33].

Recently the analysis of solid state NMR spectra has led to the conclusion that the chemical shift of the carbon atoms attached to the glycosidic oxygen are strongly dependent on the dihedral angles φ and ψ of the glycosidic bonds. Apparently there is linear correlation of the conformation of the C1-O1 and O1-Cx bond over a range of approximately 10–15 ppm [34, 35]. A similar correlation was reported from the comparison of ^{13}C-NMR chemical shifts and dihedral angles calculated by the HSEA method [36]. The regression analysis gave the dihedral angle $\varphi = 5.3 \times \Delta\delta$ [ppm] — 54.2 and $\psi = 5.1 \times \Delta\delta$ [ppm] — 56.9, where $\Delta\delta$ is the change in chemical shift caused by the glycosylation of C1' and C4 respectively.

2.3 X-Ray Analysis

X-ray analysis gives the best data of the conformation of an oligosaccharide — unfortunately in an environment which is not the same as in biological systems, i.e. the crystal structure may deviate quite strongly from the solution conformation of an oligosaccharide. Furthermore, higher oligosaccharides do not tend to crystallize, which may be due to an unfavorable packing of the structures in the solid state. Until now there has been no general way to get single crystals from molecules with more than four monosaccharides.

Nonetheless, there is one macromolecule — the Fc fragment of the human IgG_1 Immunoglobulin — with a complex carbohydrate structure attached to it which allowed the determination of the decasaccharide conformation down to atomic resolution [37, 38] (cf. below). Another system which allowed the resolution of the oligosaccharide structure is the rabbit IgG_1 Fc-fragment [39].

Some important contributions to the conformational analysis of oligosaccharides have come from the determination of the structure of oligosaccharides which form complexes with protein. In this context the complex of lysozyme with chitohexaose [40] and the phosphorylase complex with maltoheptaose [41, 42] and maltopentaose [43] have been resolved to atomic resolution.

2.4 Optical Spectroscopy

The optical rotatory dispersion (ORD)/circular dichroism (CD) family of methods including the vacuum UV technique are applicable in special cases of glycoside linkages [44, 45, 46]. The measurements of the circular dichroism of an oligosaccharide is a good indicator of the conformation of the glycosidic bond if the oligomer contains repeating units, i.e. a repeating mono- or disaccharide, and if the effect resulting from the glycoside is not overlapped by strong effects of carboxyl- or amido-groups [47]. The circular dichroism will generally give cumulative effects of all glycosidic bonds; whereas the above mentioned methods will give distinct information on each individual bond.

2.5 Theoretical Calculations

The already mentioned difficulties in the experimental determination of the conformation of glycosidic bonds have provoked high efforts towards calculating the preferred conformations of oligosaccharides. There have been many different approaches in the last few years to find solutions to the problem of the computation of oligo-saccharides. The range of methods reaches from ab initio methods to simple distance restriction maps for the evaluation of the conformation of oligosaccharides.

The following sections contain a very short description of the advantages and dis-advantages of different calculation methods. A more detailed discussion is available [48].

2.5.1 Ab Initio Calculations

Ab initio programs are based on molecular orbital theory with the atomic orbitals constructed from a number of Gaussian probability functions. One of the vital criteria for the precision of the calculation is given by the number of Gaussian functions used (basis set) in the ab initio calculation which describe the atomic orbitals. In general, ab initio calculations contain the least amount of assumptions of all types of programs discussed below. Although the most reliable results would be obtained from ab initio calculations of oligosaccharides only very few examples are reported in the literature (cf. below). The reason is the long computation time which becomes even longer if a full geometry optimization is associated with the calculation. Another restriction results from the number of orbitals allowed in the programs. Thus, only small representative fragments of oligosaccharides have been calculated by *ab initio* calculations on a high level, with large basis sets.

2.5.2 Semi Empirical Calculations

Semi empirical quantum mechanical programs are sometimes used for the calculation of oligosaccharides. Mainly the MINDO/3 [49], MNDO [50], and PCILO [51] pro-grams are used. From these, only the last two are expected to give accurate results in the calculations of oligosaccharides. In fact most calculations are performed with the PCILO program (cf. Sect. 2.5.3). The advantage of these programs in contrast to the force field programs referred to below is the relative high precision in the deter-mination of the geometry of the minimum energy conformers and the additional information on electronic parameters of the studied molecule, like the charge distri-bution or orbital coefficients. The disadvantage is the long computation time required if a full geometry optimization is carried out — even for small molecules like di-saccharides.

2.5.3 Force Field Programs

There are a large number of different force field approaches published. The following section deals only with those programs most frequently used in the calculation of oligosaccharides. The other programs deviate from the described ones usually only by either the parametrization or the use and/or the weighting of the individual func-tions which contribute to the force field.

The force field approach assumes that it is possible to describe the geometry of a molecule by a set of mechanical equivalents. Changes in bond length are, for example, represented by a modified Hook's law $V(r) = 0.5 \times k_s \times (r - r_0)^2$ for a spring or bending of a bond angle is modelled by a bending potential $V_B = k_B \times (\theta - \theta_0)^2$, where k_B, k_s, r_0, and θ_0 are parameters that are dependent on the incorporated atom types. However, the most important term in unstrained molecules describes the non bonded interactions. Non bonded atoms may have either a repulsive or an attractive interaction. Non bonded atoms are attracted to each other if they have a distance longer than the energy minimum which is at a distance a little bit shorter than the sum of the van der Waals distances of the two atoms. The non bonded interactions are approximated by Kitaigorodsky, Lennard-Jones, or Buckingham potential functions. Additionally, certain stretch-bend and torsional potentials may be used. Dipol-dipol interaction energy, hydrogen bond potentials, and partial atomic charges are sometimes included in the total energy. These functions have the problem that their evaluation requires a assessment of the dielectricity constant at a distance of a few solvent molecules. However, the dielectricity constant is inadequate because it is defined for a particular solvent only at a distance of many solvent molecules, where the orientation of any individual solvent molecule is unimportant. In contrast, usually dipols or partial charges which contribute a significant portion to the energy are only separated by some highly oriented solvent molecules.

The MM2 program [52] is by far the most widely used force field program in the area of hydrocarbon or moderately polar molecules and numerous publications have demonstrated its usefulness. Still the anomeric center had not until recently been properly defined by the force field parameters. Now there are parameters available which include the proper treatment of acetal fragments and hence reproduce the anomeric as well as the *exo* anomeric effect accurately [53]. Due to the full optimization of the geometry the MM2 program can treat only a limited number of atoms.

An extension of the MMI program [54] developed by Jeffrey et al. [55] is named MMI-CARB. This version is also adapted to the special situations which are found in oligosaccharides and reproduces the geometries of a number of model glycosides well.

A series of simplified force fields was created by Rasmussen et al. [56, 57]. The PEF400 [58] and PEF300 (potential energy function) and variants here of are used for the calculation of disaccharides and monosaccharides (cf. below). The parametrisation of this force field program includes atomic charges which (in variant PEF422) differ for the anomeric carbon and the other carbon atoms. The parametrization is obtained from ab initio calculations.

A simplified force field is developed by Lemieux et al. [23, 24] who established the secondary structure of oligosaccharides by the calculation of the non bonded interactions between the monosaccharide constituents of a disaccharide. In order to account for the exo anomeric effect an additional term is included in the force field calculation that rules the rotameric distribution at the C1-O1 bond of the glycosidic linkage E_{EA} (kcal/mol) $= A \times (1 - \cos \varphi) + B \times (1 - \cos 2\varphi) + C \times (1 - \cos 3\varphi) + D$. Parameters for this potential function which is different for the α (A = 1.58, B = -0.74, C = -0.70, and D = 1.72) and the β anomer (A = 2.61, B = -1.21, C = -1.18, and D = 2.86) are obtained from ab initio calculations of dimethoxymethane [59]. The conformations of the pyranose rings are kept at their fixed geometry as established by X-ray or neutron diffraction studies. This approach was justified by a large number of NMR spectroscopic studies which demonstrated that there is no change in the conformation of the pyranose rings. Exceptions to this rule are known only for the conformationally labile iduronic acid (cf. below). This simple procedure yields very reliable minimum energy conformations when compared to experimental data in solution or in the crystalline state [8]. The mapping of the disaccharide bonds allows an estimation of the conformational flexibility of the glacosidic bond.

An extension to this procedure for the calculation of simpler oligosaccharides was made by the introduction of the GESA program, which uses the same potential energy functions but allows the simultaneous relaxation of all relevant parameters at the same time [60]. This feature is of great importance in the calculation of linear oligosaccharides with 1–6 linkages and specially branched oligosaccharides.

2.5.4 Molecular Dynamics Calculations

A completely different method for assessing the favoured conformation is the molecular dynamics calculations [61, 62]. Until now only a few molecules have been calculated using this approach [63, 64, 65, 66] (cf. below). Recently a modified set of potential energy function based on the CHARMM force field has been published [67]. The usual problem in the optimization procedure utilized by all the programs mentioned above is the uncertainty whether the global minimum has been found yet or whether the located energetically favored conformations are just local minima with an energy above the global one. This particular problem is tackled in a different way by the molecular dynamic calculations. Using one of the potential functions as described above in the force field section, the minimum energy of a molecule is determined by simulating the thermal motion of the molecule in very short intervals in the femtosecond range. This requires that Newton's equation of motion be solved numerically. At the beginning of the calculation, all atoms are assigned random velocities and directions of motion. The magnitude of the velocities determines the temperature of the system. The next conformation of the molecules is evaluated by considering both the current motion of the atoms and the forces from the force field by which the atoms are held together. The trajectories of the molecular parameters are then followed over a certain period of time (usually several picoseconds). This allows the dynamics of molecular motion to take place in the computer. Using this procedure even higher energy varriers can be passed if a sufficiently high (artificial) temperature is selected at the beginning of the process. Thus, the general problem of finding the global minimum is attacked by following the conformational motion of the molecule over a period of time under the assumption that the minimum energy conformer will be adopted by the molecule due to Boltzmann's law. Molecular dynamics calculations have produced outstanding results for proteins where a huge number of minima prohibits standard forve field calculations to find global minima. However, in oligosaccharides we find a fairly different situation with many fewer individual minima and many fewer degrees of freedom.

3 Calculation of Small Fragments Related to Oligosaccharides

Jeffrey et al. have calculated the energy of several conformations of dimethoxymethan as these are representative examples for the glycosidic bonds in α- or β-anomers (cf. Fig. 3) [59]. The calculations were carried out on the RHF/4-31G level. The obtained potential energies for the conformations of $0°$, $90°$, $180°$, and $270°$ dihedral angles respectively revealed the presence of a strong *exo* anomeric effect in acetal moieties. Calculations on the most advanced 6-31G* level for dihydroxymethane [59] revealed a principal confirmation of the former calculations with respect to the energetical order of the staggered conformations found at the 4-31G level [68] but with decreased energy differences between the individual staggered conformations.

Calculations of the conformations of 1-*O*-acylacetalfragments — methoxymethyl-

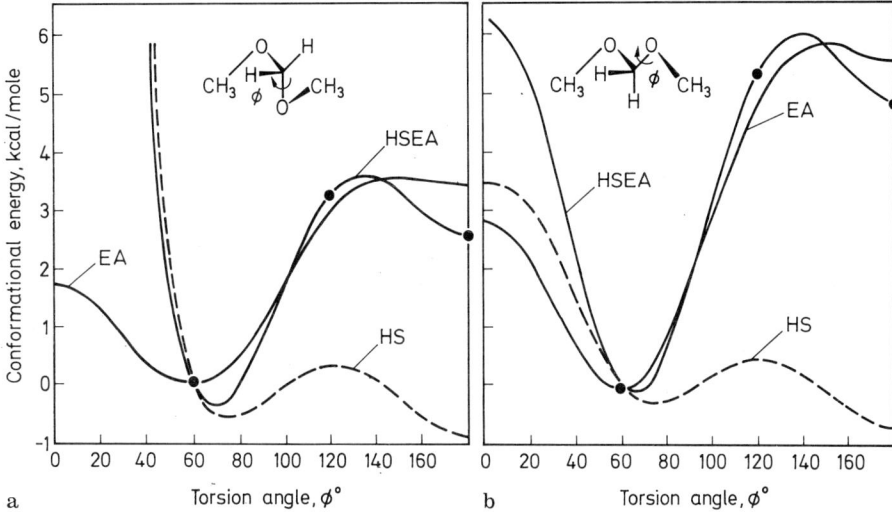

Fig. 3. Variations of the dihedral angles in dimethoxymethane to model the α- or β-anomeric center of an oligosaccharide

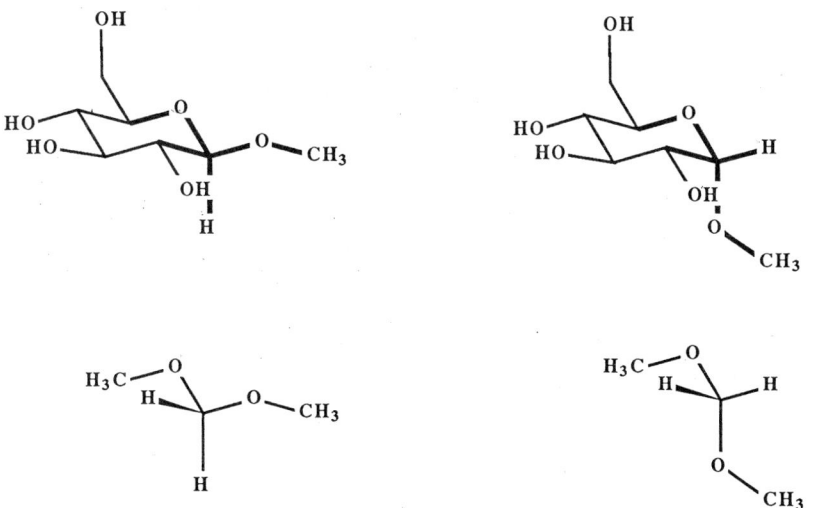

Fig. 4. Modelling of the torsion potential of the exo anomeric effect using the results of the ab initio calculation of dimethoxymethane. The difference between the non bonded terms and the ab initio calculation is fitted with the general function of torsion potentials to yield the equations for the α- and the β-anomers discussed in Sect. 2.5.3 for the HSEA method

formate and methoxymethylacetate — on the RHF/4-31G level (cf. Fig. 4) showed the presence of two minima with respect to the φ-angle (H1—C—O—C) of 90° and 180° respectively [69]. The conformations at φ = 180° are lower in energy than those with φ = 90°. The values of the φ-angle in X-ray crystal structures of α- and β-glycosidic acylated structures range from 70° to 120° with an average of the absolute value of φ = 90° for both the α- and the β-anomer. The deviation in the experimental value and theoretical prediction of the φ-angle is explained from the differences between a real pyranose ring and the models used. Pyranoses have the C2 of the ring at the position where the model compound only has a hydrogen atom.

Using the same level of approximation calculations on 1-methoxyethanol and 1,1-ethanediol [70] showed that the conformations following the *exo* anomeric effect are those with the lowest energy and thus supported the explanation given above for the discrepancies resulting from the lack of a C-2 atom in the calculation of the 1-O-acyl fragments.

It has recently been pointed out that the bond angle X—C—Y in the X-ray structures over wide range of heterocyclic acetals is larger than the tetrahedral angle if the exocyclic substituent X is axial and thus obeys the anomeric effect whereas the bond angle is smaller than the tetrahedral angle if the *exo* cyclic substituent is equatorial [71]. Glycosides with simple alcohols show values of 112° for α-anomers and 108° for β-anomers; in glycosides the angles are 113.5° for α-glycosides and 111.8° for β-glycosides. These changes are in accordance with the interactions of MOs which are used to explain the anomeric effect. Ab initio calculations on the STO 4-31G vasis set are used to show that the effects can be quantitatively reproduced.

The *exo* anomeric effect was described as being more dependent on the solvent for a β-anomer than for an α-anomer [72]. A solvent which can donate hydrogen bonds to the ring oxygen like water will strengthen the *exo* anomeric effect.

4 Conformational Aspects of Mono- and Disaccharides

4.1 Monosaccharides

In many force fields, the conformation of the pyranose ring is not varied. Thus, the ring structures of the monomers are not included in the optimization procedure. Some force field programs and the semi empirical programs optimize the ring geometry in addition to the interglycosidic bonds. Some of these data will be referred to in the following section.

The conformation of D-Glucose was calculated by Rao et al. using a potential energy procedure [73]. The pseudo rotation energy map of the ring was calculated to find the local and global minima for α-D-glucose and β-D-glucose respectively. The authors conclude that the glucose may undergo changes in its ring conformation with a rotation of up to 10° in the dihedral angles but with almost no change in energy. The $^1C_4$-conformation is calculated to have a higher energy content of approximately 3 kcal/mol more than the $^4C_1$-conformation. The anomeric ratio of many aldoses along with the preferred ring conformation were calculated [74]. The agreement with experimental data is generally good except for D-galactose where the calculation gives

$\alpha/\beta = 43/57$ and the experimental value in water is 27/73. Differences caused by the solvation of glucose are not considered in the calculation. Therefore, the result must be due to a parametrisation, which mimics the solvation effect by water quite accurately. Rasmussen [56, 75] calculated α-D-glucose and β-D-glucose with the PEF series of force fields. The anomeric ratio of D-glucose could be predicted within experimental error.

The semi empirical quantum chemical calculations of N-acetyl-β-D-glucosamine and N-acetyl-β-D-muramic acid [76, 77] have been used to determine the orientation of the side group. The results from the empirical potential energy calculations [78] are compared to the CNDO, PCILO, and MNDO results. It turned out that the different methods for calculating the preferred conformations gave quite different relative energies for the conformers. Generally, the PCILO energy agrees better with the empirical results than the CNDO or MNDO energies do. The conformation of the N-acetyl group is in accordance with the experimentally obtained conformation. The 6-hydroxymethylgroup is calculated to adopt only the *tg* orientation in the preferred conformations that are not present in solution within experimental error as determined by NMR spectroscopy [29]. PCILO calculations suggested that the N-acetyl group is important to the conformation of a neighboring β-(1–40 bond [79].

The MD (molecular dynamics) simulation of D-glucose using the parametrization of the PEF422 force field was made [63, 64]. The dihedral angles describing the ring conformation are in good agreement with the experimental values. On the other hand, the orientation of the 6-hydroxymethyl group is not properly represented by the force field. It is calculated to be mainly in the *gt* conformation, never in the *gg* and approximately 25 % in the *tg* conformation. This phenomenon is thought to be so because the hydrogen bond between OH-6 and O-4 is formed in vacuo more easily than in water. However, this does not account for the lack of the *gg* conformer.

4.2 Disaccharides

This section will present only a short selection from the numerous reports on the conformation of disaccharides. Maltose, for example, has attracted a lot of interest with numerous papers dealing with its conformational properties of which only a few have been selected.

An ab initio calculation with a minimal basis set of both α- and β-glucopyranose as well as β-maltose results in the correct prediction of the energetical order of the two glucose anomers as well as the two conformations of β-maltose [80]. But the less favored conformer contains an energy which is higher by a factor of approximately 6 compared to the results of the FF300 calculation (cf. below).

Evaluation of the coupling constants $^3J_{C4, H1'}$ and $^3J_{C1', H4}$ in dependence of the solvent was used to demonstrate changes in the conformation of the glycosidic bond in going from polar solvents (water) to unpolar solvents (dioxane) [81]. In order to be able to determine the coupling constants from the ^{13}C-NMR spectra methyl-β-maltoside was C-deuterated at positions 2, 3, 6, 2', 3', 4', and 6'. The coupling constants were measured in D_2O, DMSO-d_6, and dioxane-d_8 and changed from $^3J_{C4, H1'} = 4.0$ Hz (water) to 4.3 Hz (dioxane) and $^3J_{C1', H4} = 4.5$ Hz (water) to 5.3 Hz (dioxane) respectively. These changes were rationalized using a solvation model for

153

maltose [82] with four different conformations. The conformations used are characterized by their φ/ψ values of $-10°/5°$, $-20°/-15°$, $-40°/-30°$, and $-30°/160°$. The changes in the population of these four conformers were predicted to be up to 15% when the solvent is changed from water to dioxane. The experiment supported this calculation.

Rasmussen et al. [83] have described the calculation of β-D-maltose using the PEF300. They obtained four minima within 5 kcal/mol characterized by the angles φ/ψ of $-21°/-24°$ (0.4 kcal/mol), $17°/19°$ (1.0 kcal/mol), $-66°/-43°$ (1.5 kcal/mol), and $-29°/-168°$ (2.0 kcal/mol). The energetically most favored and that with the highest energy correlate quite well with the calculation by Tvaroska [82] mentioned above.

Lipkind et al. present their calculations of the conformational motion of maltose [84, 85] based on the assumption that only non bonded interactions and a torsion potential around the glycosidic bond are necessary to describe the energy of a glycoside. They conclude that for maltose there are four conformers with φ/ψ $-70°/-35°$ (60%), $-20°/-10°$ (27%), $30°/25°$ (10%), and $-30°/-160°$ (3%). The authors point out that $^3J_{C,H}$ coupling constants are better described by this four state approach than by a single conformation.

These findings are in contrast to the X-ray structures obtained so far from maltose derivatives which all show dihedral angles close to $-10°/-10°$. The complex of maltoheptaose with phosphorylase A also shows these particular conformations of the glycosidic bonds to be $\varphi = -15°$ and $\psi = -15°$ at the ends and $\varphi = -40°$, $\psi = -3°$ or $\varphi = -15°$, $\psi = -35°$ in the middle of the chain [42].

Methyl-β-D-lactoside was studied for its conformational preferences [86]. ^{13}C-enriched lactosides were synthesized enzymatically and then analyzed by ^{13}C- and ^{1}H-NMR methods. From coupling constants $^3J_{C4,H1'} = 3.8$ Hz and $^3J_{C2,C4'} = 3.1$ Hz the angle φ is determined to approximately $40°$ and from $^3J_{C1,H4'} = 4.9$ Hz, $^3J_{C1,C3'} = 0$ Hz, and $^3J_{C1,C5'} = 1.6$ Hz the angle ψ is determined to approximately $15°$.

Cellobiose is also in the focus of interest with respect to its conformational features. Rasmussen et al. [87] used the PEF300 force field program and they calculated for β-D-cellobiose five conformers at the glycosidic bond within 3 kcal/mol, namely

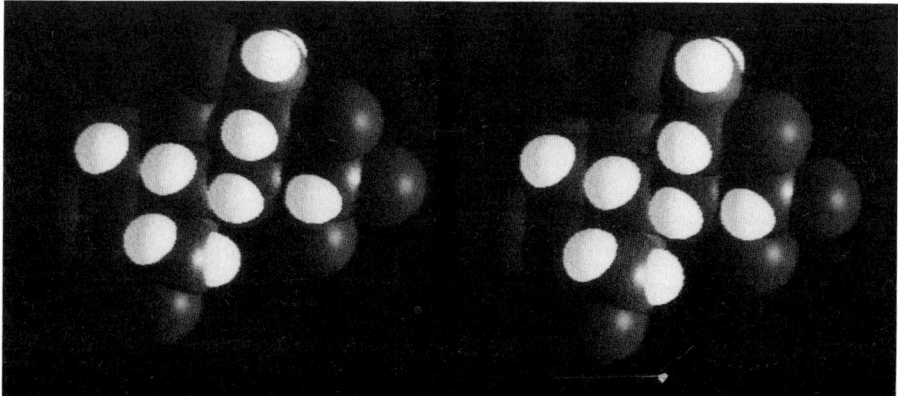

Fig. 5. Conformation of cellobiose from X-ray crystal structure analysis as a space filling stereo plot. The reducing end is to the right

$\phi/\psi = 51°/0°$ (0.4 kcal/mol), $-10°/-29°$ (0.9 kcal/mol), $164°/5°$ (1.3 kcal/mol), $21°/172°$ (1.9 kcal/mol), and $67°/-157°$ (2.8 kcal/mol).

Tvaroska showed in a model calculation of the solvation of these conformations of β-cellobiose that in water the conformer with $\phi = 51°$ and $\psi = 0°$ is the most preferred one [88]. This one is close to the X-ray crystal structure conformation. The calculated populations of the five conformers change quite a bit in going from a polar solvent (water) to an unpolar one (dioxane).

An extensive NMR study of various 1,4-linked disaccharides with D and L-mono-saccharides mixed in disaccharides has been performed [89]. Two different groups of disaccharides were defined, one containing β-DL-, β-LD-, and α-DD-disaccharides and the other containing α-DL-, α-LD-, and β-DD-disaccharides. Each group of di-saccharides exhibits specific changes in chemical shifts due to glycosidation of the hydroxygroup-4 which are different between the groups. The changes in chemical shifts were correlated to the preferred conformation of the disaccharides obtained by HSEA calculations. The results were used to predict the ^{1}H-NMR and ^{13}C-NMR chemical shifts of 1,4-linked polysaccharides.

Lipkind et al. [85, 90] used their above mentioned procedure to calculate the con-formations of methyl-β-cellobioside conformations with ϕ/ψ of $30°/-40°$ (39%), $55°/20°$ (38%), $-20°/-25°$ (10%), and $30°/175°$ (13%). They support their cal-culations with the $^3J_{C,H}$ coupling constants across the glycosidic bond. Although these authors use a torsional potential around the glycosidic C1-O1 bond, they as-sume that the *exo* anomeric effect does not exist. This was later criticized by Lemieux et al. [72] (cf. below).

The proposal of a conformation at the glacosidic bond with an upside down orient-ation of the two monomeric constituents has its support from experimental data of completely different origin. Thiem et al. were able to show that the kinetic isopropyli-denation of cellobiose in refluxing pyridine yields an isomer in low yield which has the glycosidic bond fixed via an isopropylidene bridge across the rings from OS' to O6. This implies an upside down orientation of both rings compared to its normal conformation [91, 92].

Cellobiose octaacetate and 1,6-Anhydro-β-cellobiose hexaacetate are compared with respect to their glycosidic conformation [93, 94]. For cellobiose octaacetate it was concluded that the conformation in solution is close to that one determined by X-ray crystal structure analysis to $\phi = 45°$ and $\psi = 16°$ (Fig. 5) whereas the 1,6-anhydro derivative is demonstrated by use of NOEs, relaxation data, and coupling constants $^3J_{C,H}$ to adopt torsional angles of $\phi = 25°$ and $\psi = 45°$ respectively.

The conformation of sucrose and the hydrogen bonds in aqueous solution are analyzed via HSEA calculations and NMR spectroscopic studies [95, 96]. It was found experimentally that the C1-hydroxymethyl group of the fructose shows a strongly hindered rotation due to a hydrogen bond between $O1_{frc}$ and $O2_{glc}$. These observ-ations are in full agreement with the conformation calculated by the HSEA approach. The direct experimental proof for the hydrogen bond was obtained by the isotopic shifts using a D_2O/H_2O solvent mixture. The dihedral angles for the calculated minimum energy conformer are $\phi = -20°$ and $\psi^{C1-frc} = 80°$ (Fig. 6). This study was later confirmed by a determination of the ^{13}C-NMR relaxation rates [97].

Gentiobiose octaacetate was analyzed by NMR spectroscopy and HSEA cal-culations [98]. The calculations were performed with the conformations of the hydro-

Bernd Meyer

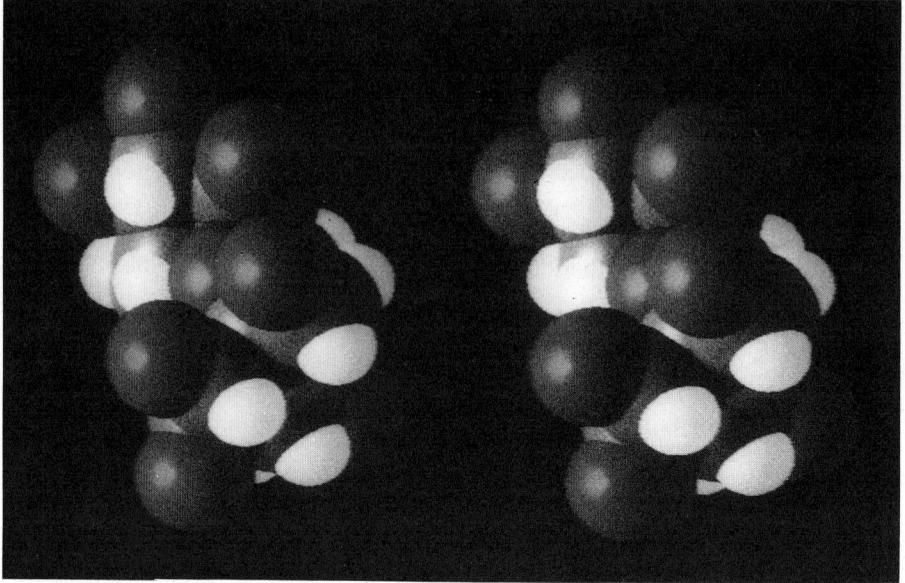

Fig. 6. Conformation of sucrose as obtained by HSEA minimization of the conformational energy in a space filling stereo plot. The β-D-fructosyl residue is to the front

xymethyl group of the reducing glucose fixed to $-60°$ (gg) and $60°$ (gt) respectively. The values for the $\varphi = 55°$ and $\psi = 190°$ were the same for both conformations of the ω-angle (cf. Fig. 7), the X-ray data are $\varphi = 61°$, $\psi = 205°$, and $\omega = -60°$ [99]. These conformations were shown to be consistent with a number of ^{1}H-NMR parameters. A previous force field calculation had yielded the dihedral angles $\varphi = 64u$, $\psi = 120°$, and $\varphi = -60°$ where especially the ψ-angle deviates from the X-ray structure [100].

Ohrui et al. [101, 102] greatly facilitated the accessibility of stereospecifically C6-deuterated hexoses with a new short and efficient synthesis. These compounds can be utilized to study the conformer equilibrium at the C5–C6 bond with high accuracy, because of the unambiguous assignment of the coupling constants to the protons pro-S H6 and pro-R H6 and because of the smaller loss of magnetization from the enhanced proton H6 to the other in an NOE experiment (cf. below).

Galactobiose, β-D-Gal-(1–6)-β-D-Gal, was shown by PEF300 and PEF400 calculations to be very flexible around the glycosidic bond [103]. The differences between the PEF300 and PEF400 force field programs led to quite different sets of populations of the individual conformers. In the PEF300 calculation there is an approximately equal distribution of the 13 conformers whereas the PEF400 calculation yields only four significant populated conformers. Furthermore the highest populated conformer in the PEF400 violates the *exo* anomeric effect and has the 6-hydroxylmethylgroup of the reducing galactose in the unfavored gg conformation, it is described by $\varphi = 159°$, $\psi = -115°$, and $\omega = -53°$. This one is chosen to model the interaction between the immunoglobulin AJ539 and the polymer galactan [104].

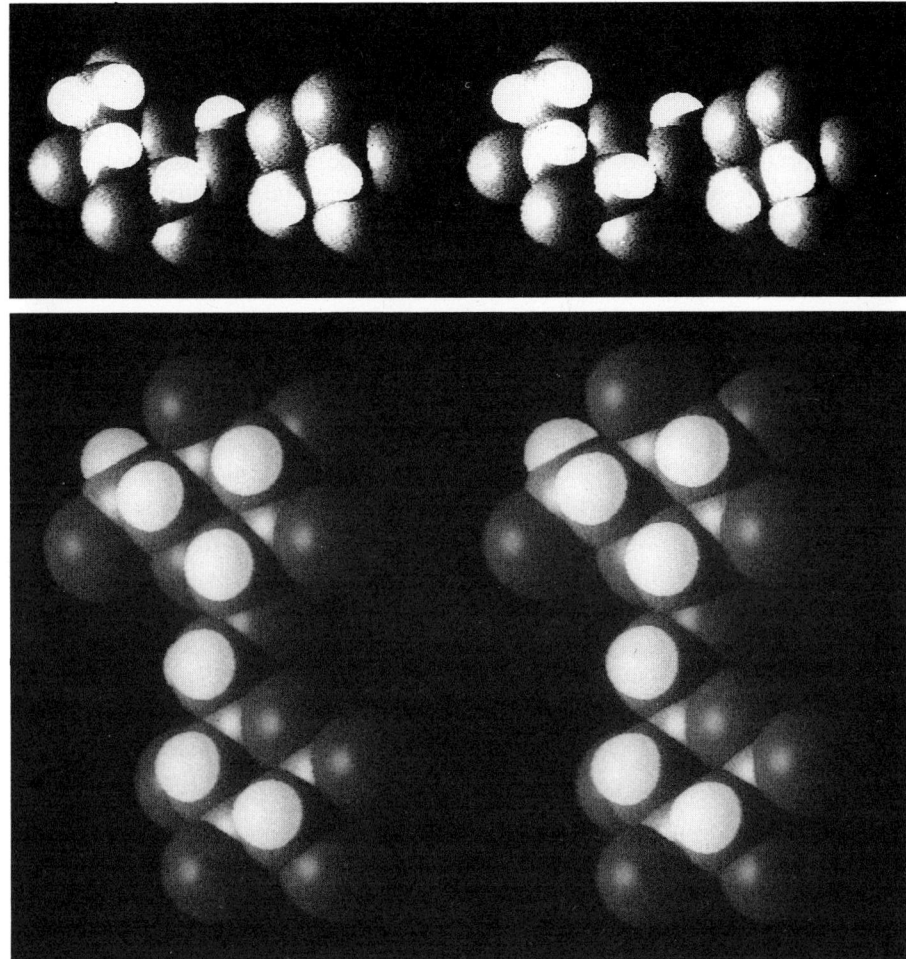

Fig. 7a, b. Conformations of gentiobiose, β-D-Glc-(1–6)-D-Glc, from HSEA optimization of the conformational energy with the ω-angle at the reducing glucose fixed to **a** the *gg* (−60°) and **b** the *gt* (60°) conformation respectively as space filling stereo plots. The reducing end is to the right

In a detailed analysis of the ^{13}C-NMR chemical shifts of several 1-1'-glycosides Pavia et al [105] have demonstrated the importance of the *exo* anomeric effect to rule also the conformations of these non reducing disaccharides.

Various analogs of disaccharides have been synthesized where the glycosidic oxygen is replaced by a CH_2-group [106, 107, 108, 109]. Analogs of cellobiose, gentiobiose, isomaltose, and simple methyl glycosides were studied. The analysis of their conformation in solution was performed by ^{1}H-NMR spectroscopy. Due to the CH_2-group instead of an oxygen the preferred conformation can easily be studied by analysis of the coupling constants across the pseudo glycosidic bond. It was concluded that the conformation of the C-glycosides follows the conformation of the normal disaccharides — even for the 1,6 linked pseudo disaccharides which show an extended

157

zigzag conformation along the interglycosidic bond. The importance of 1,3 diaxial interactions in adopting preferred conformations was emphasized [106]. The generalization, however, that stereoelectronic effects are not the "major effects ruling the conformation of glycosides" seems to be too straight forward: 1) The additional hydrogen atoms at the CH_2-group are strongly favoring the staggered conformation. 2) In simple disaccharides, the *exo* anomeric effect does not change the conformation which would be adopted, due to only steric interactions and, the effect narrows only the potential well and becomes more important in case of complex oligosaccharides where steric factors are less favorable in the conformation which follows the *exo* anomeric effect [110].

5 Blood Group Antigens

Blood group antigens are the first compounds studied with respect to their conformation both by NMR and computational methods [23, 24]. In the meantime a variety of studies have appeared on the preferred conformations of these antigens.

α-L-Fuc-1–2)-β-D-Gal-(1–3)-β-D-GlcNAc-(1–3)-β-D-Gal

H-determinant type I

β-D-Gal-(1–3)-β-D-GlcNAc-(1–3)-β-D-Gal
$\qquad\qquad\qquad$ 4
$\qquad\qquad\qquad$ |
$\qquad\qquad\qquad$ α-L-Fuc

Le^a-determinant type I

α-L-Fuc-(1–2)-β-D-Gal-(1–3)-β-D-GlcNAc-(1–3)-β-D-Gal
$\qquad\qquad\qquad\qquad\qquad\qquad$ 4
$\qquad\qquad\qquad\qquad\qquad\qquad$ |
$\qquad\qquad\qquad\qquad\qquad\qquad$ α-L-Fuc

Le^b-determinant type I

α-L-Fuc-(1–2)-β-D-Gal-(1–3)-β-D-GlcNAc-(1–3)-β-D-Gal
$\qquad\qquad\qquad\qquad\quad$ 3 $\qquad\qquad\qquad$ 4
$\qquad\qquad\qquad\qquad\quad$ | $\qquad\qquad\qquad$ |
$\qquad\qquad\qquad\qquad\quad$ α-D-GalNAc $\qquad$ α-L-Fuc

A-determinant type I

α-L-Fuc-(1–2)-β-D-Gal-(1–3)-β-D-GlcNAc-(1–3)-β-D-Gal
$\qquad\qquad\qquad\qquad\quad$ 3 $\qquad\qquad\qquad$ 4
$\qquad\qquad\qquad\qquad\quad$ | $\qquad\qquad\qquad$ |
$\qquad\qquad\qquad\qquad\quad$ α-D-Gal $\qquad\quad$ α-L-Fuc

B-determinant type I

α-L-Fuc-(1–2)-β-D-Gal-(1–4)-β-D-GlcNAc-(1–6)-β-D-Gal

H-determinant type II

β-D-Gal-(1–4)-β-D-GlcNAc-(1–6)-β-D-Gal
$\qquad\qquad\qquad\qquad\quad$ 3
$\qquad\qquad\qquad\qquad\quad$ |
$\qquad\qquad\qquad\qquad\quad$ α-L-Fuc

Le^a-determinant type II
X-determinant (without the reducing β-D-Gal)

α-L-Fuc-(1–2)-β-D-Gal-(1–4)-β-D-GlcNAc-(1–6)-β-D-Gal
$$\begin{array}{c} 3 \\ \uparrow \\ \text{α-L-Fuc} \end{array}$$

Le^b-determinant type II
Y-determinant (without the reducing β-D-Gal)

α-L-Fuc-(1–2)-β-D-Gal-(1–4)-β-D-GlcNAc-(1–6)-β-D-Gal
$$\begin{array}{cc} 3 & 3 \\ \uparrow & \uparrow \\ \text{α-D-GalNAc} & \text{α-L-Fuc} \end{array}$$

A-determinant type II

α-L-Fuc-(1–2)-β-D-Gal-(1–4)-β-D-GlcNAc-(1–6)-β-D-Gal
$$\begin{array}{cc} 3 & 3 \\ \uparrow & \uparrow \\ \text{α-D-Gal} & \text{α-L-Fuc} \end{array}$$

B-determinant type II

Lemieux et al. [23, 24] have demonstrated the powerful capabilities of the conformational analysis of blood group determinants by HSEA calculations in combination with NMR analyses. The terminal trisaccharide of the B-antigen α-L-Fuc-(1–2)-(αD-Gal-3)-β-D-Gal was shown by NMR spectroscopy to adopt a conformation which is the same as obtained from the HSEA calculation. From the ^{1}H-NMR NOEs, ^{1}H-NMR relaxation times, and an interpretation of the differential chemical shifts it was concluded that the *exo* anomeric effect is a very important factor for the prediction of the conformation of oligosaccharides. Negation of this factor would lead to a shift in the conformer equilibrium towards smaller φ angles. The φ and ψ angles describing the B-trisaccharide are α-L-Fuc-(1–2)-β-D-Gal 55°/20° and α-D-Gal-(1–2)-β-D-Gal —65°/—50° for the conformer with an energy of —3.16 kcal/mol (cf. Fig. 8). A second conformation with an energy contents of —2.02 kcal/mol was found to be present which differed in the conformation of the α-D-Gal-(1–3)-β-D-Gal linkage

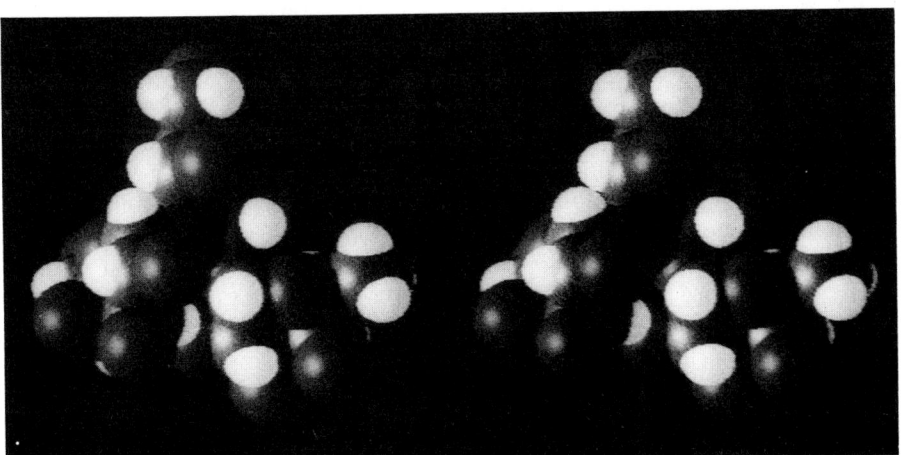

Fig. 8. Conformations of the terminal B-trisaccharide of blood group antigens, α-L-Fuc(1–2)-[α-D-Gal-(1–3)-]-β-D-Gal from HSEA optimization of the conformational energy as space filling stereo plots. The reducing end is to the right

Table 1. Calculated dihedral angles φ and ψ using the HSEA approach [23]

Type I:

	α-D-Gal-3	α-L-Fuc-2	β-D-Gal-3	α-L-Fuc-4	β-D-GlcNAc-3
Le[a]			55°/10°	55°/25°	
H			50°/10°	60°/10°	
Le[b]		45°/15°	55°/10°	55°/25°	
B	−65°/−50°	55°/20°	55°/10°	55°/25°	60°/−10°

Type II:

	α-D-Gal-3	α-L-Fuc-2	β-D-Gal-4	α-L-Fuc-3	β-D-GlcNAc-6
X			55°/10°	55°/25°	
H			55°/15°	55°/0°	
Y		50°/10°	55°/15°	50°/25°	
B	−65°/−55°	50°/15°	55°/10°	50°/25°	50°/130°

with $\varphi = -30°$ and $\psi = -45°$. The HSEA calculation of the blood group related antigens is summarized in Table 1.

The problem of the flexibility of the 1–6 glycosidic link in the type II oligosaccharides was approached by the synthesis of the β-D-Gal-(1–4)-β-D-GlcNAc-(1–6)-(6-C-CH3)-β-D-Gal trisaccharides with R and S configuration at C-6 of the reducing galactose, respectively [110]. As mentioned above the C5–C6 bond is the primary source for the flexibility of 1–6 linked glycosides. In the case of the β-D-GlcNAc-(1–6)-β-D-Gal disaccharide there is a large variety of conformations within 0.5 kcal/mol for all possible staggered conformations around the ω-angle. It could be shown that by stabilizing the *tg*-conformer through the additional C-7 methyl group, the D-epimer proved to be a good inhibitor of the anti-I Ma monoclonal antibody whereas the L-epimer is not. In fact, the D-epimer is a better substrate than the native oligosaccharide with D-galactose [111].

Rao et al [112, 113] have used a Scheraga potential [114] modified with electrostatic contributions [115]. The dihedral angles defining the minimum energy conformations are listed in Table 2. It was shown that different antigens of the same class but different type adopt a very similar overall shape and differ mainly in the orientation of the central β-D-GlcNAc residue. This feature is used to explain the cross reactions of the type I and type II antigens in certain sera. Generally the type II antigens have a much higher degree of conformational freedom as is evident from the high number of minima. Predominantly these minima are due to the flexible 1–6 bond between the β-GlcNAc-(1–6)-β-Gal. For many glycosidic bonds an agreement of the conformations obtained in this work with those from the experimentally verified HSEA calculation named is found. In a few cases, steric crowding leads to conformations with φ angles which contradict the exo anomeric effect. Furthermore, the conformations of the α-L-fucosyl residues are calculated to have different angles.

The solution conformation of the trisaccharide from the non reducing end of type II antigens was determined using ^{13}C-NMR coupling constants of ^{13}C-enriched synthetic oligomers [116, 117]. Using $^{3}J_{C,C}$ and $^{3}J_{C,H}$ coupling constants the conformations of the β-D-Gal-(1–4)-β-D-GlcNAc glycosidic linkage to $\varphi = 60°$ and $\psi = 15°$ and the α-L-Fuc-(1–2)-β-D-Gal glycosidic linkage to $\varphi = 55°$ and $\psi = 0°$.

Table 2. Calculated conformation of the blood group antigens using the modified Scheraga potential [97, 98]

Type I:

	α-D-Gal-3 or α-D-GalNAc-3	α-L-Fuc-2	β-D-Gal-3	α-L-Fuc-4	β-D-GlcNAc-3	
H		40°/24°	59°/9°		54°/15°	0.0
Le^a			59°/15°	32°/22°	55°/15°	0.0
			64°/11°	−20°/−29°	55°/15°	0.3
Le^b		35°/24°	60°/12°	−22°/−30°	55°/15°	0.0
A	−65°/−57°	40°/25°	59°/9°		54°/15°	0.0
B	−65°/−57°	40°/25°	59°/9°		54°/15°	0.0
	28°/8°	40°/25°	59°/10°		54°/15°	0.5
H		39°/24°	57°/−1°		51°/−65°/−59°	0.0
		40°/24°	57°/−1°		56°/−50°/−174°	0.6
Le^a			59°/11°	−22°/−26°	51°/−65°/−60°	0.0
			56°/12°	−22°/−27°	56°/−50°/180°	0.6
Le^b		40°/23°	61°/11°	−23°/−26°	51°/−65°/−60°	0.0
		40°/23°	61°/11°	−23°/−27°	56°/−50°/180°	0.6
A	−64°/−57°	40°/25°	57°/−2°		51°/−64°/−59°	0.0
	−65°/−56°	40°/25°	57°/−1°		56°/−51°/−175°	0.6
B	−65°/−56°	40°/25°	57°/−1°		52°/−64°/−60°	0.0
	29°/8°	40°/25°	57°/−1°		51°/−64°/−61°	0.4
	−65°/−57°	40°/25°	56°/−1°		57°/−49°/−176°	0.7

These experimental data support the values obtained from the calculated dihedral angles of various sources.

Analysis by ^{1}H-NMR techniques and different calculation methods by Bush et al. [118] showed that the minimum energy conformation obtained by the Hopfinger potential as well as the HSEA potential described the experimental data for the terminal A-blood group tetrasaccharide very well, whereas the modified Scheraga potential did not reproduce the experimental data — NOE and longitudinal relaxation times — adequately. The glycosidic bond of fucose showed calculated and experimental values of $\varphi = 60°$ and $\psi = 10°$ and that of N-acetyl-galactosamine showed $\varphi = -70°$ and $\psi = -40°$.

Using the modified Scheraga potential [119], the conformations of several oligosaccharides related to blood group substances were calculated and the conformations compared with the results from NMR and CD spectra [120]. The NOEs depend strongly on the temperature which is caused by different correlation times (τ_c) of the individual monosaccharide constituents of the oligosaccharides and not due to a change of the preferred conformations with temperature. The calculated dihedral angles for the H-type I and type II di- and trisaccharides are shown in Table 3. Distances calculated from NOE data across the glycosidic bonds were used to verify the calculated structures which are in good agreement with the experimentally determined conformations of the compounds in solution. The data obtained in this study are similar to those proposed earlier [24]. The CD data in conjunction with the $^3J_{(H-N-C-H)}$ coupling constant of the amide group were used to show the rather rigid trans orientation of the N–H bond and the C2–H2 bond. Later Bush et al. [121] confirmed that the previously established [23, 24] conformations do not change when

Table 3. Calculated dihedral φ and ψ angles for the H-type I and II di- and trisaccharides [120]

	Fuc-Gal	Gal-GlcNAc
α-L-Fuc-(1–2)-β-D-Gal	50°/10°	
β-D-Gal(1–3)-β-D-GlcNAc		50°/10°
β-D-Gal-(1–4)-β-D-GlcNAc		70°/10°
α-L-Fuc-(1–2)-β-D-Gal-(1–3)-β-D-GlcNAc	40°/20°	50°/10°
α-L-Fuc-(1–2)-β-D-Gal-(1–4)-β-D-GlcNAc	30°/30°	60°/10°

blood group H and blood group A oligosaccharides are studied by NMR in non aqueous solution.

A large number of synthetically prepared variants of several blood group substanes were analyzed with respect to their minimum energy conformations. The goal of these studies was to determine the specificity of the binding of the antigens to lectins or monoclonal antibodies. It was shown by Lemieux et al. [122, 123, 124, 125, 126, 127] that the correlation between the minimum energy conformation and the binding properties of the antigens can be explained by the relative size and the orientation of hydrophobic and hydrophilic epitopes on the surfaces of these molecules. Only the 6-hydroxymethylgroup orientation has to be changed from the preferred conformation in solution to the bound state to explain the experimental cross reaction data of the chemically modified antigens.

6 *N*-Type Oligosaccharides of Glycoproteins

N-type oligosaccharides form a large family of differently branched oligosaccharides with a common core structure linked to asparagine residues. A representative formula of the bisected tetraantennary complex type structure is given below. Most of the discussion in this section will focus on complex type oligosaccharides. The sugar units of the complex type oligosaccharides will be numbered by italic numerals in this section following the scheme below.

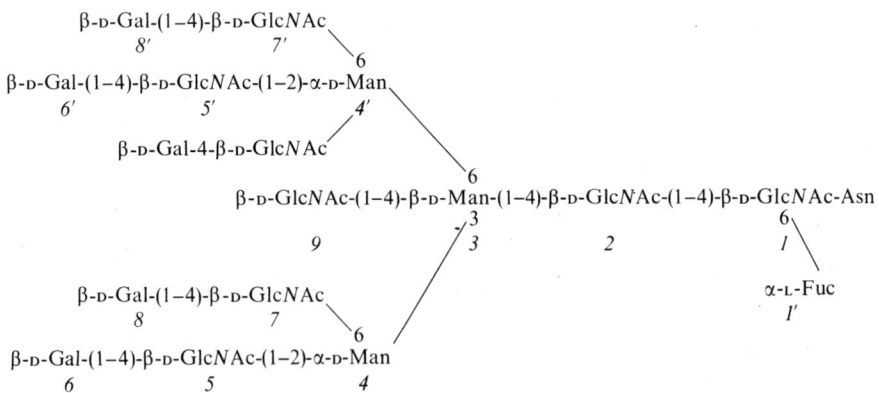

Bisected Pentaantennary *N*-linked Oligosaccharide

The terminal lactosamines of these structures may be present more than once and may carry neuramic acids at the terminal end. This family of structures has attracted a lot of interest with respect to their conformation during the last years because they are major constituents in glycoproteins. An important question is the interaction of these oligosaccharides with the protein part.

A large number of data is available which proves indirectly the conformational influences of substitution patterns at a monosaccharide. The concept of structure reporter groups for the elucidation of the structures of complex oligosaccharides using NMR spectroscopy has been shown to be extremely valuable [128]. An analogous treatment of carbon spectra has been described [129, 130]. The carbon as well as the proton spectra show very specific changes in the chemical shifts which are dependent on the type and pattern of substitution by other glycosidic bonds. This, in turn, can only be indicative if there is a certain preferred conformation of the particular fragment which remains undisturbed by remote changes in the branching pattern.

6.1 The N-Glycosidic Bond

1-N-Acetyl-β-D-glucopyranosylamine was analyzed as a model compound for the N-glycosidic bond [131]. It was shown by X-ray structure analysis and CD spectra that the H1 and the amide proton are *trans* oriented. This is the same orientation as previously described for crystalline β-D-GlcNAc-N-Asn [132, 133]. A force field calculation of the β-D-GlcNAc-N-Asn fragment carrying an asparagine residue with an N-methylamide at the carboxylgroup and a N-acetyl functional group at the amine group [134] resulted in many different conformations of the asparagine residue both with a cis and a trans relationship between the H1 and the amide proton. These cis or trans conformations are not ideal but twisted by approximately 30° from the peri-planar arrangement.

A circular dichroism measurement along with the determination of the coupling constants of the amide proton with the H1 proton of the chitobiose gives evidence that a *trans* orientation is also adopted in larger complex type oligosaccharides linked to asparagine [135]. Because the individual contributions of the different amides are independent from one another in the circular dichroism of these oligosaccharides, it is suggested that no interections between the amide groups takes place and thus an extended conformation is proposed for the molecules.

6.2 Analysis of Linear Model Oligosaccharides

From the analysis of ^{13}C NMR chemical shifts the ψ angle in chitobiose should exhibit a value of approximately 10° more positive than that of cellobiose and thus resulting in $\psi = 0°$ [136].

A trisaccharide fragment of the N-type oligosaccharides (α-D-Man-(1–3)-β-D-Man-(1–4)-β-D-GlcNAc) including the central β-D-Man has been obtained as single crystals. The dihedral angles of the glycosidic bonds which were obtained from an X-ray analysis are summarized in Fig. 9 [137]. This conformation is very closely related to the conformations calculated or determined experimentally in solution by NMR

163

Bernd Meyer

techniques (cf. below). The trisaccharide shows an extended structure with a slight bend at the central β-D-mannose.

$$-58°/-19° \qquad 48°/-1°$$
α-D-Man-(1–3)-β-D-Man-(1–4)-β-D-GlcNAc

Trisaccharide of the core region of the *N*-linked Oligosaccharide

Fig. 9. φ- and ψ-angles obtained in the X-ray structure of the trisaccharide [137]

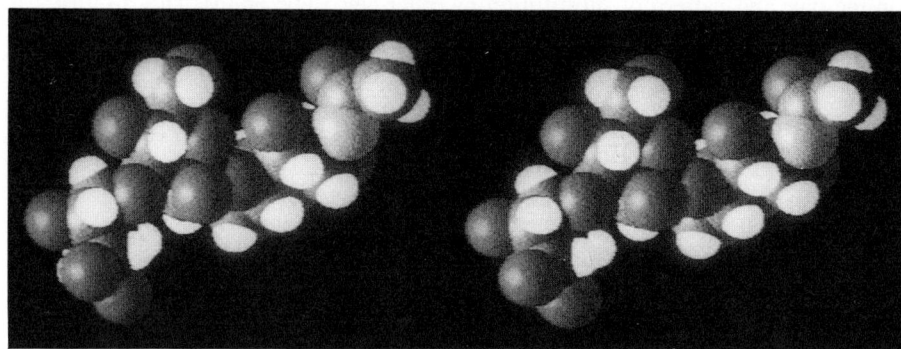

Fig. 10. Conformation of the core trisaccharide of the *N*-type oligosaccharides α-D-Man-(1–3)-β-D-Man-(1–4)-β-D-GlcNAc as obtained from the X-ray crystal analysis as space filling stereo plots. The reducing end is to the right

Table 4. Dihedral angles φ/ψ of the 3-branched trisaccharide, α-D-Man-(1–3)-β-D-Man-(1–4)-β-D-GlcNAc, and pentasaccharide, β-D-Gal-(1–4)-β-D-GlcNAc-(1–2)-α-D-Man-(1–3)-β-D-Man-(1–4)-β-D-GlcNAc, from GESA calculations and NMR spectroscopy

Glycosidic Bond	φ/ψ	φ/ψ
β-D-Gal-(1–4)-β-D-GlcNAc		55/2
β-D-GlcNAc-(1–2)-α-D-Man		53/22
α-D-Man-(1–3)-β-D-Man	−48/−13	−47/−13
β-D-Man-(1–4)-β-D-GlcNAc	56/1	56/1

GESA calculations and NMR studies of synthetic tri- and pentasaccharides show a rather rigid conformation of this part of the larger *N*-type oligosaccharides [30, 138] (cf. Table 4). The dihedral angles of the tri- and pentasaccharide of the 1–3 branch imply a stretched out conformation of this part of the *N*-type oligosaccharides. Except for these global minima, there are no other local minima found which indicates a fairly rigid conformation of the tri and the pentasaccharide (cf. Table 4). Further support comes from the fact that attachment of the 1–6 branch to the central β-D-mannose does not change the conformation of the α-(1–3) bond [138] (cf. Table 6).

164

Table 5. Dihedral angles φ/ψ or φ/ψ and ω of the trisaccharide, α-D-Man-(1–6)-β-D-Man-(1–4)-β-D-Glc*N*Ac, and pentasaccharide, β-D-Gal-(1–4)-β-D-Glc*N*Ac-(1–2)-α-D-Man-(1–6)-β-D-Man-(1–4)-β-D-Glc*N*Ac, of the 6-branch from GESA calculations [30, 138]

Glycosidic Bond	φ/ψ	φ/ψ	φ/ψ	φ/ψ	φ/ψ
β-D-Gal-(1–4)-β-D-Glc*N*Ac				55/3	55/2
β-D-Glc*N*Ac-(1–2)-α-D-Man				44/—5	55/24
α-D-Man-(1–6)-β-D-Man	−58/177	−51/188	−50/87	−52/79	−53/157
	50	75	37	38	—56
β-D-Man-(1–4)-β-D-Glc*N*Ac	53/9	57/12	59/—2	59/—6	58/6
E (kcal/mol)	0.0	0.47	1.17	0.0	1.51

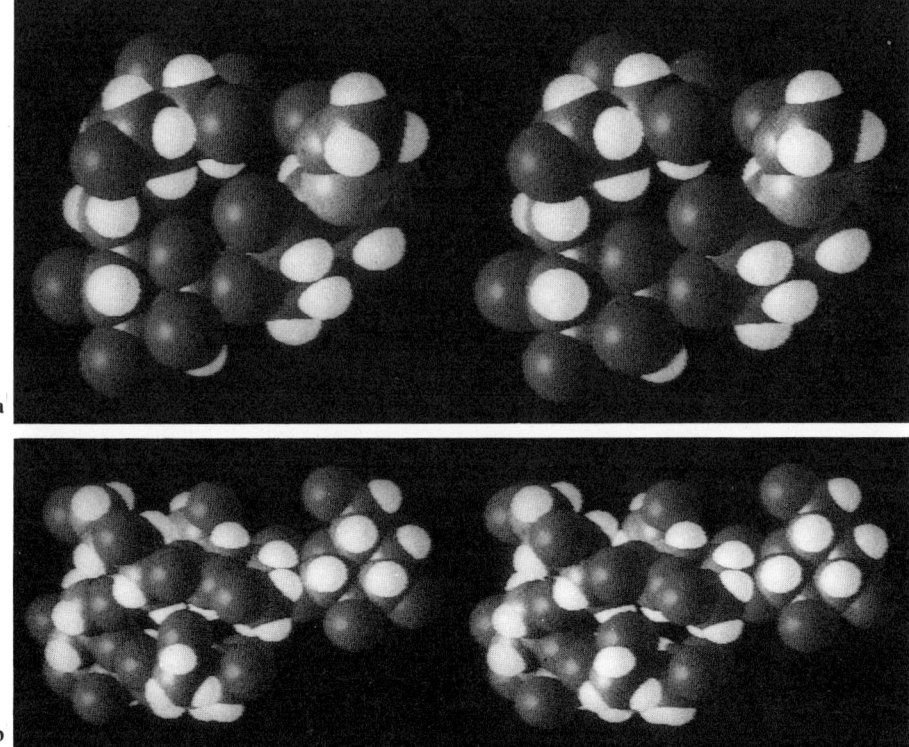

Fig. 11 a, b. Stereo plots of the space filling models of the conformations of the 1,6 linked **a** trisaccharide, α-D-Man-(1–6)-β-D-Man-(1–4)-β-D-Glc*N*Ac, and **b** pentasaccharide, β-D-Gal-(1–4)-β-D-Glc*N*Ac-(1–2)-α-D-Man-(1–6)-β-D-Man-(1–4)-β-D-Glc*N*Ac, of *N*-type oligosaccharides obtained from GESA optimization of the conformational energy. The reducing end is to the right

165

The calculated structures of minimal energy contents could be experimentally verified by extensive NMR analysis [138]. Of special interest is the change of the conformation comparing the trisaccharide to the pentasaccharide. This could be interpreted as an artifact resulting from the calculation in vacuo which would force the non bonded attractions between the branch (6-antenna) and the core to change the conformation to a backfolding of the branch towards the reducing end. But the NMR spectra of the pentasaccharide supported this conformational change [138]. The interactions between the branch and the core are mainly hydrophobic in nature and thus are still active as a conformational factor if these compounds are dissolved in a hydrophilic solvent like water.

The special problem of the conformational analysis — the orientation of the glycosylated 6-hydroxymethylgroups in these oligosaccharides — is of major importance because of the large differences in the global shape that are induced via conformational changes in this part of the molecule.

A detailed inspection of the coupling constants $^3J_{5,6'}$ and $^3J_{5,6}$ of several mannosidic oligosaccharides utilizing phase sensitive COSY spectra has been reported [29]. The population analysis based on these coupling constants shows the preference of the gg conformation over gt while tg is not found within the precision of the analysis.

The NMR analyses of the mannobiosides α-D-Man-(1–3)-α-D-Man-O-Me and α-D-Man-(1–6)-α-D-Man-O-Me from the core region have been reported [139, 140]. The glycosylation of the 6-hydroxymethylgroup of methyl α-D-mannoside causes a change

b

Fig. 12a, b. Stereo plots of the space filling models of the conformations of the 1,6 linked disaccharide α-D-Man-(1–6)-β-D-Man-O-Me proposed by Carver: **a** *gt* conformation and **b** *gg* conformation of the hydroxymethyl group at the β-D-Man residue. The reducing end is to the right

in the coupling constant $^3J_{5,6}$ from 2.2 and 6.0 Hz to 1.8 and 4.5 Hz. This is interpreted in terms of a changed rotamer population. A potential energy calculation with fixed values for the ω-angle of 60° (gt) and −60° (gg) yields φ = −60° and ψ = 150° for both C5—C6 rotamers (Fig. 12) [140].

Using the same approach for the analysis of the conformation of the α-D-Man-(1–3)-β-D-Man bond revealed that the disaccharide α-D-Man-(1–3)-β-D-Man and the trisaccharide β-D-GlcNAc-(1–2)-α-D-Man-(1–3)-β-D-Man adopt the same conformation for this bond with φ = −50° and ψ = −10° [141]. The β-(1–2) bond has calculated values of φ = 40° and ψ = 30° (Fig. 12). These dihedral angles are in good agreement with the observed NOEs of these compounds.

After criticism [142] of the interglycosidic NOEs observed in the previously described study a very precise approach towards the conformational analysis of the α-(1–3) bond was performed by NOE measurements of specifically deuterated compounds [143]. The conformation was determined earlier by several interglycosidic NOEs to protons that have their resonances in an area of high spectral overlap. The synthesis of C-deuterated di- and trisaccharides made an unequivocal assignment of the enhanced signals possible. Thus, the enhancements of H2, H3, and H4 of the β-D-mannose upon irradiation of H1$_{\alpha\text{-man}}$ as well as the enhancement of H5$_{\alpha\text{-man}}$ upon irradiation of HS$_{\beta\text{-man}}$ are indicative of the rigid conformation at the glycosidic bond and confirmed the previous study.

6.3 Biantennary Structures

The biantennary structures exhibit two branches at the central β-D-Man. In the following section several approaches to this particular conformational problem are described. The results are divided into synthetically derived substructures and the studies on biochemically obtained materials.

β-D-Gal-(1–4)-β-D-GlcNAc-(1–2)-α-D-Man-
   6'   5'   4'
                                   6
β-D-Man-(1–4)-β-D-GlcNAc-(1–4)-β-D-GlcNAc-Asn
              3   6
                3   2   1
β-D-Gal-(1–4)-β-D-GlcNAc-(1–2)-α-D-Man-
   6   5   4   α-L-Fuc
                                                                         1'

Biantennary N-linked decasaccharide

Table 6. Dihedral angles φ/ψ or φ/ψ/ω of the tetrasaccharide, α-D-Man-(1–6)-[α-D-Man-(1–3)]-β-D-Man-(1–4)-β-D-GlcNAc, calculated by the GESA program [30]

α-D-Man-6	α-D-Man-3	β-D-Man-4	E (kcal/mol)
−58°/179°/48°	−48°/−13°	57°/7°	0.0
−49°/−168°/75°	−48°/−13°	58°/11°	0.4
−47°/91°/35°	−48°/−13°	58°/−4°	1.8
−52°/171°/−53°	−47°/−13°	59°/−2°	2.0
−52°/128°/−52°	−48°/−12°	60°/−1°	3.0

Bernd Meyer

6.3.1 Synthetic Structures

Results of the conformational analysis of the biantennary tetrasaccharide by NMR and GESA calculations showed no changes for the α-(1–3) and the α-(1–6) bonds in comparison to the corresponding linear trisaccharides, which proves that the two α-mannosidic linkages are independent of each other (Table 6) [30].

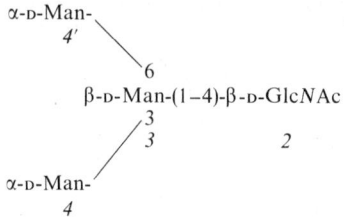

α-D-Man-
4′
⟍
6
β-D-Man-(1–4)-β-D-GlcNAc
3   3   2
╱
α-D-Man-
4

N-linked core tetrasaccharide

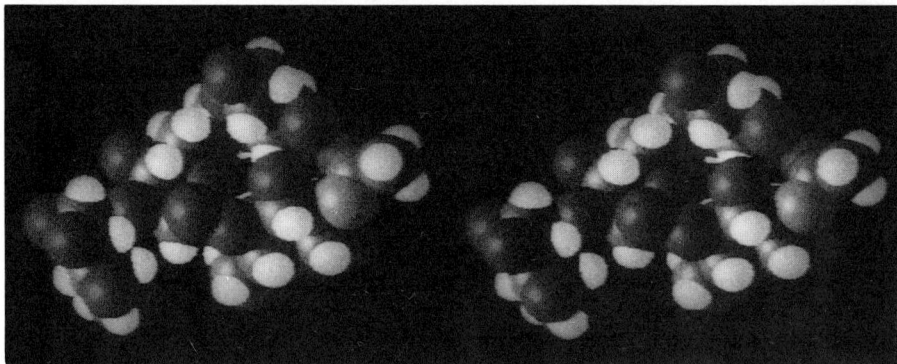

Fig. 13. Stereo plots of the space filling model of the *gt* conformation of the biantennary tetrasaccharide, α-D-Man-(1–6)-[α-D-Man-(1–3)-]-β-D-Man-(1–4)-β-D-GlcNAc, of the *N*-type oligosaccharides with lowest energy contents obtained from GESA opimization. The reducing end is to the right

In an investigation of the conformation of the synthetically obtained heptasaccharide the dihedral angle ω of the β-mannose was fixed at 60° (gt) or —60° (gg) and the other glycosidic linkages were optimized one by one using the HSEA program [144]. Each glycosidic bond was optimized in this procedure as being independent of the other. Supported by NMR chemical shifts and NOEs about 30–40% gg- and 60–70% gt-conformer are assigned (Table 7).

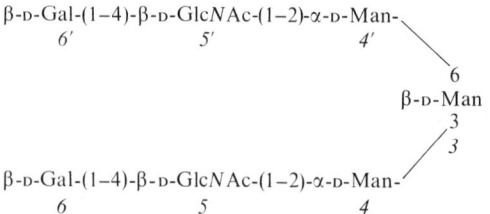

β-D-Gal-(1–4)-β-D-GlcNAc-(1–2)-α-D-Man-
6′   5′   4′
⟍
6
β-D-Man
3
╱3
β-D-Gal-(1–4)-β-D-GlcNAc-(1–2)-α-D-Man-
6   5   4

Biantennary N-linked heptasaccharide

168

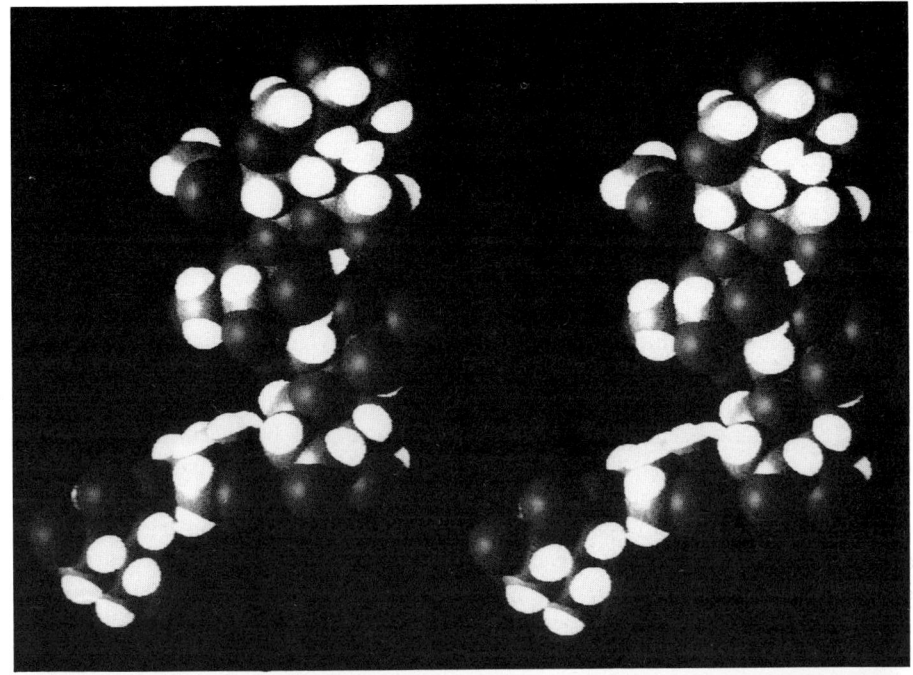

a

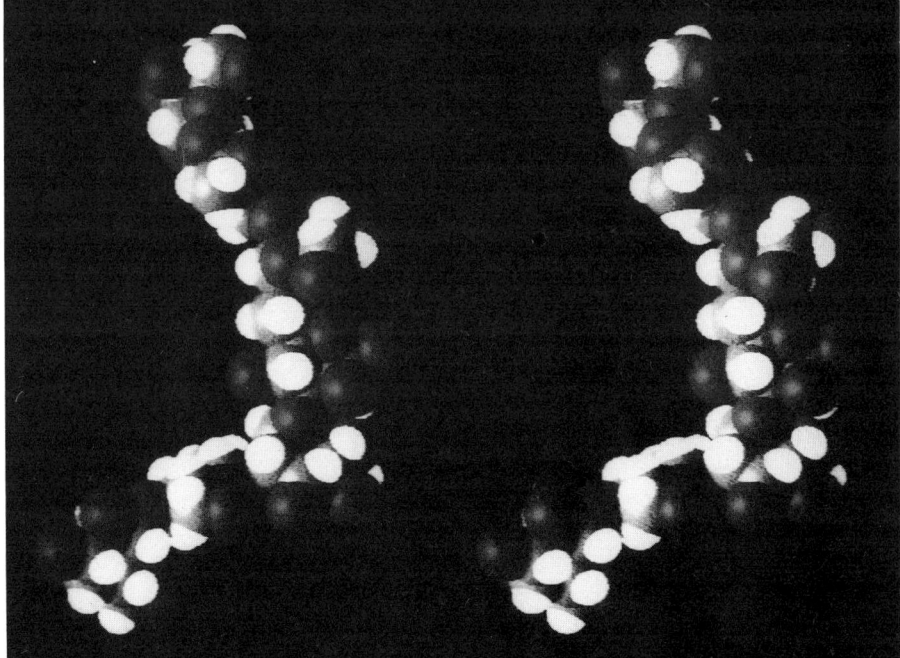

b

Fig. 14a, b. Stereo plots of the space filling models of the two conformations proposed for the bi-antennary heptasaccharide, β-D-Gal-(1,4)-β-D-GlcNAc-(1-2)-α-D-Man-(1-6)-[β-D-Gal-(1-4)-β-D-Glc-NAc-(1-2)-α-D-Man-(1-3)-]-β-D-Man, of the N-type oligosaccharides with the ω angle at the β-D-man-nose fixed to the **a** gg (−60°) and **b** gt (60°) rotamer respectively. The reducing end is to the right

Bernd Meyer

Table 7. Dihedral angles of the heptasaccharide calculated by the HSEA approach with the ω angles set to $-60°$ (gg) and $60°$ (gt) respectively [144]

	6'–5'	5'–4'	4'–3'	6–5	5–4	4–3
gg	55°/5°	55°/10°	$-50°/-170°$	55°/5°	55°/10°	$-50°/-15°$
gt	55°/5°	55°/10°	$-50°/-160°$	55°/5°	55°/10°	$-50°/-15°$

The octasaccharide which carries an additional *N*-acetylglucosamine at the reducing end in comparison to the heptasaccharide was prepared by an independent synthesis. This compound shows a different conformational behavior than the heptasaccharide [122]. The minimum energy conformation, as calculated by the GESA program, shows the trisaccharide at the 6-position of the β-mannose bent back towards the reducing end. This particular effect is in qualitative agreement with the observation of the conformational difference between the tri- and the pentasaccharide described above. The NMR analysis of this octasaccharide confirms the calculated structure. One important fact for the correct prediction of the conformational data was the simultaneous treatment of all independent parameters in the optimization procedure (Table 8).

β-D-Gal-(1–4)-β-D-GlcNAc-(1–2)-α-D-Man-
     6'   5'   4'

                                      6
                               β-D-Man-(1–4)-β-D-GlcNAc
                               3
                          3   2

β-D-Gal-(1–4)-β-D-GlcNAc-(1–2)-α-D-Man-
     6   5   4

N-linked biantennary octasaccharide

Table 8. Dihedral angles φ, ψ, and ω of the structure of the biantennary *N*-type octasaccharide from the GESA calculation [122]

		a	b
6'–5'	β-(1–4)	55°/2°	54°/2°
5'–4'	β-(1–2)	47°/0°	54°/21°
4'–3	α-(1–6)	$-54°/79°$	$-55°/147°$
6 –5	β-(1–4)	55°/2°	55°/2°
5 –4	β-(1–2)	52°/23°	52°/24°
4 –3	α-(1–3)	$-47°/-14°$	$-47°/-14°$
3 –2	β-(1–4)	57°/$-5°$	57°/7°
$ω_{β\text{-Man-3}}$		38°	$-57°$
E (kcal/mol)		0.0	1.71

170

Fig. 15. Stereo plots of the space filling models of the two conformations calculated for the biantennary octasaccharide, β-D-Gal-(1–4)-β-D-GlcNAc-(1–2)-α-D-Man-(1–6)-[β-D-Gal-(1–4)-β-D-GlcNAc-(1–2)-α-D-Man-(1–3)-]-β-D-Man-(1–4)-β-D-GlcNAc, of the N-type oligosaccharides using the GESA program. The reducing end is to the right

171

6.3.2 Asparagine Linked Oligosaccharides

The structure of the complex type decasaccharide is analyzed by X-ray crystallography of the Fc-fragment of the human immunoglobulin IgG_1 [37, 38]. This was the first example of an X-ray structure of a glycoprotein with the carbohydrate resolved to atomic resolution in the electron density map. The conformation of the bisected structure has the 6-hydroxymethyl group of the β-D-mannose in the gt-conformation with a dihedral angle of ω = 68°. This proves that the assumption of the ω-angle as being always in its ideal conformation is not justified by crystal structure analysis. The dihedral angles of the α-(1–6) bond are φ = −57° and ψ = 171° which results in an almost rectangular orientation of the trisaccharide 6'-5'-4' with respect to the core 3-2-1. The trisaccharide 6-5-4 is in an extended orientation relative to the core (Table 9). The report of an X-ray of rabbit IgG_1 subsequently showed a similar conformation [39]. The core structure of the N-type chain attached to asn-165 of the influenza virus hemagglutinin is reported [145] to adopt a conformation similar to that described by Deisenhofer and Huber for the IgG_1 [37, 38].

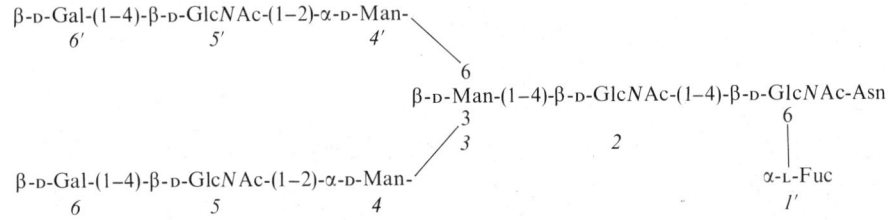

N-linked biantennary decasaccharide

The conformation of the decasaccharide linked to an asparagine residue has been determined by Carver et al. [146, 141] by NMR spectroscopy assisted with potential energy calculations. They conclude that the trisaccharide 6'-5'-4' is in two different conformations with respect to the core caused by the ω-angle of the β-mannose resulting in one conformation (gg) bent back towards the core and one (gt) with the trisaccharide almost perpendicular to the core trisaccharide. These calculations were performed with the ω-angle fixed to 60° (gt) and −60° (gg) respectively. According to the NMR spectroscopic analysis these conformations should be approximately equally populated as derived from the coupling constants $^3J_{5,6}$ and $^3J_{5,6'}$ which were interpreted from the H6-$_{β-D-Man}$ signal of the NOE difference spectrum. By contrast, the conformational energy of these two conformers differs by 4 kcal/mol which would make the gt conformer populated to approximately 0.1 %. A discrepancy between the larger observed and smaller computed NOE at the α-(1–6) branch is explained via a change in the correlation time at this rather flexible branching point. The α-(1–3) bond is demonstrated to retain the same conformation as derived for the smaller fragments of this molecule (cf. above).

Homans et al. [142, 147] have strongly criticized the determination of the conformation of the α-(1–3) linkage proposed by Carver et al. [143]. These authors in turn have ruled out the criticism by demonstration of the NOEs present in specifically deuterated model oligosaccharides where ambiguity is no longer possible

Table 9. Dihedral angles phi, psi, and omega from the isolated biantennary decasaccharide calculated by the GESA program [138] and from the x-ray structure analysis of human IgG$_1$ [33, 34].

Glycosidic Bonds		calculated structures					x-ray structure a)
		a	b	c	d	e	
6'–5'	(β 1–4)	54°/–1°	53°/4°	55°/2°	56°/0°	56°/–1°	76°/–2°
5'–4'	(β 1–2)	48°/–14°	56°/19°	52°/24°	53°/21°	53°/2°	52°/27°
4'–3	(α 1–6)	–61°/77°	–50°/59°	–57°/148°	–57°/176°	–50°/–172°	–56°/171°
6–5	(β 1–4)	54°/2°	55°/2°	55°/2°	56°/–0°	56°/0°	b)
5–4	(β 1–2)	52°/26°	52°/24°	52°/23°	53°/23°	53°/23°	71°/27°
4–3	(α 1–3)	–46°/–11°	–48°/–13°	–47°/–14°	–48°/–12°	–48°/–13°	–48°/–6°
3–2	(β 1–4)	59°/–5°	47°/–11°	57°/7°	52°/8°	56°/12°	27°/–32°
2–1	(β 1–4)	47°/0°	52°/–1°	54°/0°	54°/6°	66°/10°	44°/–35°
1'–1	(α 1–6)	52°/81°	53°/124°	55°/122°	54°/–164°	56°/–153°	61°/–164°
Omega 3		33°	47°	–56°	50°	75°	68°
Omega 1		–59°	152°	–60°	–71°	144°	–43
E (kcal/Mol)		0.0	4.7	6.3	9.6	9.9	—

a) Nonasaccharide 451–459 from the x-ray structure. The terminal sugars 6' and 5 are not well defined in the electron density map. b) Not localized due to low electron density.

173

Table 10. Calculated dihedral angles of asparagine linked oligosaccharides [130, 125]

	ω-β-Man	high mannose	bisected hybrid	complex	bisected complex
α-D-Man-6	−60°	−40°/200°	−40°/140°	−60°/120°	−60°/120°
	60°	−60°/160°	−60°/160°	−60°/180°	−60°/180°
β-D-GlcNAc-2				40°/30°	40°/30°
β-D-Gal				30°/−20°	30°/−20°
α-D-Man-3		−50°/−10°	−50°/−10°		
β-D-Man-4	−60°	60°/−20°	60°/−10°	50°/−10°	50°/−10°
	60°	50°/0°	50°/0°	30°/−50°	30°/−50°
β-D-GlcNAc-4$_{core}$	−60°			30°/−50°	30°/−50°
	60°			50°/−10°	50°/−10°
α-L-Fuc-6	−60°			60°/150°	60°/150°
	60°			60°/180°	60°/180°
β-D-GlcNAc-4$_{bisecting}$			60°/10°		60°/10°
Energy (kcal/mol)	−60°	0.8	0.4	0.0	0.0
	60°	0.0	0.0	4.0	5.0

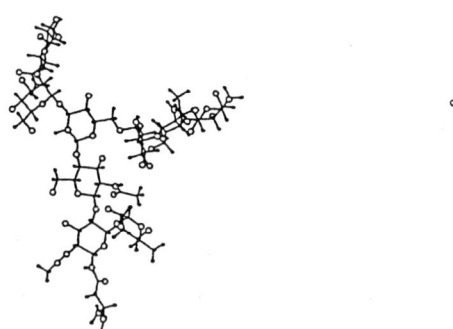
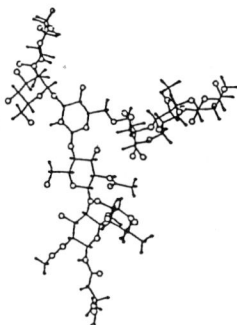

Fig. 16. Stereo plots of space filling models of the two preferred conformations calculated by a potential energy calculation for the biantennary octasaccharide, β-D-Gal-(1–4)-β-D-GlcNAc-(1–2)-α-D-Man-(1–6)-[β-D-Gal-(1–4)-β-D-GlcNAc-(1–2)-α-D-Man-(1–3)]-β-D-Man-(1–4)-β-D-GlcNAc, of N-type oligosaccharides. The conformations were obtained in keeping the ω angle of the β-D-mannose fixed to the gg (−60°) and rotamer. The reducing end is to the right

(cf. above). Due to the small NOEs obtained at the α-(1–6) linkage 4′–3 Homans et al. conclude that there is a high flexibility which leads to many conformations present in solution at this linkage [142, 148, 149]. Given all the facts described above, this interpretation would not explain the chemical shift effects observed from the anomeric configuration that are caused by the α, β ratio of the octasaccharide in solution [30, 138] (cf. above).

A more detailed analysis by the same group utilizing coupling constants showed the 6-hydroxymethylgroup of the β-D-Man to have approximately equal populations (40%:55%) of the gg and gt conformations with o-angles of 55° and 70° respectively [150, 151]. In a comparison of results from MM2 and MNDO calculations, very large deviation in the energy of the gg and the gt conformation were found. Even the comparison of the results of only MNDO calculations showed differences if the constraints in the optimization of the geometry for the rotamer population of the C5—C6 bond of the β-D-mannose were altered. The results obtained without constraints gave the gt conformation of the 6-hydroxymethyl group with 94% population, which is known to be almost entirely absent in water solution.

A completely different approach to elucidate the conformation of N-type oligosaccharides is used by Gallot et al. [152, 153] who determined the conformation of the N-type biantennary structures by linking the oligosaccharide to a lipophilic aglycon which is capable of intercalation into membranes. The lamellar structure formed in water solution was analyzed. These newly synthesized glycolipids are incorporated into double layers which are then analyzed by X-ray diffraction and freeze-fracture electron microscopy. From the periodicity of the diffraction pattern the distances of the planes of the oriented membranes with the intercalated glycolipids are obtained. Under the assumption that the distance between the membranes is mainly caused by the oligosaccharide chains, the authors obtain estimates of the longitudinal stretching of the carbohydrate part. The distances obtained at different concentrations of the glycolipid/lipid system are interpreted in terms of a changed oligosaccharide conformation. At low concentrations of the glycolipid in the lipid membrane, the "T"-shaped [154] conformation causes a distance of 22 Å and at high concentrations the "Y"-shaped [154] conformation causes a distance of 32 Å between the lipid layers.

Montreuil has proposed the terms "T"-, "Y"-, "bird" and "broken" wing shape for the description of the conformations of N-type oligosaccharides [154, 155]. These terms, although illustrative, highly simplify the conformational orientation of the different parts of the molecule with respect to the others. The more elaborate characterization of the molecule via its dihedral angles is required to describe the conformers at the glycoside bonds adequately, because within each of the families of conforma-

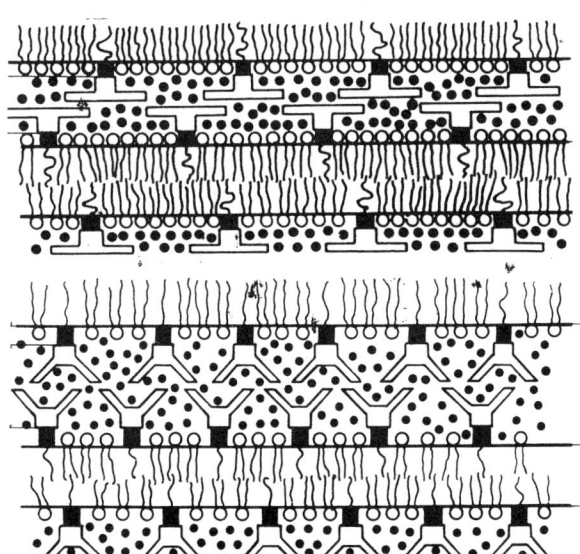

Fig. 17. Model of the synthetic glycolipid obtained from the biantennary decasaccharides incorporated in a membrane system with *a*) a high and *b*) a low concentration of the glycolipid in the membrane. Determination of the distance of the membranes yields the proposed conformations of the N-type biantennary oligosaccharides

tions defined above, variations of the three dimensional structure may occur which have significance in biological processes.

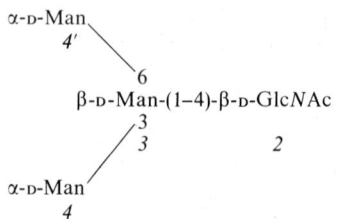

Branched tetrasaccharide

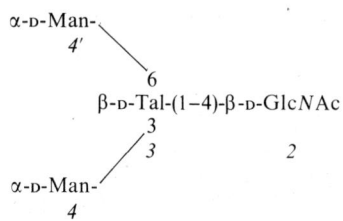

4-Epimer at the central β-mannose of the branched tetrasaccharide

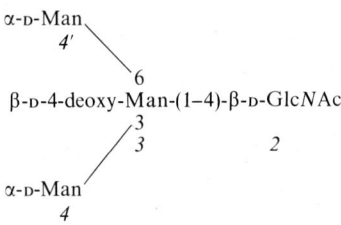

4-deoxy isomer of the branched tetrasaccharide

Modified tetrasaccharides of the branch point of the core were synthesized and their conformation analyzed by NMR spectroscopy and GESA calculations [156]. The tetrasaccharide having the β-D-mannose substituted by 4-deoxy mannose or by the 4-epimer D-talose appear to have the same overall conformation as the native substrate and have the 6-branched α-D-mannose folded back to the reducing end.

The differences between the conformation of the oligosaccharide in the X-ray structure of the immunoglobulin IgG_1 [37, 38] and the calculated and experimentally verified structure of the decasaccharide in solution [138] challenged the computation of the glycoprotein. The influence of the aglycon was analyzed by the calculation of the decasaccharide attached to the C_H2 domain of the Fc fragment of the IgG_1 immunoglobulin [157]. Because of the dimeric nature of the glycoprotein at first only one half of the glycoprotein was used in the calculation. In order to simplify the computational effort only the amino acids which are members of the C_H2-domain are treated to model the influences of the aglycon on the conformation of the oligosaccharide. The energy of that conformer which is in agreement with the X-ray study dropped from approximately 10 kcal/mol relative to the energy minimum to 6 kcal/mol when attached to the monomeric aglycon (Table 11). The modelling of the dimeric glycopro-

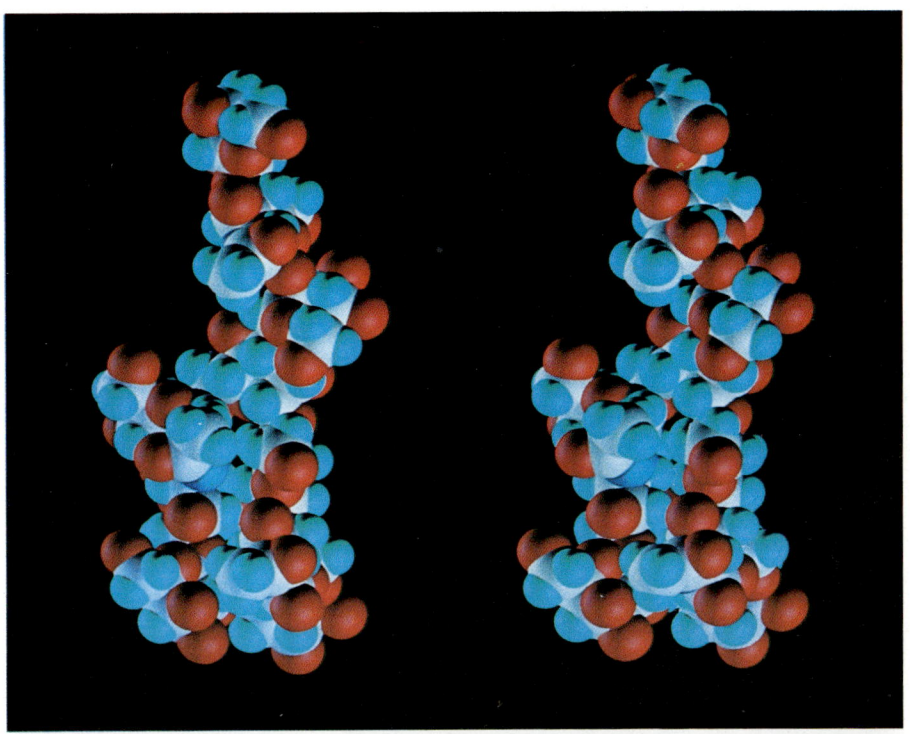

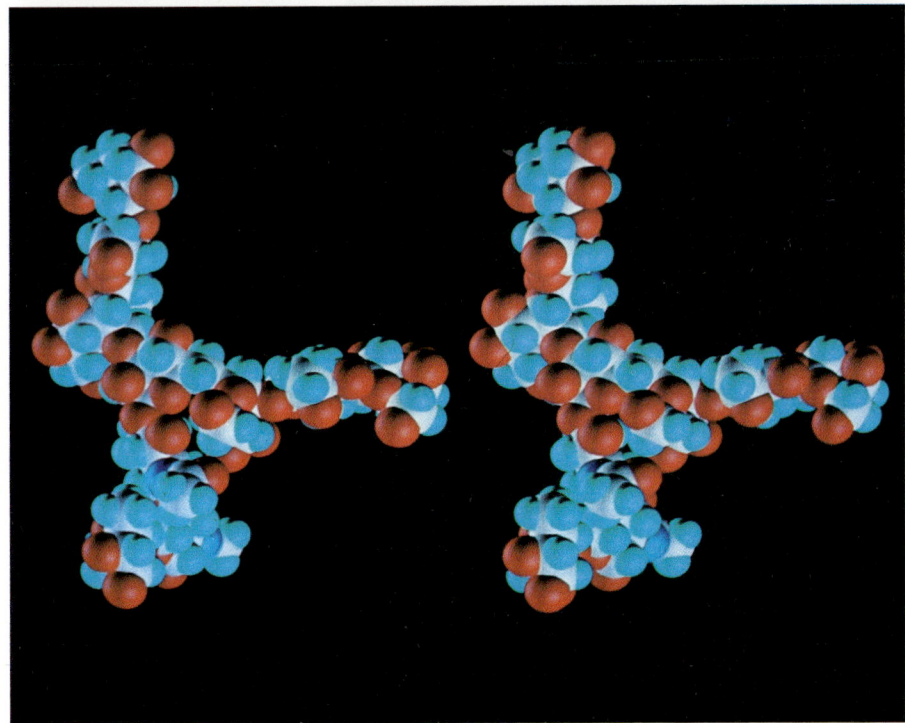

b

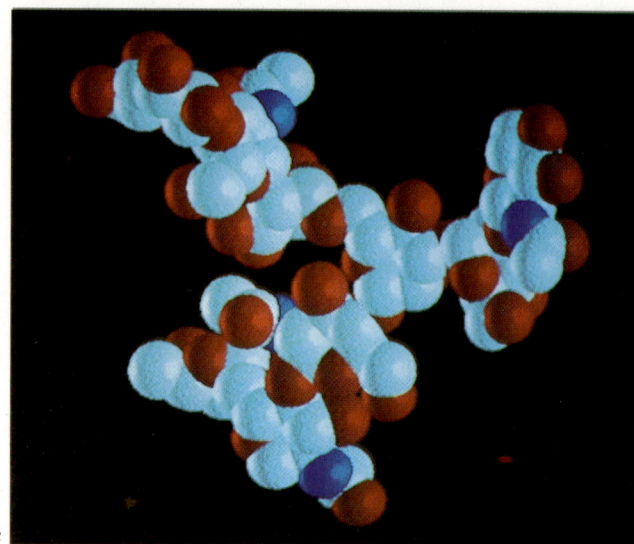

c

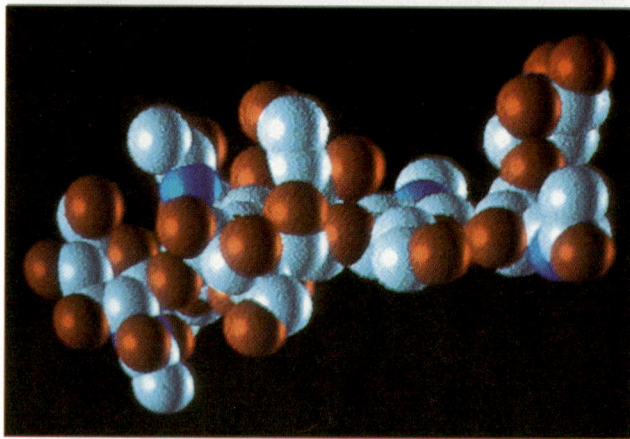

d

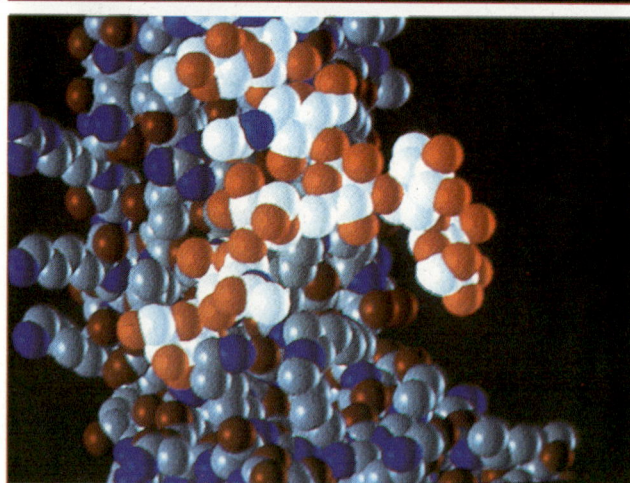

e

Fig. 18 a–g. Space filling models of the calculated and experimental conformations of the desaccharide β-D-Gal-(1–4)-β-D-GlcNAc-(1–2)-α-D-Man-(1–6)-[β-D-Gal-(1–4)-β-D-GlcNAc-(1–2)-α-D-Man-(1–3)]-β-D-Man-(1–4)-β-D-GlcNAc-(1–4)-[α-L-Fuc-(1–6)]-β-D-GlcNAc-Asn. Stereo plots of **a** the conformation with the lowest energy and **b** a relative energy of 9.6 (kcal/mol) (conformation similar to the X-ray structure (cf. text and **c**) **c** Conformation of the nonasaccharide from the X-ray structure of the IgG$_1$ with the reducing end at the bottom, central β-D-mannose viewed from the bottom, α-(1–6) trisaccharide at the top left and α-(1–3) disaccharide at the top right, **d** same conformation as in **c** viewed from the side showing the planar arrangement of the core trisaccharide with reducing end to the right, fucose to the top, α-(1–3) disaccharide points out of the plane, α-(1–6) trisaccharide goes into the plane of the paper, **e** nonasaccharide attached to the protein showing the close proximity of the α-(1–6) trisaccharide and the core of the oligosaccharide to the protein, the α-(1–3) disaccharide points away from the protein, **f** total assembly of the Fc fragment of the IgG$_1$ with the dimeric protein and two nonasaccharides attached to Asn-297, **g** same as **f** viewed from the top

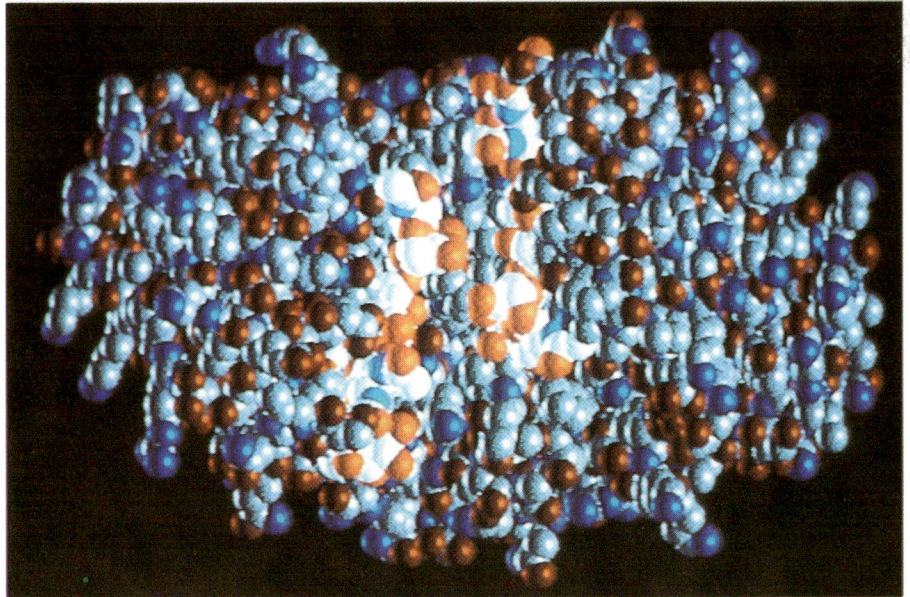

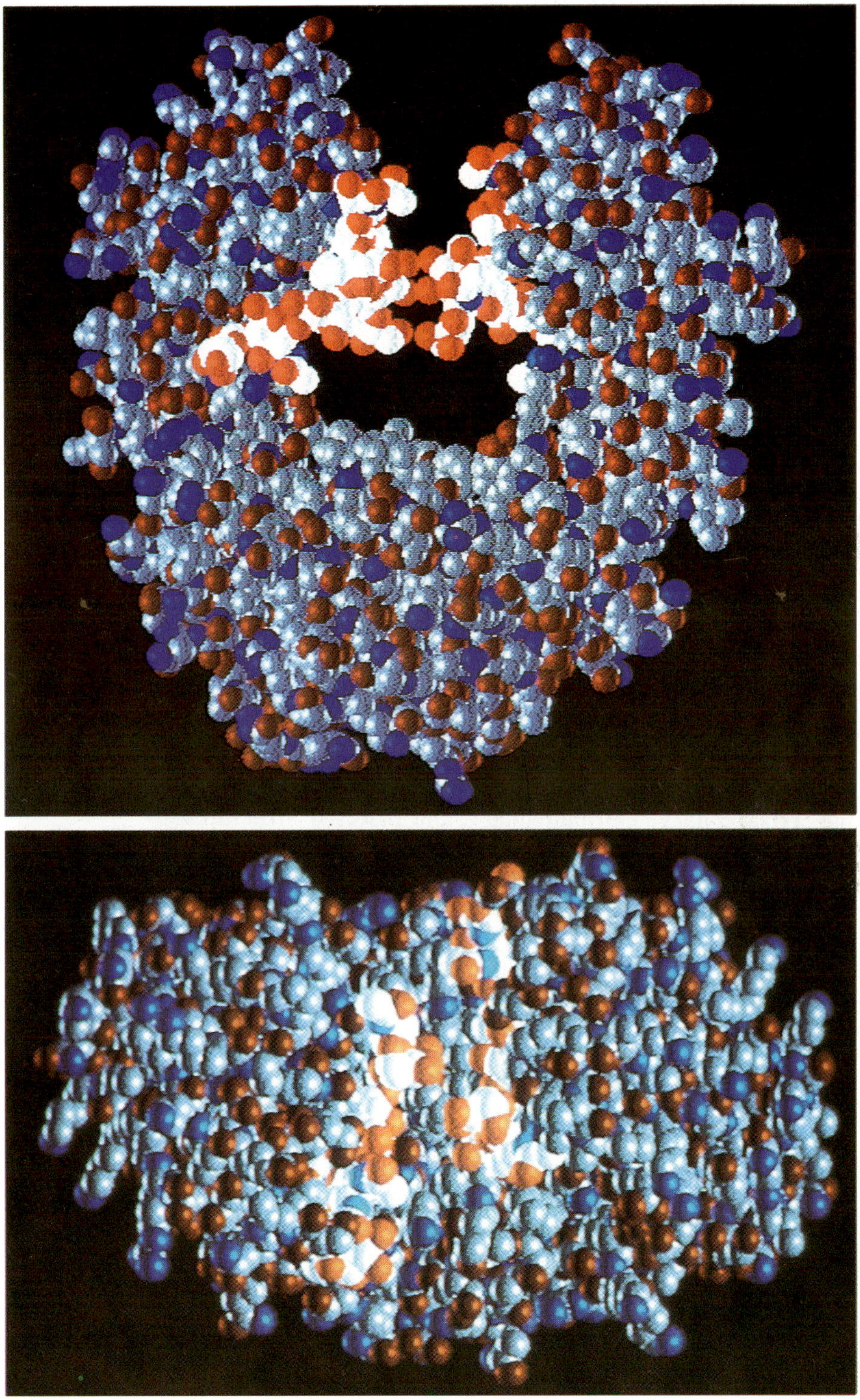

Fig. 18f, g.

Table 11. Calculation of the relative energy of the N-type decasaccharide of the IgG$_1$ attached to the monomeric protein aglycon a–d and two decasaccharides attached to the dimeric protein aglycon from the Fc fragment [138]

Glycosidic bonds	a	b	c	d	dimeric gp
6'–5' (β 1–4)	53°/1°	59°/6°	55°/1°	58°/3°	56°/—1°
5'–4' (β 1–2)	47°/—16°	60°/9°	59°/22°	56°/12°	55°/26°
4'–3 (1–6)	—58°/82°	—27°/82°	—53°/12°	—52°/—172°	—55°/172°
6 –5 (β 1–4)	56°/—0°	55°/—1°	56°/5°	56°/0°	56°/—3°
5 —4 (β 1–2)	52°/22°	52°/22°	69°/33°	52°/23°	51°/25°
4 –3 (1–3)	—48°/—13°	—47°/—14°	—42°/—22°	—48°/—14°	—39°/—3°
3 –2 (β 1–4)	57°/—1°	58°/—1°	56°/—1°	61°/14°	28°/—19°
2 –1 (β 1–4)	49°/1°	47°/5°	47°/—8°	57°/—8°	33°/—44°
1'–1 (1–6)	54°/103°	49°/94°	49°/94°	50°/94°	63°/—170°
Omega 3	33°	45°	149°	75°	73°
Omega 1	47°	49°	55°	56°	—45°
E (kcal/Mol)	0.0	1.9	2.6	5.7	—50.3

tein with two decasaccharides and the dimeric protein aglycon represented by its C$_H$2-domain amino acids yielded only one conformer with a reasonable energy which turned out to be identical in its conformation to that from the X-ray crystal structure analysis. All other conformations with a lower energy in the monomer showed energies of several thousand kcal/mol. This proved that the GESA calculations are even capable of reproducing the conformation of a carbohydrate moiety attached to a protein aglycon.

6.4 Triantennary, Tetraantennary, and Bisected Structures

An extensive NMR-study of the triantennary sugar chains of calf fetuin by 500 MHz NMR spectroscopy revealed that the conformation of the hydroxymethylgroups in GlcNAc-1 and GlcNAc-2 are different [158]. The J$_{5,6}$ and the J$_{6,6'}$ coupling constants differ for GlcNAc-2 compared to all other GlcNAc residues in these structures. This finding was explained by a conformation of the 6-branch which is folded back towards the reducing end of the structure and thus influences the conformation of the GlcNAc-2.

A study of two synthetic model compounds which are related to the branch point in triantennary and tetraantennary oligosaccharides concluded that the conformational degrees of freedom for the 1,6 linkages are restricted in the branched trisaccharide due to interactions between the two GlcNAc residues [159]. The values for the dihedral angles φ, ψ, and ω were determined by ^{1}H-NMR spectroscopy and force field calculations which included hydrogen bonding terms. For the trisaccharide two conformations were calculated to φ = 55°, ψ = 90° or 252°, and ω = 60°. The NMR spectrum showed however only one "virtual" conformation with ψ = 190°. It represents the average of the two calculated conformations and results from the slow NMR time scale. It was concluded that the conformation of the 1,2 linkage could only be correctly predicted in accordance with the NMR data when a hydrogen bonding term was included in the calculations which resulted in $\varphi = 40°$ and $\psi = -5°$. This β-(1–2) linkage is different than that found in other β-(1–2) linked GlcNAc residues [141].

β-D-GlcNAc-(1–6)-α-D-Man-OMe

Branch point disaccharide of triantennary and tetraantennary oligosaccharide

β-D-GlcNAc
          6
β-D-GlcNAc-(1–2)-α-D-Man-OMe

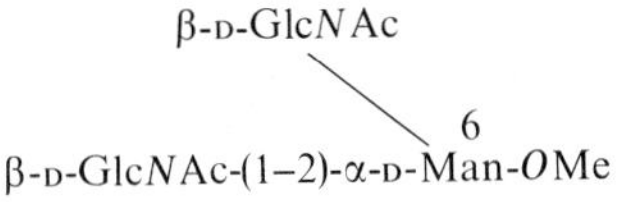

Branch point trisaccharide of triantennary and tetraantennary oligosaccharide

Cumming and Carver pointed out that it is important to consider not only the minimum energy conformers but rather a weighted average of the conformations in order to be able to interpret effects from time averaged experimental techniques like NMR spectroscopy [160, 161]. Averaging of the conformations becomes more important as the flexibility of a glycosidic bond increases and more conformational states become possible, but may also be important in glycosidic bonds to "normal" secondary alcohols [162].

The NMR analysis and GESA calculation of the bisected pentasaccharide shows that the preferred conformation of the C5—C6 bond of the β-D-mannose is gg (Table 12) [163]. Thus, the introduction of the β-D-GlcNAc at the 4-position of the β-D-Man results in a change of the preferred conformation of the C5—C6 bond from gt to gg.

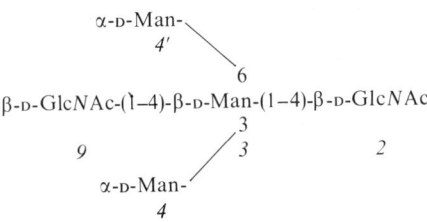

Bisected N-linked pentasaccharide

Homans et al. [150] have analyzed the conformations of bisected structures by measuring the coupling constants $^3J_{5,6'}$ and $^3J_{5,6}$. They conclude from a chemical shift analysis that the conformation of the C6$_{β\text{-D-Man}}$ of the pentasaccharide is

Table 12. Dihedral angles φ/ψ and $\varphi/\psi/\omega$ of the bisected pentasaccharide calculated by the GESA program [138]

α-D-Man-(1–6)-β-D-Man	−59/152/−66	−58/155/−67	−57/179/48
β-D-GlcNAc-(1–4)-β-D-Man	58/3	61/−4	57/−3
α-D-Man-(1–3)-β-D-Man	−48/−15	−42/2	−41/5
β-D-Man-(1–4)-β-D-GlcNAc	59/−2	58/−2	56/7
E (kcal/mol)	0.0	0.2	1.8

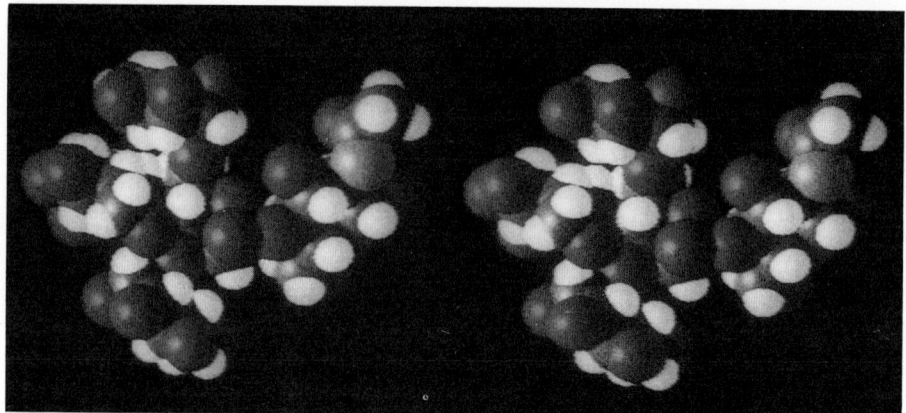

Fig. 19. Stereo plots of space filling models of the conformation of minimum energy calculated by the GESA program for the biantennary bisected pentasaccharide, α-D-Man-(1–6)-[β-D-GlcNAc-(1–4)]-[α-D-Man-(1–3)]-β-D-Man-(1–4)-β-D-GlcNAc, of the N-type oligosaccharides. The reducing end is to the right

equally populated by the gg and gt conformer whereas the bisected undecasaccharide shows coupling constants $^3J_{5,6'} = {}^3J_{5,6} = 2.0$ Hz only in accord with a single gg conformation at the β-D-Man.

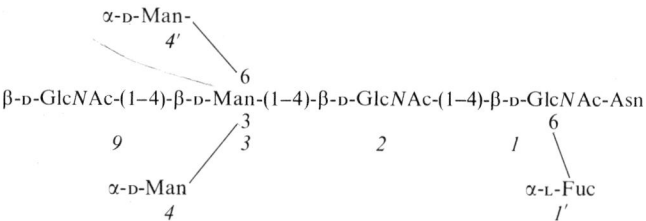

Bisected biantennary core structure of the N-linked oligosaccharides

β-D-Gal-(1–4)-β-D-GlcNAc-(1–2)-α-D-Man-
    6′   5′   4′

                6
        β-D-GlcNAc-(1–4)-β-D-Man-(1–4)-β-D-GlcNAc-(1–4)-β-D-GlcNAc-Asn
                         3   6
                  9   3   2   1

β-D-Gal-(1–4)-β-D-GlcNAc-(1–2)-α-D-Man-
    6   5   4   α-L-Fuc
                                                                  1′

Bisected biantennary undecasaccharide

From NMR spectroscopic analyses of bisected structures of the complex type [141, 164], it became clear that the relative orientation of the α-(1–3) bond is only slightly affected in the ψ-angle, which then has a value of $-20°$ after the attachment of a β-D-GlcNAc at the 4-position of the β-D-Man. The β-D-GlcNAc at the 4-position of the β-mannose has $\varphi = 60°$ and $\psi = 10°$ (Table 10).

The changes of the preferred conformation around the α-(1–6) bond at the β-mannose was studied by NOE and coupling constant analysis in the NMR spectra of different bisected structures [146]. It is concluded that the preferred conformation of the ω-angle is the gg conformation with coupling constants $^3J_{5,6'} = {}^3J_{5,6} = 2.0$ Hz. This implies an interaction between the trisaccharide 6'-5'-4'- and the bisecting β-D-GlcNAc 7.

These findings of the conformational analysis have led to an attempt to explain the substrate specificity of the glycosyl transferases during the biosynthesis [164, 165, 166]. The β-D-GlcNAc-transferases have a high selectivity for structures that do not have a bisecting β-D-GlcNAc at the 4-position of the β-mannose. Therefore, the bisecting β-D-GlcNAc acts as a stop signal during the biosynthesis [167]. A possible explanation for the high specificity to the substrate is the conformation for the trisaccharide 6'-5'-4' with respect to the core: the biantennary complex type a prefers the gt conformation and the bisected biantennary complex type prefers the gg conformation which reorients the 6-linked trisaccharide in space upon introduction of the bisecting β-D-GlcNAc 7.

It was noted that the conformation of an oligosaccharide may change depending on the protein environment where it is bound [162]. The changes in conformation may cause enzymes to exhibit different activities depending on the site where the oligosaccharide is bound. This was called "site directed" processing by enzymes.

6.5 High Mannose and Hybrid N-Type Oligosaccharides

α-D-Man-(1–2)-α-D-Man
　　　　　　　　　　＼
　　　　　　　　　　 6
α-D-Man-(1–2)-α-D-Man-(1–3)-α-D-Man
　　　　　　　　　　　　　　　　＼
　　　　　　　　　　　　　　　　 6
　　　　　　　β-D-Man-(1–4)-β-D-GlcNAc-(1–4)-β-D-GlcNAc-Asn
　　　　　　　　　　　　　　　　 3
　　　　　　　　　　　　　　　　／
α-D-Man-(1–2)-α-D-Man-(1–2)-α-D-Man

High Mannose Type Undecasaccharide

α-D-Man-(1–2)-α-D-Man
　　　　　　　　　　＼
　　　　　　　　　　 6
α-D-Man-(1–2)-α-D-Man-(1–3)-α-D-Man
　　　　　　　　　　　　　　　　＼
　　　　　　　　　　　　　　　　 6
　　　　　(β-D-GlcNAc-(1–4)-)-β-D-Man-(1–4)-β-D-GlcNAc-(1–4)-β-D-GlcNAc-Asn
　　　　　　　　　　　　　　　　 3
　　　　　　　　　　　　　　　　／
β-D-Gal-(1–4)-β-D-GlcNAc-(1–2)-α-D-Man

Hybrid Type of N-Type Oligosaccharides

In comparing complex and hybrid type structures it is assumed that the ψ angle at the α-D-Man-(1-3)-β-D-Man linkage changes only by $-10°$ [141]. Thus, the bisected hybrid type structures have dihedral angles of $φ = -50°$ and $ψ = -20°$ for the α-(1–3) bond and $φ = 60°$ and $ψ = 10°$ for the newly formed β-D-GlcNAc linkage at the 4-position of the β-mannose 3.

Homans et al. [150] conclude from the analysis of the H5, H6 coupling constants to $^3J_{5,6'} = {}^3J_{5,6} = 2.0$ Hz obtained from triple quantum filtered COSY 2D NMR spectra that the high mannose type oligosaccharides have exclusively the gg conformation at the 6-hydroxymethylgroup of the β-mannose 3. A study using combined NMR and calculational techniques including MNDO and molecular dynamics approaches was published [65]. It was shown that the heptasaccharide has two flexible 1-6 linkages and the undecasaccharide has a much more restricted conformational space for the same linkages. This is explained by interaction of the extended chains.

$$
\begin{array}{l}
\alpha\text{-}\mathrm{D}\text{-Man} \\
\qquad\qquad\diagdown 6 \\
\mathrm{D}\text{-Man-}(1\text{–}3)\text{-}\alpha\text{-}\mathrm{D}\text{-Man} \\
\qquad\qquad\qquad\qquad\diagdown 6 \\
\qquad\qquad\qquad\quad \beta\text{-}\mathrm{D}\text{-Man-}(1\text{–}4)\text{-}\beta\text{-}\mathrm{D}\text{-Glc}N\text{Ac-}(1\text{–}4)\text{-}\beta\text{-}\mathrm{D}\text{-Glc}N\text{Ac-Asn} \\
\qquad\qquad\qquad\qquad\quad 3 \\
\qquad\qquad\qquad\quad \diagup \\
\alpha\text{-}\mathrm{D}\text{-Man-}
\end{array}
$$

The enzymatic synthesis of a hybrid type oligosaccharide on the hen ovalbumin with ^{13}C-enriched galactose allowed the measurement of the ^{13}C-NMR spectra of the whole glycoprotein and additionally, the determination of the correlation times of the protein and the oligosaccharide to 25 ns and 40–80 ns, respectively [168]. This implies that the carbohydrate, at least in its terminal monosaccharide constituents, has a much higher flexibility than the protein, and its flexibility is only little impeded by the attachment to the protein.

7 *O*-Type Oligosaccharides of Glycoproteins

O-type oligosaccharides form another important group of carbohydrate structures found as parts of glycoproteins. They are either linked to serine or threonine of the protein backbone.

Pavia et al. have determined the crystal and solution conformation of α-D-mannopyranosyl-*O*-threonine [169]. This compound shows a conformation with dihedral angles $\varphi = -35°$ and $\psi = 3°$. The bond angle at the glycosidic oxygen is found to have the very small value of $\tau = 113°$. Furthermore, does the conformation of the threonine residue change upon glycosylation from a *trans* to a *gauche* relation between H_α and H_β.

A ^{13}C-NMR study of the preferred conformation of O-glycosyl serine or threonine gave for an α-glycosid of threonine angles of $\varphi = -60°$ and $\psi = 0°$ in accordance with the above mentioned X-ray structure, β-glycosides of threonine gave $\varphi = 60°$ and $\psi = 60°$ [105].

The conformation of the β-D-Glc$N$Ac-(1-6)-α-D-Gal$N$Ac bond found in *O*-type glycopeptides was determined by means of a stereospecific deuteration at the 6-position of the galactosamine [170]. The conformation of the C5—C6 bond of the galactosamine was determined by NMR spectroscopy for the trisaccharide β-D-Glc$N$Ac-(1-6)-[β-D-Gal-(1-3)]-α-D-Gal$N$Ac, the disaccharide β-D-Gal-(1-3)-α-D-Gal$N$Ac, and α-D-

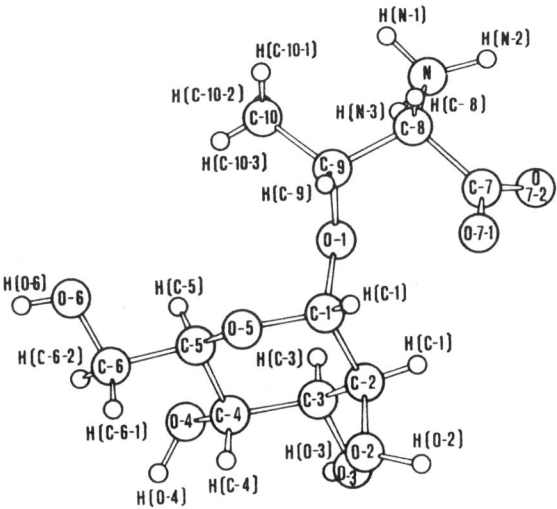

Fig. 20. Conformation of α-D-mannopyranosyl-*O*-threonine as obtained from X-ray crystallography

GalNAc. The disaccharide and the trisaccharide show no change in coupling constants and thus conformation. This finding implies that the glycosidation does not affect the conformation of the C5—C6 bond. However, the conformation of the hydroxymethyl group in α-D-GalNAc does differ from both the disaccharide and the trisaccharide in having a higher population of the tg conformation and a lower proportion of the gt conformation.

From a Monte Carlo simulation, Bush et al. [171] concluded that α-D-GalNAc-*O*-Thr has its preferred conformations at the glycosidic bond around $\varphi = -30°$ and $\psi = -60°$ which shows a qualitative agreement of the φ angle with the above named experimental results but the ψ angle deviates quite strongly.

An NMR study of α-D-GalNAc-(1-*O*)-[*N*-carbobenzoxy]-Thr-Ala-Ala-*O*Me in dimethylsulfoxide revealed that the *N*-acetyl group of the galactosamine has an important role for the conformation of the *O*-glycosidic linkage [172]. The *N*-acetyl amide hydrogen is coordinated to the carbonyl of the threonine, or the carbonyl oxygen of the *N*-acetate group is coordinated to the amide hydrogen of the threonine. The type of the coordination is dependent on the size of the protecting group at the *N*-terminus.

An NMR study and force field calculations of a disaccharide tripeptide related to the structure found in antifreeze glycoprotein (AFGP), Ala-[β-D-Gal-(1-3)-α-D-GalNAc-*O*-]Thr-Ala, in aqueous solution also suggests that the preferred conformation shows a close distance between the *N*-acetyl amide proton and the carbonyl atom of threonine [173]. The dihedral angles of the disaccharide were determined to $\varphi = 50°$ and $\psi = 40°$ for the β-linkage and to $\varphi = -40°$ and ψ (C1′—O1′—Cβ—Cα) $= 150°$ for the α-linkage. It was pointed out that the molecule has a hydrophobic and a hydrophilic side. The conformation of the AFGP was analyzed by NMR techniques. A study of the properties of fraction 8 of the AFGP [174], which contains mainly β-D-Gal-(1-3)-α-D-GalNAc-*O*-Thr, by ^{13}C-NMR, ^{1}H-NMR, and CD techniques leads to the conclusion that AFGP fraction 8 is in an extended rod-like conformation at low temperature, whereas at higher temperature it becomes a flexible coil.

The fractions 3–5 with a molecular weight of less than 23.5 kDa are subjected to an NMR analysis at 500 MHz [175]. The NOE enhancement across the glycosidic bond is stronger than the intramolecular enhancement upon irradiation of H1′ which is used as an argument for a rather rigid conformation of the glycosidic bond with a distance between H1′ and H3 of less than 3.0 Å.

The NMR spectroscopic study of two glycopeptides from calf fetuin found different chemical shifts for identical trisaccharides linked to different amino acids of the peptide [176]. One trisaccharide, α-D-NeuNAc-(2-3)-β-D-Gal-(1-3)-α-D-GalNAc, is found attached to either Glu-Ala-Pro-Ser-Ala or two trisaccharides are linked to Gly-Pro-Ser-Ala-Thr-Pro-Ala. However, the two identical sugar parts showed different chemical shifts, this was interpreted that the sugar chains either have different conformations or are interacting with the peptide or with the second oligosaccharide chain in different ways.

8 Mixed *O*- and *N*-Glycoproteins

The conformation of the glycoprotein glycophorin A has been deduced from CD spectra [177]. Setting the bisected biantennary *N*-type oligosaccharide into the conformation suggested by NMR [146], the carbohydrate covers the protein backbone from Asn-26 to Lys-30. This particular sequence is difficult to degrade by an Edman procedure. The *O*-type glycosides force the protein backbone into an extended conformation by the close packing of *O*-type glycosides. Additionally, the neuramic acid attached to the O6 of the *N*-acetylgalactosamine of the *O*-type chains is closely packed along the protein backbone. This feature is used to explain the resistance of the neuramic acid towards enzymatic cleavage by sialidase.

9 Glycolipids

Globoside was studied via NMR spectroscopy [178]. The observed inter- and intraresidue NOEs are consistent with the conformation of the glycosidic bonds as predicted by the exo anomeric effect. The interglycosidic distances of the protons neighboring the glycosidic oxygens are estimated to 2.4 + 0.1 Å.

β-D-GalNAc-(1–3)-α-D-Gal-(1–4)-β-D-Gal-(1–4)-β-D-Glc-ceramide

Globoside

In a later study from the same group, a comparison between an HSEA calculation, and an HSEA calculation combined with distance constraints implementing the distances from the NMR study was performed [179, 180]. The changes obtained in the minimum energy conformation of the three different glycosidic linkages in globoside show very different behavior on implementing the distance constraints. The β-D-GalNAc-(1-3)-α-D-Gal bond shows a change from the pure HSEA calculation from $\varphi = 60°, \psi = -10°$ to $\varphi = 40°, \psi = -50°$, the α-D-Gal-(1-4)-β-D-Gal bond changes the values from $\varphi = -40°, \psi = -10°$ to $\varphi = -50°, \psi = -10°$, and the β-D-Gal-

(1-4)-β-D-Glc bond from φ = 60°, ψ = −10° to φ = 20°, ψ = 0°. It is pointed out that the use of a single conformer for the interpretation of time averaged experimental data is an oversimplification. It was attempted to reduce this drawback by checking a two state model which takes two conformations into account and averages the observed parameters of these calculations. However, no major improvement in fitting the experimental data could be obtained.

A recent study of globoside using NMR data and a force field parametrization for the AMBER program has been published [181]. This new approach included a feature to add optional pseudo potentials to account for NMR constraints. This force field led to some changes in the torsional angles determined earlier (cf. above). The β-D-GalNAc-(1-3)-α-D-Gal bond shows φ = 44°, ψ = −62°, the α-D-Gal-(1-4)-β-D-Gal bond φ = −50°, ψ = −23°, and the β-D-Gal-(1-4)-β-D-Glc bond φ = 68°, ψ = −23°. Using distance constraints these values changed for the β-D-GalNAc-(1-3)-α-D-Gal bond to φ = 22°, ψ = −43°, for the α-D-Gal-(1-4)-β-D-Gal bond to φ = −39°, ψ = −20°, and for the β-D-Gal-(1-4)-β-D-Glc bond to φ = 53°, ψ = −13°.

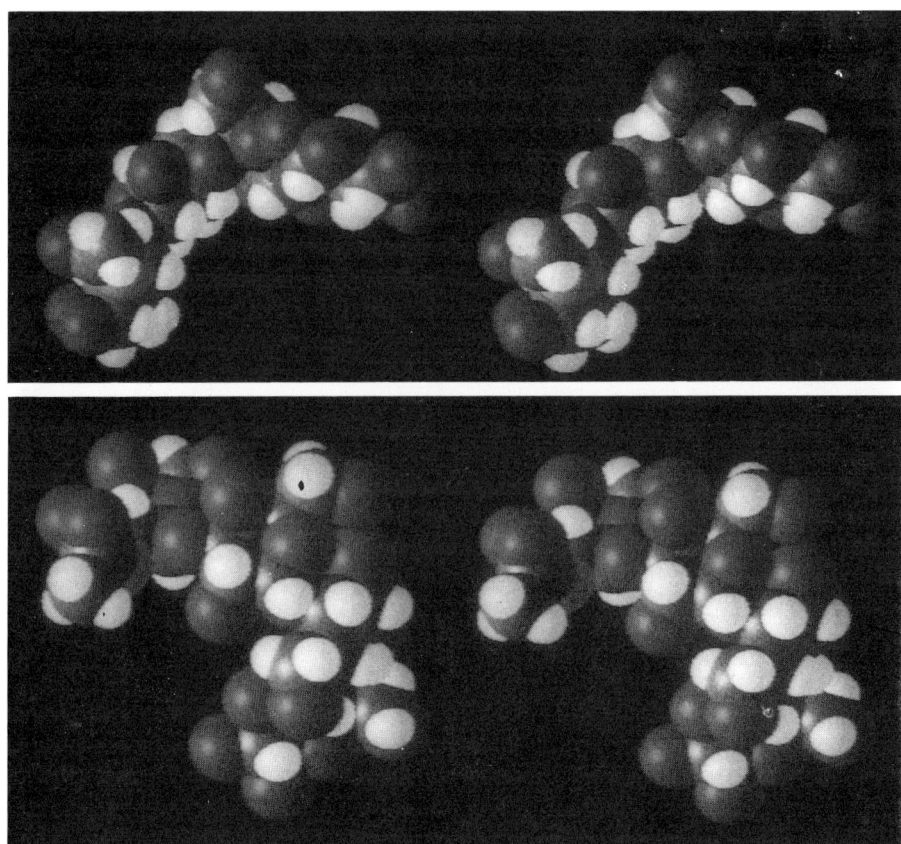

Fig. 21a, b. Conformation of globoside calculated by the HSEA program **a** without and **b** with constraint imposed via a potential function to account for the observed NOEs. Stereo plots with the reducing end to the right

187

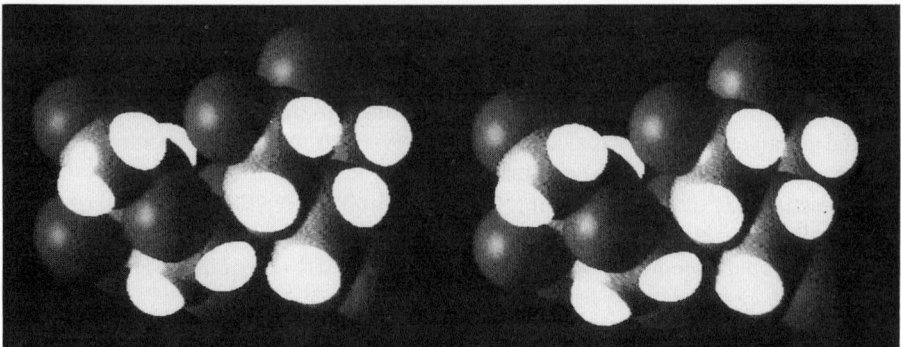

Fig. 22. Hydrophobic epitope on the convex side of the disaccharide, α-D-Gal-(1–4)-D-Gal, which is proposed to be an important factor for the binding specificity of the uropathogenic bacteria

The binding of a uropathogenic *E. coli* to the α-D-Gal-(1-4)-D-Gal sequence whether internal or terminal in the glycolipid was rationalized using HSEA and GESA calculations [182]. It could be shown that a largely hydrophobic surface lies on the convex side of this particular fragment. This part of the molecule is assumed to be responsible for the binding of the bacteria.

A PCILO calculation gives for the minimum energy conformer of the β-D-GalNAc-(1-3)-β-D-Gal disaccharide fragment $\varphi = 33°$ and $\psi = 21°$ [183]. The conformer with $\varphi = 40°$ and $\psi = 21°$ had a relative energy of 1.0 kcal/mol and that with $\varphi = -41°$ and $\psi = -13°$ already 4.4 kcal/mol.

The Forssman antigen is composed of an α-D-GalNAc linked to the 3 position of the terminal β-D-GalNAc of globoside I. The terminal disaccharide α-D-GlcNAc-(1-3)-α-D-GlcNAc of the Forssman antigen was analyzed by NMR spectroscopy, HSEA calculations, and X-ray crystallography of the peracetate [184]. The dihedral angles are obtained from the X-ray crystal structure to $\varphi = -37°$ and $\psi = -22°$, and from the HSEA calculation to $\varphi = -45°$ and $\psi = -35°$. The experimental evidence that this conformation is also adopted in solution is gained by NMR experiments.

The terminal disaccharide of the Forssman antigen was also studied using PCILO calculations [185]. The minimum energy conformation of α-D-GalNAc-(1-3)-β-D-GalNAc was found by using PCILO calculations to have $\varphi = -57°$ and $\psi = -37°$, the energetically next favorable conformer has $\varphi = -41°$ and $\psi = -50°$ and an energy content of 0.5 kcal/mol. The conformations obtained could be grouped into three distinct areas on the φ/ψ map. The second set of these areas is represented by $\varphi = -40°$ and $\psi = -30°$ for its midpoint on the surface. This conformation then agrees well with that one obtained by X-ray crystallography [186] or HSEA calculations.

Upon sialylation of *O*-type oligosaccharides there are some long range shift effects in the ^{13}C-NMR spectrum which are not easily attributable to a direct effect by the neuramic acid residue because of the long distance of the sialylated 6-hydroxygroup to the site of the shift effect [187]. The sialylation of the 6'-hydroxygroup of methyl β-D-*N*-acetyl-lactosamine causes a deshielding of 2 ppm at the C4 of the glucosamine.

It is proposed that this effect is only due to the hydrated carboxyl group of the neuramic acid.

Berman showed from a detailed analysis of the ^{13}C-NMR spectra of several sialylated oligosaccharides that the conformation of an α-(2-3) bond gives a linear extension, whereas the attachment of the α-D-NeuNAc to the 6-hydroxygroup yielded into a 'folded back' conformation of the neuramic acid [188].

Differences in the thermodynamic properties of globoside and asialo-G_{M1} are explained by conformational differences of these glycosphingolipids [189]. The globoside adopts a bent conformation due to its α-linked galactose, while the asialo G_{M1} shows a stretched out conformation. The lipid monolayers studied by differential scanning calorimetry showed for the globoside a lower transition temperature of 40.5 °C and an energy of $\Delta H = 2.0\,kcal/mol$, while the asialo G_{M1} has a transition temperature of 54.0 °C and an energy of $\Delta H = 4.2\,kcal/mol$. These data agree well with the denser packing of the asialo G_{M1} compared to globoside, which is a direct effluent of the bent conformation in the former one.

β-D-Gal-(1–3)-β-D-GalNAc-(1–4)-β-D-Gal-(1–4)-β-D-Glc ceramide

Asialo-G_{M1}-Ganglioside

From ^{13}C-NMR relaxation data of G_{D1a}, it was shown that the C1 of the ceramide is much more mobile than the adjacent parts of the molecule, both the C2—C3 of the ceramide as well as the carbohydrate part [190]. This is interpreted in terms of a hinge function of the C1 in order to facilitate the movement of the carbohydrate with respect to the ceramide and hence to the membrane.

β-D-Gal-(1–3)-β-D-GalNAc-(1–4)-β-D-Gal-(1–4)-β-D-Glc-ceramide
         3   3
         | |
         2   2
α-D-NeuNAc   α-D-NeuNAc

G_{D1a}-Ganglioside

β-D-Gal-(1–3)-β-D-GalNAc-(1–4)-β-D-Gal-(1–4)-β-D-Glc-ceramide
                                  3
                                  |
                                  2
                          α-D-NeuNAc

G_{M1}-Ganglioside

β-D-GalNAc-(1–4)-β-D-Gal-(1–4)-β-D-Glc-ceramide
                  3
                  |
                  2
          α-D-NeuNAc

G_{M2}-Ganglioside

The conformation of oligosaccharides related to G_{M1} and G_{M2} gangliosides are calculated by the HSEA method [191]. The dihedral angles for the G_{M1} ganglioside are listed in Table 13. Special focus was on the conformation of the neuramic acid glycosidic bond to the 3-position of the β-D-galactose. From calculation as well as from NOE measurements, it became clear that the neuramic acid adopts a conformation with the carboxylgroup almost antiperiplanar to the C3 of the β-galactose. The

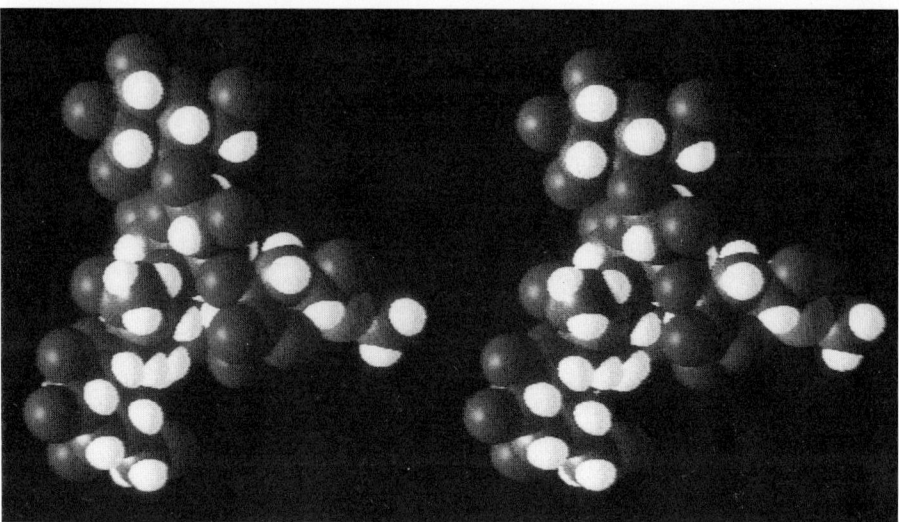

Fig. 23. Calculated minimum energy conformation of the G_{M1}-ganglioside using the HSEA program. Stereo plot with the reducing ends to the top

Table 13. Calculated dihedral angles φ and ψ for the G_{M1}-ganglioside [191]. The φ-angle of the neuramic acid glycosidic bond is defined with respect to $C1_{NeuNAc}$

	φ/ψ
β-D-Gal-(1–3)-β-D-GalNAc	55/10
β-D-GalNAc-(1–4)-β-D-Gal	55/10
β-D-Gal-(1–4)-β-D-Glc	55/0
α-D-NeuNAc-(1–3)-β-D-Gal	−165/−15

conformations at the other glycosidic bonds were also established experimentally by NMR spectroscopy and are in agreement with the calculations.

Wynn et al. have described a potential energy calculation which utilizes the non bonded interaction and a term for the electrostatic interaction with the inclusion of 'directed chargers', localized in certain atomic orbitals [192, 193]. With this potential, they obtained for the β-D-glucosyl-ceramide a conformation which is similar to the X-ray crystal structure of β-D-Gal-ceramide [194]. The calculation of the α-D-NeuNAc-(2-3)-β-D-Gal disaccharide shows a conformation with $\varphi^{C1} = 120°$ (determined from the stereoplot presented in the publication) which is inconsistent with the data obtained by Sabesan et al. [191], who were able to show that their calculated conformation with $\varphi = 165°$ is in agreement with NOE data.

The preferred conformations of the gangliosides G_{M1}, G_{M2}, G_{M3}, G_{D1a}, and G_{D1b} have been studied by the computation of their potential energies [195]. The calculated dihedral angles for the minimum energy conformations are summarized in Table 14. It is evident that the neuramic acid conformation is predicted in agreement with the

HSEA calculation and NMR data [191]. Interestingly, G_{M3} contains a different conformation of the neuramic acid compared to G_{M2} and G_{M1}, whereas the terminal neuramic acid in G_{D1b} is expected to show the same φ angle as G_{M3} and thus differing from all other neuramic acid conformations studied here. The different shapes of the molecules are discussed in terms of their specific binding of cholera and tetanus toxins.

$$\beta\text{-}\text{D-Gal-(1–3)-}\beta\text{-}\text{D-Gal}N\text{Ac-(1–4)-}\beta\text{-}\text{D-Gal-(1–4)-}\beta\text{-}\text{D-Glc-ceramide}$$
$$3$$
$$|$$
$$2$$

$$\alpha\text{-}\text{D-Neu}N\text{Ac-(2–8)-}\alpha\text{-}\text{D-Neu}N\text{Ac}$$

$$G_{D1b}\text{-Ganglioside}$$

Table 14. Calculation of the dihedral angles φ and ψ of the gangliosides G_{M1}, G_{M2}, G_{M3}, G_{D1a}, and G_{D1b} [167].

	G_{M3}	G_{M2}	G_{M1}	G_{D1a}	G_{D1b}
β-D-Gal-4-β-D-Glc	53°/4°	60°/0°	60°/0°	60°/0°	60°/0°
β-D-GalNAc-4-β-D-Gal		29°/14°	26°/14°	28°/6°	25°/14°
β-D-Gal-3-β-D-GalNAc			50°/11°	32°/—4°	50°/10°
α-D-NeuAc-3-β-D-Gal	—67°/—7°	—157°/—25°	—158°/—24°	—156°/—24°	—157°/—21°
α-D-NeuAc-3-β-D-Gal(term)				—74°/0°	
α-D-NeuAc-8-α-D-NeuAc					—70°/—6°

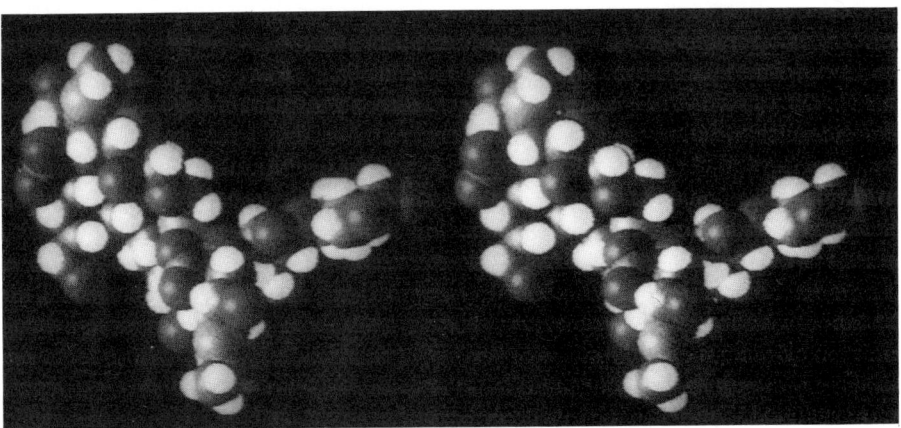

Fig. 24. Calculated minimum energy conformation of the G_{D1a}-ganglioside using a potential energy function. Stereo plot with the reducing end to the right

Bernd Meyer

10 Oligosaccharides Related to *O*-Specific Polysaccharides

Bock et al. have analyzed ten oligosaccharides related to the O-antigen of *Shigella flexneri* (Y-PS), which is a polymer with a repeating unit of [-(1-3)-α-L-Rha-(1-3)-β-D-GlcNAc-(1-2)-α-L-Rha-(1-2)-α-L-Rha-]$_n$ [23, 196]. From the detailed analysis of the NMR chemical shifts as well as the NOEs, it became evident that the conformations of the glycosidic bonds are in agreement with those predicted by HSEA calculations. The dihedral angles of the minimum energy conformer are shown in Table 15. The NMR parameters are the same for the oligosaccharides and the polymer, which demonstrates the conservation of the oligomer conformations in the polymer and thus the lack of high order phenomena in the polymer. The polymer has a helical structure with an identity period of eight monosaccharides.

The preferred conformation of the trisaccharide α-L-Rha-(1-2)-α-L-Rha-(1-2)-α-L-Rha which is common to the Streptococcal group B polysaccharide and the *Shigella flexneri* antigen referred to above was found to be similar to that one determined earlier for the *Shigella* oligosaccharides [197].

A study of the conformation of *Salmonella* O-antigens of serogroups A, B, and D$_1$ by HSEA calculations and NMR has been described [198, 199]. As can be seen from

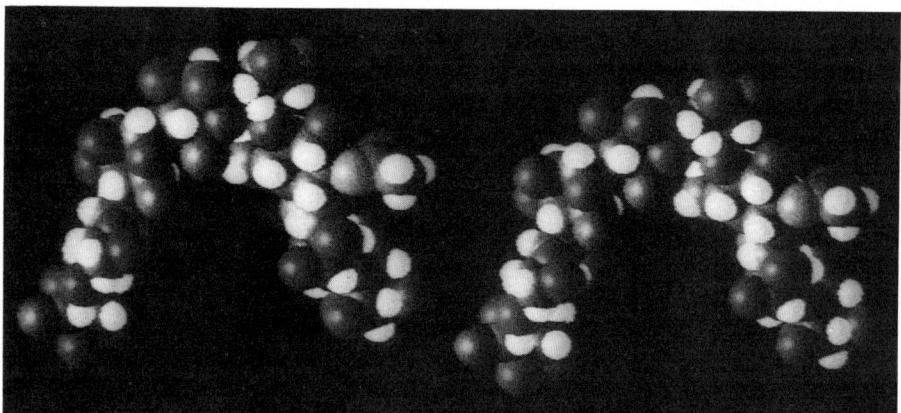

Fig. 25. Octasaccharide, [-α-L-Rha-(1–3)-β-D-GlcNAc-(1–2)-α-L-Rha-(1–2)-α-L-Rha-(1–3)-]$_2$, representing two repeating units from the *Shigella flexneri* YPS O-antigen obtained from HSEA calculations. Stereo plot with the reducing end to the top

Table 15. Dihedral angles φ and ψ of the Y-PS of *Shigella flexneri* O-antigen by HSEA calculations [196]

	φ/ψ
α-L-Rha-(1–3)-β-D-GlcNAc	40/15
β-D-GlcNAc-(1–2)-α-L-Rha	50/10
α-L-Rha-(1–2)-α-L-Rha	45/15
α-L-Rha-(1–3)-α-L-Rha	50/15

192

the dihedral angles in Table 16 by comparing the entries under disaccharides with the other columns, the conformation of the glycosidic bonds are only slightly dependent on the presence of the other glycosidic linkages, except for the case in which an α-D-Glc is attached to 4-OH of the α-D-Gal which changes the conformation of the α-L-Rha-(1-3)-α-D-Gal drastically. Thus, the backbone remains invariant among the three serogroups analyzed but changes its structure when going to the factor 12 structure. Although the minimum energy conformations, as shown in Table 16, are often not altered by additional glycosides, the degrees of freedom of motion around the glycosidic bonds may be drastically reduced as demonstrated by energy contour maps. The 3,6-dideoxy-sugars are exposed to the outer side of the helical backbone of the polymer [200]. This feature is used to explain the antigenicity of the different structures. A ^{1}H- and ^{13}C-NMR study with a discussion of differential chemical shifts and interresidue NOEs proved the calculated structures to be present in solution [199].

Using X-ray crystal structure data or, alternatively, coordinates which are generated due to geometric criteria gives, as a consequence, only an insignificant small difference with respect to the calculated minimum energy conformation. This was demonstrated later after obtaining an X-ray crystal structure of 3,6-dideoxy-α-D-*arabio*-hexo-pyranose [201] became possible.

$$[-(1-2)-\alpha\text{-}\mathrm{D}\text{-Man-}(1-4)-\alpha\text{-}\mathrm{L}\text{-Rha-}(1-3)-\alpha\text{-}\mathrm{D}\text{-Gal-}]$$

```
      3                           4(6)
    2 |             3            4 |
      |                            |
      R                         α-D-Glc

      1                           5
```

Serogroup A :   R = 3,6-dideoxy-α-D-*ribo*-hexopyranosyl (paratose)
Serogroup B :   R = 3,6-dideoxy-α-D-*xylo*-hexopyranosyl (abequose)
Serogroup D_1 : R = 3,6-dideoxy-α-D-*arabino*-hexopyranosyl (tyvelose)
Factor 12 :   R = 3,6-dideoxy-α-D-*xylo*-hexopyranosyl (abequose)

Salmonella landau serogroup N antigen is a polysaccharide with the repeating unit [-(1-2)-α-D-Per$N$Ac-(1-3)-α-L-Fuc-(1-4)-β-D-Glc-(1-3)-α-D-Gal$N$Ac-]$_n$ where Per is perosamine, 4-amino-4,6-dideoxy-D-mannose. The conformation of this polysaccharide has been calculated using the GESA program, and the conformation obtained has been compared to the experimental NOEs from one and two dimensional NMR spectroscopy [202]. The calculated structure agrees with the experimentally determined con-

Table 16. Calculated dihedral angles φ and ψ of the *Salmonella* O-antigens serogroups A, B, and D_1 in their minimum energy conformation [201]

	Disaccharides	Pentasaccharides		Hexasaccharide factor 12	
		A	B	D_1	B
4–2	−46/−19	−46/−23	−44/−19	−46/−22	−45/−18
2–3	−39/−13	−36/1	−36/−1	−36/−1	−36/−1
3–4	49/4	50/6	50/6	50/6	28/21
1–2	−48/−8	−51/−14	−51/−12	−51/−13	−51/−14
5–4	−40/−10				−36/−22

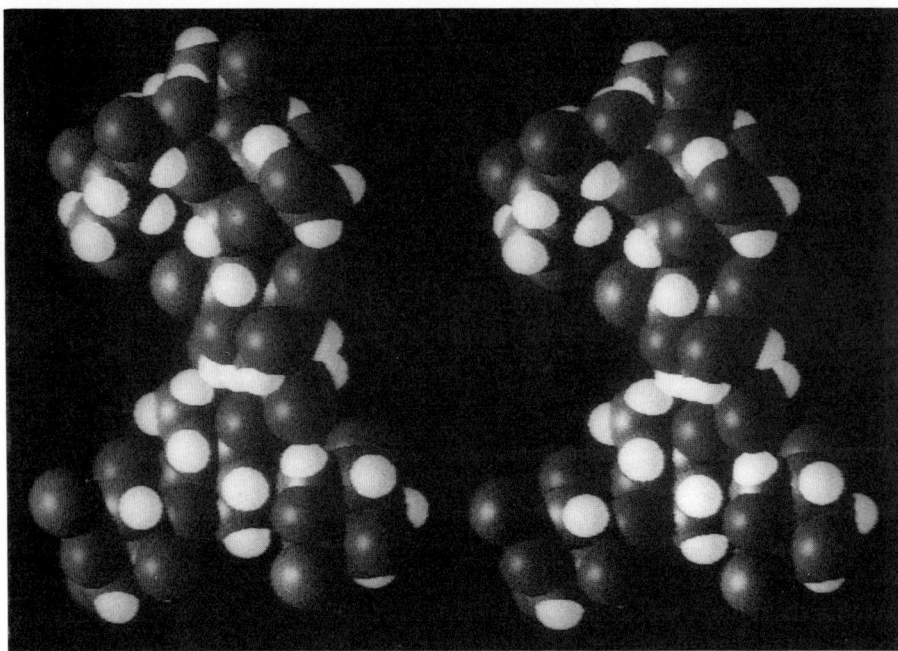

Fig. 26. Octasaccharide, [—[α-D-Abe-(1–3)]-α-D-Man-(1–4)-α-L-Rha-(1–3)-α-D-Gal-(1–2)-]₂, repres-enting two repeating units of the *Salmonella* O-antigen serogroup B. Stereo plot with the reducing ends to the bottom left. The exposure of the abequose to the outer surface is evident

formation to yield the dihedral angles (φ/ψ)α-D-Per*N*Ac-(1-3)-α-L-Fuc to $-50°/-12°$ α-L-Fuc-(1-4)-β-D-Glc to $43°/7°$, β-D-Glc-(1-3)-α-D-Gal*N*Ac to $54°/15°$, and α-D-Gal*N*Ac-(1-2)-α-D-Per*N*Ac to $-43°/-23°$.

The conformation of the backbone of the *Klebsiella* polysaccharide K$_{23}$[-(1-3)-α-L-Rha-(1-3)-β-D-Glc-]$_n$ has been calculated using a force field which includes terms for non bonded interactions, the *exo* anomeric effect, and hydrogen bonds [203]. The calculated data were checked using NMR spectroscopy. The β-D-Glc-(1-3)-α-L-Rha bond shows dihedral angles φ/ψ of $10°/40°$ and $30°/-60°$ with approximately equal energy, the α-L-Rha-(1-3)-β-D-Glc bond shows only one major conformer with φ/ψ equal to $20°/-45°$.

A conformational analysis of the *Neisseria meningitidis* serogroup B and C poly-saccharides has been performed [204]. These polysaccharides are homopolymers of *N*-acetylneuramic acid linked α-(2-8) in the group B serotype or α-(2-9) in the group C serotype, respectively. Using a force field program which calculates the energy only from short range interactions the conformation of the disaccharide bonds are deter-mined and compared to the results of NMR analyses. Due to the additional problem of the free rotation of the side C7-C8-C9 fragment of the neuramic acids, this part of the molecule attracted the major attention. From NMR analysis of the coupling con-stants, it became clear that in the case of the α-(2-9) linkage the conformation of the side chain is retained compared to the monomer with a *trans* orientation of H7 and

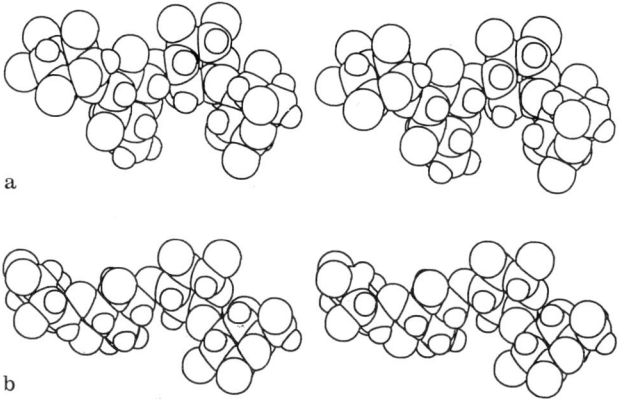

a

b

Fig. 27a, b. Tetrasaccharide with two repeating disaccharide units from *Klebsiella* capsular poly-saccharide K_{23} in its two energetically favored conformations

H8, but in case of the α-(2-8) linked polysaccharide there is a change in the conformation of the side chain orientation to a *gauche* relationship between H7 and H8 reflected by the coupling constant of $^3J_{7,8} = 3.5$ Hz compared to $^3J_{7,8} = 9.1$ Hz in the α-(2-9) linked disaccharide.

The group B meningococcal polysaccharide is an α-(2-8) linked polymer of neuramic acid. The smallest oligosaccharide that binds to antibodies was found to have a minimal length of ten residues. A study of different oligomers by NMR spectroscopy revealed that the conformation of the interglycosidic bonds changes between the internal parts of the oligomers and the two terminal bonds [205]. The internal bonds adopt a conformation close to that of colominic acid. The terminal glycosidic bonds show coupling constants of $J_{7,8} = 9.1$ and 6.7 Hz, respectively. These changes in the conformation at the cutoff of the chain were used to interpret the required minimum length of the oligomers for binding to the antibody as only the internal hexamer shows the correct conformation for binding.

The change of the side chain conformation as was found in comparing α-(2-8) and α-(2-9) linked neuramic acids (cf. above) have been supported by a study on the *E. coli* K92 capsular polysaccharide, which is a polymer made by alternating α-(2-8) and α-(2-9) linked neuramic acids [-9)-α-D-NeuNAc-(2-8)-α-D-NeuNAc-(2-]$_n$ [206]. Again, the α-(2-8) linkage causes the two protons H7 and H8 at the aglyconic part to be *gauche* and the α-(2-9) linkage causes a *trans* orientation.

Neisseria meningitidis serogroup K polysaccharide is composed of two N-acetyl-β-D-mannuronic acids, one of which is O-acetylated [-(1-4)-β-D-ManNAcA-(1-3)-β-D-(49O-Ac)-ManNAcA-]$_n$ [207]. It was shown to adopt the expected conformation with the protons across the glycosidic bonds $H1_A$ and $H3_B$ or $H1_B$ and $H4_A$ to be close together as demonstrated by large ^{1}H-NMR nuclear Overhauser enhancements.

The conformation of the core region of the lipopolysaccharides of *Citrobacter* O36 was calculated using the MM2 program and the results compared to NOE data [208]. Qualitatively, NOE data agreed with the calculated conformation. However, all sugars with a galactose configuration showed the gg conformation for the hydroxy-methyl group, which is not populated in solution.

$$
\begin{array}{ccccc}
45/16 & -64/-50 & 28/17 & -23/-45 \\
\end{array}
$$

β-D-GalNAc-(1—4)-α-D-GalNAc-(1—3)-β-D-GlcNAc-(1—4)-α-D-Glc-(1—2)-α-D-Glc

$$
\begin{array}{cc}
& 4 \\
-46/172/-54 & \diagup 58/-18 \\
\end{array}
$$

α-D-Gal-(1—6)-β-D-Glc-1⟍

11 Oligosaccharides from Proteoglycans

The polysaccharides from proteoglycans have been very extensively analyzed by means of X-ray and CD spectroscopy [209, 210]. This section will review only the newer developments based on NMR and conformational calculations.

L-Iduronic acid is a component of heparin as well as dermatan sulfate. This mono-saccharide unit has a unique feature, as its ring conformation cannot be regarded as independent of its substituents due to a small energy difference between the $^4C_1$, the $^2S_0$, and the $^1C_4$ conformer [211]. Heparin as a polymer was shown by analysis of the ^{1}H-NMR coupling constants to contain the L-iduronic acid exclusively in the $^1C_4$ conformation [212]. The antithrombin active pentasaccharide and shorter fragments of this active portion of the heparin polymer, however, adopt different conformations, as demonstrated by the unusual high coupling constants $^3J_{2,3}$ in the iduronate residues [213, 214]. The conformation that is most likely is a skew boat, $^2S_0$ for the pyranose ring of the iduronic acid. This has been shown by molecular mechanics calculations to be an energetically favoured conformer, even for an iduronic acid carrying a sulfate methyl ester or the ionic sulfate group at oxygen O2 [215, 216]. The $^2S_0$ conformation was found to be the most probable one for the AT III active pentasaccharide. This feature of the iduronic acid residue complicates the analysis of oligomers of heparin drastically because both the conformational flexibility of the glycosidic bonds and also that of the pyranose ring itself will change the overall shape of the oligomer or polymer.

In the disaccharide with α-L-iduronic-acid-2-sulfate linked to the 4-OH of 2,5-anhydro-mannose, which is obtained from heparin by nitrous acid degradation, the conformation of the iduronic acid was found to be $^1C_4$ and to be independent of temperature in the range from 20 °C to 90 °C [217].

A thorough NMR analysis of synthetic oligosaccharides related to the penta-saccharide binding to antithrombin III revealed the preferred conformations of the rings and the glycosidic bonds [218, 219]. It was concluded that the iduronate ring prefers the $^2S_0$ conformation but changes its conformation towards a $^1C_4$ conformation when a 2-sulfate is present. Furthermore, a sulfamido substituent in the 2' position of the nonreducing end of the iduronic acid tends to stabilize the $^2S_0$ conformation. The trisaccharide α-D-GlcNSO$_3$-3,6-di-O-SO$_3$-(1-4)-α-L-IduA-2-O-SO$_3$-(1-4)-α-D-GlcNSO$_3$-6-O-SO$_3$ shows at low salt concentrations a $^2S_0$ conformation which changes to a $^1C_4$ conformation in the presence of 3 M NaCl. A GESA calculation of a nonsulfated trisaccharide, α-D-GlcNAc-(1-4)-α-L-Ido-(1-4)-α-D-GlcNAc, gave φ = −37° and ψ = −31° for the $^2S_0$ conformation of the idose and φ = −41° and ψ = −36° for the $^4C_1$ conformation.

In dermatan sulfate a vacuum UV circular dichroism study showed that the iduronate ring is present only in the $^1C_4$ conformation [220]. This applies to the solution as well as to the dried films. The evidence obtained from X-ray diffraction on crystalline fibers is just the opposite. According to this study, iduronate should adopt a $^4C_1$ conformation in the solid state [221].

12 Antibiotics

The conformational analysis of antibiotics is very important in order to be able to establish a structure activity relationship (SAR) and thus gain more insight into the way the antibiotics act in the living organism.

12.1 Aminoglycosides

Mallams et al. [222, 223, 224] have determined the solution conformation of a large series of semisynthetic aminoglycoside antibiotics from the sisomycin, kanamycin and gentamycin family of compounds using ^{13}C-NMR spectroscopy. A detailed analysis of the changes in the chemical shifts gave information about the relative orientation of the glycosidic bonds. The conformations around the glycosidic bonds were not determined with the precision required to present torsion angles but still have given rise to a general description of the preferred conformation at the glycosidic centers.

The normal orientation of the glycosidic and aglyconic part of the molecules shows both C—H bond vectors adjacent to the glycosidic oxygen to be parallel. The glycosidic bonds at C-4 and C-6 of the streptamine residue show both a small clockwise rotation from the conformation described above to relieve some strain built up by dipolar or steric effects. Only in the case of 1-epi stereochemistry in the streptamine

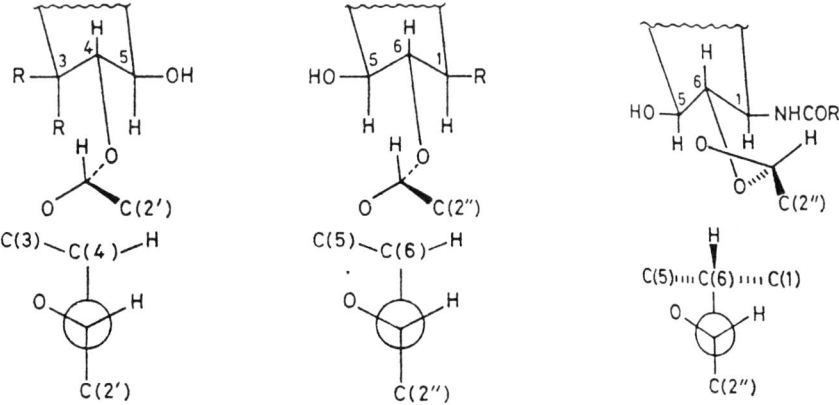

Fig. 28. Conformations around the glycosidic bonds of aminoglycoside antibiotics as proposed on the basis of ^{13}C-NMR chemical shift analysis

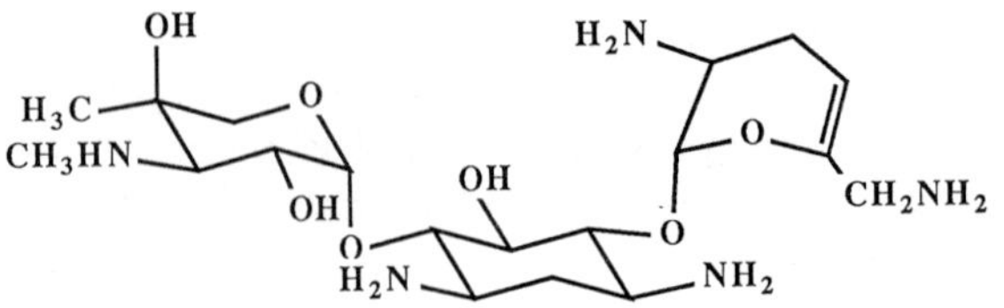

Bernd Meyer

OH
H₃C—
CH₃HN—
OH
Sisomycin

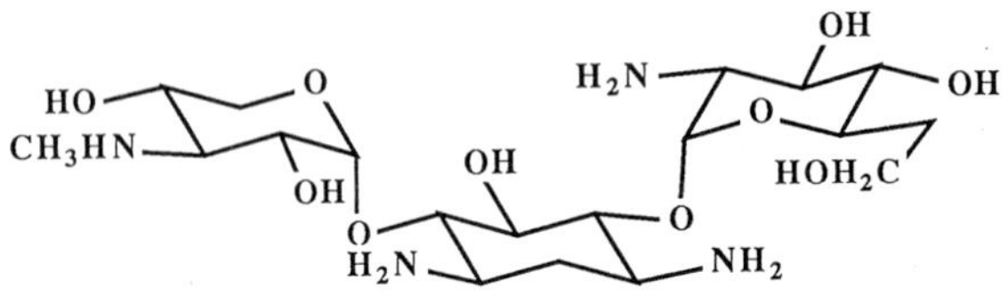

Gentamycin A

Scheme 1. Aminoglycoside antibiotics: Sisomycin and Gentamycin derivatives

the glycoside linkage to O-6 of the streptamine is rotated clockwise by 120° compared to the parallel orientation of both C-H bond vectors. All conformations deduced from ^{13}C-NMR chemical shift differences put the C1-O1 bond in the rotamer which follows the *exo* anomeric effect.

From differential shielding effects in ^{13}C-NMR spectra Wright et al. [225] concluded that the most likely conformations of the glycosidic bonds of a series of gentamycin derivatives are in accordance with the *exo* anomeric effect. Furthermore, the C-H vectors on both sides of the glycosidic bond are supposed to be in an almost parallel orientation, which is in agreement with the expectation.

The conformations of streptomycin and streptomycylamine derivatives are calculated using the Hopfinger potential functions [226]. The conformations are opti-

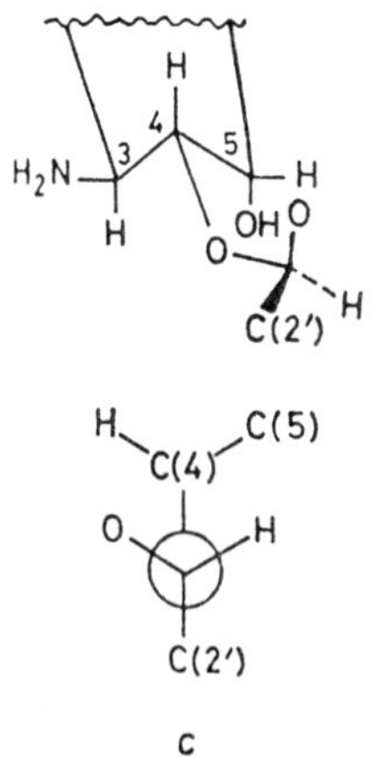

Fig. 29. Conformations around the glycosidic bonds of aminoglycoside antibiotics with 1-epi stereochemistry at the streptamine as proposed on the basis of ^{13}C-NMR chemical shift analysis

198

mized in a hydrophilic/hydrophobic phase boundary simulated via different dielectric constants in two different volumes of the space. The orientation of these conformers with respect to the negatively charged bilayers of phospholipids is simulated by the interaction of the aminoglycosides with four phosphatidylinositol molecules forming a monolayer. These data are used to show the interaction between the monolayer fragment and the carbohydrate to be both ionic and hydrophobic in nature. Experimental and calculated values of the hydrophogic/hydrophilic balance are used for comparison. The conformations obtained differ from those usually observed in glycosides dissolved in water.

12.2 Deoxy Oligosaccharides

Berman et al. have determined the conformation of chromomycin A_3 by ^{1}H-NMR spectroscopy in an unpolar solvent [227]. The conformation was determined via interresiude NOEs, which showed that there are contacts between the ends of the oligosaccharide chains, thus requesting a conformation with both chains bent to each other while the aromatic system is exposed. The mechanism of the interaction of this antibiotic with DNA was shown in a preliminary study to be consistent with an intercalation process [228]. A note added in proof states that NOEs are not in agreement with an intercalation procedure.

However, a recent study by Patel et al. found that chromomycin A_3 forms a dimer in the complex with DNA and adopts a stretched conformation [229]. Hereby the disaccharide comes close to the trisaccharide, and hence interresidue NOEs are observed. Furthermore, strong shielding effects are observed which can only be

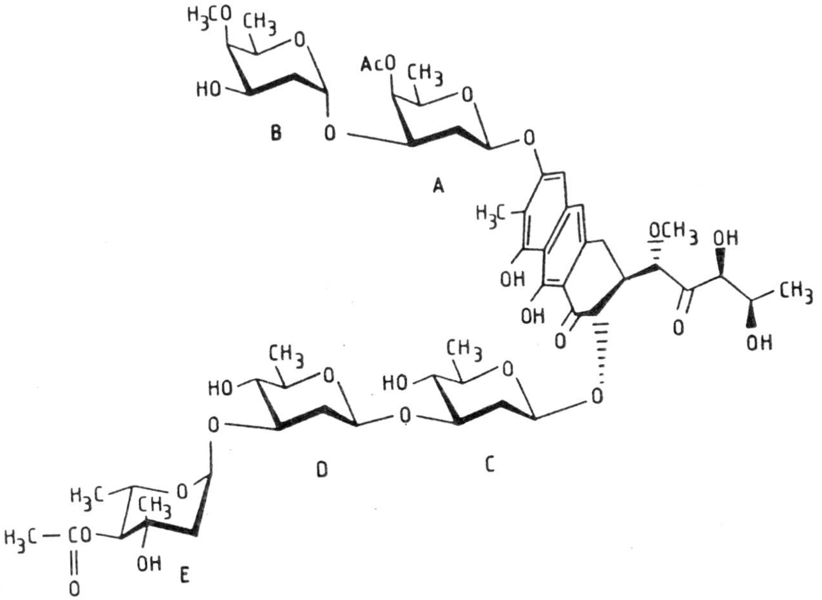

Scheme 2. Chromomycin A_3

explained by a close proximity of the sugars to the aromatic rings of the chromo-mycin. The dimerization would also explain the above mentioned NOEs for the solution structure [227].

The conformational analysis of a trisaccharide from the antibiotic kijanimycin (α-L-Dig-(1–3)-α-L-Dig-(1–3)-α-L-Dig) formed of digitoxose, 2,6-dideoxy-D-ribo-hexopyranose, by determination of the NOEs revealed u equivocally the conform-ation of this oligosaccharide because of interresidue enhancements from the reducing unit to the terminal unit [230]. The short distance between the units 1 and 3 impose a strong limitation on the range of the dihedral angles at the glycosidic bonds to $\varphi = 60°$ and $\psi = 50°$ at the non reducing unit and to $\varphi = 40°$ and $\psi = 70°$ at the inner one. The proposed conformations are in agreement with the earlier report based on the ^{13}C-NMR chemical shift analysis [231].

The conformation of the glycosidic linkage of daunomycin was determined by NMR NOE spectroscopy to $\varphi = 40°$ and $\psi = -5°$ [232].

13 Miscellaneous

The solution conformation of acarbose has been studied by HSEA calculations and NMR spectral data [233]. The valienamine shows angles of φ (H1'''-C1'''-N1'''-C4) = $= -34°$ and ψ (C1'''-N1'''-C4-H4) = $-19°$ at the pseudo glycosidic linkage to the 4-amino-4,6-dideoxy-α-D-glucose. This conformation requires short distances across the pseudo glycosidic bond which are proved to be present by NOE experiments. It is shown from a comparison with maltotetraose in its minimum energy conformation that both compounds adopt a very similar overall conformation, thus explaining the glucosidase inhibition by acarbose.

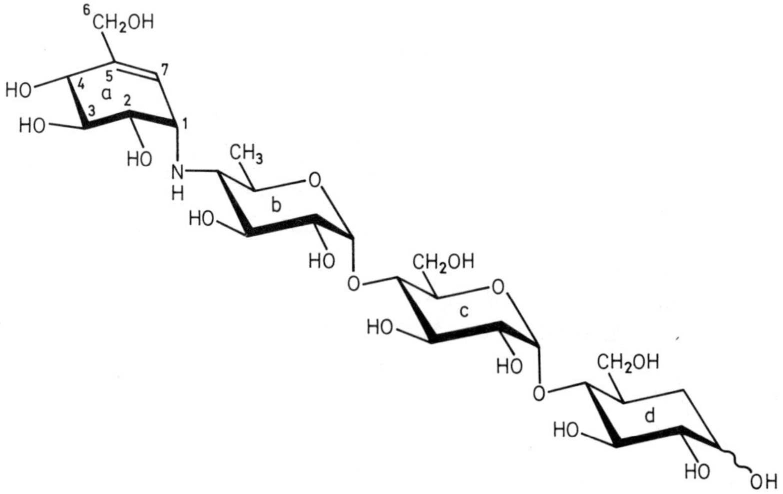

Scheme 3. Acarbose (glucosidase inhibitor)

1 Introduction

The selective protection of carbohydrates and their derivatives by alkyl groups is a subject of continuing interest (for previous reviews, see Refs. [1–3]) and has recently assumed increased importance as a result of the current widespread use of such carbohydrates as protecting units for polysaccharide, complex oligosaccharide, and glycoprotein syntheses, or as a source of chiral building blocks for the synthesis of natural products. The classical approach to these ethers, based on some kind of protection of several, usually more reactive hydroxyl groups, followed by alkylation with an excess of a reagent and deprotection, becomes now more and more displaced by direct regioselective mono-, di-, or trialkylation. No doubt, the classical three-step procedure, based on the formation of cyclic acetals, cyclic boronates, anhydro derivatives or, eventually, on a partial acylation, remains unequalled for some of the carbohydrate derivatives. The same is true for SN_2 displacement of sugar sulfonates or for an oxirane ring opening by alkoxides, for the shortening, lengthening, and other changes of an alkylated carbohydrate chain but, in most cases, the overall yield of the desired product is unsatisfactory, the procedures are tedious and complicated. The long and cumbersome 16-step synthesis [4] of methyl 3,6-di-O-methyl-β-D-galactopyranoside with an overall yield of about 1 % can serve as an example when compared with recent direct regioselective alkylation at these two positions (see, e.g., Sect. 3.3). The evolution of separation techniques plays also an important role; all of the partially methylated derivatives of methyl α-D-mannopyranoside, methyl α-D-galactopyranoside, methyl β-D-xylopyranoside can now be conveniently prepared [5, 6] by partial methylation of methyl glycosides followed by separation of the whole mixture of isomers by liquid chromatography. Only this direct, regioselective alkylation approach will be treated in this article. In addition, partial dealkylation has been included (for previous review, see Ref. [7]), as well as the reductive cleavage of cyclic acetals, because of their rapid development in the past ten years.

Of the numerous alkyl protecting groups known [8, 9], only those widely used in synthetic chemistry of carbohydrates will be discussed. Methyl ethers — with hundreds of selective methylations far exceeding the scope of the review — were excluded for their unimportant role as synthetic intermediates (great majority of partial methylations are carried out to complete the list of reference compounds for the methylation analysis of polysaccharides, or simply to prepare ethers widely occurring in biologically important molecules).

Several strategies are now employed to obtain the regioselective protection of multiple hydroxyl groups by alkylation and, in most of them, the product distribution depends on the relative reaction rate constants of the hydroxyl groups in the starting polyol as well as on the relative rate constants of the partially alkylated compound. Care must be taken in generalizing the results obtained since the reactivity of a hydroxyl group frequently changes during the course of the multi-stage alkylation, and the assessment of the relative reactivities of the polyol hydroxyl groups based on the product distribution at only one degree of substitution can lead to erroneous conclusions. For Kuhn methylation of methyl 4-O-methyl-β-D-xylopyranoside, for instance, the relative reactivity is OH–2 > OH–3 despite the fact that the 3,4-dimethyl ether can be isolated as the major partially methylated product, simply because its rate of disappearance is substantially lower compared to that of the 2,4-isomer [10]. Several methods of numerical analysis of experimental data were developed [11–17] to help with these problems, they can offer a quantitative basis for discussions on relative reactivities of hydroxyl groups with various alkylating agents.

2 Benzyl Ethers

Recent investigations related to the synthesis of complex saccharides and modified sugar analogs strongly indicate the advisability of using partially O-benzylated carbohydrate derivatives as the key intermediates. The persistent protecting benzyl ether grouping is stable and not prone to migration under the conditions of the Koenigs-Knorr reaction, it can be split off without affecting the glycosidic bond. As a result, different approaches were developed for the preparation of such suitable "aglycons" at will, with high regioselectivity. Both the selection of reaction conditions and choice of the starting carbohydrate (α- or β-anomer, glycoside or 1,6-anhydro sugar) are used to control the site of benzylation, and as the number of available techniques becomes enormous, the selective benzylation now in most cases surpasses the classical protection-benzylation-deprotection sequence.

Generally, liquid chromatography methods are used with partial benzylation, both for analytical and preparative work. All theoretically possible mono-, di-, and tri-benzyl ethers of methyl α-D-glucopyranoside have been separated by GLC via the trimethylsilyl ethers [18]. Simple and unambiguous localization of the benzyl group in the monobenzyl ethers of methyl glycosides follows from the electron-impact mass spectrometric fragmentation [19]. Benzyl-α,α-^{2}H ethers can be used to simplify [20–22] NMR spectra assignment, deuterated benzyl bromide or chloride being easily available [20, 21].

2.1 Direct Partial Benzylation

There is a broad assortment of benzylation techniques [23, 24] where the use of limited reaction time, limited amount of base or alkylating agent results in partial benzylation of hydroxyl groups. In most cases, we have to deal with a solid-liquid two-phase system with the corresponding drawbacks, and as the preparative data at only one level of substitution are usually available, generalizations are far from being easy due to differences between the hydroxyl group reactivity in the first and the second stage, respectively. For a particular carbohydrate, the regioselectivity differs from the reagent to reagent and depends frequently on the solvent used.

Benzyl Bromide — Silver Oxide — N,N-Dimethylformamide

The reaction is relatively slow and easily controled even with a large excess of reagent. The primary hydroxyl group in methyl 2,3-di-O-benzyl-α-D-galactopyranoside is benzylated as first [25, 26]. After 18 h stirring at ambient temperature, about 50% of 2,3,6- and <10% of 2,3,4-tri-O-benzyl derivative have been isolated, together with 40% of unreacted starting material [25]. The β-anomer reacts in a similar way [26]. In sucrose benzylation [27], however, the primary positions are not among the most reactive hydroxyl groups, the order being 2 > 3' > 1' > 3 (ratio of 80:10:3:1).

There seems to be a difference between the secondary equatorial hydroxyl groups

OH-2 and OH-3 in methyl 4-O-methyl-β-D-xylopyranoside [28] or alkyl 4,6-O-benzylidene-β-D-glucopyranoside [29] (cf. Ref. [30]; for discussion of the discrepancy observed, see Ref. [10]). This preponderance of position 2 over 3 appears also in methyl 4,6-O-benzylidene-α-D-mannopyranoside, the axially oriented hydroxyl group being substantially more reactive than the equatorial one [31–33]. Some 55% of 2-O-benzyl and 19% of 3-O-benzyl derivative have been isolated, together with 10% of 2,3-di-O-benzyl and 16% of the starting material [33].

Acetyl group migration or removal have been observed in the benzylation of phenyl 2,3,4-tri-O-acetyl-β-D-glucopyranoside [34] and benzyl 2-acetamido-3,6-di-O-acetyl-2-deoxy-α-D-glucopyranoside [35]. Delicate benzylation of the primary hydroxyl group in benzyl 2-acetamido-4-O-benzoyl-3-O-benzyl-2-deoxy-α-D-glucopyranoside could be realized [36] in 48% yield using benzene instead of N,N-dimethylformamide at 65 °C. Anhydrous conditions and freshly prepared silver oxide seem to be necessary precautions [24, 37, 38]. For allyl 3-O-benzoyl-2,6-di-O-benzyl-α-D-galacto-pyranoside, e.g., the benzoyl migration occurred in one instance, when silver oxide a month old was used [38, 39]. As the ageing of silver oxide as well as the purity of DMF and reagents were found to affect also the selectivity of Kuhn alkylation [10], some problems with reproducibility of this heterogeneous reaction must be calculated with.

Benzyl Bromide — Barium Oxide — N,N-Dimethylformamide

The primary hydroxyl group of various 3-O-protected N-acetyl-glucosaminides (1) has been selectively benzylated by stirring the carbohydrate with approx. 1 molar equiv. of benzyl bromide, excess barium oxide and barium hydroxide octahydrate in

R^1=OBn; R^2=H; R^3=All

R^1=OBn; R^2=H; R^3=Bn

R^1=OAll; R^2=H; R^3=Bn

R^1=H; R^2=OAll; R^3=Bn

R^1=H; R^2=OC$_6$H$_4$NO$_2$-p; R^3=Bn

(1)

DMF in the dark at room temperature for many hours [40–43] (for this compound, a two-step procedure based on regioselective tosylation of OH-6 followed by SN$_2$ displacement with sodium benzoxide proved more practical [44, 45]). Barium carbonate [46] or sodium hydroxide [43] as a base have also been used. Without protection of OH-3, the order of reactivity seems to be OH-6 > OH-3 > OH-4; 40% of 3,6-di-O-benzyl and 20% of 3,4,6-tri-O-benzyl derivative have been isolated after 4 days stirring with 2 molar equiv. of benzyl bromide [41]. Ultrasonication increases the rate of this benzylation substantially [47].

Partial benzylation of methyl 4,6-O-benzylidene-β-D-mannopyranoside with 1.2 molar equiv. of benzyl bromide yielded 3-O-benzyl (66%), 2-O-benzyl (16%), and 2,3-di-O-benzyl (10%) derivative, together with 8% of unreacted starting material [33].

213

Under the same conditions, the reactivity of hydroxyl groups is OH-2 > OH-3 for methyl 4,6-O-benzylidene-α-D-glucopyranoside, and OH-3 > OH-2 for the β-anomer [48] or for the corresponding galactopyranoside [49]. The hydroxyl group in position 2 is the most reactive in sucrose benzylation [27]. It seems that the *cis*-OR substituent activates the adjacent equatorial hydroxyl group [33] in benzylations in the presence of barium oxide.

Benzyl Bromide — Sodium Hydride — *N,N*-Dimethylformamide or Dimethyl Sulfoxide

Alkylations with sodium hydride in dipolar aprotic solvents usually proceed very efficiently and with high yields [50]. Dibenzyl ether, the major byproduct when alkali hydroxide is used, is not formed even under forcing conditions necessary for benzylation of sterically hindered hydroxyl groups [51]. Under limiting conditions of base, the rate of formation, the stability, and the rate of further reaction of competing alkoxides determine the distribution of products.

Monomolar benzylation of methyl 2,3-di-*O*-benzyl-α-D-galactopyranoside in DMF gave 2,3,6-tri-*O*-benzyl derivative in 73% yield [52]. The primary 6-benzyl ether also forms the major part of the monobenzyl fraction obtained from methyl α-D-galactopyranoside or from its β-anomer [53]. Interestingly, position 6 becomes less reactive than position 2 if 3,4-*O*-isopropylidene acetals is used to protect the other two secondary hydroxyl groups. The ratio of 2- and 6-benzyl ethers was found to be 11:1 in the α-anomer and 2.5:1 in the β-anomer [53] (see also, Ref. [54]). Uridine [55], cytidine [56], and 4-(methylthio)uridine [56] also prefer OH-2′ over the primary position when benzylated in dimethyl sulfoxide (for other benzylations in this solvent, see Refs. [35, 57]).

The hydroxyl group OH-4 seems to be the most reactive among the secondary ones in methyl α-D-galactopyranoside, whereas OH-3 seems to be prefered in the β-anomer [53]; the yields are, however, too low for any conclusions. Axial hydroxyl group at C-4 is more reactive than the equatorial OH-3 in both methyl 2,6-di-*O*-benzyl-α-D-galactopyranoside and the β-anomer [52] (see also, Ref. [58]), but the overall reactivity of the latter compound is substantially lower; more than 50% of unreacted methyl 2,6-di-*O*-benzyl-β-D-galactopyranoside were recovered after reaction with a slight molar excess of sodium hydride and benzyl bromide in DMF for 3 h at ambient temperature, whereas only small amounts of the α-anomer remained unreacted under the same conditions [52]. Surprisingly, no 4-benzyl ether was isolated from the reaction of the corresponding 6-deoxy glycoside. This paucity of 4-benzyl ether in the selective monobenzylation of methyl α-L-fucopyranoside can only partly be explained by its rapid conversion to 2,4-di-*O*-benzyl derivative, because the isolated amount of the dialkylated product (14%) is too low compared to the yield of the other two monobenzylated compounds (32% of 2-*O*-benzyl and 16.5% of 3-*O*-benzyl derivative [59]. The effect of C-5 substituent (methyl versus benzyloxymethyl) can play a more important role here. Dimolar benzylation affords methyl 2,4-di-*O*-benzyl- and 2,3-di-*O*-benzyl- and 2,3-di-*O*-benzyl-α-L-fucopyranoside in approx. 3:1 ratio. These and other results by Flowers [59] are consonant with the lack of tendency for formation of vicinal dibenzyl ethers by this method. Thus, the ratio of 3:2 for 4- to 3-benzyl ether was observed for methyl 2-*O*-benzyl-α-L-fucopyranoside [59] as well as for 3-*O*-(2-*O*-benzyl-α-L-fucopyranosyl)-1,2:5,6-di-*O*-isopropylidene-α-D-glucofuranose [60].

4,6-Di-O-benzyl-D-galactal was the main product of the dimolar benzylation of D-galactal with benzyl bromide, whereas 3,6-di-O-benzyl derivative was obtained with benzyl chloride from 1,2,O-(1-methoxyethylidene)-α-D-galactopyranose in 56% yield [61]. Reexamination of the results of partial benzylation [62] of methyl 4,6-O-benzylidene-α-D-mannopyranoside revealed [33, 63] that the 2-benzyl ether is in fact the dominant product, with the 2- to 3-ratio of 1.86:1. Approximately equal proportions of monobenzyl derivatives have been obtained from 1,2-O-isopropylidene-3,6-di-O-allyl-*myo*-inositol [64].

Attempted sodium hydride mediated benzylation of methyl 3-O-benzoyl-4,6-O-benzylidene-β-D-galactopyranoside failed due to a benzoyl migration [65]. The acyl group migration and removal are also responsible for only 62% yield of benzyl 2-acetamido-3,6-di-O-acetyl-4-O-benzyl-2-deoxy-α-D-glucopyranoside. Thallium ethoxide instead of sodium hydride or alkoxide successfully restrained the acetyl group migration in this reaction [35].

Benzyl Halide — Sodium Hydride — Aprotic Solvent

Oxolane [66–69], 1,4-dioxane [70] or, eventually, an excess of benzyl bromide [71] or chloride [72, 73] are the solvents of choice complementing the dipolar aprotic solvents mentioned above. A catalytic effect of tetrabutylammonium iodide has been observed [66, 69] for this reagent system; 1,2:5,6-di-O-isopropylidene-α-D-glucofuranose needs 24 h boiling with excess benzyl bromide and sodium hydride in oxolane for complete benzylation of the hindered secondary hydroxyl group, but 165 min at 20 °C are sufficient when 0.01 equiv. of the catalyst is present. Ion-pair formation is probably responsible for the increase of the alkoxide reactivity [66].

Methyl 2,3-di-O-benzyl-α-D-glucopyranoside gives 58% of 2,3,6- and 21% of 2,3,4-tri-O-benzyl ether on reaction with 1.4 equiv. of sodium hydride and benzyl bromide [71]. In a remarkably selective reaction, 62% of methyl 2,4,6-tri-O-benzyl-α-D-glucopyranoside result from the unprotected methyl α-D-glucopyranoside [74]. Benzyl chloride has been used for this transformation, as well as for the efficient synthesis of methyl 2,4-di-O-benzyl-α-D-xylopyranoside [75]. As expected, OH-2 in methyl 4,6-O-benzylidene-α-D-glucopyranoside is more reactive [71] than OH-3.

Activation of the tertiary hydroxyl group by *cis*-oriented vicinal oxygen atom can be responsible for the difference in 3-C-(hydroxymethyl)-1,2-O-isopropylidene-β-L-threofuranose (2) and -α-D-erythrofuranose (3) benzylation. The yield of 1′-monobenzyl ether is much lower (4.5% versus 43%) in the latter compound 3 where such an interaction is possible, the 1′,3-dibenzyl ether being the main product [70].

(2)   (3)

A difference between benzene and oxolane as a solvent has been observed in attempts for O-benzylation of allyl 2-acetamido-4,6-di-O-benzyl-2-deoxy-β-D-glucopyranoside [72, 73]. Extensive formation of the N-benzylated side products could be circumvented in the latter solvent provided that the reaction time is carefully controlled [76]. N-Trityl-5'-O-trityladenosine gave a mixture of 2'-O- and 3'-O-benzyl derivatives in 80% yield when acetonitrile-1,4-dioxane (1:1) with 1–2 molar equiv. of water was used, whereas, with a dry solvent mixture, monobenzyl ethers could not be traced [77]. Attempted di-O-benzylation of 3'-O-trityluridine yielded the 2'-benzyl and 5'-benzyl ethers only, together with some N,O-dibenzylated products. An interesting side-product, 2',3'-O-diphenylmethyleneuridine, was formed under the reaction conditions by an attack of O-2' alkoxide at the quaternary carbon atom of the trityl residue [78]. The amount of oxolane used has been made responsible [69] for the incompleteness of benzylation of 2,4-O-ethylidene-D-erythritol showing the preferential reaction at the secondary position. It seems, however, that a more probable reason is the shorter reaction time and a lower temperature compared to those used for complete benzylation, even though the purity of solvents is also of importance. No benzyl group migration has been observed [68] in the reaction of 3-O-benzoyl-6-O-(*tert*-butyldimethylsilyl)-D-glucal with potassium hydride and benzyl bromide in oxolane at 0 °C.

Benzyl Chloride — Potassium Hydroxide — Aprotic Solvent

Partial benzylation with powdered potassium hydroxide as a base and toluene as a solvent was used some 50 years ago for the preparation of 1,6-anhydro-2,4-O-benzyl-β-D-glucopyranose [79]. Since that time, other solvents, such as benzene [80–82], 1,4-dioxane-toluene mixtures [83, 84], or excess benzyl chloride [82, 85] were used as well, with apparent effects on the regioselectivity. Thus, the axially oriented secondary hydroxyl group of 1L-1,2,3,4-tetra-O-benzyl-*chiro*-inositol is more reactive than the equatorial one using benzyl chloride alone (ratio of 79:21), whereas the opposite is true (35:65) in benzene as a solvent [82]. Benzylation of *myo*-inositol derivatives in the latter solvent was also described [80, 81, 86].

Approx. a 1:1 mixture of 2,3- and 2,4-di-O-benzyl derivatives has been prepared from methyl 2-O-benzyl-α-L-fucopyranoside using 1,4-dioxane-toluene solvent mixture [83]. Surprisingly, no trace of 2,3-dibenzyl ether could be detected in the mixture of dibenzyl derivatives obtained from the unprotected methyl α-L-fucopyranoside, the 2,4- to 3,4- ratio being about [83] 3:2. Partial benzylation of allyl α-L-fucopyranoside gave similar results [84]. The isolation of 38% of 2,2',3,6,6'-penta-, 30% of 2,2',6,6'-tetra-, and 7% of 2',3,6,6'-tetra-O-benzyl derivative from the reaction of benzyl 3',4'-O-isopropylidene-β-lactoside with benzyl bromide and potassium hydroxide suggests that the high nucleophilicity of OH-2' may arise from the intramolecular hydrogen bond [85]. The effect of the alkali hydroxide used is apparent from the results of tribenzylation of methyl α-D-galactopyranoside. The use of lithium hydroxide gave the 2,3,6-ether, whereas the use of potassium hydroxide and rubidium hydroxide gave the 2,4,6-ether as the main product [87]. No differences in the reactivity of OH-2 and OH-3 in methyl 5-O-triphenylmethyl-β-D-ribofuranoside were observed if potassium hydroxide in toluene was used [87a].

216

Benzyl Trichloroacetimidate — Trifluoroacetic Acid

Though not yet used for direct partial benzylation, this mild and efficient procedure [88, 89] is very important as it is compatible with both acid- and alkali-sensitive groups in the molecule. For instance, a mixture of methyl 3-*O*-benzoyl-4,6-*O*-benzylidene-β-D-galactopyranoside, 2 molar equiv. of benzyl trichloroactimidate, and a catalytic amount of trifluoroacetic acid in dry dichloromethane — hexane (1:2) was quenched after 12 h at room temperature by the addition of triethylamine to give the 2-benzyl ether in good yield. Phase-transfer or sodium hydride-mediated benzylation failed in this case due to benzoyl migration [65]. *p*-Nitrophenyl 2-*O*-benzyl-4,6-*O*-cyclohexylidene-α-D-mannopyranoside could also be prepared with this reagent only (via the 3-*O*-benzoyl derivative), as the direct partial benzylation of the diol with benzyl bromide and silver oxide failed to give a recognizable product [90].

Miscellaneous Benzylation Methods

Excess benzyl bromide and triethylamine in the presence of tin(II)chloride catalyst were used [91] for partial benzylation of methyl α-L-rhamnopyranoside and its 4-*O*-benzyl ether. Complexation of tin(II)chloride with vicinal *cis*-disposed hydroxyl groups permitted selective benzylation at the 3- and 2-positions. Methyl 3-*O*-benzyl- or 3,4-di-*O*-benzyl-α-L-rhamnopyranoside, respectively, could be prepared in good yield by this method. No benzylation took place when benzyl chloride was used instead of benzyl bromide, or when other solvents than ethyl acetate or acetonitrile were tested.

Benzylation of nucleosides with diazo(phenyl)methane in the presence of a catalytic quantity of tin(II)chloride dihydrate in a 1,2-dimethoxyethane-methanol mixture also showed a high regioselectivity for the vicinal diol system. Approximately a 1:1 mixture of 2'- and 3'-benzyl ethers was obtained for adenosine, cytidine, uridine, etc., without any attack at the primary hydroxyl group [92]. Similarly, glycerol gives 2-*O*-benzylglycerol only, together with some unreacted starting material [93]. The yield was lower and the product much more difficult to purify when the reagent was prepared by oxidation of benzaldehyde hydrazone instead of from α-diazobenzyl phenyl ketone [93]. In the presence of boron trifluoride etherate, diazo(phenyl)methane converted benzyl 2-acetamido-3,6-di-*O*-acetyl-2-deoxy-α-D-glucopyranoside into the 4-benzyl ether in 59% yield, without any acetyl group migration [35].

Partial benzylation of methyl 4,6-*O*-benzylidene-α-D-glucopyranoside by sequential treatment with the Vilsmeier salt $Me_2N^+ = CClPh_2$ Cl^- and sodium telluride has been reported, with unsignificant regioselectivity [94]. Benzylation of D-mannose with benzyl bromide in dimethyl sulfoxide in the presence of potassium hydroxide gave more than 95% of crystalline benzyl tetra-*O*-benzyl-β-D-mannopyranoside [95]. Perbenzyl derivatives of other hexoses and pentoses were prepared by this method [95], first used for the benzylation of glycosides [96]. Cellulose can be completely benzylated in one step by using powdered sodium hydride and benzyl chloride in the sulfur dioxide — diethylamine — dimethyl sulfoxide system [97].

Benzyl trifluoromethanesulfonate has been shown to be a powerful benzylating agent (for review, see Ref. [98]). Benzylations with the in situ prepared reagent in the presence of 2,4,6-trimethylpyridine or similar bases proceed readily at −60 °C, wherein neither an acetyl group migration nor a loss of acetyl groups occurs [99, 100] (see also, Ref. [29]).

In the reaction of benzyl 3-azido-5-O-benzoyl-3,6-dideoxy-α-L-talofuranoside (4) with (diethylamino)sulfur trifluoride (DAST) the anomeric alkoxyl group migrated to the C-2 position and a fluorine atom entered into C-1 from the β-side to give the 2-benzyl ether 5 having β-L-*galacto* configuration [101]. Methyl 2-O-benzoyl-5-O-

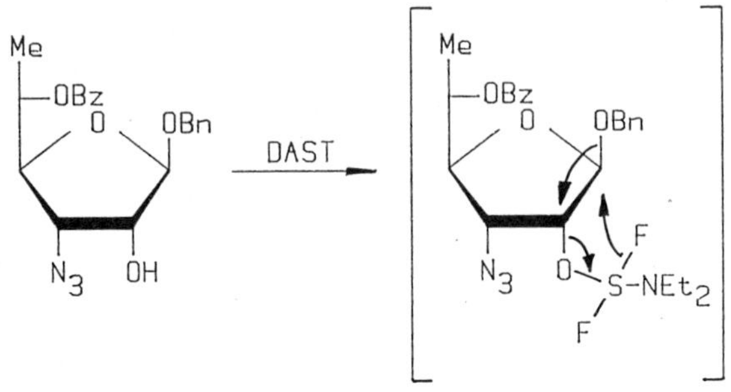

(4)   (5) (79%)

benzyl-3,6-dideoxy-α-D-*arabino*-hexofuranoside was obtained [102] by treatment of methyl 3,6-dideoxy-α-D-*arabino*-hexopyranoside with benzaldehyde in the presence of aluminium trichloride in an excellent, 95% yield. The nine-membered bis-acetal A formed from the equilibrium mixture of pyranosides and furanosides is supposed to be the intermediate of this surprisingly regioselective reaction consisting in either *inter* or *intra* molecular hydrogen transfer, as depicted in formula B.

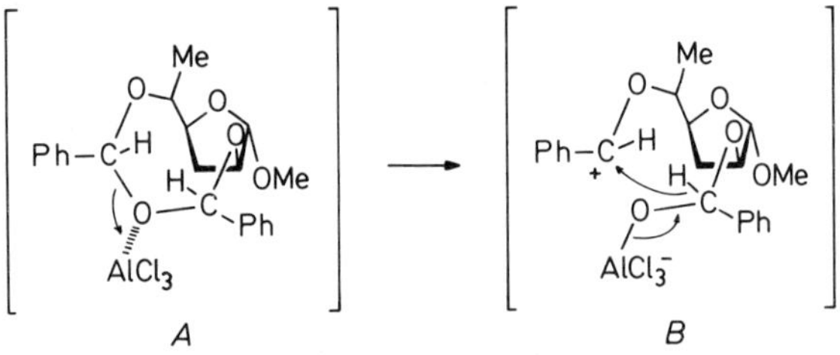

A   B

2.2 Phase-Transfer Catalyzed Benzylation

The phase-transfer technique is a simple and efficient tool for the benzylation of carbohydrates. With benzyltriethylammonium chloride or tetrabutylammonium bromide as a catalyst, a mixture of aqueous, 50% sodium hydroxide and benzyl bromide or chloride in benzene or dichloromethane solution gives a good yield of the fully protected product [103, 104], such as methyl 2,3-di-O-benzyl-4,6-O-benzylidene-α-D-glucopyranoside, when stirred at room temperature for several hours. The latter catalyst is slightly more efficient. Dichloromethane has been observed to produce methylene acetals from *cis* vicinal diols under comparable conditions [103].

At a lower, ca. 5% base concentration the reaction is substantially slower and enables the isolation of monobenzylated products in much higher yields than corresponds to statistical product composition [105]. Supposing the alkoxide formed from the alcohol in the aqueous phase forms an ion-pair with the catalyst cation which is then transferred into the organic phase and reacts there to produce an ether, the higher tendency of the monobenzyl derivative to remain in the organic phase (relative to the starting polyol) has been utilized to explain this preparatively important decrease of rate constants to the second stage of benzylation [105]. As the fully unprotected saccharides, however, are not easily alkylated under phase-transfer catalytic conditions (DMSO as a solvent can help in such a case [106]), other factors, such as the change of sterical demands of a neighbouring substituent (hydroxyl versus benzyloxy group) may be more important in the formation and reaction of bulky tetraalkylammonium alkoxide.

Regioselective benzylation at the primary position of the 4,6-diol has been observed in α- (Ref. [107]) and β-D-galacto [108, 109] series (see also, Ref. [110]). Benzyl 2,3,6-tri-O-benzyl-β-D-galactopyranoside was obtained crystalline first by this method in 81% yield [108]. The preference is lower for D-gluco series [45, 105], benzyl 2-acetamido-3-O-benzyl-2-deoxy-β-D-glucopyranoside gave [45] the desired 3,6-di-O-benzyl isomer in only 43% yield, together with 31.5% of 4,6- and 14.8% of 3,4,6-tri-O-benzyl derivative.

Among the secondary hydroxyl groups, the 2-position displays the highest reactivity because of a higher acidity. About 50% of methyl 2-O-benzyl-4,6-O-benzylidene-β-D-glucopyranoside [105], its α-anomer [105, 111], or trichloroethyl [112] or benzyl [113] analog have been isolated after several days boiling at 40 °C. Benzylation of methyl 4,6-O-benzylidene-α-D-manno- and -galactopyranoside has also been described [114, 115]. 1,6-Anhydro-4-O-benzyl-β-D-mannopyranose [116, 116a], methyl 3,6-di-O-allyl-α-D-mannopyranoside [117], and methyl 4,6-di-O-benzyl-α-D-mannopyranoside [118, 119] (see also, Ref. [120]) were all preferentially etherified at OH-2. Methyl 4-O-benzyl-α-L-rhamnopyranoside gives a mixture of 71% 2,4-di-O-benzyl and 9% 3,4-di-O-benzyl derivatives [121, 122] (for the D-enantiomer, see Ref. [123]), the 2,3-dibenzyl ether is the dominant product in the case of methyl 3-O-benzyl-α-L-rhamnopyranoside benzylation [124].

Benzyl 2,6-dideoxy-α-L-ribo-hexopyranoside yielded 61% of 4-benzyl ether and 20% of 3-benzyl ether [125], but the alkylation of methyl 2-O-benzyl-α-L-fucopyranoside [59, 109] and of some dicyclohexylidene-myo-inositols [126] was non-selective. Benzyl 2,6-di-O-benzyl-4-O-(2,3,4,6-tetra-O-benzyl-β-D-galactopyranosyl)-β-D-glucopyranoside has been obtained [127] in 26% yield by the phase-transfer catalytic procedure from benzyl β-lactoside or, more conveniently, from its hepta-O-acetyl derivative. The OH-2' is the secondary hydroxyl group preferentially benzylated here, and also in the 3',4'-O-isopropylidene derivative, probably due to the existence of an intramolecular hydrogen bond [127]. A large excess of benzyl bromide, tetrabutylammonium hydrogensulfate, and the partially protected 1,6-anhydro-maltose 6 gave

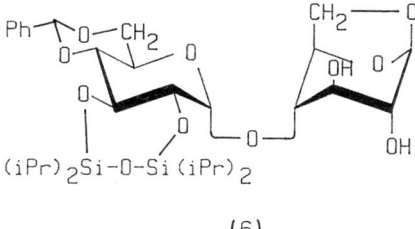

(6)

the 2,3-di-*O*-benzyl and probably 2-*O*-benzyl derivatives in 31% and 21% yield, respectively [29], when vigorously stirred in a mixture of toluene and 50% aqueous sodium hydroxide during 2.3 h at −10 °C. The allylic hydroxyl group of 1,5-anhydro-2,6-dideoxy-L-*arabino*-hex-1-enitol (L-rhamnal) is less reactive [127a] than a hydroxyl group in position 4, the ratio of 3-benzyl and 4-benzyl derivatives being about 1:2. The reversed order of reactivity was observed [127a] for the isomer having the L-lyxo configuration, i.e. for L-fucal.

2.3 Dibutyltin Oxide Method

Cyclic dibutylstannylene derivatives are formed when a suitable polyol is heated with an equimolar amount of dibutyltin oxide in methanol [128–130] or in benzene with azeotropic removal of water [58]. Such a stannylation enhances the nucleophilicity of one of the bound oxygen atoms, so that a high yield of mono-benzylated products is obtained by the action of benzyl bromide either in DMF [58, 109, 124, 128–137] or by

working in toluene or other apolar solvent in the presence of a quarternary ammonium halide as a catalyst [54, 136, 138–145]. The latter method seems to be more convenient (for a review see Ref [145a]). For instance, benzylation of the stannylene derivative of benzyl 2,3-di-*O*-benzyl-α-D-glucopyranoside in DMF at 100 °C for 3 h yields the 2,3,6-tribenzyl ether only as a minor component (36%) in an untractable mixture of products [135], whereas the other technique gives this compound in 80% yield (16 h, 80 °C) [54]. Coordination of the halide catalyst or the solvent to the tin atom seems to be necessary for the nucleophilicity enhancement. Instead of DMF, 1,4-dioxane [131], acetonitrile [131], or even an excess of benzyl bromide [146] were also efficient, whereas no reaction occured in benzene and other apolar solvents [54].

Substitution of a six-membered stannylene derivative involving the primary position always occurs at that position. Various 6-*O*-benzyl derivatives of D-*gluco* and D-*galacto* configuration have been prepared by this method in good yields [54, 135]. Treatment of allyl 2-acetamido-3-*O*-benzyl-2-deoxy-4,6-*O*-dibutylstannylene-α-D-galactopyranoside with benzyl bromide in the presence of tetrabutylammonium iodide furnished 75% of the 6-benzyl ether [141].

Five-membered stannylene derivatives of furanosides exhibit almost no regioselectivity. A 1:1 mixture of 2'- and 3'-*O*-benzyl derivatives has been obtained from uridine [128]. Similarly, methyl 5-*O*-benzoyl-β-D-ribofuranoside gave a 2:3 mixture of 2-benzyl and 3-benzyl ethers in 70% yield [147], a slightly better result (ratio of 1:3) was obtained for the parent methyl β-D-ribofuranoside [147a]. The reversed ratio (4:1) was found for the α-anomer, methyl 2-*O*-benzyl-α-D-ribofuranoside being the major product [147a].

In contrast, the five-membered cyclic dibutylstannylene derivatives resulting from the reaction of dibutyltin oxide with pyranoid *cis* vicinal diols are substituted with high regioselectivity for the equatorial oxygen atom. Allyl 2,6-di-*O*-benzyl-α-D-galactopyranoside gave essentially one product, allyl 2,3,6-tri-*O*-benzyl-α-D-galactopyranoside, isolated in 72% yield [129, 130]. Selective benzylation has been accomplished also with benzyl 6-*O*-allyl-2-*O*-benzyl- [58], allyl 6-*O*-benzyl-2-*O*-(2-butenyl)- [132], and methyl 2,6-di-*O*-benzyl-α-D-galactopyranosides [109], benzyl 2,6-di-*O*-benzyl-β-D-galactopyranoside [54], methyl 2-*O*-benzyl-α-L-fucopyranoside [59, 109], and with some fucopyranosyl disaccharides [60]. In the synthesis of the blood-group specific glycoprotein, stannylene derivative of benzyl 2-acetamido-4-*O*-(2,6-di-*O*-benzyl-β-D-galactopyranosyl)-3,6-di-*O*-benzyl-2-deoxy-α-D-glucopyranoside gave again the monosubstituted products almost exclusively, with very high regioselectivity for the reaction at the equatorial oxygen [133]. The regioselectivity is not influenced by the anomeric configuration [148]; dideoxy glycoside (7) was converted to the corresponding ether (8) in 95% yield [145], the same yield being observed for the α-anomer [148, 149].

(7) (8) (95%)

Methyl 4,6-*O*-benzylidene-2,3-di-*O*-butylstannylene-α-D-mannopyranoside reacts with benzyl bromide in an analogous fashion, it is benzylated almost exclusively at the equatorial O-3 [129, 137]. Higher than 50% yields of the 3-*O*-benzyl derivative were obtained also from methyl 4-*O*-methyl-α-L-rhamnopyranoside [134] and its 4-*O*-benzyl analog [122, 146], and 4-azido-4-deoxy [149 a] analog.

Stereoselective formation of the glycosidic bond can be achieved by this procedure. The dibutylstannylene complex of 3,4,6-tri-*O*-benzyl-D-mannose was converted into benzyl 3,4,6-tri-*O*-benzyl-β-D-mannopyranoside, useful intermediate for the synthesis of 2-deoxy-2-[^{18}F]fluoro-D-glucose [138].

In the synthesis of D-tagatose from the more common D-fructose, 1-*O*-benzoyl-2,3-*O*-isopropylidene-β-D-fructopyranose afforded two products identified as 1-*O*-benzoyl-5-*O*-benzyl-2,3-*O*-isopropylidene-β-D-fructopyranose (97%) and its 4-*O*-benzyl isomer (2.8%). The skew-boat $^{6}S_{4}$(D) conformation with an oxygen atom at C-5 adopting a quasi-equatorial position is responsible for the unexpected regioselectivity observed [136]. Conformational equilibria may also be a reason for the non-exclusive, though preferential substitution at O-3 of benzyl 4-*O*-benzyl-6-deoxy-α-L-talopyranoside (9) [142]. Even in this case, however, no tri-*O*-benzyl derivative was formed and no starting material 9 remained, the total isolated yield of 10 and 11 being 87%.

If the five-membered dibutylstannylene ring spans two equatorial positions, the outcome of the reaction is more complex. Benzyl 4,6-O-benzylidene-β-D-galacto-

(9)   (10) (55%)   (11) (32%)

pyranoside yields [54] 65% of a mixture of 3-O-benzyl and 2-O-benzyl derivatives in the ratio of 5.5:1, methyl 4,6-O-benzylidene-β-D-glucopyranoside yields 90% of a mixture with the ratio of 2.2:1 [140], and its α-anomer again 90% but 1:3.5 ratio [111]. Methyl 2,6-dideoxy-3-O-benzyl-α-L-*arabino*-hexopyranoside can be synthesized from the corresponding diol in 70% yield [148].

It should be noted that not only a hydroxyl group, but also an alkoxyl group in a *cis* relationship at an adjacent position, can activate the equatorial hydroxyl group via stannylation [131]. Negligible activation has been observed for isolated hydroxyl groups such as in the case of benzyl 3,6-di-O-benzyl-β-D-galactopyranoside [54].

Compounds with three or four free hydroxyl groups can also be used in the reaction with equimolar amount of dibutyltin oxide. Building the six-membered stannylene derivative does not occur when a five-membered ring is possible. Both benzyl β-D-galactopyranoside and its 2-benzyl ether are substituted regioselectively at the equatorial OH-3, leaving the primary hydroxyl group free [54]. Similarly methyl β-D-galactopyranoside [144] and its α-anomer [143] yielded 64% of 3-O-benzyl derivative (see also, Ref. [131]). The 3-position of methyl α-L-rhamnopyranoside [124] and methyl α-L-fucopyranoside [59] has been protected by this method as well. Interestingly, the result from benzyl 6-O-benzyl-β-D-galactopyranoside was very complicated [54], thus indicating some role of the C-5 substituent. The six-membered stannylene derivative of 1,6-anhydro-β-D-glucopyranose enables a regioselective substitution of the 1,3-diaxially related pair of hydroxyl groups. Only 4-O-benzyl and 2-O-benzyl derivatives of this triol were isolated (61% yield) in a ratio [149b] of about 2:1.

From methyl α-D-glucopyranoside, 55% of 2-O-benzyl and 18% of 3-O-benzyl derivatives could be obtained using 1,4-dioxane as a solvent. No monobenzyl derivative was observed under standard DMF solvent conditions [131]. Conformational equilibria $^4C_1 \rightleftharpoons {}^1C_4$ play a role in the results of alkylation of all-equatorial methyl β-D-xylopyranoside and methyl β-D-glucopyranoside [131].

Benzyl 4',6'-O-benzylidene-β-lactoside with five free hydroxyl groups was converted to the dibutylstannylene intermediate by azeotropic removal of water from its mixture with 2.5 molar equiv. dibutyltin oxide in benzene, the reaction with benzyl bromide in the presence of tetrabutylammonium bromide then gave the 2,3'-di-O-benzyl derivative in 52% yield [139]. When the 3',4'-O-isopropylidene analog was treated with 1.2 molar equiv. only, the 2-O-benzyl derivative was the main product [150].

2.4 Bis(tributyl)tin Oxide Method

Polyhydroxy compounds react with a limited amount of bis(tributyl)tin oxide in boiling toluene with continuous removal of water to give a mixture of tributylstannyl derivatives. As the tin atom of trialkylstannyl alkoxides is known to form a coordination bond with a suitably oriented neighboring oxygen atom (OH, OMe, etc.), a relatively high regioselectivity of this partial stannylation can be expected, with the corresponding regioselective enhancement of the nucleophilicity of carbohydrate oxygens. Even though the conditions necessary for the subsequent benzylation are rather vigorous (heating with benzyl bromide at ca. 90 °C for several days) so that the equilibration of partially stannylated intermediates occurs, the regioselectivity of this benzylation sequence [151] is remarkably good (for a review see Ref. [145a]. A catalytic effect of tetrabutylammonium bromide described first for allylation [152] (see Sect. 3.4) has been frequently used to improve the benzylation step, and N-methylimidazole seems to be even a better catalyst [150]. It should be noted that tributylstannyl intermediates are immediately hydrolyzed when spotted on plates of silica gel, and cannot be detected as such.

$$\text{OH} \quad \xrightarrow{(Bu_3Sn)_2O} \quad \text{OH} \quad \xrightarrow{BnBr} \quad \text{OH}$$

(HO, OH) → (O····OH with Bu—Sn—Bu, Bu) → (BnO, OH)

Primary hydroxyl groups are benzylated first, followed by equatorial hydroxyl groups having a *cis* oriented oxygen atom in vicinal position. Tributylstannylation of methyl α-D-glucopyranoside with 1.5 molar equiv. of bis(tributyl)tin oxide followed by alkylation in benzyl bromide for 2 days at 80–90 °C gave a mixture of 48.6% 6-O-benzyl and 30.5% 2,6-di-O-benzyl derivatives. Minor amounts of 3,6- (4.5%) and 4,6- (6.0%) were isolated as well [151, 153]. Methyl β-D-galactopyranoside gives 6-O- and 3,6-di-O-benzyl derivatives [151, 154], just as methyl, benzyl, or p-nitrophenyl α-D-mannopyranoside [151, 155, 156]. As no alternative approach for the regioselective introduction of two ether linkages directly at O-3 and O-6 seems to be available, this one is of preparative significance, despite its moderate yield.

Systematic study of the selective benzylation of conformationally rigid 1,6-anhydro-β-D-hexopyranoses has been published recently [156a]. High regioselectivity has been observed for all configurations having *cis* orientation of vicinal hydroxyl groups, that is for six of the eight possible isomers. Benzylation of 1,6-anhydro-β-D-glucopyranose and 1,6-anhydro-β-D-idopyranose was not selective. Best results were obtained with N-methylimidazol or tetrabutylammonium fluoride as a catalyst [156a].

Treatment of benzyl 3',4'-O-isopropylidene-β-lactoside with 3 molar equiv. of bis(tributyl)tin oxide and subsequent reaction with benzyl bromide gave a complex mixture of products. When benzylation was carried out in the presence of tetrabutylammonium bromide or N-methylimidazole, 50% of the 2-O-benzyl derivative was obtained after some hours. After 7 days, the 2,6,6'-tri-O-benzyl derivative was the main product (38%) [150].

When applied to carbohydrates with only two hydroxyl groups, the reaction schemes becomes quite simple. Methyl 4,6-di-O-benzyl-α-D-mannopyranoside is converted to 3,4,6-tri-O-benzyl derivative in 73% yield [119]. Methyl 4-O-benzyl-α-L-rhamnopyranoside [122] or methyl 2-O-benzyl-α-L-fucopyranoside [109] are attacked at O-3, too. Methyl 4,6-O-benzylidene-α-D-glucopyranoside gives 59.1% of 2-O- and 31.9% of 3-O-benzyl derivative [111]. Derivatives of lactose [157] or maltose [158] were selectively protected by this method at a primary hydroxyl group OH-6′ leaving the secondary OH-4′ free. Various glycosides of N-acetylated 3-O-benzyl-D-glucosamine were benzylated at a primary position [45, 159–162] in order to obtain a key intermediate of glycoprotein oligosaccharide synthesis. However, attempts to shorten the multi-step synthesis by the application of the stannylation-benzylation sequence to the triol (12) showed a poor regioselectivity. Only 9.3% of the desired 3,6-di-O-benzyl derivative could be isolated [163] after several days heating with benzyl bromide at 100–105 °C.

(12)

The fact that methyl 2,3- and 2,6-di-O-benzyl galactopyranosides give the same 2,3,6-trisubstituted compound has been utilized in the synthesis of the trisaccharide moiety of gangliotriosyl-ceramide. The intermediate methyl lactoside having only the axial OH-4 of the galactose unit free resulted from a mixture of both penta-O-benzyl lactosides, thus avoiding the separation of the 4,6- and 3,4-O-isopropylidene precursors [164].

2.5 Copper Chelates

Regioselective deactivation of the dianions of carbohydrate diols by complexing with copper salts (see Sect. 3.5) has been employed for their effective monobenzylation [165, 166]. When benzyl iodide was used, the yield of monosubstituted product was higher than 85% in almost every instance, the rest being accounted for an unreacted starting material. In no case was any disubstituted product isolated. Benzyl bromide gave <20% of alkylation in a reasonable time, benzyl chloride showed no reaction at all.

The less acidic hydroxyl groups being more reactive in this alkylation sequence [166], the 3-benzyl ethers dominate in the reaction of 2,3-diols, and 4-benzyl ethers in the case of 4,6-diols. Thus, 95% of methyl 3-O-benzyl-4,6-O-benzylidene-α-D-mannopyranoside together with 5% of unreacted starting material have been obtained from methyl 4,6-O-benzylidene-α-D-mannopyranoside. Methyl 2,3-O-isopropylidene-α-D-mannopyranoside can be quantitatively transformed to the corresponding 4-O-benzyl derivative when treated with 2 molar equiv. sodium hydride in 1,2-dimethoxyethane with the following addition of copper(II)chloride and boiling with 5 molar equiv. benzyl iodide for 24 h. Even for the unfavorable axial position in 1,6-anhydro-4-*O*-benzyl-β-D-mannopyranose (13) the yield of the expected ether (14) is very good.

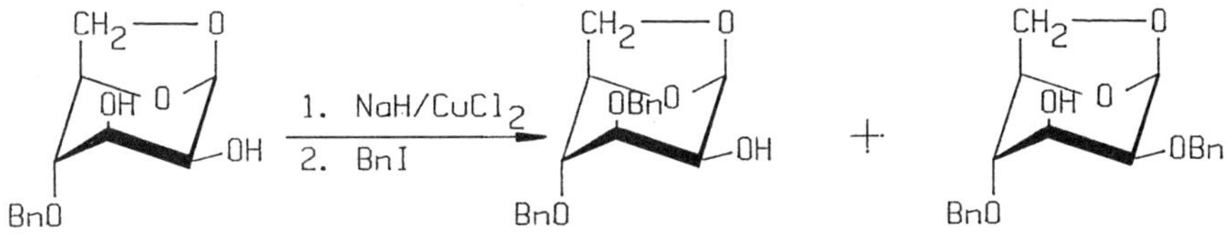

$$\text{(13)} \qquad \text{(14) (64\%)} \qquad \text{(15) (36\%)}$$

Almost no regioselectivity has been observed for 3-*O*-benzyl-1,2-*O*-isopropylidene-α-D-glucofuranose, the total yield of monosubstituted products being, however, imposantly high (88 %) [166].

2.6 Reductive Cleavage of Benzylidene Acetals

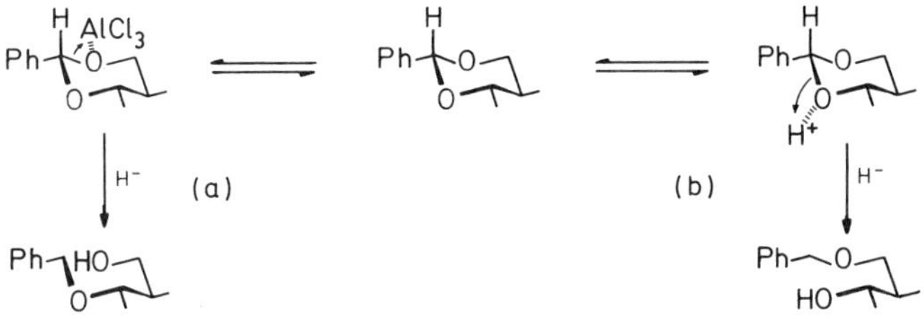

The dioxane and dioxolane-type benzylidene acetals can be reductively cleaved to give a partially benzylated diol. Lithium aluminium hydride — aluminium chloride, sodium cyanoborohydride — hydrogen chloride, and borane — trimethylamine — aluminium chloride are the reagents used with carbohydrate acetals. The regioselectivity in the reductive ring openings is the same with the all three reagents for dioxolane benzylidene acetals, but strongly varies with the reagent when the dioxane rings of 4,6-*O*-benzylidene acetals of protected hexopyranosides are to be cleaved. For sodium cyanoborohydride — hydrogen chloride, the benzyl group in the product is at O-6 and hydroxyl group OH-4 is free [167, 168]. The yields are generally high for all configurations tested [45, 47, 113, 167–173]. The corresponding reductions with lithium aluminium hydride and aluminium chloride, on the other hand, tend to give the opposite regioselectivity, particularly if the O-3 protective group is bulky [174–181]. The difference in the opening 4,6-*O*-benzylidene acetals using the two methods may be explained by steric factors. The greater steric demand of a Lewis acid, as compared to a proton, directs the reductive opening using LiAlH$_4$–AlCl$_3$ to take path (a). In NaBH$_3$CN–HCl reductions, however, the steric demand of the electrophile is much smaller, and the direction is governed by acidities of OH-4 and OH-6 (path b). The regioselectivity achieved with the third reagent, borane — trimethylamine — aluminium chloride, depends on the solvent. The reaction in oxolane produces the 6-benzyl ethers in good yields, whereas the use of toluene gives the 4-benzyl ethers, in addition to degradation products. Stronger solvation of cationic intermediates in the former solvent has been used for explanation of this difference in regioselectivity [182].

The 4-O-benzyl compound was isolated as the main product of cleavage (> 65%) of each of the alkyl 2,3-di-O-benzyl-4,6-O-benzylidene-gluco-, -manno-, and -galacto-pyranosides with LiAlH$_4$–AlCl$_3$ reagent [176, 177, 180] (for comparison with the alternative tritylation sequence, see Ref. [183]). The yields are high even for 4,6-O-benzylidene derivatives having large substitutents, such as an 18-crown-6 moiety, at positions 2 and 3 [184], and for 3-O-allyl derivatives [185, 186]. The less bulky group at C-3, such as methoxyl or hydroxyl group, tends to favor the formation of 6-O-benzyl compounds [176, 177, 180, 187]. The direction of cleavage is independent of the anomeric configuration, character of the aglycon moiety, and substitution at O-2 [176].

The reaction selectively distinguishes between the 5- and 6-membered acetal rings, the former being cleaved more rapidly [188, 189]. Benzyl 2,3(R):4,6(R)-di-O-benzyli-dene-α-D-mannopyranoside gives the 3-benzyl ether of 4,6-O-benzylidene glycoside in good yield [63, 189]. On prolonged treatment, of course, benzyl 3,4-di-O-benzyl-α-D-mannopyranoside can be prepared in a 52% yield [190].

Cleavage of the dioxolane is selective and dependent on the configuration at the acetal carbon atom. The 3-O-benzyl derivative is formed from the 2,3(R)-, i.e., *exo* isomer mentioned above, whereas the other, *endo* isomer yields the 2-benzyl ether [63, 189]. Steric reasons are responsible for the validity of this rule that the *exo* dioxolane provides benzyl ether containing free axial hydroxyl group [188–197]. Even the 1,2-O-benzylidene derivatives (16) and (18) give the benzyl glycoside (19) or 2-O-benzyl-aldose (17) according to the configuration at the acetal atom [198]. The eventual presence of the other benzyl ether as a by-product is caused by the *endo-exo* isomerization of dioxolane-type benzylidene acetals under the conditions of LiAlH$_4$–AlCl$_3$ acetal ring cleavage [199].

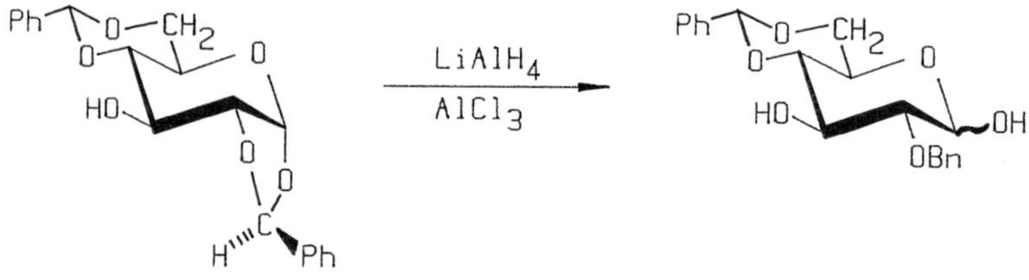

(16)   (17)

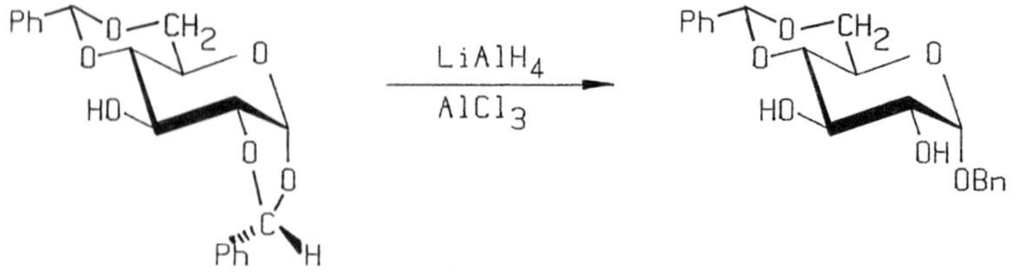

(18)   (19)

The hydrogenolysis of methyl 3,5-O-benzylidene-α- and -β-D-xylofuranoside derivatives gave 5-benzyl ethers as main products. In some cases the attack of the reagent occurred at the ring oxygen of the furanoside skeleton to yield 5-O-benzyl-1-O-methylxylitol derivatives [200].

For sodium cyanoborohydride — hydrogen chloride reagent the direction of the reductive opening of the dioxolane acetals obeys the same rule, it depends on the stereochemistry at the asymmetric, benzylidene acetal carbon [168, 169]. Methyl *exo*-2,3:4,6-di-O-benzylidene-α-D-mannopyranoside is cleaved in oxolane solution to give 50% of the 3,6-di-O-benzyl derivative, which is also the major product in the reaction of methyl 3-O-benzyl-4,6-O-benzylidene-α-D-mannopyranoside. The 2-O-benzyl isomer (20) was cleaved nonselectively, indicating again the effect of the bulk of the C-3 substituent [169]. A compatibility of this reagent, as well as of borane — tri-

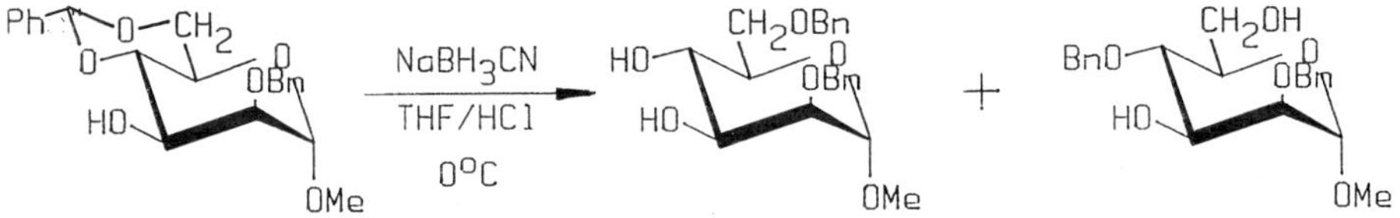

| (20) | (21) (32%) | (22) (30%) |

methylamine — aluminium chloride one, with the presence of ester protecting groups should prove useful in synthetic carbohydrate chemistry [171, 182].

2.7 Debenzylation

Of the variety of debenzylation methods (for review see Refs. [7, 23, 24]), acetolysis is the most frequently used for the preparation of partially benzylated saccharides. As observed for hexitol derivatives first [201], primary benzyl ethers are split more rapidly than the secondary ones. Thus, 2,3,4,6-tetra-O-benzyl-D-glucopyranose was acetolyzed with acetic anhydride — acetic acid — sulfuric acid reagent to give 1,6-di-O-acetyl-2,3,4-tri-O-benzyl-α,β-D-glucopyranose in a 95% yield [202, 203]. When compared to tritylation-benzylation-detritylation sequence, the complete benzylation followed by selective acetolysis at the primary position has been found more convenient also with D-mannose and D-galactose derivatives, either in the form of hemiacetal, or glycoside, or orthoester [202, 204, 204a]. The detailed results obtained [205] with methyl 2,3,4,6-tetra-O-benzyl-α-D-glucopyranoside and methyl 2,4,6-tri-O-benzyl-α-D-glucopyranoside (23) show that the ease of cleavage of the benzyl ethers

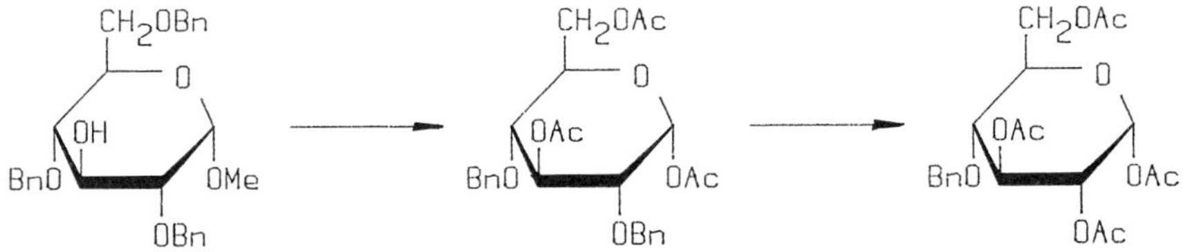

| (23) | (24) | (25) |

of D-glucopyranose follows the order C-6 > C-3 > C-2 > C-4. Depending on the reaction time, 2,4-di-O-benzyl or 4-O-benzyl derivative (24 or 25) could be isolated in higher than 50% yield. With this in mind, the high reactivity of the 4-O-benzyl group (C-6 > C-4 > C-3 > C-2) under acetolysis using ferric chloride — acetic anhydride reagent claimed recently [206] for D-*gluco* configuration seems suspicious, moreover without sufficient experimental support. Acetolysis of 3,4,6-tri-O-benzyl-1,2-O-(1-methoxyethylidene)-β-D-mannopyranose with 2% sulphuric acid at 0–5 °C for 2 h gave 1,2,6-tri-O-acetyl-3,4-di-O-benzyl-α-D-mannopyranose in good yield [207, 207a]. On prolonged treatment at ambient temperature, the 3-benzyl ether was the major product [207]. Similar suceptibility of the 6-O-benzyl group on a mannopyranosyl residue towards acetolysis was observed in the synthesis of oligosaccharides [116a] or in the case of methyl 2,4-di-O-acetyl-3,6-di-O-benzyl-α-D-mannopyranoside which on acetolysis afforded 1,2,4,6-tetraacetate in 74.8% yield [207a]. The steric hindrance caused by *cis*-vicinal neighboring substituent may be responsible for the slow reaction at O-3 of the D-*manno* derivatives, just as described for selective acetolysis of benzyl ethers of *myo*-inositol [208].

As slight variations in acidity are sufficient to markedly alter the rate, some problems with the reproducibility of the acid catalyzed acetolysis may be encountered. Careful monitoring of the reaction by ^{1}H NMR spectroscopy, in preference to TLC, is recommended [202].

Treatment of benzyl 2,3,4,6-tetra-O-benzyl-β-D-glucopyranoside with a limited amount of sodium in liquid ammonia furnished 2,3,4,6-tetra-O-benzyl-D-glucopyranose as a principal product [209].

Heterogeneous catalytic transfer hydrogenolysis with palladium catalysts and formic acid [210], ammonium formate [210a, 210b], 2-propanol [211, 212, 212a], and cyclohexene [213] as a hydrogen donor provides a highly efficient and experimentally facile means for the removal of O-benzyl groups of carbohydrate derivatives. The reaction seems to be structure sensitive, 20% of the 2-O-benzyl and 20% of the 3-O-benzyl derivative have been obtained [211] from methyl 2,3-di-O-benzyl-β-D-glucopyranoside using palladium on alumina in 2-propanol. Benzyl group in position 2 of methyl 2,3-di-O-benzyl-4,6-O-benzylidene-α-D-glucopyranoside was regioselectively cleaved [210b] using palladium on carbon and ammonium formate in methanol, the product, methyl 3-O-benzyl-4,6-O-benzylidene-α-D-glucopyranoside being isolated in 83% yield.

Complete series of 1,6-anhydro-2,3,4-tri-O-benzyl-β-D-hexopyranoses was used to follow the stereochemical aspects of this heterogeneous, catalytic transfer hydrogenolysis. When adjacent *cis*-disposed O-benzyl groups are present, one of them may act as hydrogen donor and partially O-benzoylated and O-benzylated derivatives can be obtained. For instance, a 1:1 mixture of 3,4-di-O-benzyl and 3-O-benzoyl-2-O-benzyl derivatives results from the hydrogenolysis of 1,6-anhydro-2,3,4-tri-O-benzyl-β-D-galactopyranose. On prolonged treatment, 70% of 1,6-anhydro-3-O-benzoyl-β-D-galactopyranose can be isolated, together with some D-galactosan [212, 212a]. The benzyloxy group in 1,3-diaxial relationship to the anomeric methoxyl group could be hydrogenolyzed on Pearlman's catalyst in ethanol even without cyclohexene added. Simple benzyl ethers, such as 1-benzyloxy-9-decene are not cleaved with this mixture. The methoxyl group might be acting as a second ligand to stabilize binding to palladium [214].

Chemoselective catalytic hydrogenolysis (10 % Pd/C) of the glycosidic benzyl group of a 2-amino-2-deoxy-D-glucosederivative that carries another benzyl group protecting a 3-hydroxytetradecanoyl substituent at O–3 has been applied in an efficient synthesis of lipid Y [215]. Benzyl ether is cleaved selectively in the presence of p-methoxybenzyl ether by Raney nickel hydrogenolysis [216].

Some other mild deprotection methods, such as ozone treatment followed by debenzoylation [217], cleavage with alkylthiotrimethylsilane [218], ethanethiol — boron trifluoride etherate [219], lithium —ethylamine [220], or anodic oxidation [221] could appear to be of potential value for partial deprotection. Important selectivity has been observed in the reaction of methyl 4-O-benzyl-6-deoxy-2,3-O-isopropylidene-α-D-mannopyranoside with dibromomethyl methyl ether and zinc bromide, 4-O-benzyl-6-deoxy-2,3-O-isopropylidene-α-D-mannopyranosyl bromide being the only product (90 %) after 5.5 h at ambient temperature. Prolonged treatment (21 h) is necessary to cleave the benzyl protecting group from this useful synthon [221a]. Similarly, 3-O-benzyl group of methyl 2,4,6-tri-O-benzoyl-3-O-benzyl-β-D-galactopyranoside is not cleaved with excess of dichloromethyl methyl ether in the presence of fused zinc chloride catalyst under carefully controled conditions. The corresponding glycosyl chloride (80 % yield), an important intermediate in the synthesis of oligosaccharides, can of course be debenzylated on further treatment with the reagent [222].

Iodotrimethylsilane in acetonitrile converts [222a] benzyl 2,3-di-O-acetyl-4,6-di-O-benzyl-α-D-glucopyranoside in 15 min into benzyl 2,3-di-O-acetyl-4-O-benzyl-α-D-glucopyranoside (yield 68 %). Some 5 hours are necessary to cleave the second benzyl group in position 4, benzyl 2,3-di-O-acetyl-α-D-glucopyranoside being isolated in 61 % yield [222a].

After hydrogenolysis of 1,6:2,3-dianhydro-4-O-benzyl-β-D-mannopyranose over 10 % Pd/C in ethanol, the reductive cleavage of an oxirane as well as of the 1,6-anhydro ring has been observed to give 1,5-anhydro-2-deoxy-D-*arabino*-hexitol as the final product [223].

3 Allyl Ethers

Allyl ethers are useful temporary protecting groups in the chemical synthesis of carbohydrate analogues and complex oligosaccharides since they can be removed in the presence of other protecting groups (e.g., acyl, benzyl, or acetal) and can also act as a nonparticipating substituent to effect, for instance, α-glycosylation. Their facile rearrangement to easily removable, labile prop-1-enyl ethers forms a basis of their increasing, convenient use [224, 225]. Allyl groups may be used as TLC tracers as they are selectively detected by potassium permanganate spray [43]. The advantage of the use of the perdeuterioallyl analogs for the ^{1}H-NMR structure elucidation of complex oligosaccharides was illustrated for the 2,6-dideoxy-L-*ribo*-hexose series [226]. Perdeuterioallyl bromide is easily prepared from propargylic acid [226].

Jan Staněk

3.1 Direct Partial Allylation

The great majority of experimental variants discussed for benzylation (see Sect. 2.1) has also been applied for the preparation of allyl ethers. Among them, the alkylation with allyl bromide and sodium hydride in a dipolar aprotic solvent is most frequently used for complete allylation. Reaction with the methylsulfinyl carbanion in DMSO to form an alkoxide, followed by the reaction with allyl bromide provides a convenient high-yield route to 2,3,6-tri-O-allyl-amylose [227]. With the limited amount of reagent, 35% of methyl 2-O-allyl-3,6-dideoxy-α-D-xylo-hexopyranoside was synthesized from the corresponding glycoside [228]. The 2-allyl ether was the major product (43% yield) of the reaction of methyl 4,6-O-benzylidene-α-D-glucopyranoside with allyl bromide and 1.1 equiv. of sodium hydride in benzene [71].

Sodium hydroxide-induced partial allylation was used for preferential reaction at the equatorial OH-4 in 1,6-anhydro-2-O-benzyl-β-D-galactopyranose [229]. A similar reaction of allyl 2,6-di-O-benzyl-α-D-galactopyranoside with allyl bromide and sodium hydroxide in benzene at room temperature was not as regioselective, the persubstituted 3,4-di-O-allyl derivative dominated in the reaction product [73]. A high, 87% yield of the monoallyl derivative was obtained from 1,2:5,6-di-O-isopropylidene-D-mannitol after the treatment with allyl bromide and sodium hydroxide in aqueous acetone [230].

Silver oxide in DMF was successfully applied for the regioselective reaction of (26) with allyl bromide to give compound (27) in 74% yield [231]. Hydroxyl groups in positions 2 and 6 in cycloheptaamylose were selectively protected by reaction with allyl bromide, barium oxide, and barium hydroxide octahydrate in a DMF-DMSO mixture [232].

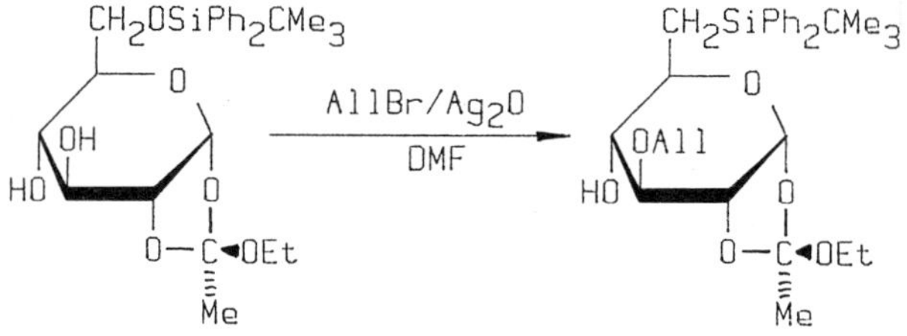

(26)   (27) (74%)

Acid-catalyzed allylation with allyl trichloroacetimidate was used to protect the hydroxyl group without migration of any acyl group present in the molecule of saccharide [89]. Thus, methyl 3,4-di-O-allyl-2,6-di-O-benzoyl-α-D-mannopyranoside was obtained from the corresponding dibenzoate in 68% yield using trifluoromethanesulfonic acid as a catalyst.

3.2 Phase-Transfer Catalyzed Allylation

The phase-transfer technique of partial alkylation introduced by Garegg in the carbohydrate field (see Sect. 2.2) can be used also for the preparation of partially allylated derivatives. It was found that the reaction of benzyl 4-O-benzyl-α-L-rhamnopyranoside with allyl bromide under catalysis by tetrabutylammonium bromide gave 73 % of the 2-allyl ether. Only 9 % of the 3-*O*-allyl and less than 5 % of 2,3-di-*O*-allyl derivative were formed, and negligible amounts of the starting diol remained unreacted [121, 233] under conditions similar to those of benzylation. About 75 % of the 2-allyl ether was obtained also from the corresponding methyl 4-*O*-benzyl-α-L-rhamnopyranoside [121] and from methyl 4,6-*O*-benzylidene-α-D-mannopyranoside [234], even though, in the latter case, the reported conditions (50 % sodium hydroxide, 20 h boiling) seem to be sufficient for complete allylation of both hydroxyl groups.

When benzyl 6-*O*-trityl-α-D-mannopyranoside was allowed to react with allyl bromide under phase-transfer catalysis, the 3-allyl and 2-allyl ethers and the starting material were obtained in yields of 36 %, 23 %, and 24 %, respectively [152]. The absence of the protective group at C-4 might be responsible for this difference in selectivity.

3.3 Dibutyltin Oxide Method

Following the methodology of regioselective enhancement of the nucleophilicity of equatorial hydroxyl groups through dibutylstannylation discussed thoroughly in Sect. 2.3, allyl 2,6-di-*O*-benzyl-α-D-galactopyranoside was convered [39, 129] via its 3,4-di-*O*-dibutylstannylene derivative, into allyl 3-*O*-allyl-2,6-di-*O*-benzyl-α-D-galactopyranoside in 79 % yield. Allyl iodide in DMF for 1 h at 100 °C was necessary for complete monoallylation, as allyl bromide reacted too sluggish [129]. Similarly, methyl 3-*O*-allyl-4,6-*O*-benzylidene-α-D-mannopyranoside has been prepared from the corresponding 2,3-diol [137]. The *myo*-inositol derivative with unprotected vicinal *cis* diol grouping was also allylated at the equatorial position [64]. 3,4,6-Tri-*O*-benzyl-D-mannopyranose (28) was quantitatively converted into allyl β-D-mannopyranoside (29) having a free axial OH-2 [235].

Reaction of benzyl 4-*O*-benzyl-α-L-talopyranoside [142] and methyl 4,6-*O*-benzylidene-β-D-glucopyranoside [140] was less regioselective, giving 54 % and 66 %, respectively, of the 3-allyl ethers. No diallyl ethers were formed and no starting material remained, the rest being the isomeric 2-allyl ether. In both cases, tetrabutylammonium bromide was used to catalyze the reaction of the 2,3-*O*-dibutylstannylene intermediate with allyl bromide.

Benzyl 3-*O*-allyl-β-D-galactopyranoside with three free hydroxyl groups was converted into the 3,6-di-*O*-allyl derivative in 60 % yield, the fully unprotected benzyl β-D-galactopyranoside gave 85 % of benzyl 3-*O*-allyl-β-D-galactopyranoside [54]. All these results confirm that stannylation selectively activates an equatorial hydroxyl group which has an oxygenated function in a *cis* relationship at an adjacent position, even in the presence of a more reactive primary hydroxyl group. Methyl β-D-galactopyranoside [131, 236], methyl α-D-galactopyranoside [131], and methyl α-D-mannopyranoside [235] behave similarly, they all produce only the 3-allyl ether in high yield. Other nonprotected methyl or phenyl hexopyranosides were monoallylated by this method with excellent regioselectivity [131]. Solvent variations can be used to modify the outcome of the reaction in some cases. Thus, the dibutylstannylated methyl α-D-glucopyranoside gave 74.8% of both 2- and 3-allyl ether in the ratio of 79:21 when 1,4-dioxane was used, whereas in acetonitrile the 2-O-allyl derivative was the only monoallylated product isolated in 46% yield [131]. One step functionalization of methyl β-lactoside through its dibutylstannylene complex afforded methyl 3'-O-allyl-β-lactoside with high regioselectivity [237].

3.4 Bis(tributyl)tin Oxide Method

The highly regioselective, partial tributylstannylation of carbohydrates resulting in the nucleophilicity enhancement of the corresponding hydroxyl groups (see Sect. 2.4) was successfully applied in the efficient preparation of partially allylated mono-saccharides. Stannylation of methyl α-D-mannopyranoside with 1.5 molar equiv. of bis(tributyl)tin oxide afforded methyl 3,6-bis-O-tributylstannyl-α-D-mannopyrano-side as an unstable oil which was directly transformed into methyl 3,6-di-O-allyl-α-D-mannopyranoside. However, 7 days heating with allyl bromide at 80 °C were necessary to achieve a 70.9% yield, together with some 12.7% of the 3-allyl ether [151, 238] (for the allylation of the corresponding benzyl glycoside, see Ref. [152]). When benzyl 6-O-triphenylmethyl-α-D-mannopyranoside (30) was treated under the same conditions, the reaction was even slower; only 31% of the desired 3-O-allyl

(30)   (31) (62%)   (32) (15%)

derivative (31) could be isolated from the mixture still containing 38% of the starting triol [152]. Fortunately, it was found that the reaction of allyl bromide with the stan-nylated intermediate can be strongly accelerated by the addition of 0.1–0.3 equiv. of tetrabutylammonium bromide. After 48 h at 80 °C, 62% of (31) nad 15% of (32) were isolated from the mixture, no more starting material being recovered [152]. Similar results were obtained for allyl 6-O-triphenylmethyl-α-D-mannopyranoside [239]. Tetrabutylammonium iodide was found to be a more efficient catalyst but is not used since the presence of free iodine makes the isolation more difficult [152].

Partial stannylation of methyl β-D-galactopyranoside leads to the formation of the 3,6- and 2,3,6-stannylated derivatives which react with allyl bromide preferentially at O–6 and O–3 to give 50.7% of the 3,6-di-O-allyl and 11.3% of 6-O-allyl derivative [151, 154]. N-Methylimidazole can be used to catalyze the reaction [240]. Methyl α-D-glucopyranoside was converted into methyl 2,6-di-O-allyl-α-D-glucopyranoside in a 41.8% yield, accompanied with 23% of the 6-allyl and 8% of the 3,6-diallyl ether [151, 153].

3.5 Copper Chelates

Carbohydrate derivatives having vicinal hydroxyl groups (eq-eq or ax-eq) or those having free HO-4 and OH-6 are able to form copper complexes when mixed with sodium hydride and copper(II)chloride in the molar ratios of 1:2:1. The resulting dark-green solutions in oxolane or 1,2-dimethoxy-ethane react with excess of boiling allyl iodide for several hours to give excellent yield of monosubstituted products [165, 166]. The more acidic of the two hydroxyl groups generated an anion which bonds more tightly with copper, making its electrons less available for nucleophilic attack on allyl iodide, so that the other one reacts preferentially. Consequently, the 3-allyl ethers [e.g. (34)] are the major products obtained from 2,3-diols [e.g. (33)], and the 4-allyl ethers from the 4,6-diols [166]. No disubstitution is observed, although both alkoxides generally react. Monoalkylation and the formation of one copper-iodide bond evidently decrease the reactivity of the second alkoxide [166]. In 3,4,6-

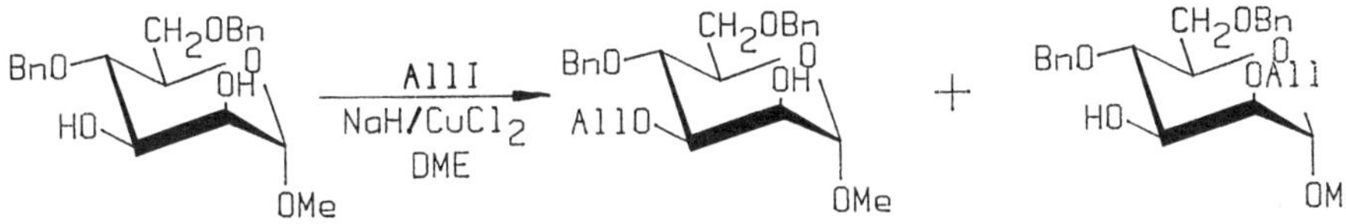

(33)   (34) (76%)   (35) (19%)

tri-O-benzyl-D-glucopyranose both free hydroxyl groups are acidic, i.e., both anions are deactivated to such a degree that neither react with allyl iodide.

The regioselectivity is influenced by coordination of solvent molecules to copper central atom of complexes. This effect of solvent seems to be variable and not yet predictable but, nevertheless, it is of preparative value. Thus, methyl 2,3-di-O-benzyl-α-D-glucopyranoside gave 61% of the 4-O-allyl and 15% of the 6-O-allyl derivative in 1,2-dimethoxyethane, whereas only 36% of the former and 31% of the latter were formed in oxolane, the rest being the unreacted starting material in both cases. Mono-O-allylated sucroses were prepared from the corresponding copper complexes in high yield using dimethyl sulfoxide as a solvent [241].

3.6 Reductive Cleavage of Prop-2-enylidene Acetals

Reductive cleavage of 4,6-O-prop-2-enylidene acetals, such as (36), with sodium cyanoborohydride — hydrogen chloride in oxolane gives the corresponding 6-O-allyl derivatives [e.g. (37)] in good yield [242]. The reaction is compatible with the

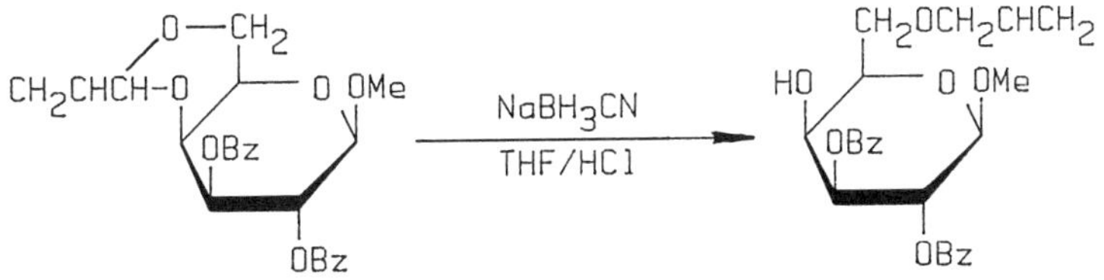

(36)   (37)

presence of ester or acetamido groups at other positions of the hexopyranoside. Opening of these acetals with lithium aluminium hydride — aluminium chloride which gives the alternative regioselectivity (see Sect. 2.5) proceeds only in low yield [243].

3.7 Deallylation

Of the methods described for the removal of the allyl protecting groups (for review see Refs. [7, 225]), the rearrangement of allyl ethers to labile prop-1-enyl ethers followed by dilute acid hydrolysis [244], or by mercury(II)chloride and mercury(II)oxide treatment [245], or by oxidative cleavage [246] is the most frequently used in the carbohydrate field. This rearrangement, readily achieved by treatment with potassium tert-butoxide in DMSO [246, 247], tris(triphenylphosphine)rhodium(I)chloride in the presence of 1,4-diazabicyclo[2.2.2]octane [248, 249], palladium on carbon [250], palladium chloride and sodium acetate [251], diethylazodicarboxylate [252], or with trans-[Pd(NH$_3$)$_2$Cl$_2$] [253] has a great yet unexplored potential for the preparation of partially allylated carbohydrates, because its rate is expected to depend strongly on the structure. For example, allyl 6-O-allyl-α-D-galactopyranoside was quantitatively converted with the last-mentioned reagent into the corresponding prop-1-enyl analog in 60 min, whereas only a 30% isomerization could be observed for the more sterically hindered 6-O-allyl-1,2:3,4-di-O-isopropylidene-α-D-galacto-pyranose after 4 h [253]. It should be noted that the prolonged refluxing with this reagent in tert-butanol causes not only isomerization but also cleavage of the prop-1-enyl grouping formed. The higher stability of the prop-1-enyl grouping at the anomeric center enables a partial removal of allyl groups, without deprotection of the anomeric hydroxyl group [253]. In methyl 2,3-di-O-allyl-4,6-O-benzylidene-α-D-glucopyrano-side the 2-O-allyl group was found to rearrange more rapidly [245]. Tris(triphenylphosphine)rhodium(I)chloride isomerizes the allyl group significantly faster than the analogous but-2-enyl group [249]. Selective cyclization to acetals of the prop-1-enyl ethers having a hydroxyl group in a suitable position (see, e.g. the conversion [245] of 38 to 39) presents further possibilities of partial deprotection [245, 246, 253].

(38)   (39) (57%)

4 Triphenylmethyl Ethers

The triphenylmethyl (trityl) group remains the most frequently used for the temporary protection of primary hydroxyl groups (for review see Refs. [2, 3, 254]) which can be conveniently regenerated from the corresponding ethers by mild acid treatment or by hydrogenolysis. Moreover, good crystallizing properties of trityl ethers, easy analytical control of the course of tritylation, the steric bulk of this alkyl group, and other advantages made this protective group attractive also for secondary hydroxyl groups. Assignment of trityl residues in derivatized carbohydrates is simple using NMR data, especially with a two-dimensional, heteronuclear shift-correlated NMR technique [255].

Problems may appear in the chromatographic purification of trityl ethers; depending on the type and activity of the sorbent, the silica gel-catalyzed detritylation has been observed [256–259]. In 1962, Buchanan and Schwarz found it necessary to pre-treat with ammonium hydroxide the silica gel used for the chromatographic purification of trityl derivatives [258].

In addition to standard syntheses in solution, even the solid-phase version has been developed, using a polymer-bound tritylation reagent [260–263]. With such temporary protection of the primary hydroxyl group, methyl 2,3,4-tri-*O*-benzoyl-α-D-gluco-pyranoside has been prepared from methyl α-D-glucopyranoside [260, 261].

4.1 Triphenylmethyl Chloride — Pyridine

The classical method [2, 3] for the preparation of triphenylmethyl (trityl) ethers in-volves the reaction of the carbohydrate with trityl chloride in the presence, preferably as solvent, of pyridine at temperatures ranging from room temperature up to 100 °C. Only the tertiary hydroxyl group, such as in 1,2-*O*-isopropylidene-3-*C*-methyl-α-D-erythrofuranose [264], is resistant under these conditions. However, the selectivity for the primary position over the secondary one is fairly good due to a large steric requirement of *N*-triphenylmethylpyridinium cation from which the alkyl group is transferred to the alcohol. Hunreds of such applications with higher than 50% yields of the isolated products partially tritylated at the primary positions can be found in the literature and will not be treated here. In suitable cases, such as the derivatives of β-D-*galacto* configuration [265, 266] like benzyl 2-azido-2-deoxy-β-D-galacto-pyranoside (40) [267] the almost quantitative yield of the product (e.g. 41) can be obtained. In unsuitable cases, of course, the reactivity of the primary hydroxyl

$$\text{(40)} \qquad\qquad\qquad \text{(41) (97\%)}$$

groups can be very low. For example, methyl 4-*O*-methanesulfonyl-2,3-di-*O*-methyl-α-D-galactopyranoside is almost impossible to be tritylated with the excess of reagent [268].

The observed differences [3] in the reactivity of primary hydroxyl groups were preparatively exploited in the structural modifications of oligosaccharides. The monomolar tritylation of phenyl α-maltoside gives the 6'- to 6- ratio of about 1.3, whereas 2.9 is observed for phenyl β-maltoside [269]. Generally, the orientation of the anomeric group has an influence on the reactivity of OH-6 (see also, Refs. [270–272]). The 6'-O-trityl derivative dominates over the 6-O-trityl isomer also in the mono-molar tritylation of maltose [273, 274], In dimolar tritylation, of course, the 6,6'-di-O-trityl derivative is the major product [274, 275]. Small amounts of trityl 6'-O-trityl-β-maltoside, 2,6'-di-O-trityl-α-maltose, and trityl 6,6'-di-O-trityl-β-maltoside could also be isolated using an appropriate separation technique [275]. In malto-triose, as well as in its 1,6-anhydro derivative, the primary hydroxyl group at C-6'' seems to be the most reactive [276].

Tritylation of sucrose has been explored intensively [277–282] to show the low reactivity of the primary hydroxyl group at C-1 of the β-D-fructofuranosyl moiety. Monotritylation of raffinose afforded the 6'-trityl and 6''-trityl ethers in 1.7 and 9.4% yields [283], whereas the 1',6',6''-tri-O-trityl derivative was the major product (20%) of the trimolar tritylation [284].

Attempted regioselective monotritylation of methyl β-sophoroside was unsuccess-ful, it gave a mixture that was inseparable by column chromatography or fractional crystallization [285]. Similar problems were encountered for tritylation of laminari-biose [286] and cellobiose [287–289].

The hemiacetal hydroxyl group is more reactive than the other secondary ones. Thus, trityl 5-O-tritylpentopyranosides are obtained from pentoses when an excess of trityl chloride [290] or elevated temperature [291] is used. For D-lyxose, the ditrityl derivative is formed considerably even at room temperature with only one molar equiv. of the reagent. Under such conditions, the other three pentoses give the 5-trityl ethers in ca. 60% yields [291]. Further re-investigation of the monotritylation of D-xylose at 50 °C revealed the 5-, 4-, 3-, and 1- substitution in the ratio [292] of 100:25:0:36.

Differences in the reactivity of secondary hydroxyl groups of nucleosides were studied very thoroughly. In 1934, Levene and Tipson prepared [293] the first ditrityl ether starting from 5'-O-trityluridine. It was later shown to be [294] the 2',5'-ditrityl ether, and both 2',5'- and 3',5'-di-O-trityl derivatives were then isolated in comparable yields from the tritylation of uridine by several research groups [295–301]. Further tritylation of the cis vicinal secondary hydroxyl group in these compounds was found not to be precluded from the sterical reasons [296–299] (cf. Ref. [295]), pertrityluridine (42) being isolated in a 1–4% yield. Similar results were obtained for 5'-O-acetyl-uridine [302], N-acetylcytidine [303], N-benzoylcytidine [304], and other nucleosides [300, 305, 306]. However, an analogous treatment of 6-azauridine [295] or 6-azacyti-dine [307] gave the 5'-monotrityl ethers only. Tritylation of 1-(5'-O-trityl-β-D-ara-

(42)

binofuranosyl)uracil yielded the 3′,5′-ditrityl ether as the sole partially alkylated product [308, 309]. Partial tritylation of the anomeric mixture of methyl D-xylo-furanosides gave the 2,5-di-O-trityl derivative [310].

Just as in the case of nucleosides, the older results were re-investigated with novel methods also for tritylation of pyranosides. All the monotrityl ethers of methyl β-D-xylopyranoside and methyl α-D-xylopyranoside have been isolated recently [311] to complement the pioneering work by Hockett and Hudson [312]. The monotrityl fraction obtained [313] from methyl α-L-fucopyranoside was now found [314] to consist of 2-O- and 3-O-trityl derivatives in the ratio of 3:2 (D-enantiomer has been actually studied).

Selective tritylation of methyl α-L-rhamnopyranoside yielded 57% of the 3-O-trityl derivative, together with 1% of the 2-O- and 3% of 4-O-trityl isomers. For methyl β-L-rhamnopyranoside, 34% of the 3-trityl and 17% of the 4-trityl ether were obtained [315]. Similarly, the 3,6-ditrityl ether is the major product of ditrityl-ation of methyl and benzyl α-D-mannopyranosides [316]. The corresponding α-D-glucopyranosides yielded the 2,6-di-O-trityl derivative, whereas both the 2,6- and 3,6-ditrityl ethers were isolated in the case of α-D-*galacto* configuration [316].

The bulky triphenylmethyl group can be introduced also at the sterically hindered axial position of 1,6-anhydro sugars. With 3 molar equiv. of the reagent at 85–90 °C for 24 h, more than 8% of 2-acetamido-1,6-anhydro-2-deoxy-3-O-trityl-β-D-gluco-pyranose (45) has been obtained [317] from (43). The 4-trityl ether [i.e., (44)] was, of course, the major product, just as in the case of tritylation of 1,6-anhydro-2-O-*p*-toluenesulfonyl-β-D-glucopyranose [318] or, 1,6-anhydro-2-O-benzyl-β-D-glucopyra-nose [229].

(43)   (44) (66.9%)   (45) (8.3%)

4.2 Miscellaneous Tritylation Techniques

The application of tritylpyridinium tetrafluoroborate, easily obtained from trityl tetrafluoroborate and pyridine, presents definitive advantages [319, 320] over the classical procedure discussed in the previous section. Acetonitrile solvent simplifies the workup procedure, the substantially shorter reaction times enable the use of only a slight excess of the reagent. Preferential tritylation of the primary hydroxyl group in carbohydrates is especially convenient with this method.

A combination of trityl tetrafluoroborate or perchlorate with 2,4,6-tri(*tert*-butyl)-pyridine instead of pyridine enables an almost quantitative tritylation of the secondary hydroxyl group in 1,2:5,6-di-O-isopropylidene-α-D-glucofuranose in 10 min at room temperature [321]. The more readily available 2,6-di(*tert*-butyl)-4-methylpyridine was used [322] in combination with trityl perchlorate for the selective tritylation of methyl 2,6-di-O-acetyl-α-D-galactopyranoside at the equatorial OH-3. Combination of trityl perchlorate with 2,4,6-collidine presents another attractive alternative [323–325].

The electron-donating effect of pyridine substituents becomes especially evident with 4-dimethylaminopyridine. Methyl α-D-glucopyranoside reacts with 1.1 molar equiv. of trityl chloride, 1.5 molar equiv. of triethylamine, and a catalytic amount of 4-dimethylaminopyridine in DMF to give 88% of the 6-trityl ether after only several hours at room temperature [326]. With the classical method, the yield was 61% and some difficulties with the formation of addition complexes of the product with pyridine were encountered [326]. Surprisingly, pyridine still remains frequently used as a solvent in combination with this effective catalyst [327, 328]. Of the two secondary hydroxyl groups of methyl 4,6-O-benzyl-α-D-mannopyranoside [119] or methyl 4-O-benzyl-α-L-rhamnopyranoside [122], the equatorial OH-3 was protected first. Tritylation of methyl 2-O-benzyl-α-L-fucopyranoside (46) proceeded readily to give (47) in 70% yield [109].

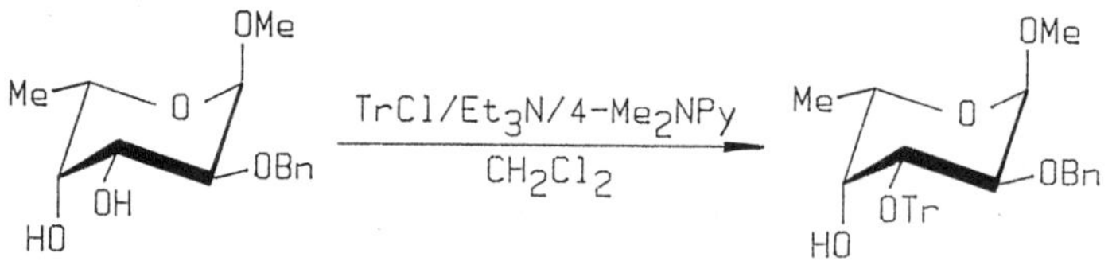

(46)   (47) (70%)

Monotritylation of D-xylose in HMPT with trityl chloride in the presence of silver acetate gave 5-, 4-, 3-, and 1-trityl ethers in the ratio of 100:57:49:0, quite different of that of the classical procedure [292]. Nucleosides were partially tritylated with

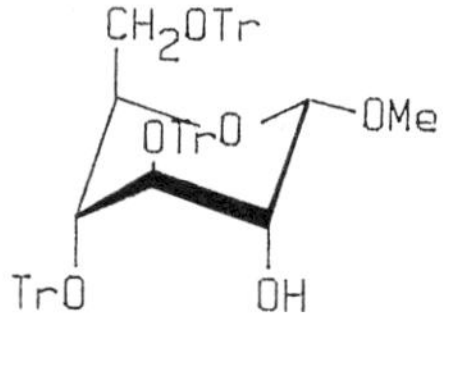

(48)

trityl chloride in dichloromethane with 4-Å molecular sieve as a neutral acid scavenger [329]. The selectivity of this method can be increased by the addition of pyridine [329].

Activation of hydroxyl groups via tributylstannylation (see Sect. 2.4) was used in the partial tritylation of methyl α-D-glucopyranoside with trityl chloride [153]. Stannylation effectively enhances the reactivity of OH-6 and OH-2 to give 53.6% of the corresponding ditrityl ether. Surprisingly, 37.3% of methyl 3,4,6-tri-O-trityl-α-D-glucopyranoside were also isolated, the $^1C_4$(D) conformation (48) being more stable in deuterochloroform than the expected $^4C_1$(D). A sterically demanding electrophile trityl chloride shows different regioselectivity than benzyl or allyl bromide; the treatment of the partially stannylated methyl β-D-galactopyranoside with trityl chloride for 35 h at 55–60 °C afforded 71.6% of the 2,6-ditrityl ether [154] (cf. Sects. 2.4, 3.4).

4.3 Detritylation

As indicated by a large number of procedures described, detritylation seems to be a problematic operation. Low yields, formation of by-products, glycoside cleavage (see, e.g., purine deoxynucleoside derivatives), and especially the acyl migration, often accompanying the acid-catalyzed detritylations (80% acetic acid at reflux [330–333], hydrogen chloride in methanol [334] and other solvents [335–338], hydrogen bromide in acetic acid [339–342], etc. [3]), are attracting permanent interest of researchers. Detritylation with aqueous trifluoroacetic acid [343], sometimes diluted with butanol [344], copper(II)sulfate in boiling benzene [345], powdered anhydrous zinc bromide in dichloromethane [329], boron trifluoride etherate and methanol in aprotic solvents [346], aqueous tetrafluoroboric acid in acetonitrile [347], or iodotrimethylsilane [222a, 348] were developed and used successfully to avoid hydrolysis or migration of acetyl groups, the last two methods being superior. The easily available pyridinium perchlorate in nitromethane-methanol mixtures is another effective catalyst for detritylation of otherwise acetylated glycosides [349]. It should be noted that the acidity of pyridinium chloride is also sufficient for complete detritylation [350]. For reductive detritylation, sodium or lithium in liquid ammonia were applied [351–353]. These reagents are, however, unattractive for use with nucleoside derivatives since the thymine ring is reduced as well. In such cases, the reaction with naphthalene radical-anion in HMPT has been effective [354].

For partial detritylation, however, only the classical methods were used. 3',5'-Di-O-trityluridine was partially detritylated to about the same extent at C-3' and C-5' by heating with 80% acetic acid at 50 °C for 1 h. By contrast, 2',5'-di-O-trityluridine was not detritylated under these conditions, but required a temperature of 80 °C to yield 20% of the 2'-trityl ether and traces of 5'-trityl ether [355]. Different reactivity of trityl groups was also observed for 3',5'-di-O-tritylthymidine, resulting in 38% of the 3'- and 6% of the 5'-trityl ether [356]. Removal of trityl groups from deoxynucleosides is much slower at the 3'- than at 5'-position also with finely powdered zinc bromide catalyst, probably due to the absence of the neighboring oxygen-chelating site [357]. Selective loss of the 5'-O-trityl group of 2',5'-di-O-trityluridine occurred on treatment with methyltriphenoxyphosphonium iodide in DMF-containing pyridine [358].

Catalytic hydrogenolysis of trityl ethers [3, 335, 359, 360] on platinum, palladium, or Raney nickel has been used for the preparation of partially tritylated carbohydrate derivatives. Partial hydrogenolysis of 3',5'-di-O-trityluridine gave 5'-O-trityluridine as the only monotritylated product [302]. Similarly, formation of the 5'-trityl ether was favored in the reaction of 3',5'-di-O-tritylthymidine [356]. 2'-O-Trityluridine has been obtained, albeit in poor yield, from 2',5'-di-O-trityluridine [302]. Some selectivity for the primary C-5 position was observed also for trityl 2(or 3),5-di-O-trityl-D-riboside [361]. Catalytic hydrogen transfer cleavage by using palladium and 2-propanol as hydrogen donor may offer some selectivity in the removal of this protecting group [211].

5 Miscellaneous Arylalkyl Ethers

5.1 Arylmethyl Ethers

The substituted arylmethyl group can offer various advantages over the traditional benzyl-protecting group. Thus, 2-nitrobenzyl ethers of carbohydrates are of interest because they are readily removed by photolysis [362, 363] in methanol and DMF (see also, the cleavage of 2-nitrobenzyl glycosides [363 a, 363 b]), the 4-methoxybenzyl group can be cleaved in the presence of benzyl ethers by DDQ oxidation [216] or by using cerium(IV)ammonium nitrate in aqueous acetonitrile [364], solid 4-chloro-benzyl chloride is less lachrimatory than benzyl chloride and the corresponding ethers are more readily crystalline [365], etc. Methods of preparation of such partially alkylated carbohydrates parallel those discussed in Sect. 2.

Mixtures of photolabile 2'- and 3'-O-(2-nitrobenzyl)ribonucleosides were obtained in good yields when adenosine, uridine, cytidine, and inosine were treated with 2-nitrophenyldiazomethane in the presence of tin(II)chloride [362]. The dibutylstan-nylene approach has been used for the preparation of 2'-(4-nitrobenzyl)uridine [366]. 2-Nitrobenzyl chloride and sodium hydride in DMF gave the 2'-ethers in 26–37% yield, depending on the starting nucleoside [363, 367].

Treatment of adenosine with 4-methoxybenzyl bromide and sodium hydride yielded the 2'-ether selectively, whereas both 2'- and 3'-ethers resulted from the tin(II)chloride catalyzed reaction with 4-methoxyphenyldiazomethane [368]. A 1:1 mixture of 2'- and 3'-O-(4-methoxybenzyl)uridine has been obtained by the dibutylstannylene approach [369]. Either 6- or 4-O-(4-methoxybenzyl) derivatives were obtained from Garegg's sodium cyanoborohydride redutive cleavage (see Sect. 2.6) of methyl 4,6-O-(4-methoxybenzylidene)-α-D-gluco- and galactopyranosides, depending on the type of catalyst and solvent. Trifluoroacetic acid catalyzes the formation of primary 6-O-(4-methoxybenzyl) derivative, whereas the sterically more demanding chloro-trimethylsilane activates preferentially O–6 to give 4-O-(4-methoxybenzyl) isomer [364]. Similarly, the reductive cleavage of allyl glycoside 49 gave the 4-ether 50 [370].

(49)   (50)

Methyl 4-O-(4-methoxybenzyl)2,3-di-O-methyl-α-D-glucopyranoside has been prepared [371] by Liptak's procedure (see Sect. 2.6). This lithium aluminium hydride — aluminium trichloride method was also used in the synthesis of 4-hydroxy-3-methoxy-benzyl [372], 4-hydroxy-3,5-dimethoxybenzyl [372], and 1-phenylethyl [373] ethers from the 4,6-acetals derived from vanillin, syringealdehyde, and acetophenone. Various vinylbenzyl ethers were prepared by the reaction of carbohydrates with vinylbenzyl chloride, and copolymerized with styrene [374].

The remarkable regioselective cleavage of the 4-chlorobenzyl group at O–2 with the simultaneous inversion of the anomeric configuration of methyl 2,3,5-tris-O-(4-chlorobenzyl)-β-D-ribofuranoside (51) with tin(IV)chloride gave a partially protected glycoside 52 in 83 % yield [374a]. The relative orientation of the 4-chlorobenzyloxy groups plays an important role, the corresponding arabinofuranoside and lyxofuranoside do not react in this way.

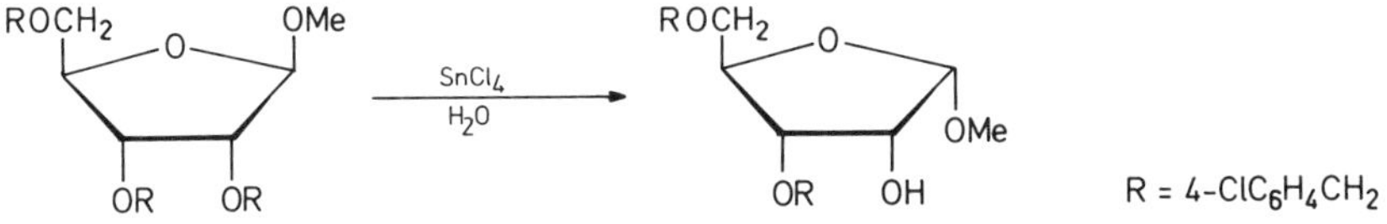

(51) (52) (83 %) R = 4-ClC$_6$H$_4$CH$_2$

5.2 Diarylmethyl Ethers

The diphenylmethyl protecting group has been introduced to complement the benzyl — trityl series. Diphenylmethylation of carbohydrate hydroxyl groups may be effected by the uncatalyzed thermal reaction with diazo(diphenyl)methane in benzene or acetonitrile solution. No marked selectivity for any particular hydroxyl group was observed under these conditions [375]. However, in the presence of catalytic amounts of tin(II)chloride, methyl 4,6-O-benzylidene-α-D-mannopyranoside (53) gave a high yield of 3-O-diphenylmethyl derivative (54). The more convenient and easily available reagents diazo[bis(4-methylphenyl)]methane, diazo[bis(4-chlorophenyl)]methane, diazo[bis(4-methoxyphenyl)]methane, and 9-diazofluorene gave similar results. It should be noted that methyl 4,6-O-isopropylidene-α-D-mannopyranoside showed less pronounced selectivity of alkylation [376].

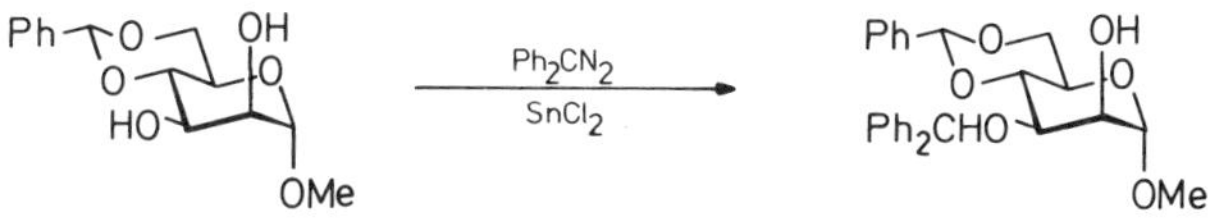

(53) (54) (70 %)

5.3 Triarylmethyl Ethers

The introduction of methoxy substituents increases the ease of removal of trityl groups under acid conditions, but also decreases the selectivity for the primary hydroxyl group [377]. As a compromise, the monomethoxytrityl and dimethoxytrityl groups found widespread use for the protection of primary positions of glycosides (see, e.g. Ref. [378]) and, especially for the OH-5′ function of nucleosides [377]. The use of 6-nitroquinoline instead of pyridine was described [379] to improve the selectivity of monomethoxytritylation of nucleosides and nucleotides bearing free amino groups. A general and rapid procedure was developed for the preparation and isolation of 5′-O-dimethoxytrityl derivatives [380].

The selectivity can be further improved by turning to the more bulky triarylmethyl groups, such as bis(4-benzyloxyphenyl)phenylmethyl [381], 4-bromophenacyloxyphenyldiphenylmethyl [381], or tris(4-benzoyloxyphenyl)methyl [382] group. Thus, methyl α-D-glucopyranoside carrying the latter group at O–6 could be isolated in 81 % yield after some 20 min treatment with 1.2 molar equiv. of the corresponding tri-arylmethyl bromide in pyridine — triethylamine mixture at 65 °C. Even higher yields were obtained for various nucleoside derivatives [382]. Under acid conditions, this protecting group is ca. 5 times more stable than the trityl group, but it can be easily removed by treatment with 0.5 M sodium hydroxide [382]. For the removal of the 4-bromophenacyloxyphenyldiphenylmethyl group, mild reduction with zinc and 20 % acetic acid is efficient [381].

5.4 Miscellaneous Ethers

The use of the p-nitrophenylethyl group, a very versatile blocking group in nucleo-sides, has been reviewed recently [383, 384]. The 2,3-diphenylcyclopropen-1-yl group has been introduced in the carbohydrate field in the search for an ether group which cleaves to form a stable cation, like the trityl group, but in addition can be introduced readily at secondary hydroxyl groups [385]. Thus, benzyl 2-acetamido-3,6-di-O-acetyl-2-deoxy-α-D-glucopyranoside (55) treated with a slight excess of 2,3-diphenyl-1-cyclopropen-1-ylium perchlorate and 2,4,6-trimethylpyridine in acetonitrile or benzene at room temperature yielded 53 % of ether 56. Mixtures of 2'- and 3'-protected ribonucleosides were obtained from the alkylation of nucleosides with 1-oxido-3-

(55)　　　　　(56) (53 %)

methyl-2-pyridyldiazomethane in the presence of tin(II)chloride [386]. 2-Pyridyl-methyl, 2-quinolylmethyl, and similar ethers were prepared by the reaction of cor-responding chlorides under phase-transfer conditions [387].

6 Silyl Ethers

The role of silyl ethers in carbohydrate synthesis increased enormously in the last decade with the introduction of novel types of protecting groups, especially with the introduction of the *tert*-butyldimethylsilyl group [388, 389]. High selectively and manipulative convenience of the reagent, ca. 10^4 times increased stability of ethers against hydrolysis compared to the traditional trimethylsilyl group, and good crystal-lizing properties of the products are responsible for this success. From similar reasons, some authors prefer the *tert*-butyldiphenylsilyl ethers which are easily detected due to a strong UV absorbance and are unaffected by conditions that hydrolyze the *tert*-butyldimethylsilyl group [390]. The high selectivity for the primary hydroxyl group has been observed in carbohydrates also for other, less readily available reagents, such as chlorotriisopropylsilane [391–394], *tert*-butylchloromethoxyphenylsilane [395], and others [392, 393]. The easily accessible chlorodimethyl-(1,1,2-trimethyl-propyl)silane can find wide use in carbohydrate synthesis as well [395a]. Preparative applications of the trimethylsilyl group for a partial protection of hydroxyl groups are scarce, its domain remains in GLC analysis of volatile persilylated derivatives.

6.1 Trimethylsilyl Ethers

Despite the fact that the differencies in the rates of reaction of sterically hindered and unhindered hydroxyl groups towards various trimethylsilylating agents are well established [396–398], the numer of preparative partial trimethylsilylation results is very low in carbohydrates. The ethers are too susceptible to solvolysis in protic media and, consequently, partial silylation is not easily controlled, the results can vary from run to run [399].

For aldohexoses, OH-1, OH-2, and OH-6 seem to be particularly reactive [399, 400]. Nevertheless, the initial product of the treatment of D-*altro*-3-heptulose (coriose) with chlorotrimethylsilane in pyridine is claimed to have a free hemiacetal hydroxyl group [401]. 1,6-Anhydro-β-D-glucopyranose yielded 87% of the 2,4-disilyl ether [402].

Selective hydrolysis can be used to remove the O–6 protecting group of methyl 2,3,4,6-tetrakis-*O*-(trimethylsilyl)-α-D-glucopyranoside [403, 404]. Under properly controlled conditions, selective cleavage at one or both of the primary positions of pertrimethyl-silylated α,α-trehalose has been achieved [405]. The simple heteronuclear $^1H^{29}Si$ chemical shift correlated two-dimensional NMR spectroscopy enables the assignment of positions of trimethylsilyl substituents [406]. Interestingly, octakis-*O*-(trimethylsilyl)sucrose is a stable, crystalline material which has been even proposed as an internal standard for GLC [407].

6.2 *Tert*-butyldimethylsilyl Ethers

Both primary and secondary hydroxyl groups react rapidly [389, 408–412] under classical Corey conditions [388] with *tert*-butylchlorodimethylsilane in DMF in the presence of imidazole catalyst. The primary hydroxyl group reacts preferentially, the ratio of products being dependent on the amount of imidazole present. For thymidine, the highest, ca. 75% yield of 5′-ether has been obtained [389, 408] with a ratio of nucleoside : *tert*-butylchlorodimethylsilane : imidazole near 1:1.1:2.2. Similar results were obtained with other deoxyribonucleosides [389, 408], as well as with uridine [410, 411, 413], adenosine [410, 411], cytidine [414], guanosine [414], and their

(57)   (58) (98%)

derivatives [411, 413–415]. Methyl 2-deoxy-3-C-methyl-α-D-*ribo*-hexopyranoside (57) yielded [416] 98% of the silyl ether (58), the selective protection of the primary OH-6 was also achieved for other hexopyranosides [417, 418].

Among the secondary hydroxyl groups of ribonucleosides, the 2'-OH is slightly more reactive than 3'-OH towards this reagent [411, 414, 419, 420]. The higher re-activity of OH-2 has been observed also for methyl α-D-glucopyranoside [421, 421a], its β-anomer [421, 421a], and 4,6-O-benzylidene derivative [422]. With 3.2 molar equiv. of the reagent, 84% of methyl 2,6-bis-O-(*tert*-butyldimethylsilyl)-α-D-galacto-pyranoside and 14% of 2,3,6-tris-O-(*tert*-butyldimethylsilyl)-α-D-galactopyranoside were isolated [421a]. A synthetically useful yield of the last mentioned compound (57%) can be achieved with 4.3 molar equiv. of the reagent. The 2,6-disilyl ether dominates over the 3,6-disilyl isomer also in the reaction product of methyl α-D-mannopyranoside (50% and 33%), but the ratio is reversed (35% and 43%) for methyl β-D-galactopyranoside [421a]. 1',2:4,6-Di-O-isopropylidenesucrose yielded 86% of 3',4',6'-trisilyl ether, leaving thus the OH-3 of D-glucose unit unprotected [423].

The reagent reacts selectively with the allylic hydroxyl group of 6-deoxyglycals; 95% of 3-O-(*tert*-butyldimethylsilyl)-L-rhamnal were isolated after 2 h reaction at room temperature, in addition to 3% of the 4-silyl and 2% of 3,4-disilyl ethers [424]. Reinvestigation of this reaction by Thiem and coworkers revealed that temperatures below −20 °C are necessary in order to achieve such a high regioselectivity. At room temperature, a complicated mixture of products was obtained [127a]. Derivatives of N-acetylneuraminic acid such as 59 were also partially silylated with efficiency

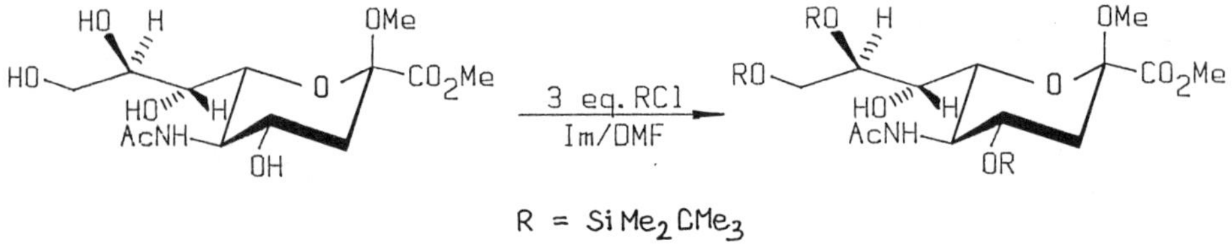

$$3 \text{ eq. RCl} \quad \text{Im/DMF}$$

$$R = Si Me_2 CMe_3$$

(59)   (60) (70%)

[425, 426]. Substitution of 1,2,4-triazole for imidazole as the activating agent can be advantageous in some special cases as it prevents the simultaneous acyl group mi-gration [427].

On prolonged treatment with *tert*-butylchlorodimethylsilane and imidazole in DMF, silyl group migrations between vicinal *cis* [424, 428, 429] and even *trans* [430]

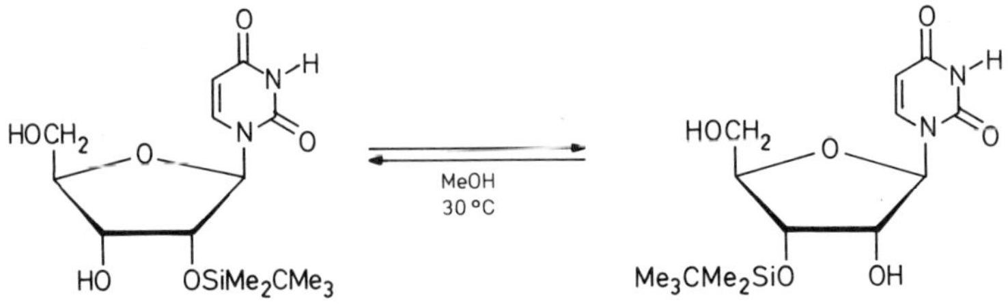

$$\text{MeOH} \quad 30 °C$$

(61) (57%)   (62) (43%)

hydroxyl groups have been observed. About 20% of 2',5'-bis-O-(tert-butyldimethyl-silyl)uridine were found to isomerize to 3',5'-disilyl ether in 24 h. The isomerization is almost negligible in dry aprotic solvents like pyridine, oxolane, acetonitrile, but it is much faster in protic solvents like methanol [428]. In the latter solvent, equilibrium state is reached in only several hours, with the 2'-isomer (e.g., 61) generally preponderating over the 3'-isomer (e.g., 62) in nucleosides [411, 414, 415, 419, 420, 428]. Migration is slower for pyranosides than in the ribonucleoside field, no changes were detected in methanol during 24 h. However, considerable 3 → 2 migration occured when methyl 3,6-bis-O-(tert-butyldimethylsilyl)-α-D-mannopyranoside was treated with imidazole — DMF [421a]. Methyl 3,6-bis-O-(tert-butyldimethylsilyl)-β-D-glucopyranoside isomerized on treatment with diethylazodicarboxylate and triphenylphosphine to give 83% of 4,6- and 10% of the 2,6-isomer [421].

The silylation with tert-butylchlorodimethylsilane in pyridine as a solvent and catalyst is substantially slower but gives much more selectivity [411, 414, 431, 432]. For example, 2',5'-disilyluridine is obtained in 65% yield after a 48-h reaction in pyridine, but in only 46% after a 2-h reaction in DMF — imidazole mixture [414]. With one molar equiv. of the reagent, methyl α-D-glucopyranoside gave the 6-silyl ether in virtually quantitative yield, its secondary hydroxyl groups reacted slowly even with a large excess of tert-butylchlorodimethylsilane [432]. Primary hydroxyl groups in D-glucal [431] or in L-iditol derivative [433] were selectively protected as well. Similarly, 3.5 molar equiv. of the reagent converted sucrose into 1',6,6'-tris-O-(tert-butyldimethylsilyl)sucrose almost exclusively [432]. For this disaccharide, the use of a limited amount of tert-butylchlorodimethylsilane in pyridine have led to the 6,6'-disilyl ether isolable in 36% yield without chromatography [434]; the reaction starts with OH-6' of the D-fructose unit [432].

High selectivity of tert-butylchlorodimethylsilane for the primary hydroxyl group is observed with 4-dimethylaminopyridine catalyst [435–438]. About 89% of 5'-O-(tert-butyldimethylsilyl)uridine resulted from the application of this method in DMF using triethylamine to regenerate the catalyst [435]. No significant differentiation between secondary hydroxyl groups could be traced from the results of dimolar silylation of common methyl glycopyranosides [421a]. The selectivity achieved with N,O-bis(tert-butyldimethylsilyl)acetamide or tert-butyldimethylsilylimidazole was also slightly better when compared to classical imidazole catalyzed technique [391, 392). The polymeric reagent tert-butylchlorodimethylsilane — polyvinylpyridine was effective for the selective protection of the primary hydroxyl group in methyl 2,3-dideoxy-α-D-erythro-hex-2-eno-pyranoside [439].

(63)   (64) (95%)

Silver nitrate displayed a pronounced effect on the selectivity of silylation [380, 392, 440, 441]. The treatment of uridine (63) with 2.2 molar equiv. of each *tert*-butylchloro-dimethylsilane and silver nitrate in oxolane led to the exclusive formation of 5'-*O*-silyl derivative 64 in 95% isolated yield [380]. Under analogous conditions, 1,2-*O*-isopropylidene-α-D-glucofuranose was selectivity silylated at O-6, and 1,2-*O*-isopropylidene-α-D-xylofuranose at O-5, with 94–96% yield [442]. Combination of silver nitrate with pyridine resulted in disilylation of nucleosides at O-5' and O-2' with excellent selectivity. Interestingly, the selectivity for secondary hydroxyl groups is reversed with a combination of silver nitrate and 1,4-diazabicyclo[2.2.2]octane as a base. Such a disilylation led to 3',5'-*bis*-*O*-(*tert*-butyldimethylsilyl) derivatives in as high yields as 94%. Other soluble silver salts, e.g., silver perchlorate, behaved similarly [380].

For the removal of this protecting group, tetrabutylammonium fluoride in oxolane is the most frequently used [388, 389, 409–411]. A much simpler reagent to prepare, potassium fluoride — crown ether, has been introduced for the same purpose [427]. Silyl group at O-2' of nucleosides is cleaved more rapidly [411] than at O-5'. Acyl migrations occurred under the tetrabutylammonium fluoride-catalyzed desilylation [432, 434, 443]. Differencies between the primary and secondary position were also observed for acid- or base-catalyzed solvolysis [391, 409–412]. 5'-*O*-(*Tert*-butyldimethylsilyl)nucleosides are much more labile towards acid than either 2'- or 3'-silyl ethers [391, 410–412], whereas the situation is reversed for base hydrolysis [411]. *N*-Bromosuccinimide in aqueous DMSO is another alternative for the removal of this type of silyl group [444].

6.3 *Tert*-butyldiphenylsilyl Ethers

To selectively block the primary hydroxyl group, the *tert*-butyldiphenylsilyl group is now frequently chosen instead of the traditional trityl group because of the relative ease of preparation of the required derivative in high yield, its stability under mild acid conditions, and the ease and convenience with which the group may be removed on treatment with fluoride ion [390]. For instance, the reaction of methyl α-D-gluco-pyranoside with 1.1 molar equiv. of *tert*-butylchlorodiphenylsilane in DMF containing 2.2 equiv. of imidazole for 4 h at 25 °C afforded 80% of the corresponding 6-*O*-(*tert*-butyldiphenylsilyl) derivative [390]. Similar conditions were used for the 6,6'-disilyl ether of methyl 2-*O*-benzyl-3-*O*-α-D-mannopyranosyl-α-D-mannopyranoside [445], whereas the 6,6'-disilyl ether of sucrose was prepared in 78% yield by the action of 3 molar equiv. of *tert*-butylchlorodiphenylsilane in pyridine and 4-dimethyl-aminpyridine [446]. The fact that *tert*-butyldiphenylsilyl group is fairly stable under benzyl bromide — silver oxide — *N*,*N*-dimethylformamide benzylating conditions (note, that the base catalyzed hydrolysis of this group is an unavoidable side reaction during benzylation in the presence of sodium hydride or barium oxide) makes the conversion of methyl β-D-galactopyranoside (65) in its 2-*O*-benzyl derivative 68 via 66 and 67 quite feasible [446a].

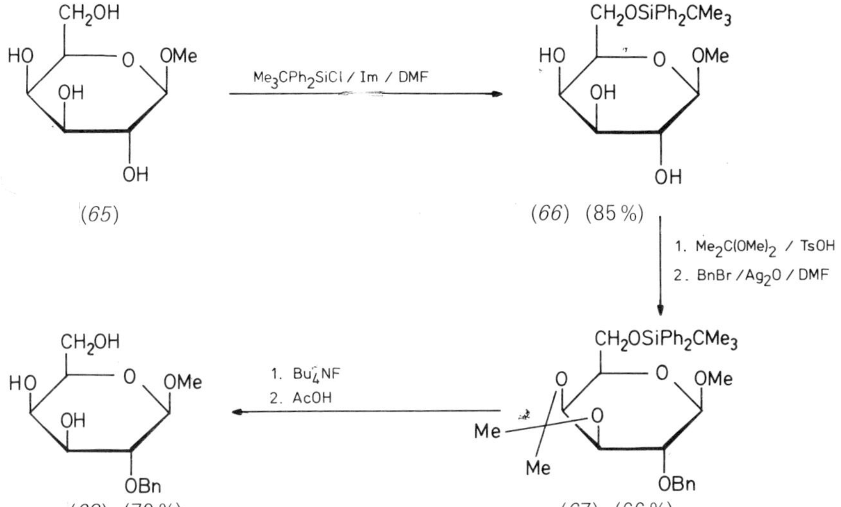

Secondary hydroxyl groups also react if sufficient amount of *tert*-butylchlorodiphenylsilane — imidazole — DMF reagent is used. The less hindered OH-2 of 1,4-anhydro-5,6-*O*-isopropylidene-D-glucitol was silylated in more than 78% yield [447]. Of the two secondary hydroxyl groups of the 4',6'-*O*-isopropylidene-α-D-glucopyranosyl part of a kinetic acetonation product of maltose, that one at C-3' was selectively protected [448].

Imidazole or pyridine mediated silylation of 1,2-*O*-[1-*exo*-ethoxy)ethylidene]-α-D-glucopyranose failed to give a high yield of the 6-silyl ether due to some polymerization and side reactions. Activation of hydroxyl groups via a tributylstannyl intermediate followed by the tetrabutylammonium bromide catalyzed reaction with *tert*-butylchlorodiphenylsilane was more successful [231], the 6-*O*-silyl derivative being isolated in 87% yield. The lability of this protecting group under benzylation with benzyl bromide and sodium hydride at 0 °C has been observed [449].

7 References

1. Sugihara JM (1953) Adv Carbohydr Chem 8: 1
2. Haines AH (1976) Adv Carbohydr Chem Biochem 33: 11
3. Helferich B (1948) Adv Carbohydr Chem 3: 79
4. Penman A, Rees DA (1973) J Chem Soc Perkin I: 2188
5. Mega T, Nishikawa A, Ikenaka T (1983) Carbohydr Res 112: 313
6. Staněk J (jr), Jeřábková J, Jarý J (1984) Collect Czech Chem Commun 49: 2922
7. Haines AH (1981) Adv Carbohydr Chem Biochem 39: 13
8. Greene TW (1981) Protective groups in organic synthesis, Wiley, New York
9. McOmie JFW (1973) Protective groups in organic chemistry, Plenum Press, New York
10. Staněk J (jr), Balcarová Š, Jurý J (in press) Carbohydr Res

11. Staněk J (jr), Chuchvalec P, Čapek K, Kefurt K, Jarý J (1974) Carbohydr Res 36: 273
12. Spurlin HM (1954) in: Ott E, Spurlin HM, Grafflin MW (eds) Cellulose and cellulose derivatives, II, High polymers, vol V, Interscience, New York
13. Garegg PJ (1963) Acta Chem Scand 17: 1343
14. Spurlin HM (1939) J Am Chem Soc 61: 2222
15. Marek M, Chuchvalec P, Kefurt K, Jarý J (1978) Collect Czech Chem Commun 43: 115
16. Lehrfeld J (1975) Carbohydr Res 39: 364
17. Luby P (1975) J Chem Soc Perkin II: 1196
18. Küster JM, Luftmann H, Dyong I (1976) Chem Ber 106: 2223
19. Mihálov V, Kováčik V, Toman R, Tvaroška I (1983) Carbohydr Res 119: 13
20. Paulsen H, Röben W, Heiker FR (1980) Tetrahedron Lett 3679
21. Zamojski A, Grzeszczyk B, Banaszek A, Babiński J, Bordas X, Dziewiszek K, Jarosz S (1985) Carbohydr Res 142: 165
22. Paulsen H, Bünsch A (1981) Liebigs Ann Chem 2204
23. Closkey CM (1957) Adv Carbohydr Chem 12: 137
24. Wing RE, BeMiller JN (1972) Methods Carbohydr Chem 6: 368
25. Gorin PAJ (1982) Carbohydr Res 101: 13
26. Borén HB, Garegg PJ, Kenne L, Pilotti Å, Svensson S, Swahn CG (1973) Acta Chem Scand 27: 2740
27. Reinefeld E, Heincke KD (1971) Chem Ber 104: 265
28. Reicher F, Corrêa JBC, Gorin PAJ (1984) Carbohydr Res 135: 129
29. Ichikawa Y, Manaka A, Kuzuhara H (1985) Carbohydr Res 138: 55
30. Borén HB, Garegg PJ, Kenne L, Maron L, Svensson S (1972) Acta Chem Scand 26: 644
31. Borén HB, Garegg PJ, Wallin NH (1972) Acta Chem Scand 26: 1082
32. Alfredsson G, Borén HB, Garegg PJ (1972) Acta Chem Scand 26: 3431
33. Kondo Y, Noumi K, Kitagawa S, Hirano S (1983) Carbohydr Res 123: 157
34. Seib PA (1968) Carbohydr Res 8: 101
35. Bovin NV, Zurabyan SE, Khorlin AYa (1980) Izv Akad Nauk SSSR, Ser Khim 199; (1980) CA 92: 215673
36. Jacquinet JC, Petit JM, Sinaÿ P (1974) Carbohydr Res 38: 305
37. Berry JM, Dutton CGS (1974) Carbohydr Res 38: 339
38. Nashed MA, Anderson L (1976) Carbohydr Res 51: 65
39. Nashed MA, Anderson L (1977) Carbohydr Res 56: 419
40. Jacquinet JC, Sinaÿ P (1977) J Org Chem 42: 720
41. Rollin P, Sinaÿ P (1977) J Chem Soc Perkin I: 2513
42. Jacquinet JC, Sinaÿ P (1976) Carbohydr Res 46: 138
43. Warren CD, Jeanloz RW (1977) Carbohydr Res 53: 67
44. Petit JM, Jacquinet JC, Sinaÿ P (1980) Carbohydr Res 82: 130
45. Rana SS, Vig R, Matta KL (1982/83) J Carbohydr Chem 1: 261
46. Ekborg G, Sone Y, Glaudemans CPJ (1982) Carbohydr Res 110: 55
47. Farkaš J, Ledvina M, Brokeš J, Ježek J, Zajíček K, Zaoral M (1987) Carbohydr Res 163: 63
48. Kondo Y (1975) Agric Biol Chem 39: 1879
49. Grishkovets VI, Chirva VYa (1982) Khim Prirod Soedin 28 (1982); CA 96: 218137
50. Brimacombe JS (1972) Methods Carbohydr Chem 6: 376
51. Tate ME, Bishop CT (1963) Can J Chem 41: 1801
52. Flowers HM (1975) Carbohydr Res 39: 245
53. Flowers HM (1982) Carbohydr Res 100: 418
54. David S, Thieffry A, Veyrières A (1981) J Chem Soc Perkin I: 1796
55. Imura N, Tsuruo T, Ukiki T (1968) Chem Pharm Bull 16: 1105
56. Kikugawa K, Sato F, Tsuruo T, Imura N, Ukita T (1968) Chem Pharm Bull 16: 1110
57. Iwashige T, Saeki H (1967) Chem Pharm Bull 15: 1803
58. Augé C, David S, Veyrières A (1976) J Chem Soc, Chem Commun 375
59. Flowers HM (1982) Carbohydr Res 99: 170
60. Flowers HM (1983) Carbohydr Res 119: 75
61. Bovin NV, Zurabyan SE, Khorlin AYa (1983) J Carbohydr Chem 2: 249
62. Srivastava HC, Srivastava VK (1977) Carbohydr Res 58: 227

63. Lipták A, Czégéni I, Harangi J, Nánási P (1979) Carbohydr Res 73: 327
64. Gigg J, Gigg R, Payne S, Conant R (1985) Carbohydr Res 140: C1
65. Kihlberg J, Frejd T, Jansson K, Magnusson G (1986) Carbohydr Res 152: 113
66. Czernecki S, Georgoulis C, Provelenghiou C (1976) Tetrahedron Lett 3535
67. Lubineau A, Thieffry A, Veyrières A (1976) Carbohydr Res 46: 143
68. Ireland RE, Wrets PGM, Ernst B (1981) J Am Chem Soc 103: 3205
69. Wróblewski AE (1984) Carbohydr Res 131: 325
70. Ezekiel AD, Overend WG, Williams NR (1971) J Chem Soc C: 2907
71. Küster JM, Dyong I (1975) Liebigs Ann Chem 2179
72. Gent PA, Gigg R, Conant R (1973) J Chem Soc Perkin I: 1858
73. Gent PA, Gigg R, Conant R (1972) J Chem Soc Perkin I: 1535
74. Koto S, Takebe Y, Zen S (1972) Bull Chem Soc Jpn 45: 291
75. Morishima N, Koto S, Kusuhara C, Zen S (1982) Bull Chem Soc Jpn 55: 631
76. Van der Vleugel DJM, Vliegenthart JFG (1982) Carbohydr Res 105: 168
77. Myles A, Pfleiderer W (1972) Chem Ber 105: 3327
78. Kowollik G, Otto A, Etzold G, Langen P (1973) Z Chem 13: 217
79. Angyal I (1938) Magyar Biol Kutatóintézet Munkái 10: 449; (1939) CA 33: 4963
80. Angyal SJ, Russell AF (1969) Aust J Chem 22: 391
81. Angyl SJ, Tate ME (1965) J Chem Soc 6949
82. Angyal SJ, Stewart TS (1966) Aust J Chem 19: 1683
83. Dejter-Juszynski M, Flowers HM (1973) Carbohydr Res 28: 61
84. Garegg PJ, Norberg T (1976) Carbohydr Res 52: 235
85. Küster JM, Dyong I, Schmeer D (1976) Chem Ber 109: 1253
86. Klyashchitskii BA, Krylova EB, Shvets VI (1972) Zh. Obsch Khim 42: 2586
87. Morishima N, Koto S, Oshima M, Sugimoto A, Zen S (1983) Bull Chem Soc Jpn 56: 2849
87a. Schmidt RR, Gohl A (1979) Chem Ber 112: 1689
88. Iversen T, Bundle DR (1981) J Chem Soc Chem Commun 1240
89. Wessel HP, Iversen T, Bundle DR (1985) J Chem Soc Perkin I: 2247
90. Ekborg G, Glaudemans CPJ (1985) Carbohydr Res 142: 213
91. Toman R, Rosik J, Zikmund M (1982) Carbohydr Res 103: 165
92. Christensen LF, Broom AD (1972) J Org Chem 37: 3398
93. Chittenden GJF (1981) Carbohydr Res 91: 85
94. Barrett AGM, Read RW, Barton DHR (1980) J Chem Soc Perkin I: 2184
95. Decoster E, Lacombe JM, Strebler JL, Ferrari B, Pavia AA (1983) J Carbohydr Chem 2: 329
96. Fügedi P, Nánási P (1981) J Carbohydr Nucl 8: 547
97. Isogai A, Ishizu A, Nakano J (1984) J Appl Polym Sci 29: 2097
98. Binkley RW, Ambrose MG (1984) J Carbohydr Chem 3: 1
99. Lemieux RU, Kondo T (1974) Carbohydr Res 35: C4
100. Berry JM, Hall LD (1976) Carbohydr Res 47: 307
101. Hasegawa A, Goto M, Kiso M (1985) J. Carbohydr Chem 4: 627
102. Florent JC, Monneret C (1980) Carbohydr Res 81: 225
103. Di Cesare P, Gross B (1976) Carbohydr Res 48: 271
104. Zhdanov YuA, Alekseev YuE, Doroshenko SS, Bogdanova GV, Sudareva TP, Alekseeva VG (1978) Dokl Akad Nauk SSSR 238: 1102; (1978) CA 89: 43992
105. Garegg PJ, Iversen T, Oscarson S (1976) Carbohydr Res 50: C12
106. Nouguier R, Lambert C, Azria O (1985) Tetrahedron Lett 26: 5769
107. Rana SS, Barlow JJ, Matta KL (1980) Carbohydr Res 84: 353
108. Takeo K, Nakaji T, Shinmitsu K (1984) Carbohydr Res 133: 275
109. Rana SS, Piskorz CF, Barlow JJ, Matta KL (1980) Carbohydr Res 83: 170
110. Sarkar AK, Roy N (1987) Carbohydr Res 163: 285
111. Ogawa T, Kaburagi T (1982) Carbohydr Res 103: 53
112. Doboszewski B, Zamojski A (1987) Carbohydr Res 164: 470
113. Hoffman J, Theander O, Lindberg B, Norberg T (1985) Carbohydr Res 137: 265
114. Verez Bencomo V, Diaz Rodriguez P, Garcia Fernandez G, Basterrechea Rey MCM (1981) Rev Cubana Farm 15: 26; (1982) CA 96: 52594
115. Verez Bencomo V, Saenz de la C GA, Garcia Fernandez G, Basterrechea MCM, Coll Manchado F (1981) Rev Cubana Farm 15: 169; (1982) CA 96: 85867

116. Paulsen H, Lebuhn R, Lockhoff O (1982) Carbohydr Res 103: C7
116a Paulsen H, Lebuhn R (1983) Liebigs Ann Chem: 1047
117. Paulsen H, Heume M, Györgydeak Z, Lebuhn R (1985) Carbohydr Res 144: 57
118. Kong F, Schuerch C (1983) Carbohydr Res 112: 141
119. Handa VK, Barlow JJ, Matta KL (1979) Carbohydr Res 76: C1
120. Awad LF, El Ashry ESH, Schuerch C (1983) Carbohydr Res 122: 69
121. Poszgay V (1979) Carbohydr Res 69: 284
122. Handa VK, Piskorz CF, Barlow JJ, Matta KL (1979) Carbohydr Res 74: C5
123. Fang Y, Kong F, Wang Q (1987) J Carbohydr Chem 6: 169
124. Rana SS, Barlow JJ, Matta KL (1980) Carbohydr Res 85: 313
125. Köpper S, Thiem J (1987) J Carbohydr Chem 6: 57
126. Garegg PJ, Iversen T, Johansson R, Lindberg B (1984) Carbohydr Res 130: 322
127. Fernandez-Mayoralas A, Martin-Lomas M, Villanueva D (1985) Carbohydr Res 140: 81
128. Wagner D, Verheyden JPH, Moffatt JG (1974) J Org Chem 39: 24
129. Nashed MA, Anderson L (1976) Tetrahedron Lett 3503
130. Nashed MA, Anderson L (1977) Carbohydr Res 56: 325
131. Haque ME, Kikuchi T, Yoshimoto K, Tsuda Y (1985) Chem Pharm Bull 33: 2243
132. Slife CW, Nashed MA, Anderson L (1981) Carbohydr Res 93: 219
133. Jacquinet JC, Duchet D, Milat ML, Sinaÿ P (1981) J Chem Soc Perkin I: 326
134. Hong N, Funabashi M, Yoshimura J (1981) Carbohydr Res 96: 21
135. David S, Thieffry A (1979) J Chem Soc Perkin I: 1568
136. Cubero II, López-Espinosa MTP (1986) J Carbohydr Chem 5: 299
137. Nashed MA (1978) Carbohydr Res 60: 200
138. Dessinges A, Olesker A, Lukacs G, Ton That Thang (1984) Carbohydr Res 126: C6
139. Takeo K, Tei S (1985) Carbohydr Res 141: 159
140. Takeo K, Shibata K (1984) Carbohydr Res 133: 147
141. Nashed MA, El-Sokkary RI, Rateb L (1984) Carbohydr Res 131: 47
142. Aspinall GO, Takeo KI (1983) Carbohydr Res 121: 61
143. Kováč P, Glaudemans CPJ, Taylor RB (1985) Carbohydr Res 142: 158
144. Kováč P, Glaudemans CPJ (1985) Carbohydr Res 138: C10
145. Martin A, Païs M, Monneret C (1983) Carbohydr Res 113:21
145a. David S, Hanessian S (1985) Tetrahedron 41: 643
146. Poszgay V, Neszmélyi A (1980) Carbohydr Res 85: 143
147. Kosma P, Christian R, Schulz G, Unger FM (1985) Carbohydr Res 141: 239
147a. Schmidt RR, Gohl A, Karg T (1979) Chem Ber 112: 1705
148. Monneret C, Gagnet R, Florent JC (1987) J Carbohydr Chem 6: 221
149. Brimacombe JS, Rahman KMM (1985) J Chem Soc Perkin I: 1067
149a. Eis MJ, Ganem B (1988) Carbohydr Res 176: 316
149b. Grindley TB, Thangarasa R (1988) Carbohydr Res 172: 311
150. Fernandez-Mayoralas A, Martin-Lomas M (1986) Carbohydr Res 154: 93
151. Ogawa T, Matsui M (1978) Carbohydr Res 62: C1
152. Alais J, Veyrières A (1981) J Chem Soc Perkin I: 377
153. Ogawa T, Takahashi Y, Matsui M (1982) Carbohydr Res 102: 207
154. Ogawa T, Nukada T, Matsui M (1982) Carbohydr Res 101: 263
155. Arnarp J, Lönngren J (1981) J Chem Soc Perkin I: 2070
156. Arnarp J, Baumann H, Lönn H, Lönngren J, Nyman H, Ottosson H (1983) Acta Chem Scand B37: 329
156a. Cruzado MC, Martin-Lomas M (1988) Carbohydr Res 175: 193
157. Koike K, Sugimoto M, Sato S, Ito Y, Nakahara Y, Ogawa T (1987) Carbohydr Res 163: 189
158. Takahashi Y, Ogawa T (1987) Carbohydr Res 164: 227
159. Ogawa T, Kitajima T, Nukada T (1983) Carbohydr Res 123: C5
160. Veyrières A (1981) J Chem Soc. Perkin I: 1626
161. Ogawa T, Nakabayashi S (1981) Carbohydr Res 97: 81
162. El Sadek MM, Warren CD, Jeanloz RW (1984) Carbohydr Res 131: 166
163. Ogawa T, Nakabayashi S, Sasajima K (1981) Carbohydr Res 96: 29
164. Wessel HP, Iversen T, Bundle DR (1984) Carbohydr Res 130: 5
165. Eby R, Schuerch C (1982) Carbohydr Res 100: C41

166. Eby R, Webster KT, Schuerch C (1984) Carbohydr Res 129: 111
167. Garegg PJ, Hultberg H (1981) Carbohydr Res 93: C10
168. Garegg PJ, Hultberg H, Wallin S (1982) Carbohydr Res 108: 97
169. Noumi K, Kitagawa S, Kondo Y, Hirano S (1984) Carbohydr Res 134: 172
170. Madiyalakan R, Chowdhary MS, Rana SS, Matta KL (1986) Carbohydr Res 152: 183
171. Garegg PJ, Oscarson S (1983) Carbohydr Res 114: 322
172. Takeo K, Okushio K, Fukuyama K, Kuge T (1983) Carbohydr Res 121: 163
173. Rana SS, Matta KL (1983) Carbohydr Res 116: 71
174. Bhattacharjee SS, Gorin PAJ (1969) Can J Chem 47: 1195
175. Bhattacharjee SS, Gorin PAJ (1969) Can J Chem 47: 1207
176. Lipták A, Jodál I, Nánási P (1975) Carbohydr Res 44: 1
177. Nánási P, Lipták A (1974) Magy Kem Foly 80: 217
178. Fügedi P, Lipták A, Nánási P, Neszmélyi A (1980) Carbohydr Res 80: 233
179. Lipták A, Pekár F, Jánossy L, Jodál, Fügedi P, Harangi J, Nánási P, Szejtli J (1979) Acta Chim Hung 99: 201
180. Fügedi P, Lipták A, Nánási P, Szejtli J (1982) Carbohydr Res 104: 55
181. Lipták A (1982) Carbohydr Res 107: 300
182. Ek M, Garegg PJ, Hultberg H, Oscarson S (1983) J Carbohydr Chem 2: 305
183. Alais J, Veyrières A (1981) Carbohydr Res 92: 310
184. Bako P, Fenichel L, Töke L, Tóth G (1986) Carbohydr Res 147: 31
185. Lipták A, Fügedi P, Nánási P (1979) Carbohydr Res 68: 151
186. Takeo K, Tei S (1986) Carbohydr Res 145: 293
187. Gorin PAJ, Finlayson AJ (1971) Carbohydr Res 18: 269
188. Lipták A, Imre J, Harangi J, Nánási P, Neszmélyi A (1982) Tetrahedron 38: 3721
189. Lipták A, Bobák Á, Nánási P (1977) Acta Chim Hung 94: 261
190. Arnarp J, Haraldsson M, Lönngren J (1981) Carbohydr Res 97: 307
191. Lipták A, Fügedi P, Nánási P (1976) Carbohydr Res 51: C19
192. Rollin P, Sinaÿ P (1977) CR Acad Sci C284: 65
193. Lipták A (1978) Carbohydr Res 63: 96
194. Lipták A, Fügedi P, Nánási P (1978) Carbohydr Res 65: 209
195. Lipták A, Jánossy L, Imre J, Nánási P (1979) Acta Chim Hung 101: 81; (1980) CA 92: 129254
196. Lipták A, Neszmélyi A, Wagner H (1979) Tetrahedron Lett 741
197. Lipták A, Nánási P, Neszmélyi A, Wagner H (1980) Tetrahedron 36: 1261
198. Lipták A, Imre J, Harangi J, Nánási P (1983) Carbohydr Res 116: 217
199. Harangi J, Lipták A, Oláh VA, Nánási P (1981) Carbohydr Res 98: 165
200. Lipták A, Neszmélyi A, Kováč P, Hirsch J (1981) Tetrahedron 37: 2379
201. Allerton R, Fletcher HG (jr) (1954) J Am Chem Soc 76: 1757
202. Eby R, Sondheimer SJ, Schuerch C (1979) Carbohydr Res 73: 273
203. Kováč P, Yeh HJC, Jung GL (1987) J Carbohydr Chem 6: 423
204. Rana SS, Barlow JJ, Matta KL (1981) Carbohydr Res 88: C20
204a. Shah RN, Baptista J, Pardomo GR, Carver JP, Krepinsky JJ (1987) J Carbohydr Chem 6: 645
205. Sakai JI, Takeda T, Ogihara Y (1981) Carbohydr Res 95: 125
206. Kartha KPR, Dasgupta F, Singh PP, Srivastava HC (1986) J Carbohydr Chem 5: 437
207. Ponpipom MM (1977) Carbohydr Res 59: 311
207a. Ogawa T, Sasajima K (1981) Carbohydr Res 93: 67
208. Angyal SJ, Randall MH, Tate ME (1967) J Chem Soc C: 919
209. Holick SA, Anderson L (1974) Carbohydr Res 34: 208
210. Rao VS, Perlin AS (1980) Carbohydr Res 83: 175
210a. Bieg T, Szeja W (1985) Synthesis: 76
210b. Beaupere D, Boutbaiba I, Demailly G, Uzan R (1988) Carbohydr Res 180; 152
211. Rao VS, Perlin AS (1983) Can J Chem 61: 652
212. Cruzado MC, Martin-Lomas M (1986) Tetrahedron Lett 27: 2497
212a. Cruzado MC, Martin-Lomas M (1987) Carbohydr Res 170: 249
213. Hanessian S, Liak TJ, Vanasse B (1981) Synthesis 396
214. Prugh JD, Rooney CS, Deana AA, Ramjit HG (1985) Tetrahedron Lett 26: 2947
215. Ikeda K, Nakamoto S, Takahashi T, Achiwa K (1986) Carbohydr Res 145: C5
216. Oikawa Y, Tanaka T, Horita K, Yonemitsu O (1984) Tetrahedron Lett 25: 5397

217. Angibeaud P, Defaye J, Gadelle A, Utille JP (1985) Synthesis 1123
218. Hanessian S, Guidon Y (1980) Tetrahedron Lett 21: 2305
219. Fuji K, Ichikawa K, Node M, Fujita E (1979) J Org Chem 44: 1661
220. Monneret C, Florent JC, Kabore I, Khong-Huu Qui (1974) J Carbohydr Nucl 1: 161
221. Czernecki S, Gorson G (1978) Tetrahedron Lett: 4113
221a. Thiem J, Gerken M (1982/3) J Carbohydr Chem 1: 229
222. Kováč P, Whittaker NF, Glaudemans CPJ (1985) J Carbohydr Chem 4: 243
222a. Klemer A, Bieber M, Wilburs H (1983) Liebigs Ann. Chem: 1416
223. Van Rijsbergen R, Anteunis MJO, De Bruyn A (1982) Carbohydr Res 105: 269
224. Gigg R (1977) ACS Symp Ser 39: 253
225. Monthorpe PA, Gigg R (1980) Methods Carbohydr Chem 8: 305
226. Thiem J, Mohn H, Heesing A (1985) Synthesis 775
227. Lee KS, Gilbert RD (1981) Carbohydr Res 88: 162
228. Classon B, Garegg PJ, Norberg T (1984) Acta Chem Scand B38: 195
229. Gent PA, Gigg R, Penglis AAE (1976) J Chem Soc Perkin I: 1395
230. Bladon P, Owen LN (1950) J Chem Soc 591
231. Liu CM, Warren CD, Jeanloz RW (1985) Carbohydr Res 136: 273
232. Bergeron RJ, Meeley MP, Machida Y (1976) Bioorg Chem 5: 121
233. Fujiwara T, Aspinall GO, Hunter SW, Brennan PJ (1987) Carbohydr Res 163: 41
234. Winnik FM, Brisson JR, Carver JP, Krepinski JJ (1982) Carbohydr Res 103: 15
235. Srivastava VK, Schuerch C (1979) Tetrahedron Lett 3269
236. Kohata K, Abbas SA, Matta KL (1984) Carbohydr Res 132: 127
237. Alais J, Maranduba A, Veyrières A (1983) Tetrahedron Lett 24: 2383
238. Ogawa T, Katano K, Sasajima K, Matsui M (1981) Tetrahedron 37: 2779
239. Ogawa T, Yamamoto H (1985) Carbohydr Res 137: 79
240. David S, Fernandez-Mayoralas A (1987) Carbohydr Res 165: C11
241. Avela E, Aspelund S, Holmborn B, Melander B, Jalonen H, Peltonen C (1977) ACS Symp Ser 41: 62
242. Garegg PJ, Hultberg H, Oscarson S (1982) J Chem Soc Perkin I: 2395
243. Gelas J (1981) Adv Carbohydr Chem Biochem 39: 71
244. Gigg R, Warren CD (1965) J Chem Soc 2205
245. Gigg R, Warren CD (1968) J Chem Soc C: 1903
246. Gigg J, Gigg R (1966) J Chem Soc C: 82
247. Price CC, Snyder WH (1961) J Am Chem Soc 83: 1773
248. Corey EJ, Suggs JW (1973) J Org Chem 38: 3224
249. Gent PA, Gigg R (1974) J Chem Soc Chem Commun 277
250. Boss R, Scheffold R (1976) Angew Chem Int Ed Engl 15: 558
251. Ogawa T, Nakabayashi S (1981) Carbohydr Res 93: C1
252. Ho TL, Wong CM (1974) Synth Commun 4: 109
253. Bieg T, Szeja W (1985) J Carbohydr Chem 4: 441
254. Agarwal KL, Yamazaki A, Cashion PL, Khorana HG (1972) Angew Chem Int Ed Engl 11: 451
255. Appleton ML, Cottrell CE, Behrman EJ (1986) Carbohydr Res 158: 227
256. Lehrfeld J (1967) J Org Chem 32: 2544
257. Gros EG, Gruñeiro EM (1970) Carbohydr Res 14: 409
258. Buchanan JG, Schwarz JCP (1962) J Chem Soc 4770
259. Buchanan JG, Fletcher R (1965) J Chem Soc 6316
260. Fréchet JM, Haque KE (1975) Tetrahedron Lett 3055
261. Fréchet JM, Nuyens LJ (1976) Can J Chem 54: 926
262. Leznoff CC, Fyles TM, Weatherston J (1977) Can J Chem 55: 1143
263. Fyles TM, Leznoff CC (1976) Can J Chem 54: 935
264. Ezekiel AD, Overend WG, Williams NR (1971) Carbohydr Res 20: 251
265. Kováč P, Glaudemans CPJ, Guo W, Wong TC (1985) Carbohydr Res 140: 299
266. Ekborg G, Vranešić B, Bhattacharjee AK, Kováč P, Glaudemans CPJ (1985) Carbohydr Res 142: 203
267. Paulsen H, von Deessen U, Tietz H (1985) Carbohydr Res 137: 63
268. Foster AB, Hems R, Westwood JH (1970) Carbohydr Res 15: 41
269. Takeo K, Kato S, Kuge T (1974) Carbohydr Res 38: 346

270. Helferich B, Speicher W (1953) Liebigs Ann Chem 579: 106
271. Durette PL, Hough L, Richardson AC (1974) J Chem Soc Perkin I: 97
272. Durette PL, Hough L, Richardson AC (1974) J Chem Soc Perkin I: 88
273. Wolfrom ML, Koizumi K (1967) J Org Chem 32: 656
274. Koizumi K, Utamura T (1974) Carbohydr Res 33: 127
275. Koizumi K, Utamura T (1983) Chem Pharm Bull 31: 1260
276. Takeo K, Matsunami T, Kuge T (1976) Carbohydr Res 51: 73
277. Otake T (1970) Bull Chem Soc Jpn 43: 3199
278. Otake T (1972) Bull Chem Soc Jpn 45: 2895
279. Hough L, Mufti KS, Khan R (1972) Carbohydr Res 21: 144
280. Otake T (1974) Bull Chem Soc Jpn 47: 1938
281. Khan RA (1977) U.S. Patent 4,002,609; (1977) CA 87: 6297
282. Buchanan JG, Cummerson DA, Turner DM (1972) Carbohydr Res 21: 283
283. Suami T, Otake T, Ikeda T, Adachi R (1977) Bull Chem Soc Jpn 50: 1612
284. Hough L, Salam MA, Tarelli E (1977) Carbohydr Res 57: 97
285. Takeo K (1983) Carbohydr Res 112: 73
286. Takeo K (1981) Carbohydr Res 93: 157
287. Takeo K, Fukatsu T, Yasato T (1982) Carbohydr Res 107: 71
288. Utamura T, Koizumi K (1980) Yakugaku Zasshi 100: 307; (1980) CA 93: 132742
289. Kozumi K, Utamura T (1978) Yakugaku Zasshi 98: 327; (1978) CA 89: 44014
290. Zeile K, Kruckenberg W (1942) Ber Dtsch Chem Ges 75: 1127
291. Kam BL, Oppenheimer NJ (1979) Carbohydr Res 69: 308
292. Dupeyre D, Uttile JP, Vottero PJA (1979) Carbohydr Res 72: 105
293. Levene PA, Tipson RS (1934) J Biol Chem 105: 419
294. Yung NC, Fox JJ (1961) J Am Chem Soc 83: 3060
295. Žemlička J (1964) Collect Czech Chem Commun 29: 1734
296. Blank HU, Pfleiderer W (1967) Tetrahedron Lett 869
297. Blank HU, Pfleiderer W (1970) Liebigs Ann Chem 742: 1
298. Cook AF, Moffatt JG (1967) J Am Chem Soc 89: 2697
299. Codington JF, Fox JJ (1967) Carbohydr Res 3: 124
300. Cook AF, Moffatt JG (1970) U.S. Patent 3,491,085; (1970) CA 72: 111785
301. Johnston GAR (1968) Aust J Chem 21: 513
302. Baker J, Jarman M, Stock JA (1973) J Chem Soc Perkin I: 665
303. Brodbeck U, Moffatt JG (1970) J Org Chem 35: 3552
304. Hutzenlaub W, Pfleiderer W (1973) Chem Ber 106: 665
305. Japan Kokai 7654599; (1977) CA 86: 5768
306. Blank HU, Frahne D, Myles A, Pfleiderer W (1970) Liebigs Ann Chem 742: 34
307. Beránek J, Pitha J (1964) Collect Czech Chem Commun 29: 625
308. Codington JF, Doerr IL, Fox JJ (1964) J Org Chem 29: 564
309. Codington JF, Cushley RJ, Fox JJ (1968) J Org Chem 33: 466
310. McIlroy RJ (1946) J Chem Soc 100
311. Otake T, Sonobe T (1983) Bull Chem Soc Jpn 56: 3187
312. Hockett RS, Hudson CS (1931) J Am Chem Soc 53: 4456
313. Hockett RS, Hudson CS (1934) J Am Chem Soc 56: 945
314. Otake T, Sonobe T (1976) Bull Chem Soc Jpn 49: 1050
315. Otake T, Sonobe T, Suami T (1979) Bull Chem Soc Jpn 52: 3109
316. Koto S, Morishima N, Yoshida T, Uchino M, Zen S (1983) Bull Chem Soc Jpn 56: 1171
317. Itoh Y, Tejima S (1982) Chem Pharm Bull 30: 3383
318. Černý M, Pacák J, Staněk J (1965) Collect Czech Chem Commun 30: 1151
319. Hanessian S, Staub APA (1973) Tetrahedron Lett 3555
320. Hanessian S, Staub APA (1976) Methods Carbohydr Chem 7: 63
321. Wozney YV, Kochetkov NK (1977) Carbohydr Res 54: 300
322. Wozney YV, Backinowsky LV, Kochetkov NK (1979) Carbohydr Res 73: 282
323. Betaneli VI, Ovchinnikov MV, Backinowsky LV, Kochetkov NK (1979) Carbohydr Res 76: 252
324. Malysheva NN, Kochetkov NK (1982) Carbohydr Res 105: 173
325. Kochetkov NK, Zhulin VM, Klimov EM, Malysheva NN, Makarova ZG, Ott AYa (1987) Carbohydr Res 164: 241

326. Chaudhary SK, Hernandez O (1979) Tetrahedron Lett 95
327. Roy R, Jennings HJ (1983) Carbohydr Res 112 : 63
328. Suami T, Tadano K, Suga A, Ueno Y (1984) J Carbohydr Chem 3 : 429
329. Kohli V, Blöcker H, Köster H (1980) Tetrahedron Lett 21 : 2683
330. Kuhn R, Rudy H, Weygand F (1936) Ber Dtsch Chem Ges 69 : 1543
331. McKeown GE, Serenius RSE, Hayward LD (1957) Can J Chem 35 : 28
332. Lemieux RU, Barrette JP (1958) J Am Chem Soc 80 : 2243
333. Suami T, Otake T, Ogawa S, Shoji T, Kato N (1970) Bull Chem Soc Jpn 43 : 1219
334. Helferich B, Speidel PE, Toeldte, W (1923) Ber Dtsch Chem Ges 56 : 766
335. Micheel F (1932) Ber Dtsch Chem Ges 65 : 262
336. Choy YM, Unrau AM (1971) Carbohydr Res 17 : 439
337. Verkade PE, Van der Lee J, Meerburg W (1935) Rec Trav Chim Pays-Bas 54 : 716
338. Dewar J, Fort G (1944) J Chem Soc 496
339. Helferich B, Klein W (1926) Liebigs Ann Chem 450 : 219
340. Roy N, Timell TE (1968) Carbohydr Res 7 : 82
341. Roy N, Timell TE (1968) Carbohydr Res 6 : 475
342. Barker GR (1963) Methods Carbohydr Chem 2 : 168
343. Baker CW, Whistler RL (1974) Carbohydr Res 33 : 372
344. MacCoss M, Cameron DJ (1978) Carbohydr Res 60 : 206
345. Randazzo G, Capasso R, Cicala MR, Evidente A (1980) Carbohydr Res 85 : 298
346. Dax K, Wolflehner W, Weidmann H (1978) Carbohydr Res 65 : 132
347. Albert R, Dax K, Pleschko R, Stütz AE (1985) Carbohydr Res 137 : 282
348. Kováč P, Glaudemans CPJ (1985) Carbohydr Res 140 : 313
349. Kochetkov NK, Dmitriev BA, Bairamova NE, Nikolaev AV (1978) Izv Akad Nauk SSSR, Ser Khim 652 : (1978) CA 89 : 24655
350. Wolfrom ML, Burke WJ, Waisbrot SW (1939) J Am Chem Soc 61 : 1827
351. O'Donell GW, Richards GM (1972) Aust J Chem 25 : 407
352. Inouye S, Tsuruoka T, Ito T, Niida T (1968) Tetrahedron 24, 2125
353. Kováč P, Bauer Š (1972) Tetrahedron Lett 2349
354. Greene GL, Letsinger RL (1975) Tetrahedron Lett 2081
355. Kowollik G, Gaertner K, Langen P (1972) Tetrahedron Lett 3345
356. Davies LC, Farmer PB, Jarman M, Stock JA (1980) Synthesis 75
357. Matteuci MD, Caruthers MH (1980) Tetrahedron Lett 21 : 3243
358. Verheyden JPH, Mofatt JG (1970) J Org Chem 35 : 2868
359. Verkade PE, Cohen WD, Vroege AK (1940) Rec Trav Chim Pays-Bas 59 : 1123
360. Kenner J, Richards GN (1955) J Chem Soc 1810
361. Bredereck H, Greiner W (1953) Chem Ber 86 : 717
362. Bartholomew DG, Broom AD (1975) J Chem Soc Chem Commun 38
363. Ohtsuka E, Tanaka S, Ikehara M (1977) Chem Pharm Bull 25 : 949
363a. Zehavi U, Amit B, Patchornik A (1972) J Org Chem 37 : 2281
363b. Zehavi U, Patchornik A (1972) J Org Chem 37 : 2285
364. Johansson R, Samuelsson B (1984) J Chem Soc Perkin I : 2371
365. Koto S, Inada S, Morishima N, Zew S (1980) Carbohydr Res 87 : 294
366. Ohtsuka E, Tanaka S, Ikehara M (1974) Nucleic Acids Res 1 : 1351
367. Ohtsuka E, Tanaka S, Ikehara M (1977) Synthesis 453
368. Takahaku H, Kamaike K (1982) Chem Lett 189
369. Takahaku H, Kamaike K, Tsuchiya H (1984) J Org Chem 49 : 51
370. Kloosterman M, Kuyl-Yeheskiely E, Van Boom JH (1985) Rec Trav Chim Pays-Bas 104 : 291
371. Joniak D, Košiková B, Kosáková L (1978) Collect Czech Chem Commun 43 : 769
372. Joniak D, Košiková B, Kosáková L (1980) Collect Czech Chem Commun 45 : 1959
373. Lipták A, Fügedi P (1983) Angew Chem 95 : 245
374. Busfield WK, Franke FP, Guthrie RD (1978) Aust J Chem 31 : 2559
374a. Martin OR, Kurz MG, Rao SP (1987) J Org Chem 52 : 2922
375. Jackson G, Jones HF, Petursson S, Webber JM (1982) Carbohydr Res 102 : 147
376. Petursson S, Webber JM (1982) Carbohydr Res 103 : 41
377. Smith M, Rammler DH, Goldberg IH, Khorana HG (1962) J Am Chem Soc 84 : 430

378. van Boeckel CAA, Beetz T, Vos JN, de Jong AJM, van Aelst SF, van den Bosch RH, Mertens JMR, van der Vlugt FA (1985) J Carbohydr Chem 4: 293
379. Okupniak J, Adamiak RW, Wiewiorowski M (1981) Pol J Chem 55: 679; (1982) CA 96: 143252
380. Hakimelahi GH, Proba ZA, Ogilvie KK (1982) Can J Chem 60: 1106
381. Taunton-Rigby A, Kim YH, Crosscup CJ, Starkovsky NA (1972) J Org Chem 37: 956
382. Sekine M, Hata T (1983) J Org Chem 48: 3011
383. Pfleiderer W, Schwarz M, Schirmeister H (1986) Chem Scr 26: 147
384. Pfleiderer W, Himmelsbach F, Charubala R, Schirmeister H, Beiter A, Schulz B, Trichtinger T (1985) Nucleosides Nucleotides 9: 81
385. Khorlin AYa, Nesmeyanov VA, Zurabayan SE Carbohydr Res 33: C1
386. Mizuno Y, Endo T, Ikeda K (1975) J Org Chem 40: 1385
387. Zhdanov Yu, Polenov VA, Alekseeva VG, Korol EL, Popov II (1985) Dokl Akad Nauk SSSR 283: 402
388. Corey EJ, Venkateswarlu A (1972) J Am Chem Soc 94: 6190
389. Ogilvie KK (1973) Can J Chem 51: 3799
390. Hanessian S, Lavallee P (1975) Can J Chem 53: 2975
391. Ogilvie KK, Thompson EA, Quilliam MA, Westmore JB (1974) Tetrahedron Lett 2865
392. Ogilvie KK, Beaucage SL, Entwistle DW, Thompson EA, Quilliam MA, Westmore JB (1976) J·Carbohydr Nucl 3: 197
393. Corey EJ, Pan BC, Hua DH, Deardorff DR (1982) J Am Chem Soc 104: 6816
394. Nishino S, Nagato Y, Yamamoto H, Ishido Y (1986) J Carbohydr Chem 5: 199
395. Guindon Y, Fortin R, Yoakim C, Gillard JW (1984) Tetrahedron Lett 25: 4717
395a. Wetter H, Oertle K (1985) Tetrahedron Lett 26: 5155
396. Sakauchi E, Horning EC (1971) Anal Lett 4: 41
397. Chambaz EM, Horning EC (1969) Anal Biochem 30: 7
398. Schneider HJ (1972) J Am Chem Soc 94: 3636
399. Kim SM, Bentley R, Sweeley CC (1967) Carbohydr Res 5: 373
400. Birkofer L, Ritter R, Bentz F (1964) Chem Ber 97: 2196
401. Okuda T, Konishi K, Saito S (1974) Chem Pharm Bull 22: 1624
402. Černý M, Gut V, Pacák J (1961) Collect Czech Chem Commun 26: 2542
403. Hurst DT, McInnes AG (1965) Can J Chem 43: 2004
404. McInnes AG (1965) Can J Chem 43: 1998
405. Gensler WJ, Alam I (1977) J Org Chem 42: 130
406. Schraml J, Petráková E, Pelnář J, Kvíčalová M, Chvalovský V (1985) J Carbohydr Chem 4: 393
407. Bentley R (1977) Carbohydr Res 59: 274
408. Ogilvie KK, Iwacha DJ (1973) Tetrahedron Lett 317
409. Kraska B, Klemer A, Hagedorn H (1974) Carbohydr Res 36: 398
410. Ogilvie KK, Sadana KL, Thompson EA, Quilliam MA, Westmore JB (1974) Tetrahedron Lett 2861
411. Ogilvie KK, Beaucage SL, Schifman AL, Thèriault NY, Sadana KL (1978) Can J Chem 56: 2768
412. Seela F, Hissmann E, Ott J (1983) Liebigs Ann Chem 1169
413. Ohtsuka E, Yamane A, Ikehara M (1983) Chem Pharm Bull 31: 1534
414. Ogilvie KK, Schifman AL, Penney CL (1979) Can J Chem 57: 2230
415. Köhler W, Pfleiderer W (1979) Liebigs Ann Chem 1855
416. Lagrange A, Olesker A, Lukacs G, Ton That Thang (1982) Carbohydr Res 110: 165
417. Costa SS, Lagrange A, Olesker A, Lukacs G (1980) J Chem Soc Chem Commun 721
418. Jain RK, Dubey R, Abbas SA, Matta KL (1987) Carbohydr Res 161: 31
419. Flockerzi D, Silber G, Charubala R, Schlosser W, Varma RS, Creegan F, Pfleiderer W (1981) Liebigs Ann Chem 1568
420. Ogilvie KK, Theriault NY (1979) Tetrahedron Lett 2111
421. Brandstetter HH, Zbiral E (1978) Helv Chim Acta 61: 1832
421a. Halmos T, Mortserret R, Filippi J, Antonakis K (1987) Carbohydr Res 170: 57
422. Tulshian DB, Tsang R, Fraser-Reid B (1984) J Org Chem 49: 2347
423. Binder TP, Robyt JF (1986) Carbohydr Res 147: 149
424. Horton D, Priebe W, Varela O (1985) ibid 144: 325

255

425. Brandstetter HH, Zbiral E, Schulz G (1982) Liebigs Ann Chem 1
426. Zbiral E, Schmid W (1985) Monatsh Chem 116: 253
427. Chavis C, Dumont F, Wightman RH, Ziegler JC, Imbach JL (1982) J Org Chem 47: 202
428. Ogilvie KK, Entwistle DW (1981) Carbohydr Res 89: 203
429. Jones SS, Reese CB (1979) J Chem Soc Perkin I 2762
430. van Boeckel CAA, van Aelst SF, Beetz T (1983) Rec Trav Chim Pays-Bas 102: 415
431. Blackburne ID, Fredericks PM, Guthrie RD (1976) Aust J Chem 29: 381
432. Franke F, Guthrie RD (1977) ibid 30: 639
433. Usui T, Takagaki Y, Tsuchiya T, Umezawa S (1984) Carbohydr Res 130: 165
434. Franke F, Guthrie RD (1978) Aust J Chem 31: 1285
435. Chaudhary SK, Hernandez O (1979) Tetrahedron Lett 99
436. Martin IR, Szarek WA (1984) Carbohydr Res 130: 195
437. Molino BF, Cusmano J, Mootoo DR, Faghih R, Fraser-Reid B (1987) J Carbohydr Chem 6: 479
438. Liang D, Schuda AD, Fraser-Reid B (1987) Carbohydr Res 164: 229
439. Cardillo G, Orena M, Sandri S, Tomasini C (1983) Chem Ind (London) 643
440. Hakimelahi GH, Proba ZA, Ogilvie KK (1981) Tetrahedron Lett 22: 4775
441. Hakimelahi GH, Proba ZA, Ogilvie KK (1981) Tetrahedron Lett 22: 5243
442. Ogilvie KK, Hakimelahi GH (1983) Carbohydr Res 115: 234
443. Dodd GH, Golding BT, Ioannou PV (1975) J Chem Soc Chem Commun 249
444. Batten RJ, Dixon AJ, Taylor RJK, Newton RF (1980) Synthesis 234
445. Dubey R, Jain RK, Abbas S, Matta KL (1987) Carbohydr Res 165: 189
446. Karl H, Lee CK, Khan R (1982) Carbohydr Res 101: 31
446a. Dubey R, Reynolds D, Abbas SA, Matta KL (1988) Carbohydr Res 183: 155
447. Hanessian S, Guindon Y, Lavallée P, Dextraze P (1985) Carbohydr Res 141: 221
448. Glushka JN, Gupta DN, Perlin AS (1983) Carbohydr Res 124: C12
449. Sadeh S, Warren CD, Jeanloz RW (1983) Carbohydr Res 123: 73

Chemistry of Pseudo-Sugars

Tetsuo Suami

Department of Chemistry, Faculty of Science and Engineering, Meisei University, Hino, Tokyo 191, Japan

Table of Contents

"Pseudo-sugar" is the name of a class of compounds in which a ring-oxygen of a hexopyranoid sugar is replaced by a methylene group. There are thirty-two theoretically possible stereoisomers in the pseudo-sugar family, including anomer-like compounds. The first pseudo-sugar was synthesized by G. E. McCasland and his coworkers in 1966. The most accessible starting material for the synthesis of pseudo-sugars was the Diels-Alder adduct of furan and acrylic acid. The Diels-Alder adduct was readily resolved by means of optically active amines as resolution reagents into the two antipodes, which were then used as starting compounds for the synthesis of enantiomeric pseudo-sugars. Also, a chiral synthesis from true sugars has been used for the preparation of enantiomeric pseudo-sugars.

Topics in Current Chemistry, Vol. 154
© Springer-Verlag Berlin Heidelberg 1990

1 Introduction

The term "pseudo-sugar" is the name of a class of compounds in which a ring-oxygen of a sugar is replaced by a methylene group. Therefore, there are two types of pseudo-sugars: pseudo-pyranose and pseudo-furanose, but in this review article the term pseudo-sugar is restricted to pseudo-pyranoses especially pseudo-hexopyranose, since pseudo-hexopyranoses and their derivatives have been extensively studied up to the present, while only a little is known on pseudo-furanose. The term "pseudo-sugar" was first proposed by G. E. McCasland in 1966 when he and his coworkers synthesized a first pseudo-sugar, namely pseudo-α-DL-talopyranose [1]. They prepared two more pseudo-sugars: pseudo-β-DL-gulopyranose [2] and pseudo-α-DL-galactopyranose [3] in 1968. He suggested in his paper [1] that pseudo-sugars may possess some biological activities, due to their structural resemblances to natural true sugars. Hope has been expressed that, in some cases, a pseudo-sugar will be accepted by enzymes or biological systems in place of a corresponding true sugar, and thus may serve to inhibit the growth of malignant or pathogenic cells.

In fact, several biologically active pseudo-sugars and their derivatives have been found in Nature in the last two decades. It was especially notable that a first naturally occurring pseudo-sugar: pseudo-α-D-galactopyranose (*1*) was discovered as an antibiotic in 1973 [4] seven years after McCasland's prediction. Prior to the discovery of pseudo-galactose *1*, pseudo-trisacchariddc validamycin antibiotic had been discovered in 1970 in a fermentation broth of *Streptomyces hygroscopicus* var. *limoneus* [5]. Validamycin A (*2*), a main component of the validamycin complex, exhibited strong inhibitory activity against a sheath blight of rice plants and a damping off of cucumber seedlings which were caused by an infection of *Pellicularia sasakii* and *Rhizoctonia solani* [5] (Scheme 1, 2).

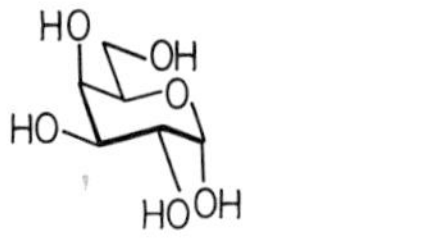

α-D-galactopyranose

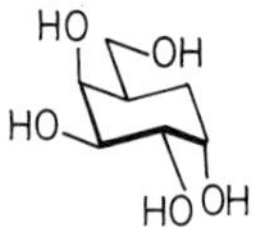

pseudo-α-D-galactopyranose

1

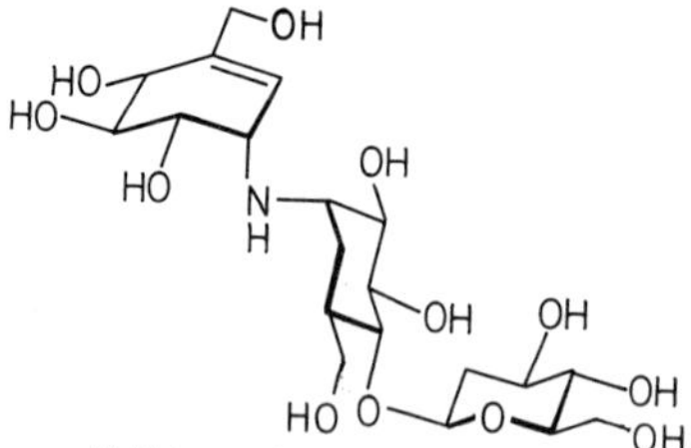

Validamycin A

2

Schemes 1, 2

Recently, pseudo-oligosaccharidic enzyme-inhibitors: acarbose (*3*) [6], trestatins (*4*) [7], adiposins (*5*) [8], S-AI [9, 10], and oligostatins [11, 12] have been found in fermentation beers (Scheme 3, 4, 5).

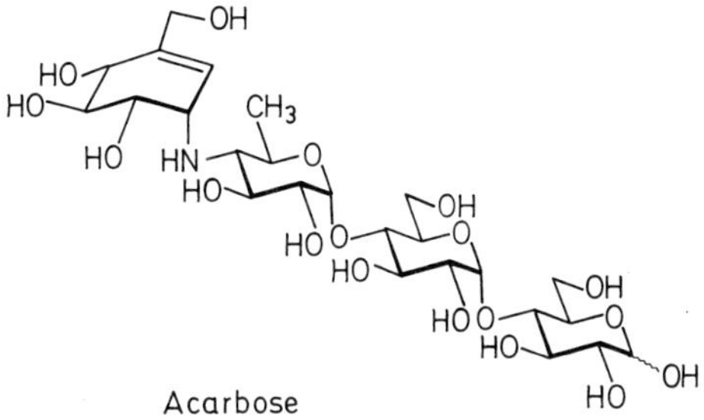

Acarbose
3

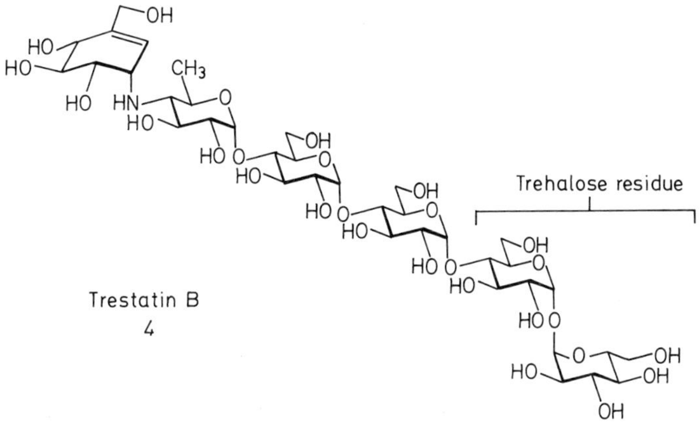

Trestatin B
4

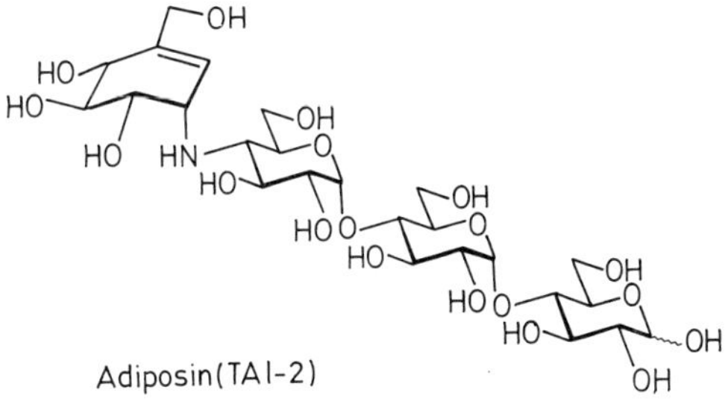

Trehalose residue

Adiposin(TAI-2)
5

Schemes 3, 4, 5

In the pseudo-sugar family there are 32 theoretically possible stereoisomers, including anomer-like compounds, and up to the present time, all the predicted sixteen racemic pseudo-sugars have been prepared, as well as ten optically active enantiomers. In the present review, preparations of pseudo-sugars, pseudo-β-fructopyranoses and biological effects of pseudo-sugars will be described.

2 Synthesis of Racemic Pseudo-sugars

The first McCasland's pseudo-sugar, pseudo-α-DL-talopyranose (9) was synthesized from 4-acetoxy-2,3-dihydroxy-5-oxo-cyclohexanecarboxylic acid (6) as follows [13]. Reduction of the *oxo* acid 6 with sodium borohydride and subsequent esterification with methanol and trifluoroacetic acid, followed by acetylation gave methyl (1, 2, 3, 4/5)-2, 3, 4, 5-tetraacetoxycyclohexanecarboxylate (7). Hydrogenation of 7 with lithium aluminium hydride and successive acetylation yielded pseudo-α-DL-talopyranose pentaacetate* (8). Hydrolysis of 8 in ethanolic hydrochloric acid gave pseudo-α-DL-talopyranose 9 in 23% overall yield from 6 [1] (Scheme 6).

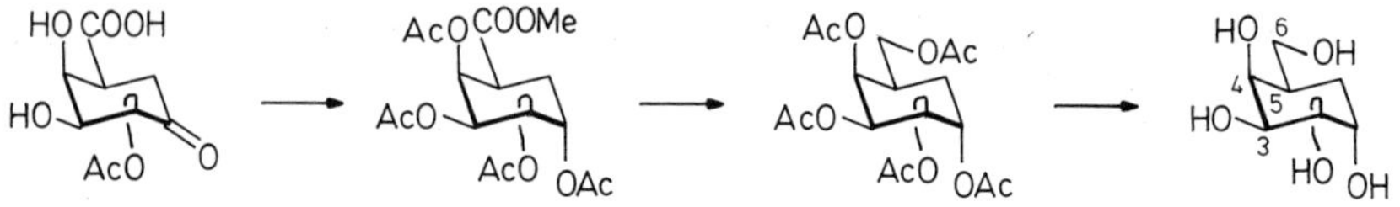

Scheme 6

Then, pseudo-β-DL-gulopyranose (14) was synthesized by hydroxylation of 2,5-di-hydroxy-3-cyclohexene-1-methanol triacetate (12), which was prepared by Diels-Alder cycloaddition of 1,4-diacetoxy-1,3-butadiene (10) and allyl acetate (11), with osmium tetroxide and hydrogen peroxide and successive acetylation as the penta-acetate (13). Analogous hydrolysis of 13 in ethanolic hydrochloric acid afforded the free pseudosugar 14 in 33% yield from 12 [2] (Scheme 7).

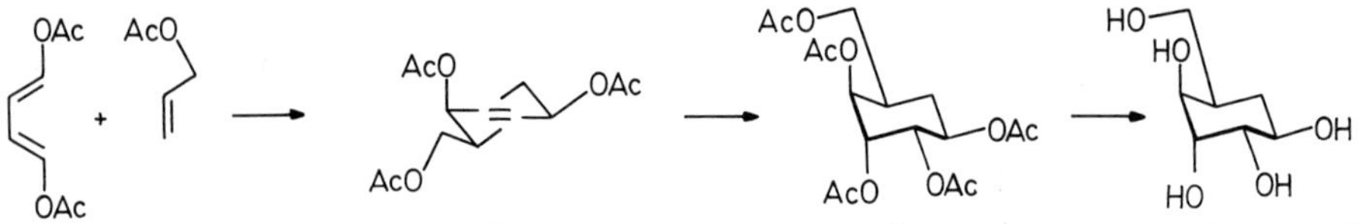

Scheme 7

When the first pseudo-sugar, pseudo-α-DL-talopyranose pentaacetate 8 was heated in 95% acetic acid containing a small volume of concentrated sulfuric acid, it underwent an epimerization on C-2 through an intermediary cyclic acetoxonium ion (15) which was formed by an anchimeric assistance of the neighboring acetoxy group, pseudo-α-DL-galactopyranose pentaacetate (16) was obtained in 14% yield, after conventional acetylation. Hydrolysis of the pentaacetate 16 gave pseudo-α-DL-galactopyranose (17) in 71% yield [3] (Scheme 8).

---

* For the sake of convenience, the numbering of a pseudo-sugar has been used tentatively in this article in the analogous numbering of a true hexopyranose.

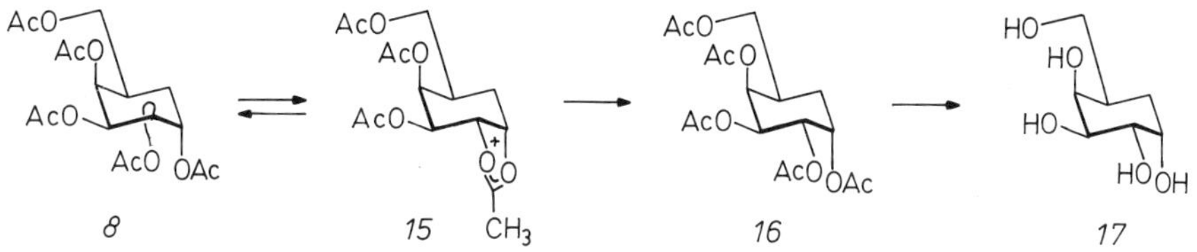

Scheme 8

Two other pseudo-sugars pseudo-β-DL-galactopyranose (28) and pseudo-α-DL-altropyranose (29) were prepared as pentaacetates from *myo*-inositol (18) [14]. *O*-Cyclohexylidenation of *myo*-inositol 18 gave 1,2-*O*-cyclohexylidene-*myo*-inositol (19), which was converted into 1,2-*O*-cyclohexylidene-3-*O*-*p*-toluenesulfonyl-*myo*-inositol (20) by preferential tosylation. Exclusive epoxidation of 20 with methanolic sodium methoxide afforded 1,2-anhydro-5,6-*O*-cyclohexylidene-*chiro*-inositol (21) and hydrogenation of 21 with lithium aluminium hydride in tetrahydrofuran (THF) gave 1,2-*O*-cyclohexylidene-5-deoxy-*chiro*-inositol (22) in 32% overall yield from 19 [15]. *O*-Isopropylidenation of 22 with 2,2-dimethoxypropane and successive Pfitzner-Moffatt oxidation with dimethylsulfoxide (DMSO) and acetic anhydride gave 5,6-*O*-cyclohexylidene-2-deoxy-3,4-*O*-isopropylidene-*chiro*-inosose-1 (23). Reactions of the ketone 23 with diazomethane gave 1,1'-anhydro-5,6-*O*-cyclohexylidene-2-deoxy-1-*C*-hydroxymethyl-3,4-*O*-isopropylidene-*chiro*-inositol (24) and its stereoisomer (25) in 82 and 9% yields, respectively. Nucleophilic opening of the oxirane ring of the major product 24 with hydriodic acid gave 1,2,3,4,6-penta-*O*-acetyl-5-deoxy-6-*C*-(iodomethyl)-*chiro*-inositol (26), after acetylation. Beta-elimination reactions of 26 by heating with zinc powder in glacial acetic acid resulted in a formation of the exocyclic olefin (27). Hydroboration of 27, followed by oxidation with hydrogen peroxide, and successive acetylation afforded pseudo-β-DL-galactopyranose pentaacetate (28), pseudo-α-DL-altropyranose pentaacetate (29) and 6-deoxy-pseudo-α-DL-altropyranose tetraacetate (30) in 13, 17 and 13% yields, respectively [14] (Scheme 9).

Since a preparation of pseudo-sugars from *myo*-inositol seemed to be unexpectedly laborious, as well as having a long reaction process, Diels-Alder cycloaddition has been attempted to obtain a more convenient starting material. The cycloaddition of furan (31) as an electron-rich diene and acrylic acid (32) as an electron-deficient dienophile has been carried out in the presence of hydroquinone as a polymerization-inhibitor to give known *endo*-7-oxabicyclo[2.2.1]hept-5-ene-2-carboxylic acid (33) in a good yield (45%) [16] which had been obtained in a poor yield by a previous reaction [17, 18]. Compund 33 was found to be the most important starting compound for a synthesis of pseudo-sugars by the following reactions.

Hydroxylation of 33 with hydrogen peroxide in formic acid resulted in a formation of *exo*-9-hydroxy-2,7-dioxatricyclo[4.2.1.0^{1,8}]nonan-3-one (34) by a spontaneous lactonization of an initially formed glycol. Reduction of 34 with lithium aluminium hydride in THF and successive acetylation gave *exo*-5, *endo*-6-diacetoxy-*endo*-2-(acetoxymethyl)-7-oxabicalo[2.2.1]heptane (35). Cleavage of the 1,4-epoxide linkage of 35 was carried out by heating in a mixture of acetic acid, acetic anhydride and sulfuric acid in a sealed tube, and a crude product was acetylated to give an approximately 1:1 mixture of two pseudo-sugar polyacetates. The mixture was successfully separated by column chromatography to give pseudo-α-DL-galactopyranose pentaacetate 16 and

261

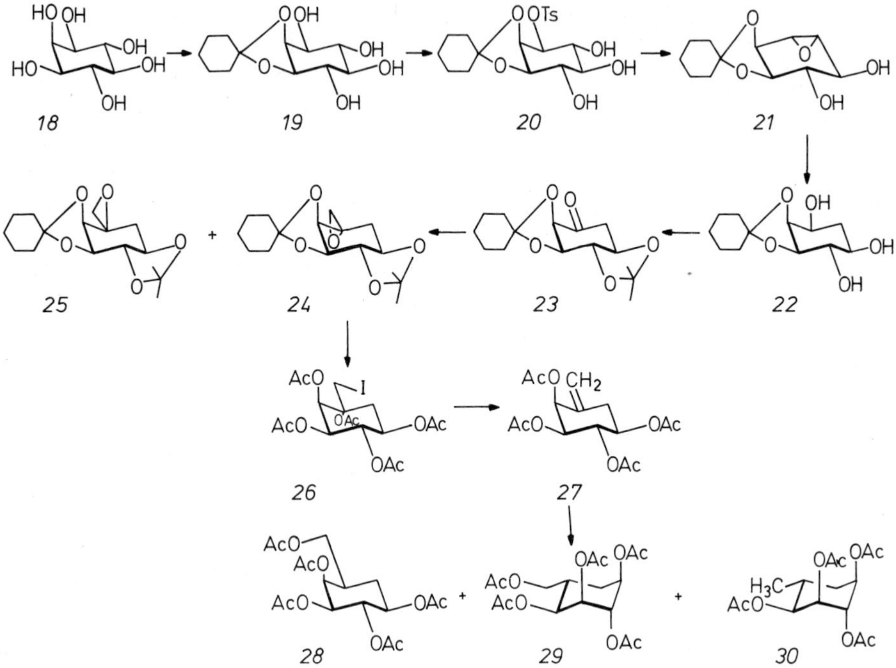

Scheme 9

pseudo-β-DL-glucopyranose pentaacetate (36) in 19 and 18% yields, respectively. Hydrolysis of the pentaacetate 16 in methanolic sodium methoxide gave paeudo-α-DL-galactopyranose 17, and an analogous hydrolysis of 36 gave pseudo-o-DL-gluco-pyranose (37) in quantitative yields [16, 20] (Scheme 10).

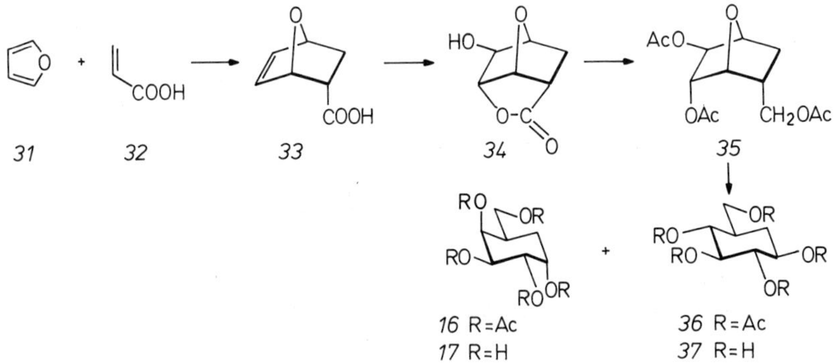

16 R=Ac   36 R=Ac
17 R=H   37 R=H

Scheme 10

Four new pseudo-sugars having α-ido, α-manno, β-altro and β-manno configurations have been synthesized from the Diels-Alder adduct 33 by the following method. Bromolactonization of 33 with hypobromous acid gave 2-exo-bromo-4,8-dioxatricyclo[4.2.1.0³,⁷]nonan-5-one (38). Reduction of 38 with lithium aluminium hydride, followed by acetylation afforded *endo*-2-acetoxy-*endo*-6-(acetoxymethyl)-*exo*-3-bromo-7-oxabicyclo[2.2.1]heptane (*39*) [20]. Acetolysis of *39* in a mixture of acetic acid, acetic anhydride and sulfuric acid in a sealed tube gave 2-bromo-2-deoxy-pseudo-α-DL-galactopyranose tetraacetate (*40*) and 2-bromo-2-deoxy-pseudo-β-DL-glucopyranose tetraacetate (*41*) in 31 and 13% yields, respectively. Nucleophilic displacement of *40* with an acetate ion and successive acetylation gave pseudo-α-DL-idopyranose pentaacetate (*42*) in 31% yield and known pseudo-α-DL-galactopyranose pentaacetate *16* in 10% yield. The displacement reaction involved a formation of an intermediary 2,3-cyclic acetoxonium ion. An analogous displacement reaction of *41* with an acetate ion gave pseudo-α-DL-mannopyranose pentaacetate (*44*) in 29% yield and pseudo-β-DL-altropyranose pentaacetate (*46*) in 27% yield. A displacement reaction of *41* by a benzoate ion, instead of the acetate ion, proceeded in a direct S_N2 mechanism afforded pseudo-β-DL-mannopyranose pentaacetate (*48*) in a yield of 49%. Hydrolysis of *42*, *44*, *46* and *48* in methanolic sodium methoxide gave pseudo-α-DL-idopyranose (*43*), pseudo-α-DL-mannopyranose (*45*), pseudo-β-DL-*altro*pyranose (*47*) and pseudo-β-DL-mannopyranose (*49*), respectively [20] (Scheme 11).

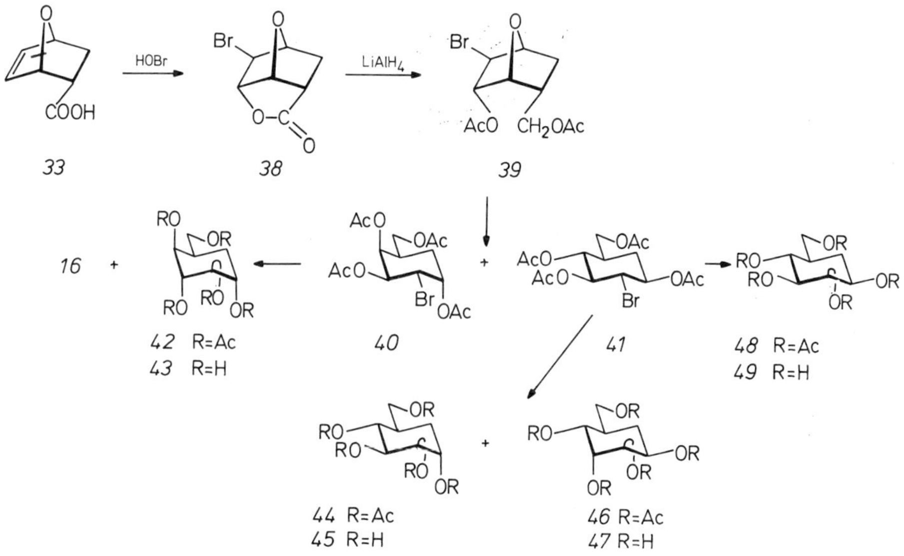

Scheme 11

On the other hand, when *39* was heated with hydrogen bromide in acetic acid, 1,2-di-*O*-acetyl-(1,3/2,6)-3,4-dibromo-6-(bromomethyl)-1,2-cyclohexanediol (*50*) was obtained, which was converted into 1,2-di-*O*-acetyl-(1,3/2)-3(bromomethyl)-5-cyclohexene-1,2-diol (*51*) by debromination with zinc dust in glacial acetic acid [21]. Hydroxylation of *51* with osmium tetroxide, and successive acetylation yielded 1,2,3,4-tetra-*O*-acetyl-6-bromo-6-deoxy-pseudo-α-DL-glucopyranose (*52*). Nucleophilic substitution reactions of *52* with sodium acetate gave pseudo-α-DL-glucopyranose pentaacetate (*53*), which gave pseudo-α-DL-glucopyranose (*54*) by usual hydrolysis [22]. Alternatively, the pentaacetate *53* was obtained as a minor component in a poor yield by nucleophilic substitutions of 2,3,4-tri-*O*-acetyl-1,6-dibromo-1,6-dideoxy-pseudo-β-DL-glucopyranose (*55*) with sodium benzoate and subsequent exchange of protective groups from benzoyl to acetyl groups [23] (Scheme 12).

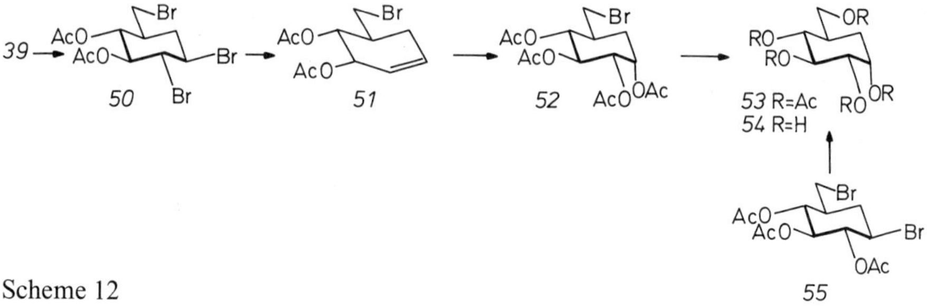

Scheme 12

Pseudo-α-DL-allopyranose (*61*) has been prepared from *54* by epimerization of the C-3 configuration as follows. *O*-Isopropylidenation of *54* with 2,2-dimethoxypropane gave 1,2:4,6-di-*O*-isopropylidene-pseudo-α-DL-glucopyranose (*56*). On oxidation with ruthenium tetroxide and sodium metaperiodate, *56* gave the 3-oxo derivative (*57*), which was converted into 1,2:4,6-di-*O*-isopropylidene-pseudo-α-DL-allopyranose (*58*) exclusively by catalytic hydrogenation under the presence of Raney nickel. Conventional acetylation of *58* furnished the 3-*O*-acetyl derivative (*59*). Hydrolysis of *59* with aqueous acetic acid, followed by acetylation afforded pseudo-α-DL-allopyranose pentaacetate (*60*), which gave the free pseudo-sugar *61* on usual alkaline hydrolysis [22] (Scheme 13).

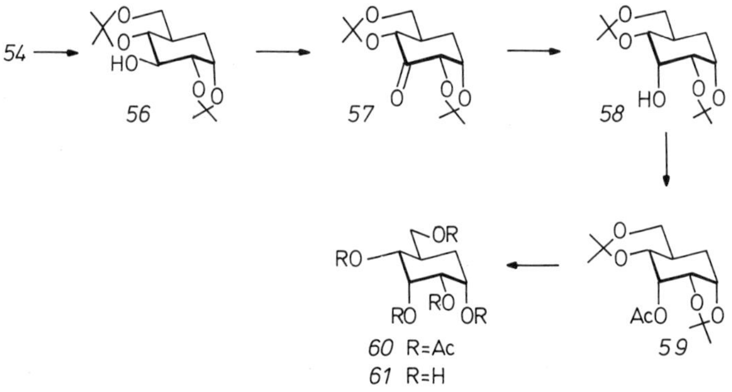

Scheme 13

Similarly, *O*-isopropylidenation of pseudo-β-DL-glucopyranose *37* [20] gave 1,2:4,6-di-*O*-isopropylidene-pseudo-β-DL-glucopyranose (*62*) and 2,3:4,6-di-*O*-isopropylidene-pseudo-β-glucopyranose (*63*) in 26 and 39% yields, respectively. Oxidation of the former compound *62* with ruthenium tetroxide and sodium metaperiodate gave the 3-oxo derivative (*64*), which gave 1,2:4,6-di-*O*-isopropylidene-pseudo-β-DL-*allo*pyranose (*65*) on analogous catalytic hydrogenation. Acetylation of *65* afforded the 3-*O*-acetyl derivative (*66*). Hydrolysis of the isopropylidene group in aqueous acetic acid, followed by acetylation gave pseudo-β-DL-allopyranose pentaacetate (*67*). Usual hydrolysis of *67* gave pseudo-β-DL-allopyranose (*68*) [22] (Scheme 14).

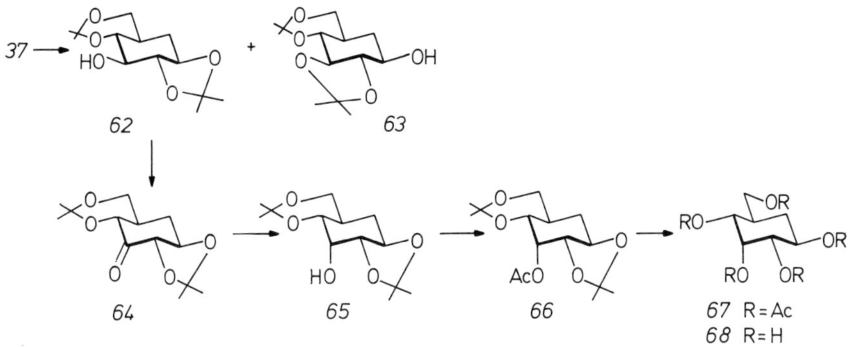

Scheme 14

While *O*-benzylidenation of pseudo-α-DL-galactopyranose *17* [3] with α,α-dimetoxy-toluene gave two stereoisomers (*69*) and (*70*) of 1,2:4,6-di-*O*-benzylidene-pseudo-α-DL-galactopyranose in 25 and 22% yields, respectively. Concerning a configuration of the phenyl group in 1,2-*O*-benzylidene group, *69* and *70* were tentatively assigned as two stereoisomers by their ^{1}H NMR spectra. Oxidation of the former isomer *69* with ruthenium tetroxide and sodium metaperiodate gave the 3-oxo compound (*71*), which was converted into 1,2:4,6-di-*O*-benzylidene-pseudo-α-DL-gulopyranose (*72*) by reduction with sodium borohydride. Hydrolysis of *72* in aqueous acetic and successive acetylation afforded pseudo-α-DL-gulopyranose pentaacetate (*73*), which on hydrolysis provided pseudo-α-DL-gulopyranose (*74*) [22] (Scheme 15).

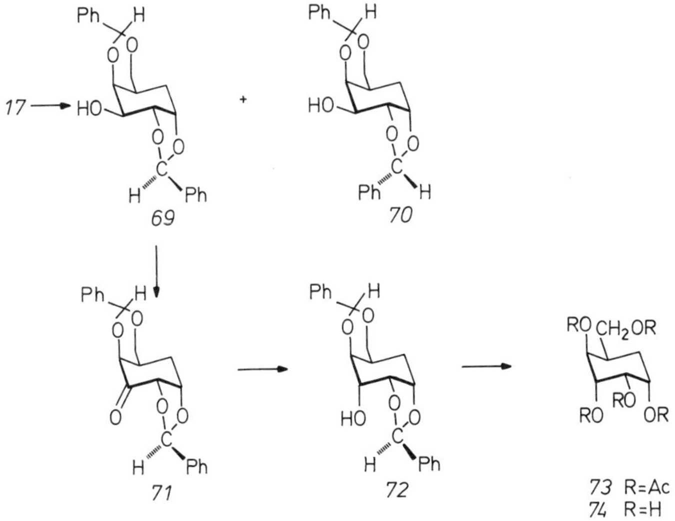

Scheme 15

Furthermore, one of the hitherto unknown two pseudo-sugars, pseudo-β-DL-idopyranose (*77*) was synthesized from readily accessible (1,2,4/3)-5-(hydroxymethyl)-5-cyclohexene-1,2,3,4-tetrol (*75*) [24]. Catalytic hydrogenation of *75* in the presence of platinum catalyst gave pseudo-β-DL-idopyranose pentaacetate (*76*) after acetylation.

Hydrolysis of *76* in methanolic sodium methoxide gave pseudo-β-DL-idopyranose *77* [22] (Scheme 16).

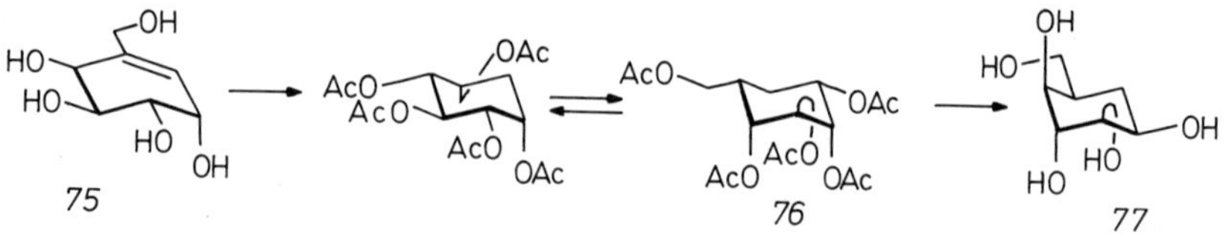

75   76   77

Scheme 16

The last unknown pseudo-sugar, pseudo-β-DL-talopyranose was prepared from 2-bromo-2-deoxy-pseudo-α-DL-galactopyranose tetraacetate (*78*) [20]. When *78* was heated with methanolic sodium methoxide and subsequently acetylated, 1,6-anhydro-pseudo-α-DL-galactopyranose triacetate (*80*) was obtained. The reaction involved an intramolecular rearrangement with a migration of an epoxide ring from an initially formed 2,3-epoxide (*79*) to the 1,6-ether compound *80*. In fact, the 2,3-epoxide *79* has been isolated from the reaction mixture at an early stage of the reaction. As the reaction progressed, the originally formed *79* disappeared, which then transformed into an intermediary 1,2-epoxide, and finally, concentrated in the 1,6-anhydro derivative.

Deacetylation of *80* in methanolic sodium methoxide and successive *O*-isopropylidenation with 2,2-dimethoxypropane gave 2-*O*-acetyl-3,4-*O*-isopropylidene-1,6-anhydro-pseudo-β-DL-galactopyranose (*81*), after acetylation. Removal of the acetyl group of *81*, followed by oxidation with ruthenium tetroxide and sodium metaperiodate afforded the 2-oxo derivative (*82*). Catalytic hydrogenation of *82* under the presence of platinum catalyst and acetolysis in a mixture of acetic acid, acetic anhydride and sulfuric acid gave pseudo-β-DL-talopyranose pentaacetate (*83*) [25] (Scheme 17).

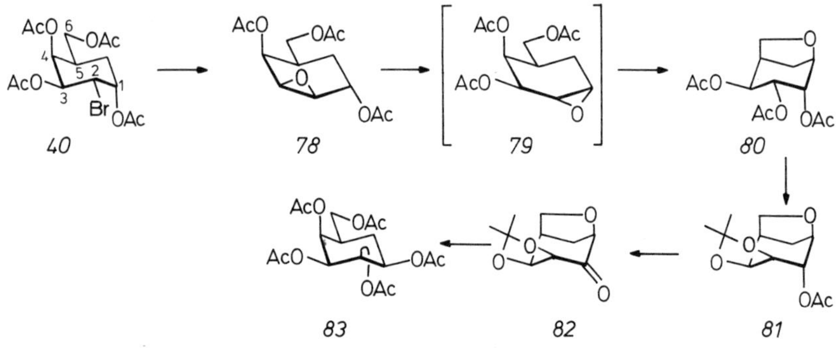

40   78   79   80

83   82   81

Scheme 17

Now, all the predicted sixteen stereoisomers of racemic pseudo-sugars have been synthesized and their physical constants are listed in Table 1. The first three pseudo-sugars *9*, *14* and *17* have been synthesized by McCasland and his coworkers. The two pseudo-sugar pentaacetates *28* and *29* were prepared from *myo*-inositol. All the remaining eleven pseudo-sugars have been synthesized from the Diels-Alder adduct

Table 1. Physical constants of DL-pseudo-sugars

Configuration	Pentaacetate M.p. (C)	Free M.p. (C)	References
α-Allopyranose	120–121 °	syrup	[22]
β-Allopyranose	110–111 °	185–186 °	[22]
α-Altropyranose	115–116 °	–	[14]
β-Altropyranose	106–107 °	143.5–144.5 °	[20]
α-Galactopyranose	147–148 °*	173–174 °*	[3]
	137–138 °	167–168 °	[20]
β-Galactopyranose	123–124 °	–	[14]
α-Glucopyranose	110–111 °	146–147 °	[21], [22], [23]
β-Glucopyranose	111–112 °	syrup	[16], [20]
α-Gulopyranose	109–110 °	139–140 °	[22]
β-Gulopyranose	132–133 °*	syrup	[2]
α-Idopyranose	106–107 °	154.5–156°	[20]
β-Idopyranose	syrup	syrup	[22]
α-Mannopyranose	99.5–100 °	syrup	[20]
β-Mannopyranose	123–125 °	198–199 °	[20]
α-Talopyranose	109–110 °*	160–162 °*	[1]
β-Talopyranose	117–119 °	–	[25]

M.ps. were measured in a capillary tubes in a liquid bath, and m.p. marked with asterisk was determined on a hot-plate.

of furan and acrylic acid. Therefore, the Diels-Alder adduct *33* was the most useful starting material for the synthesis of a wide variety of pseudo-sugars.

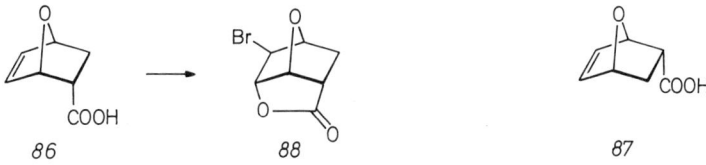

86   88   87

Scheme 18

3 Synthesis of Enantiomeric Pseudo-sugars

As the Diels-Alder cyclic adduct *33* has been recognized as the most accessible starting compound for the synthesis of racemic pseudo-sugars, a resolution of *33* has been attempted to prepare enantiomeric pseudo-sugars, starting from an optically active antipode of *33*. It has been revealed that *33* was readily separated into the enantiomers by using optically active α-methylbenzylamines as resolution reagents.

When *33* was mixed with an equimolar amount of (R)-(+)-α-methylbenzylamine in ethanol, a mixture of two crystalline salts, [(+)-amine$^+$ (−)-adduct$^-$] and [(+)-amine$^+$ (+)-adduct$^-$] was obtained. Fractional crystallization of the mixture from ethanol gave the former salt [(+)-amine$^+$ (−)-adduct$^-$] (*84*) in 42 % yield as an optically pure salt, m.p. 137.5–138.5 °C, $[\alpha]_D^{27}$ −66.4° (methanol). Similarly, by using (S)-(−)-α-methylbenzylamine, an optically pure salt of [(−)-amine$^+$ (+)-adduct$^-$] (85), m.p. 137.5–138.5 °C, $[\alpha]_n^{27}$ +66.7° (methanol) was obtained.

Optically active adducts: (−)-adduct (86), m.p. 97–98.5 °C, $[\alpha]_D^{22}$ −111.8° (ethanol), and (+)-adduct (87), m.p. 97–98.5 °C, $[\alpha]_D^{22}$ +110.7° (ethanol), can be recovered from the respective salts by treating an each solution with an ion-exchange resin, Dowex 50W X2 [26] (Scheme 18).

Before we use these optically active adducts as starting materials for a synthesis of enantiomeric pseudo-sugars, their absolute configurations must be learned. Fortunately, bromolactonization of the (−)-adduct 86 with hypobromous acid gave a nice crystalline bromolactone (88), m.p. 117–118.5 °C, $[\alpha]_D^{25}$ +92° (chloroform), whose absolute configuration has been established by a X-ray crystal structure analysis as (3S)-(+)-exo-2-bromo-4,8-dioxatricyclo[4.2.1.0^{3,7}]nonan-5-one (Fig. 1). Accordingly, it has been discovered that the (−)-adduct 86 corresponds to the D-series of pseudo-sugars, and the antipode (+)-adduct 87 belongs to the L-series [26, 27].

Starting from the (−)-adduct 86, two enantiomeric pseudosugars: pseudo-α-D-galactopyranose 1, and pseudo-β-D-glucopyranose (93) have been synthesized by analogous reactions employed in the preceding synthesis of the corresponding racemic pseudo-sugars [16]. That is, hydroxylactonization of 96 with hydrogen peroxide in formic acid gave the tricyclic hydroxylactone (89), m.p. 107–108 °C, $[\alpha]_D^{22}$ +47.9° (ethanol). Hydrogenation of 89 with lithium aluminium hydride and successive acetylation afforded the bicyclic triacetate (90), which was converted into pseudo-α-D-galactopyranose pentaacetate (91), m.p. 143–144 °C, $[\alpha]_D^{20}$ +43.2° (chloroform), and pseudo-β-D-glucopyranose pentaacetate (92), m.p. 115–116 °C, $[\alpha]_D^{20}$ +13.8° (chloroform), in 27 and 34% yields, respectively, by usual acetolysis. Hydrolysis of 91 in methanolic sodium methoxide gave pseudo-α-D-galactopyranose 1, m.p. 161.5 to 162.5 °C, $[\alpha]_D^{23}$ +66.3° (water) [27] that was identical with an authentic sample in

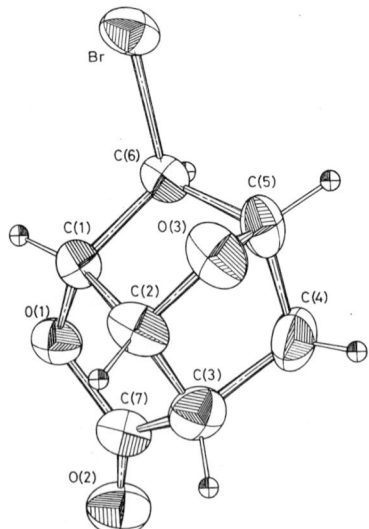

Fig. 1. Molecular structure of the bromo lactone (2)

all respects [4]. Analogous hydrolysis of *92* gave pseudo-β-D-glucopyranose (*93*) as a syrup, $[\alpha]_D^{20}$ +13.0° (water) [27] (Scheme 19).

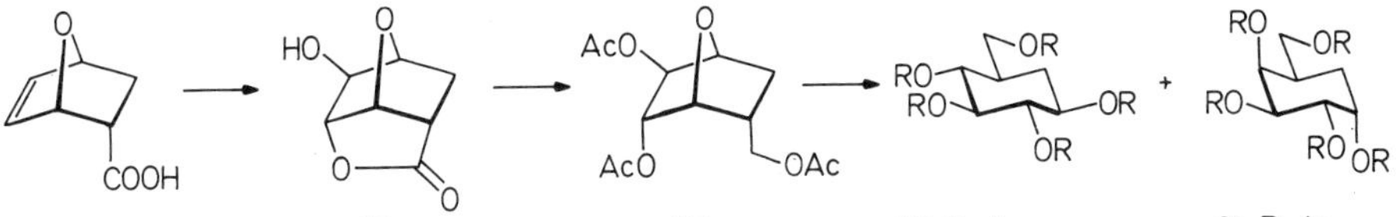

Scheme 19

Furthermore, Paulsen and his coworkers [28] synthesized compound *1* by another synthetic route, starting from naturally occurring quebrachitol: 1L-2-*O*-methyl-*chiro*-inositol (*94*). *O*-Demethylation of 1L-3,4:5,6-di-*O*-isopropylidene-2-*O*-methyl-*chiro*-inositol (*95*) with boron tribromide gave 1L-*chiro*-inositol (*96*), which was converted into 1L-1,2:3,4:5,6-tri-*O*-isopropylidene-*chiro*-inositol (*97*), m.p. 216.5 °C, $[\alpha]_D^{20}$ +35.96° (chloroform). Restricted hydrolysis of *97* with 95% aqueous acetic acid yielded 1L-1,2:5,6-di-*O*-isopropylidene-*chiro*-inositol (*98*), m.p. 153.2 °C, $[\alpha]_D^{20}$ −25.8° (chloroform). Mono-*O*-benzylation of *98* with benzyl bromide, tetraethyl-ammonium iodide and 20% sodium hydroxide solution gave 1L-4-*O*-benzyl-1,2:5,6-di-*O*-isopropylidene-*chiro*-inositol (*99*), m.p. 72.5 °C, $[\alpha]_D^{20}$ +43.4° (chloroform). Oxidation of *99* with dimethyl sulfoxide and oxalyl chloride afforded 1L-4-*O*-benzyl-1,2:5,6-di-*O*-isopropylidene-*chiro*-inosose-3 (*100*) as a syrup, $[\alpha]_D^{20}$ −77.15° (chloroform). Wittig reaction of *100* with methyl(triphenyl)phosphonium bromide and *n*-butyllithium gave the exocyclic olefin (*101*), $[\alpha]_D^{20}$ −84.0ë (chloroform). Hydroboration of *101* with boron hydride, followed by a reaction with hydrogen peroxide gave 1L-4-*O*-benzyl-3-deoxy-3-*C*-(hydroxymethyl)-1,2:5,6-di-*O*-isopropylidene-*allo*-inositol (*102*) as a syrup, $[\alpha]_D^{20}$ +0.7° (chloroform). Protection of the primary hydroxyl group with (2-methoxyethoxy)methyl chloride, followed by catalytic hydrogenolysis over palladium charcoal gave 1L-3-deoxy-1,2:5,6-di-*O*-isopropylidene-3-*C*-[(2-methoxy-ethoxy)methoxymethyl]-*allo*-inositol (*103*), $[\alpha]_D^{20}$ −30.0° (chloroform). By a reaction with carbon disulfide and methyl iodide, *103* was converted into the *S*-methyl-dithio-carbonate derivative (*104*), $[\alpha]_D^{20}$ −71.34° (chloroform). Deoxygenation of *104* with tri-*n*-butyl tinhydride afforded 1,2:3,4-di-*O*-isopropylidene-6-*O*-(2-methoxyethoxy)methyl-pseudo-α-D-galactopyranose (*105*), $[\alpha]_D^{20}$ −73.53° (chloroform). Hydrolysis of *105* with methanolic hydrogen chloride and successive acetylation gave pseudo-α-D-galactopyranose pentaacetate *91*, m.p. 144 °C, $[\alpha]^{20}$ +35.16° (chloroform). Hydrolysis of *91* gave pseudo-α-D-galactopyranose *1*, m.p. 161 °C, $[\alpha]_D^{20}$ + 47,9° (methanol) [28] (Scheme 20).

Also, psudo-β-D-mannopyranose (*115*) has been synthesized from *99* by the following reactions [28]. Halogenation of *99* with triphenylphosphine, imidazole and iodine gave 1L-4-*O*-benzyl-3-deoxy-3-iodo-1,2:5,6-di-*O*-isopropylidene-*allo*-inositol (*106*), m.p. 77.4 °C, $[\alpha]_D^{20}$ −30.1° (chloroform). Treatment of *106* with lithium aluminium hydride resulted in a formation of two endocyclic olefins (*107*) and (*108*) in an approximately 1:2 ratio. Oxidation of *108* with dimethyl sulfoxide and oxalyl chloride gave the enone (*109*) as a syrup, $[\alpha]_D^{20}$ −68.11° (chloroform). Stereoselective

269

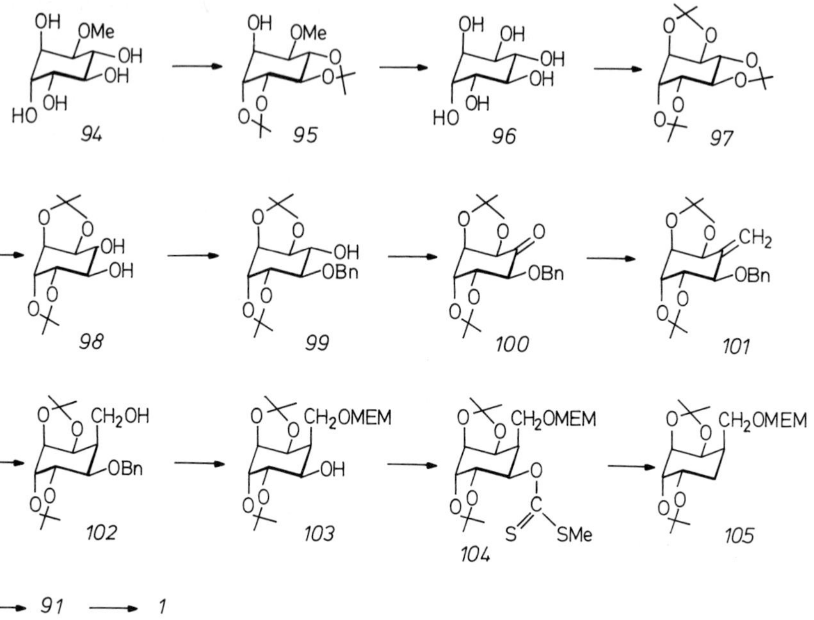

Scheme 20

alkylation of *109* with ethyl 2-lithio-1,3-dithian-2-carboxylate afforded the 2-(ethoxy-carbonyl)-1,3-dithian-2-yl derivative (*110*), m.p. 71.5 °C, $[\alpha]_D^{20}$ +28.9° (chloroform). Reduction of *110* with lithium aluminium hydride and successive acetylation gave the compound (*111*), m.p. 106 °C, $[\alpha]_D^{20}$ +66.71° (chloroform). Deprotection of *111* with mercuric oxide and mercuric chloride, reduction with sodium borohydride and successive acetylation gave a mixture of two stereoisomers (*112*). *O*-Deacetylation of *112* with methanolic sodium methoxide, followed by oxidative cleavage with sodium metaperiodate gave an aldehyde, which was further converted into 1,6-di-*O*-acetyl-4-*O*-benzyl-2,3-*O*-isopropylidene-pseudo-β-D-mannopyranose (*113*) by reduction with sodium borohydride and subsequent acetylation. Exhaustive deprotection of *113* and successive acetylation gave pseudo-β-D-mannopyranose pentaacetate (*114*), m.p. 119 °C, $[\alpha]_D^{20}$ +2.9° (chloroform). *O*-Deacetylation of *114* gave pseudo-ô-D-manno-pyranose (*115*), m.p. 217 °C, $[\alpha]_D^{20}$ +11.9° (methanol) [28] (Scheme 21).

Six other enantiomeric pseudo-sugars having α-D-gluco, β-L-altro, α-L-altro, β-L-allo, α-D-manno and β-L-manno configurations have been prepared by a chiral synthesis, starting from an appropriate true sugar. The first two pseudo-sugars were synthesized from L-arabinose, the next three pseudo-sugars were from D-glucose, and the last one was from D-ribose.

Selective *O*-tritylation of a primary hydroxyl group of L-arabinose diethyl dithio-acetal (*116*) with trityl chloride and 4-dimethylaminopyridine (DMAP), successive benzylation with benzyl bromide, and *O*-detritylation with *p*-toluenesulfonic acid, followed by tosylation gave 2,3,4-tri-*O*-benzyl-5-*O*-tosyl-L-arabinose diethyl dithio-acetal (*117*), $[\alpha]_D^{18}$ −2° (chlorŒoform). Cleavage of the diethyl dithioacetal group with

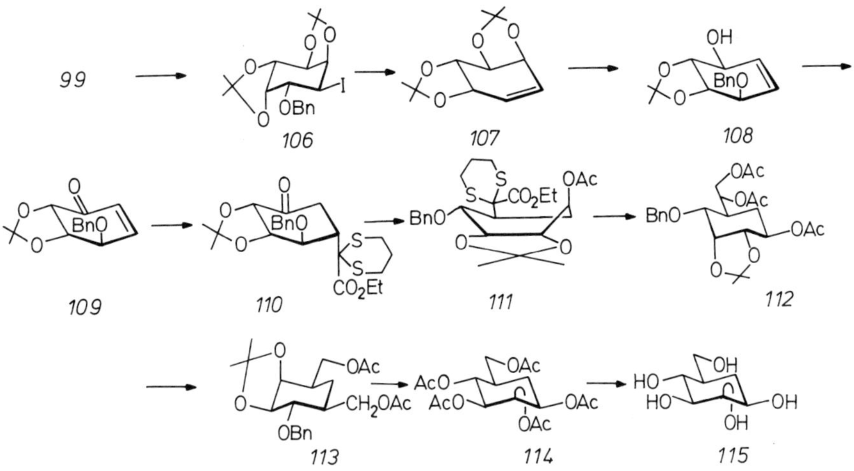

99 → 106 → 107 → 108 →

109 → 110 → 111 → 112 →

→ 113 → 114 → 115

Scheme 21

mercuric chloride and calcium carbonate gave the parent aldehyde (*118*), which was converted into 2,3,4-tri-*O*-benzyl-5-deoxy-5-iodo-L-arabinose (*119*), $[\alpha]_D^{20}$ +9° (chloroform), by a reaction with sodium iodide. Cyclization of *119* with dimethyl malonate and sodium hydride, and subsequent acetylation provided the cyclohexane derivative (*120*), $[\alpha]_D^{19}$ +10.7° (chloroform) and the tetrahydropyran derivative (*121*), $[\alpha]_D^{19}$ +7.7° (chloroform), in 43 and 33% yields, respectively. The crucial cyclization involved a nucleophilic attack of the malonic ester carbanion to the aldehyde group at the first step of the reaction. And then, a substituted malonic ester carbanion attacked on the C-5 of the sugar moiety to give *120*, or a hydroxyl group formed in the first step attacked on the C-5 to give *121*.

Thermal decarboxylation of *120* yielded the cyclohexene derivative (*122*) $[\alpha]_D^{17}$ +64° (chloroform), by the beta-elimination. Hydrogenation of *122* with lithium aluminium hydride gave the compound (*123*). Hydroboration of *123* with diborane gave the pseudo-α-D-glucopyranose derivative (*124*), m.p. 98–100 °C, $[\alpha]_D^{21}$ +20° (chloroform), and the pseudo-β-L-altropyranose derivative (*127*), $[\alpha]_D^{20}$ −24° (chloroform), in 34 and 35% yields, respectively, after oxidation with hydrogen peroxide and subsequent acetylation. *O*-Debenzylation of *124* with sodium in liquid ammonia, and successive acetylation gave pseudo-α-D-glucopyranose pentaacetate (*125*) as a syrup, $[\alpha]_D^{22}$ +37° (chloroform), which was converted into pseudo-α-D-glucopyranose (*126*), $[\alpha]_D^{25}$ +30° (methanol), on hydrolysis. Analogous reactions of *127* gave pseudo-β-L-altropyranose pentaacetate (*128*), α_D^{21} +7° (chloroform) and its free form (*129*), $[\alpha]_D^{24}$ −49.5° (methanol) [29] (Scheme 22).

Another synthetic reaction for *126* has been reported by a French research group [30], starting from D-glucose, but a detail of the reaction has not been described.

More recently, a facile synthesis of *126* and pseudo-α-L-glucopyranose (*137*) has been described [32]. When the bromolactone *38* was heated with glacial acetic containing hydrogen bromide and subsequently acetylated, DL-(1,3,5/2,4)-2,3-diacetoxy-4,5-dibromocyclohexane-1-carboxylic acid (*130*) was obtained [31]. Resolution of *130* with optically active α-methylbenzylamines provided the enantiomer (*131*), m.p.

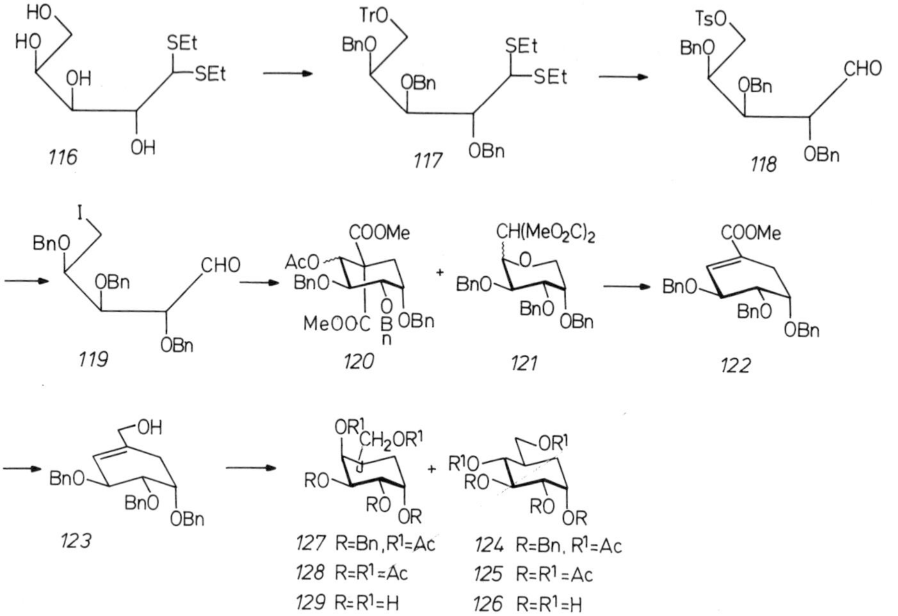

Scheme 22

180–181 °C, $[\alpha]_D^{19}$ −5.1° (ethanol), and the antipode (132), m.p. 180–181 °C, $[\alpha]_D^{21}$ +5.2° (ethanol), in 49 and 44% yields, respectively. Compound 131 was found to be a good precursor for pseudo-sugars of D-series, by the fact that 131 was obtainable from 88 whose absolute configuration had been unequivocally established by the X-ray crystal structure analysis [27]. Esterification of 131 with methanol gave methyl (1S)-(−)-(1,3,5/2,4)-2,3-diacetoxy-4,5-dibromocyclohexane-1-carboxylate (133), m.p. 140–141 °C, $[\alpha]_D^{20}$ −15.5° (chloroform). Elimination reaction of 133 with zinc dust in acetic yielded methyl (1S)-(−)-(1,3/2)-2,3-diacetoxy-4-cyclohexene-1-carboxylate (134), m.p. 45–47 °C, $[\alpha]_D^{18}$ −22° (chloroform). Hydroxylation of 134 with osmium tetroxide and hydrogen peroxide, followed by acetylation afforded methyl (1S)-(+)-(1,3/2,4,5)-2,3,4,5-tetraacetoxycyclohexane-1-carboxylate (135), m.p. 79–80 °C, $[\alpha]_D^{23}$ +56° (chloroform). Reduction of 135 with lithium aluminium hydride and successive acetylation gave pseudo-α-D-glucopyranose pentaacetate 125 as a syrup, $[\alpha]_D^{21}$ +57° (chloroform). Hydrolysis of 125 gave 126, $[\alpha]_D^{21}$ + 67° (methanol) [32].

Analogous reactions from 132 gave pseudo-α-L-glucopyranose pentaacetate (136), $[\alpha]_D^{20}$ −56° (chloroform), which was converted into pseudo-α-L-glucopyranose 137, $[\alpha]_D^{21}$ −67° (methanol), on hydrolysis [32] (Scheme 23).

Pseudo-α-L-altropyranose (154) was synthesized from D-glucose. D-Glucose was converted into 1,2:5,6-di-O-isopropylidene-α-D-ribo-hexofuranos-3-ulose (138) by a known procedure [33]. Wittig reaction of 138 with acetylmethylenetriphenyl phosphorane gave known 3-C-(acetylmethylene)-3-deoxy-1,2:5,6-di-O-isopropylidene-α-D-ribo-hexofuranose [34] (139). Catalytic hydrogenation of 139 over Raney nickel and successive oxidation with pyridinium chlorochromate (PCC) gave 3-C-(acetylmethyl)-3-deoxy-1,2:5,6-di-O-isopropylidene-α-D-allofuranose (140), m.p. 36.5–37.0 °C, $[\alpha]_D^{20}$ +82.1° (chloroform). Selective hydrolysis of 140 with aqueous acetic acid gave the

272

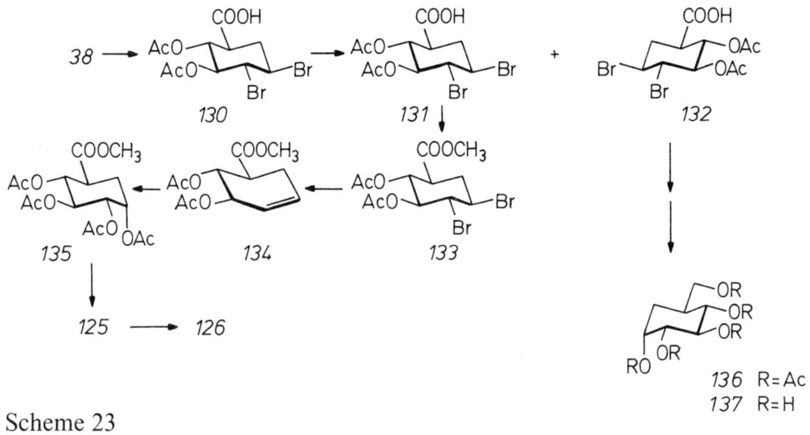

$38 \longrightarrow$ **130** $\longrightarrow$ **131** $+$ **132**

COOH (130) AcO— AcO— Br Br

COOH (131) AcO— AcO— Br Br

COOH (132) Br OAc OAc Br

135 $\longleftarrow$ **134** $\longleftarrow$ **133**

COOCH₃ (135) AcO AcO AcO OAc

COOCH₃ (134) AcO AcO

COOCH₃ (133) AcO AcO Br Br

125 $\longrightarrow$ **126**

OR OR OR RO OR

136 R = Ac
137 R = H

Scheme 23

1,2-*O*-isopropylidene derivative (*141*), m.p. 69.0–69.5 °C, $[\alpha]_D^{24}$ +79.7° (chloroform). Periodic acid oxidation of *141* afforded the aldehyde (*142*), which was converted into 8,9-*O*-isopropylidene-3-oxo-7-oxabicyclo[4.3.0]non-4-ene-8,9-diol (*143*), m.p. 48.0 to 49.0 °C, $[\alpha]_D^{17}$ −45.3° (chloroform), by an intramolecular aldol condensation with 1,8-diazabicyclo[5.4.0]undec-7-ene (DBU) and successive dehydration with acetic anhydride and pyridine. Epoxidation of *143* with hydrogen peroxide gave the epoxide (*144*) as a major product (96% yield) and another epoxide (*145*) as a minor product (3% yield). Reduction of *144* with sodium borohydride afforded two compounds (*146*), syrup, $[\alpha]_D^{27}$ +26.1° (chloroform), and (*147*), m.p. 141–143 °C, $[\alpha]_D^{27}$ +63.2° (chloroform) in a 4:1 ratio. Nucleophilic opening of the oxirane ring of *146* by hydroxide ions gave exclusively 8,9-*O*-isopropylidene-7-oxabicyclo[4.3.0]nonane-3,4,5,8,9-pentol (*148*), m.p. 159–161 °C, $[\alpha]_D^{25}$ −6.6° (methanol), by migration of the oxirane ring and subsequent *trans* diaxial opening. The reaction involved a migration of the oxirane ring with an anchimeric assistance of the vicinal hydroxyl group. The compound *148* was also obtained from *147* by an analogous reaction.

O-Benzylation of *148* with benzyl bromide in the usual manner yielded the tri-*O*-benzyl derivative (*149*), $[\alpha]_D^{27}$ −0.8° (chloroform). On hydrolysis with aqueous acetic acid, *149* gave the compound (*150*), which was further converted to the compound (*151*) by sodium borohydride reduction. Periodic acid oxidation of *151* and successive sodium borohydride reduction gave 5,6-di-*O*-acetyl-2,3,4-tri-*O*-benzyl-pseudo-α-L-altropyranose (*152*), $[\alpha]_D^{26}$ −25.7° (chloroform), after conventional acetylation. Reductive cleavage of *152* with sodium in liquid ammonia and subsequent acetylation afforded pseudo-α-L-altropyranose pentaacetate (*153*), m.p. 84–85 °C, $[\alpha]_D^{26}$ −13.7° (chloroform). Hydrolysis of *153* gave *154*, $[\alpha]_D^{25}$ −43.6° (methanol) [35] (Scheme 24).

Two pseudo-sugars of β-L-allo and α-D-manno have been prepared from the preceding intermediate *143* as follows. Reduction of *143* with diisobutyl aluminium hydride afforded the 3-hydroxyl derivative (*155*), which gave the 3-O-acetyl derivative (*156*) on acetylation. Hydroxylation of *156* with osmium tetroxide and hydrogen peroxide gave the compound (*157*), m.p. 154–155 °C, $[\alpha]_D^{21}$ +71.5° (chloroform), as a main component (53% yield) and an unidentified compound, m.p. 191–193 °C, $[\alpha]_D^{22}$ −7.9° (chloroform), as a minor component (6% yield). On acetylation, *157* was converted into the tri-*O*-acetyl derivative (*158*), m.p. 66–67 °C, $[\alpha]_D^{23}$ +41.8° (chloroform).

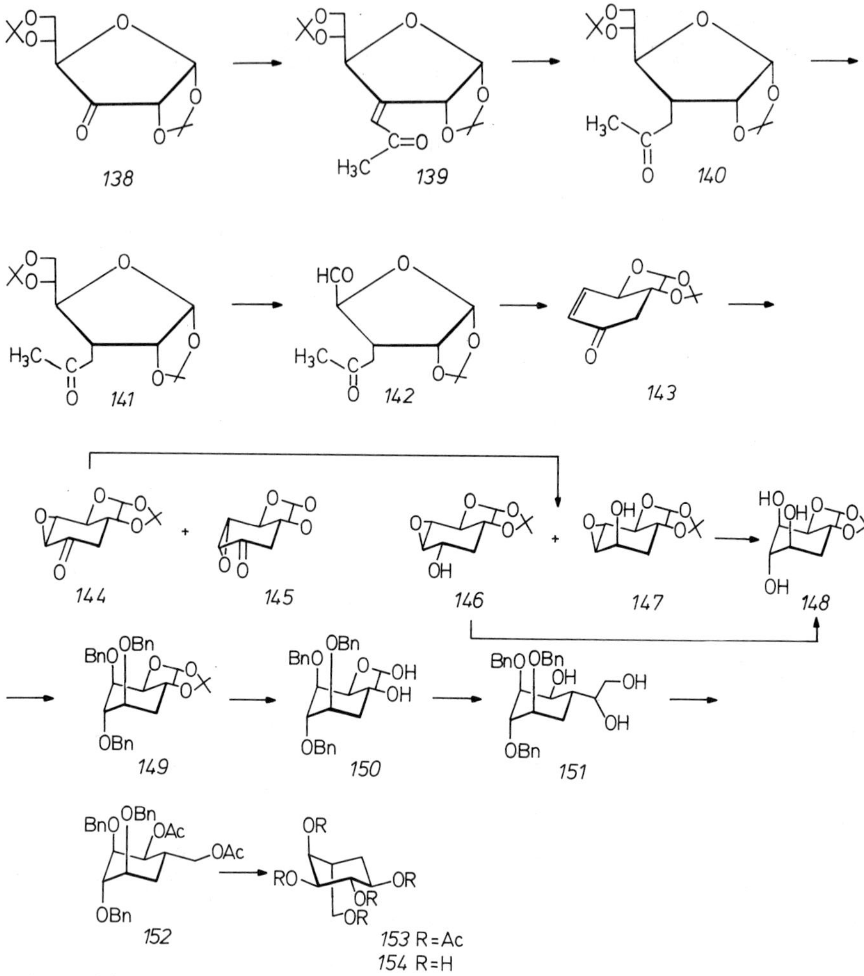

Scheme 24

O-Deacetylation of *158* and successive O-benzylation gave the tri-O-benzyl derivative (*159*), $[\alpha]_D^{21}$ +36.7° (chloroform). When *159* was hydrolyzed with aqueous acetic acid and subsequently hydrogenated with sodium borohydride, the compound (*160*), $[\alpha]_D^{22}$ +4.4° (chloroform), was obtained. Periodic acid oxidation of *160* and reduction with sodium borohydride afforded 4,6-di-O-acetyl-1,2,3-tri-O-benzyl-pseudo-β-L-allopyranose (*161*), $[\alpha]_D^{18}$ —40.2° (chloroform), after acetylation. Catalytic hydrogenolysis of *161* and successive acetylation gave pseudo-β-L-allopyranose pentaacetate (*162*), m.p. 135–136 °C, $[\alpha]_D^{18}$ +3.7° (chloroform) [36]. (Scheme 25).

The aldehyde (*163*), which was prepared from *160* by periodic acid oxidation, was further converted into pseudo-α-D-*manno*pyranose as follows. Dehydration of *163* with mesyl chloride and pyridine, and subsequent reduction with lithium aluminium hydride gave (3S, 4R, 5S)-3,4,5-tris(benzyloxy)-1-cyclohexene-1-methanol (*164*). Hydroxylation of *164* with diborane and hydrogen peroxide yielded 4,6-di-O-acetyl-1,2,3-tri-O-

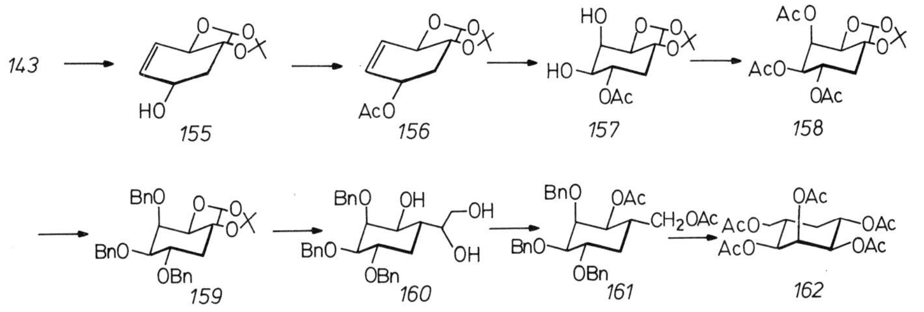

Scheme 25

benzyl-pseudo-α-D-mannopyranose (165), after usual acetylation. Catalytic hydro-genolysis of 165 in the presence of palladium catalysts, and successive acetylation gave pseudo-α-D-mannopyranose pentaacetate (166), m.p. 80–81 °C, $[\alpha]_D^{19}$ +27.8° (chloroform) [36] (Scheme 26).

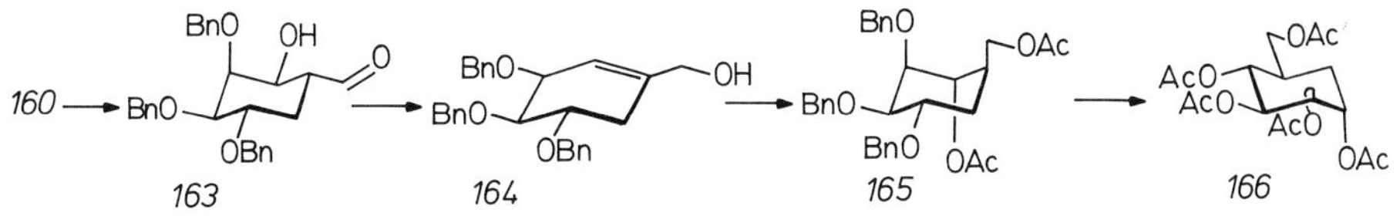

Scheme 26

The last hitherto known pseudo-sugar is pseudo-β-L-mannopyranose, which has been synthesized from D-ribose. 2,3,4-Tri-O-bezyl-5-O-(t-butyldiphenylsilyl)-D-ribose diethyl dithioacetal (168), $[\alpha]_D^{26}$ −3.3° (chloroform), was obtained from D-ribose di-ethyl dithioacetal (167) [37] by sequential reactions of O-tritylation of the primary hydroxyl group, O-benzylation, O-detritylation and O-silylation with t-butyldiphenyl-chlorosilane. Cleavage of the dithioacetal group of 168 gave the aldehyde (169) by a reaction with mercuric chloride in acetonitrile. When 169 reacted with dimethyl malo-nate in a mixture of acetic anhydride and pyridine, the unsaturated condensation product (170), $[\alpha]_D^{29}$ −5.6° (chloroform), was obtained. Catalytic hydrogenation of 170 over Raney nickel, and removal of the silyl group with tetraammonium fluoride afforded the compound (171), $[\alpha]_D^{29}$ −7.2° (chloroform). Oxidation of 171 with PCC gave an intermediary aldehyde, which underwent an intramolecular aldol condensa-tion to give the cyclohexane derivative (172), m.p. 134–136 °C, $[\alpha]_D^{32}$ +12.1° (chloro-form), after acetylation. Thermal decarboxylation of 172 gave the cyclohexene deri-vative (173), $[\alpha]_D^{29}$ −53.3° (chloroform). Pseudo-β-L-mannopyranose pentaacetate (174), $[\alpha]_D^{19.5}$ −1.1° (chloroform), was prepared from 173 by sequential reactions of reduction with lithium aluminium hydride, hydroxylation with lithium aluminium hydride, hydroxylation with diborane and hydrogen peroxide, catalytic O-debenzyla-tion over palladium black and usual acetylation [38] (Scheme 27).

Thus, ten optically active pseudo-sugars are known and their physical constants are listed on Table 2.

Table 2 Physical constants of enantiomeric pseudo-sugars

Configuration	Pentaacetate		Free		References
	M.p. (C)	$[\alpha]_D$	M.p. (C)	$[\alpha]_D$	
β-L-Allopyranose	135–136°	$[\alpha]_D^{18}$ +3.7° (CHCl$_3$)	—	$[\alpha]_D^{25}$ −43.6° (m)	[36]
α-L-Altropyranose	84–85°	$[\alpha]_D^{26}$ −13.7° (CHCl$_3$)	syrup	$[\alpha]_D^{24}$ −49.5° (m)	[35]
β-L-Altropyranose	syrup	$[\alpha]_D^{21}$ +7° (CHCl$_3$)	syrup		[29]
α-D-Galactopyranose	143–144°	$[\alpha]_D^{20}$ +43.2° (c)	161.5–162.5°	$[\alpha]_D^{23}$ +66.3° (w)	[27]
	144°	$[\alpha]_D^{20}$ +35.16° (c)	164°	$[\alpha]_D^{23}$ +61.5 ± 4° (w)	[4]
α-D-Glucopyranose	syrup	$[\alpha]_D^{22}$ +37° (c)	161°	$[\alpha]_D^{20}$ +47.9° (m)	[28]
	syrup	$[\alpha]_D^{21}$ +57° (c)	syrup	$[\alpha]_D^{25}$ +30° (m)	[29]
β-D-Glucopyranose	115–116°	$[\alpha]_D^{20}$ +13.8° (c)	syrup	$[\alpha]_D^{21}$ +67° (m)	[35]
β-D-Glucopyranose	115–116°	$[\alpha]_D^{20}$ +13.8° (c)	syrup	$[\alpha]_D^{20}$ +13.0° (w)	[27]
α-L-Glucopyranose	syrup	$[\alpha]_D^{20}$ −56° (c)	syrup	$[\alpha]_D^{20}$ +13.0° (w)	[27]
α-D-Mannopyranose	80–81°	$[\alpha]_D^{19}$ +27.8° (c)	syrup	$[\alpha]_D^{21}$ −67° (m)	[35]
	119°	$[\alpha]_D^{20}$ +2.9° (c)	—		[36]
β-D-Mannopyranose	syrup	$[\alpha]_D$ −1.1° (c)	217°	$[\alpha]_D^{20}$ +11.9° (m)	[28]
β-L-Mannopyranose			—		[38]

(m) = methanol, (w) = water.

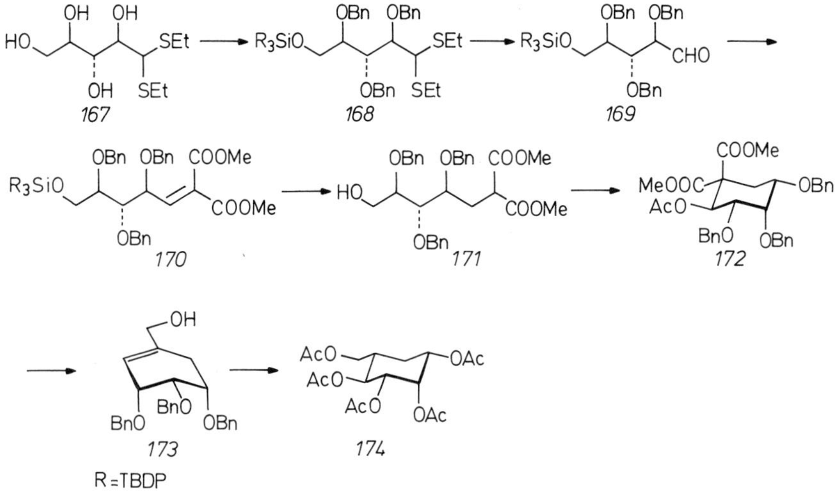

Scheme 27

4 Pseudo-β-fructopyranose

It has been revealed that pseudo-β-DL-glucopyranose *37* and pseudo-α-DL-galacto-pyranose *17* are almost as equally sweet as D-glucose and D-galactose, respectively [39]. This is not surprising, because a common molecular feature of a sweet tasting compound has been established as a bifunctional entity of AH (a proton donor) and B (a proton acceptor) components, together with a third hydrophobic component (X) [40, 41, 42]. The required distance between AH and B is approximately 3.0 Å, and the X component is located 3.5 Å from AH and 5.5 Å from B [41, 43]. In the D-glucopyranose molecule, the OH group on C-4 is assigned to be AH, the OH group on C-3 is ascribed to B, and the CH$_2$ group on C-6 is assigned to be X [44]. The geometry of a pyranose ring is practically same as that of a cyclohexane ring, and hence, the

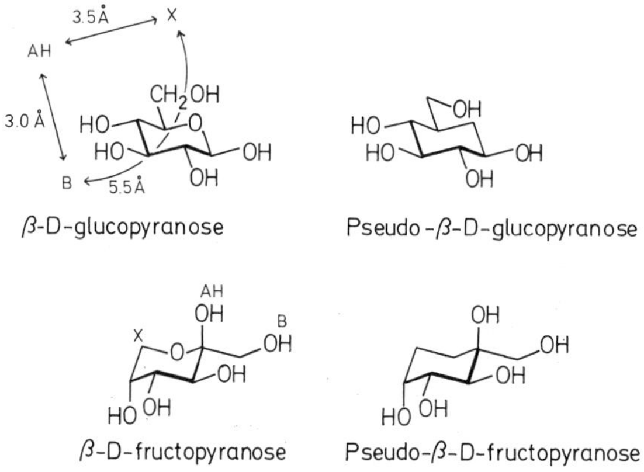

β-D-glucopyranose   Pseudo-β-D-glucopyranose

β-D-fructopyranose   Pseudo-β-D-fructopyranose

Scheme 28

277

sweetness eliciting components may exist on some pseudo-sugars. Also, the relative sweetness of L-glucose is about the same as that of D-glucose [45] and hence, pseudo-L-glucopyranose may taste as sweet as its D-enantiomer.

Among naturally occurring carbohydrates, D-fructose is the most sweet sugar, and pseudo-β-fructopyranose might be as equally sweet as fructose, since the intense sweetness of D-fructose arises only from a β-D-fructopyranose form [40, 46, 47]. The sweetness eliciting tripartite, AH, B and X, in the β-D-fructopyranose is assigned to the OH group on C-2, the OH group on C-1 and the CH$_2$ group on C-6, respectively. On the pseudo-β-fructopyranose molecule, the tripartite may exist in the same positions (Scheme 28).

4.1 Synthesis of Racemic Pseudo-β-fructopyranose

To substantiate the above-mentioned prediction, pseudo-β-DL-fructopyranose (180) has been synthesized by the following two routes. Debromination of DL-(1,3/2)-1,2-diacetoxy-3-bromomethyl-5-cyclohexene 51 [21] with DBU afforded the diene (175). Preferential epoxidation of the exocyclic double bond with m-chloroperbenzoic acid (mCPBA) gave the spiro epoxide (176). Opening of the oxirane ring of 176 with acetate ions, and subsequent acetylation with acetic anhydride and 4-dimethylaminopyridine (DMAP) gave the tetra-O-acetyl derivative (177). O-Deacetylation of 177 and successive epoxidation with mCPBA, followed by acetylation yielded the epoxide (178). The epoxidation proceeded stereoselectively with a cis-directing effect of the vicinal hydroxyl group [48] to give 178 as the primary product.

Reductive cleavage of the oxirane ring of 178 with lithium aluminium hydride, followed by acetylation with acetin anhydride and DMPA afforded pseudo-β-DL-fructopyranose pentaacetate (179), m.p. 147–148 °C, in an overall yield of 6%. O-Deacetylation of 179 gave pseudo-β-DL-fructopyranose 180 as a syrup [49, 50] (Scheme 29).

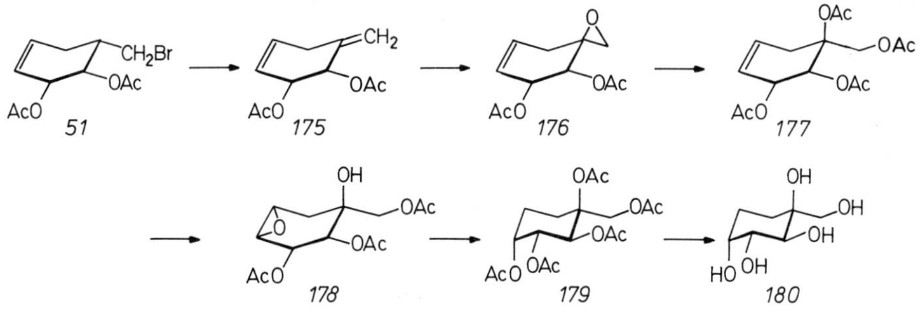

Scheme 29

A facile synthesis of 180 with a better yield has been developed as follows. O-Deacetylation of 51 with hydrochloric acid gave DL-(1,3/2)-3-bromomethyl-5-cyclohexene-1,2-diol (181). Stereospecific epoxidation of 181 with mCPBA and subsequent acetylation gave the epoxide (182), which afforded the exocyclic methylene derivative (183) by dehydrobromination with silver fluoride [51]. Reductive cleavage of the oxirane ring of 183 with lithium aluminium hydride, followed by acetylation yielded

278

the compound (*184*). Epoxidation of *184* with nCPBA gave the spiro epoxide (*185*) selectively. Opening of the oxirane ring by acetate ions and successive acetylation with acetic anhydride and pyridine gave the tetra-O-acetyl derivative (*186*), or with acetic anhydride and DMAP gave *179*. Compounds *186* and *179* were readily converted into *180* by usual hydrolysis. The overall yield was 21 % [49, 50] (Scheme 30).

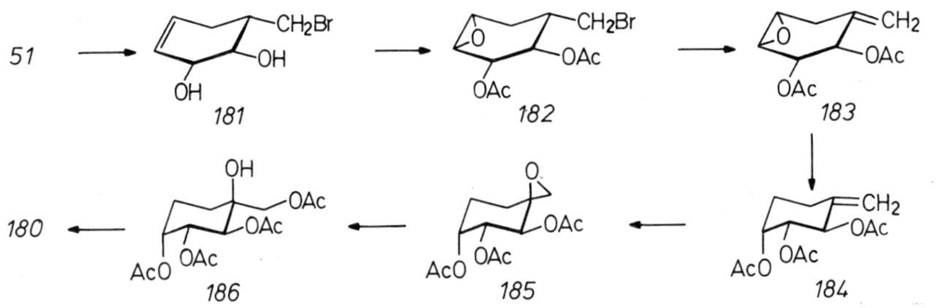

Scheme 30

4.2 Synthesis of Enantiomeric Pseudo-β-fructopyranoses

The above-mentioned facile synthesis of *180* can be applied for a preparation of enantiomeric pseudo-β-fructopyranoses [52], starting from the optically active antipodes of the Diels-Alder adduct, as follows.

Bromolactonization of the (−)-Diels-Alder adduct *86* [26] with hypobromous acid gave (3S)-(+)-*exo*-2-bromo-4,8-dioxatricyclo[4.2.1.0^{3,7}]nonan-5-one *88* [27]. Reduction of *88* with lithium aluminium hydride and successive acetylation gave (1S)-*endo*-3-acetoxy-*endo*-5-(acetoxymethyl)-*exo*-2-bromo-7-oxabicyclo[2.2.1]heptane (*187*), m.p. 76–78 °C, $[\alpha]_D^{22}$ +54° (ethanol) [53]. Cleavage of the anhydro linkage with hydrogen bromide in acetic acid afforded (1R)-(1,3,5/2,4)-3,4-diacetoxy-1,2-dibromo-5-(bromomethyl)cyclohexane (*188*), m.p. 126–128 °C, $[\alpha]_D^{24}$ −1.4° (chloroform). Debromination of *188* with zinc dust in acetic acid gave (1R)-(1,3/2)-2,3-diacetoxy-1-(bromomethyl)cyclohex-4-ene (*189*), m.p. 56–58 °C, $[\alpha]_D^{19}$ −2.2° (chloroform) [53]. O-Deacetylation of *189* with hydrochloric acid in ethanol yielded (1R)-(1/2,6)-6-(bromomethyl)-3-cyclohexene-1,2-diol (*190*), m.p. 76–77 °C, $[\alpha]_D^{27}$ +29° (chloroform). Expoxidation of *190* with mCPBA gave (1R)-(1,2,3,5/4)-3,4-di-O-acetyl-1,2-anhydro-5-(bromomethyl)cyclohexane (*191*), m.p. 125–126 °C, $[\alpha]_D^{23}$ −3.6° (chloroform). Regioselective reduction of *191* with diborane and sodium borohydride, followed by acetylation gave (1R)-(1,2,4/3)-1,2,3-triacetoxy-4-(bromomethyl)cyclohexane (*192*), m.p. 114–115 °C, $[\alpha]_D^{24}$ +3.8° (chloroform). Dehydrobromination of *192* with silver fluoride and pyridine afforded (1R)-(1,2/3)-1,2,3-triacetoxy-4-methylenecyclohexane (*193*), m.p. 71–72.5 °C, $[\alpha]_D^{24}$ −47° (chloroform). Stereoselective epoxidation of *193* with mCPBA gave (1R)-(1,2/3,4)-1,2,3-tri-O-acetyl-4,7-anhydro-4-(hydroxymethyl) cyclohexane-1,2,3,4-tetrol (*194*), syrup, $[\alpha]_D^{23}$ −43° (chloroform). Opening of the oxirane ring of *194* with acetate ions, followed by conventional acetylation gave 1,3,4,5-tetra-O-acetyl-pseudo-β-D-fructopyranose (*195*), m.p. 111–112 °C, $[\alpha]_D^{21}$ −50°

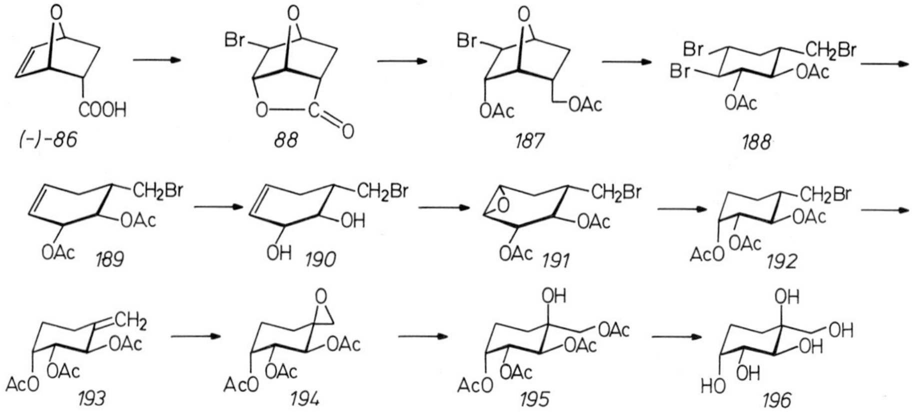

Scheme 31

Compounds in scheme: (-)-86, 88, 187, 188, 189, 190, 191, 192, 193, 194, 195, 196

(chloroform). O-Deacetylation of *195* gave pseudo-β-fructopyranose (*196*), syrup, $[\alpha]_D^{22}$ −57° (methanol) [54] (Scheme 31).

(3R)-(−)-*exo*-2-Bromo-4,8-dioxatricyclo[4.2.1.0^{3,7}]nonan-5-one (*197*), m.p. 114 to 115 °C, $[\alpha]_D^{30}$ −105° (water), was prepared from the (+)-Diels-Alder adduct [26] *87* by analogous bromolactonization [54]. The antipode of *187* (*198*), m.p. 73.5–74.5 °C, $[\alpha]_D^{28}$ −69° (chloroform), was prepared from *197* by analogous reduction. The antipode of *188* (*199*), m.p. 130–131 °C $[\alpha]_D^{28}$ −69° (chloroform), was prepared from *197* by analogous reduction. The antipode of *188* (*199*), m.p. 130–131 °C, $[\alpha]_D^{27}$ −0.8ë (chloroform), was obtained by analogous cleavage of the anhydro linkage of *198* [54]. The antipode of *189* (*200*), m.p. 52–53 °C, $[\alpha]_D^{21}$ +2.5° (chloroform), was obtained by analogous debromination of *199*. The antipode of *190* (*201*), m.p. 76–77 °C, $[\alpha]_D^{21}$ −36° (chloroform) was prepared by analogous O-deacetylation of *200*. Similarly, the antipode of *191* (*203*), m.p. 125–126 °C, $[\alpha]_D^{27}$ +1.5° (chloroform), was obtained from *201* by epoxidation. The antipode of *192* (*203*), m.p. 115–116 °C, $[\alpha]_D^{27}$ −3.2° (chloroform) was obtained by analogous reduction of *202*. The antipode of *193* (*204*), m.p. 70.5–72.5 °C, $[\alpha]_D^{22}$ +38° (chloroform), was obtained from *203* by analogous dehydrobromination. The antipode of *194* (*205*), syrup, $[\alpha]_D^{23}$ +37° (chloroform), was obtained from *204* by analogous epoxidation. The L-enantiomer: 1,3,4,5-tetra-O-acetyl-β-L-fructopyranose (*206*), m.p. 113.5–114.5 °C, $[\alpha]_D^{24}$ +46° (chloroform), was obtained from *205* by analogous opening of the oxirane ring and subsequent acetylation. Pseudo-β-L-fructopyranose (*207*), syrup, $[\alpha]_D^{23}$ +57° (methanol) was obtained from *206* by analogous O-deacetylation [54] (Scheme 32).

Pseudo-β-DL-fructopyranose *180* was found to be nearly as sweet as D-fructose. The D-enantiomer *196* and the L-antipode *207* were also as sweet as D-fructose, but *196* was somewhat sweeter than *207*. The small but observable difference in sweetness of *196* and *207* might be due to a stereogeometrical deformation of interrelations between the sweetness eliciting tripartite and a sweet receptor.

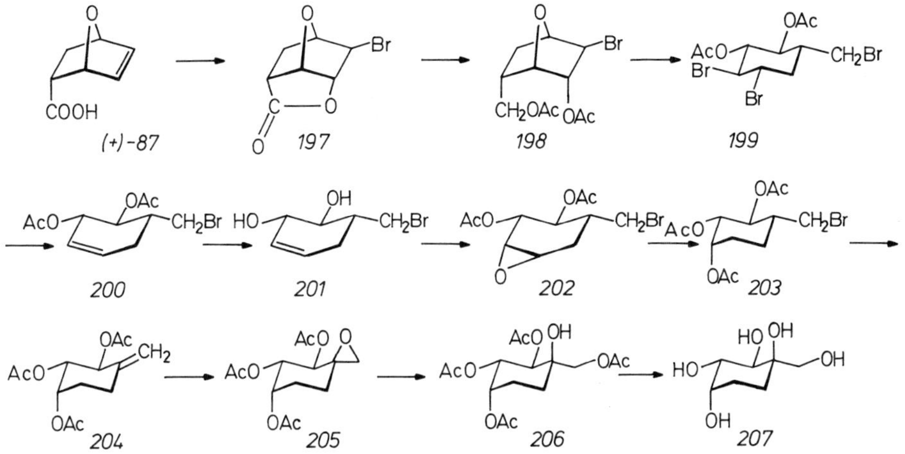

Scheme 32

5 Biological Effect of Pseudo-sugars

Besides the sweetness of pseudo-sugars, a pseudo-sugar may have a biological activity, as expected earlier, owing to its structural close resemblance to a true sugar.

As mentioned earlier in this article, pseudo-α-D-galactopyranose *1* has been found in a fermentation broth of *Streptomyces* sp. MA-4145 as an antibiotic [4]. The potency of the antibiotic was rather low. A concentration of about 125 ug/ml is required to produce a standard inhibition zone of 25 mm (diameter) against *Klebsiella pneumoniae* MB-1264, using 13 mm assay discs in a discplate assay. A sample of the synthetic pseudo-α-DL-galactopyranose *17* [3] showed to be about half as potent as the natural product *1* in the same assay system, thus indicating that the L-enantiomer is probably inactive.

An inhibition of glucose-stimulated insulin release has been studied by using pseudo-α-DL-glucopyranose *54* as a glucokinase inhibitor. That is, *54* and pseudo-β-DL-glucopyranose *37* were used as synthetic analogs of D-glucose anomers to study the mechanism of glucose-stimulated insulin release by pancreatic islets. And it was found that pseudo-sugar was neither phosphorylated by liver glucokinase, nor stimulated an insulin release from islets. Incubation of the islets with *54* resulted in an accumulation of the pseudo-sugar, probably the D-enantiomer, in the islets. Compound *54* inhibited both glucose-stimulated insulin release (44% inhibition at 20 mM) and islet glucokinase activity (36% inhibition at 20 mM), but *37* did not show any activity.

These results strongly suggested that the inhibition of glucose-stimulated insulin relase by *54* was due to the inhibition of islet glucokinase by the pseudo-sugar, providing an additional evidence for the essential role of islet glucokinase in glucose-stimulated insulin [55].

6 Conclusion

All the theoretically predictable sixteen racemic pseudo-sugars have been synthesized, as well as the ten enantiomers. The most accessible starting material for the synthesis was the Diels-Alder adduct *33* of furan and acrylic acid. Furthermore, the adduct *33* was readily resolved by means of optically active α-methylbenzylamines into the two antipodes *86* and *87*, which were also used for the preparation of enatiomeric pseudo-sugars. A chiral synthesis from true sugars was another prominent method for the preparation of the enantiomers. The remaining twenty two unknown enantiomeric pseudo-sugars will be prepared by either one of the two methods in the near future.

The chemistry of pseudo-sugars is quite a newly opened field of chemistry, and their biological effects have not been well studied yet. But sweetness, antibiotic activity, and inhibition of a glucose-stimulated insulin release have been observed in pseudo-sugars. Concerning their sweetness, some pseudo-sugars were equally as sweet as corresponding true sugars, and non mutarotation in pseudo-sugars is of great advantage for the study of the molecular mechanism of sweetness.

The molecular features of pseudo-sugars closely resemble those of true sugars, and hence, a sugar or an amino sugar moiety of an antibiotic, such as an aminocyclitol antibiotic, might be replaced by a pseudo-sugar or a pseudo-amino sugar without any detrimental effect to their biological activity, thus, providing new antibiotics that might be active against a resistant strain of a microorganism.

7 Acknowledgments

The author thanks Professor S. Ogawa, Drs. K. Tadano, T. Toyokuni, and N. Chida for their faithful collaboration in this research. The financial support of the work by the Grant-in-Aid for Scientific Research from the Japanese Ministry of Education, Science and Culture, and by the Asahi Glass Foundation for Industrial Technology is gratefully acknowledged.

8 References

1. McCasland GE, Furuta S, Durham LJ (1966) J. Org. Chem. 31: 1516
2. McCasland GE, Furuta S, Durham LJ (1968) J. Org. Chem. 33: 2835
3. McCasland GE, Furuta S, Durham LJ (1968) J. Org. Chem. 33: 2841
4. Miller TW, Arison BH, Albers-Schonberg G (1973) Biotech. and Bioeng. 15: 1075
5. Iwasa T, Yamamoto H, Shibata M (1970) J. Antibiot., 32: 595
6. Schmidt DD, Frommer W, Junge B, Muller W, Wingender W, Truscheit E, Schafter D (1977) Naturwissenschaften 64: 535
7. Yokose K, Ogawa S, Suzuki Y, Buchschacher P The 23rd Symp. on Natural Org. Compounds, Nagoya, Japan, Oct., 1980
8. Seto H, Orihata K, Otaka N, Namiki S, Kamigori K, Hara H, Mizokami K, Kimura S The 223rd Meeting of Japan Antibiot. Res. Assoc., Tokyo, Japan, March, 1981
9. Murao S, Ohyma K (1975) Biol. Chem. 39: 2271
10. Murao S, Ohyama K (1979) Agric. Biol. Chem. 43: 679
11. Itoh J, Omoto S, Shomura T, Ogino H, Iwamatsu K, Inouye S (1981) J. Antibiot., 34: 1424
12. Omoto S, Itoho J, Ogino H, Iwamatsu K, Nishizawa N, Inouye S (1981) J. Antibiot., 34: 1429

13. Daniels R, Doshi M, Smissman EE The 145th National Meeting of ACS, New York, N.Y., Sept., 1963
14. Suami T, Ogawa S, Ishibashi T, Kasahara I (1976) Bull. Chem. Soc. Jpn. 49: 1388
15. Suami T, Ogawa S, Ueda T, Uchino H (1972) Bull. Chem. Soc. Jpn. 45: 3226
16. Suami T, Ogawa S, Nakamoto K, Kasahara I (1977) Carbohydr. Res. 58: 240
17. Kunstman MP, Tarbell DS, Autrey RL (1962) J. Am. Chem. Soc. 84: 4115
18. Nelson WL, Allen DR (1972) J. Heterocycl. Chem. 9: 561
19. Ogawa S, Kasahara I, Suami T (1979) Bull. Chem. Soc. Jpn., 52: 118
20. Ogawa S, Ara M, Kondoh T, Saitoh M, Masuda R, Toyokuni T, Suami T (1980) Bull. Chem. Soc. Jpn. 53: 1121
21. Ogawa S, Toyokuni T, Kondoh T, Hattori Y, Iwasaki S, Suetsugu M, Suami T (1981) Bull. Chem. Soc. Jpn. 54: 2739
22. Ogawa S, Tsukiboshi Y, Iwasaki Y, Suami T (1985) Carbohydr. Res. 136: 77
23. Ogawa S, Chida N, Suami T (1980) Chem. Lett. 1559
24. Toyokuni T, Abe Y, Ogawa S, Suami T (1983) Bull. Chem. Soc. Jpn. 56: 505
25. Ogawa S, Kobayashi N, Nakamura K, Saitoh M, Suami T Carbohydr. Res. (1986) 153: 25
26. Ogawa S, Iwasawa Y, Suami T (1984) Chem. Lett. 355
27. Ogawa S, Iwasawa Y, Nose T, Suami T, Ohba S, Ito M, Saito Y (1985) J. Chem. Soc., Perkin Trans I. 903
28. Paulsen H, Deyn WV, Röben W (1984) Liebigs Ann. Chem. 433
29. Suami T, Tadano K, Kameda Y, Iimura Y (1984) Chem. Lett. 1919
30. Cleophax J, Gero SD, Krausz P, Machado AS, Philippe M, Sire B, Tachdjian C, Vass G The 3rd Europ. Symp. Carbohyd., Grenoble, France, Sept. 1985, Abst., B2–5
31. Ogawa S, Yato Y, Nakamura K, Takata M, Takagaki T (1986) Carbohydr. Res., 148: 249
32. Ogawa S, Nakamura K, Takagaki T (1986) Bull. Chem. Soc. Jpn., 59: 2956
33. Onodera K, Hirano S, Kashimura N (1968) Carbohydr. Res 6: 276
34. Tronchet JMJ, Cottet C, Gentile B, Mihaly E, Zumwald JB (1973) Helv. Chem. Acta 56: 181
35. Suami T, Tadano K, Ueno Y, and Fukabori C (1985) Chem. Lett., 1557
36. Miyazaki M: Master Thesis, "Synthesis of enantiomeric pseudo-sugars", Keio University, 1986
37. Kenner GW, Rodda HJ, and Todd AR (1949) J Chem. Soc. 1613
38. Tadano K, Maeda H, Hoshino M, Iimura Y, and Suami T (1986) Chem. Lett., 1081
39. Suami T, Ogawa S, and Toyokuni T (1983) Chem. Lett., 611
40. Shallenberger RS (1978) Pure Appl. Chem., 50, 1409
41. Shallenberger RS, and Acree TE (1967) Nature (London), 216, 480
42. Deutsch EW, and Hansch C (1966) Nature (London), 211, 75
43. Kier LB (1972) J. Pharm. Sci., 61, 1394
44. Birch GG, Lee CK, and Rolf E (1970) J. Sci. Food Agric., 21, 650
45. Shallenberger RS, Acree TE, and Lee CY (1969) Nature (London), 221, 555
46. Lindley MG, and Birch GG (1975) J. Sci. Food Agric., 26, 117
47. Martin OR, Tommola SK and Szareck WA (1982) Can. J. Chem., 60, 1857
48. Henbest HB and Wilson RA (1957) J. Chem. Soc., 1958
49. Suami T, Ogawa S, Takata M, Yasuda K, Suga A, Takei K, and Uematsu Y (1985) Chem. Lett., 719
50. Suami T, Ogawa S, Takata M, Yasuda K, Takei K, and Suga A (1986) Bull. Chem. Soc. Jpn., 59, 819
51. Helferich B and Himmen E (1928) Ber., 61, 1825
52. Suami T, Ogawa S, Uematsu Y, and Suga A (1986) Bull. Chem. Soc. Jpn., 59, 1261
53. Ogawa S and Takagaki T (1985) J. Org. Chem., 50, 2356
54. Ogawa S, Uematsu Y, Yoshida S, Sasaki N, and Suami T (1987) J. Carbohyd. Chem., 6, 471
55. Miwa I, Hara H, Okuda J, Suami T, and Ogawa S (1985) Biochem. Intern., 11, 809

Syntheses of Deoxy Oligosaccharides

Joachim Thiem and Werner Klaffke

Institut für Organische Chemie, Universität Hamburg, Martin-Luther-King-Platz 6, D-2000 Hamburg 13, FRG

Table of Contents

This paper focusses on stereoselective syntheses of deoxy oligosaccharides which are an integral part of a number of antibiotics and cytostatics. Following a short review on the biosynthesis of these sugars, the general techniques including classical methods as well as recent stereospecific procedures for the preparation of deoxy glycosides are outlined.

For these oligosaccharide-containing pharmaceuticals, brief explanations of their current employment in medical applications are given. The key problems of glycosylations and detailed discussions of selectively adjusted methods will be addressed.

Topics in Current Chemistry, Vol. 154
© Springer-Verlag Berlin Heidelberg 1990

Joachim Thiem and Werner Klaffke

1 Introduction

Complex glycosylated compounds like macrolides, anthracyclines, aureolic acids, cardiac glycosides, orthosomycines and tetronic acids are of considerable scientific as well as pharmaceutical interest. Obviously, each of them is responsible for certain therapeutic effects with respect to different diseases. Anthracyclines and aureolic acids are applied in cancer chemotherapy, orthosomycines are active as antibiotics against Gram-positive bacteria, and the cardiac glycosides are used in the treatment of cardiac insufficiency.

In all cases, 2,6-dideoxy sugars of the D- or L-series are common and important parts of these various molecules. In general, the specific therapeutic effect is thought to be caused by the aglycone, and the sugar residue to be responsible mainly for regulating the pharmacokinetics. Thus, parameters like bioavailability, resorption, distribution, or therapeutic width are influenced by the carbohydrate moieties. By modification of the carbohydrate moiety it is, e.g., possible to enhance the efficacy of unspecific aglycones like anthracyclines or aureolic acids, or also to reduce possible side effects. Such as approach is followed in the field of class-I and -II anthracyclines in order to decrease their considerable cardiotoxicity.

Carbohydrate chemistry is engaged in the synthesis and variation of deoxy sugar chains, where a wide set of protective groups and stereoselective glycosylation techniques are required. This contribution centers on stereoselective syntheses of mono- and oligosaccharides in the field of 2,6-dideoxy- and, in particular cases, branched-chain sugars, and summarizes modern synthetic glycosylation reactions which have been developed for this special kind of carbohydrate chemistry.

2 Biosynthesis of Deoxy Sugars

Natural occurring 2,6-dideoxy oligosaccharides of the D-series are β-glycosidically linked, those of the L-series represent α-glycosides. The absolute configuration at the anomeric center, however, is similar in both cases.

Recently, some knowledge was acquired concerning nature's approach to the synthesis of dideoxy sugars. 6-Deoxygenation of glucose seems to follow a redox pathway via the nucleotide glucoside dTDP-glucose with oxidoreductase which leads to a 6-deoxy-4-uloside [1].

As shown in the biosynthesis of granaticin, a hydride shift occurs intramolecularly. This process is mediated by an enzyme-bond pyridine nucleotide. A concerted abstraction of H-4 as a hydride in **1a** and a C-5 deprotonation in **2a** leads to the 4,5-enol ether **3a**. The reduced form of the pyridine nucleotide transfers the hydride to C-6, simultaneously releasing a hydroxide to give **4a**. Final tautomerization yields the dTDP-4-keto-6-deoxy-sugar in D-*xylo* configuration **4a**. In other enzymes of the oxidoreductase type, the active site may show a different configuration. Thus, the intermediate **3a** can be protonated from "above" at C-5 to yield the L-*arabino* isomer of **4a** [2]. The stereochemistry of this mechanism was demonstrated by double labelling (cf. **1–4b** series), and as a net result proved a suprafacial 4→6 hydride shift.

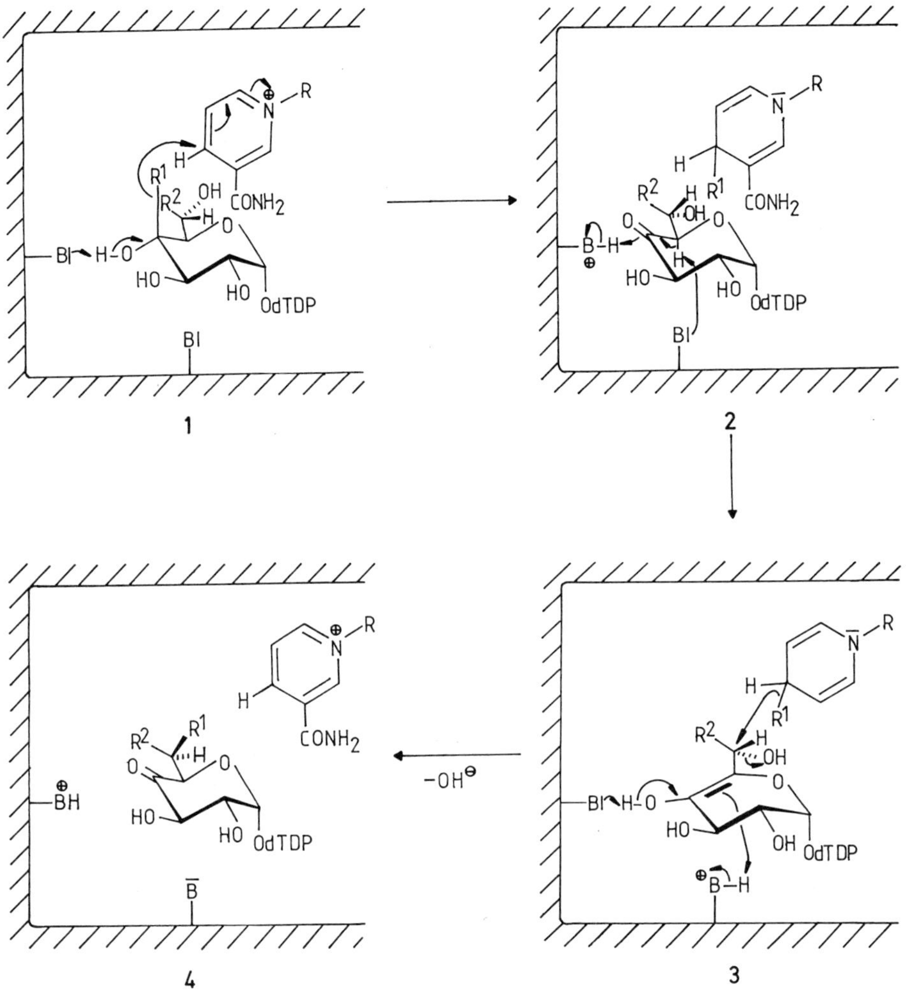

Scheme 1

In contrast to previous studies [3], recent research shows that there are two or probably more ways for 2-deoxygenation, one by inversion (chlorothricin) and another one by retention (granaticin) of the C-2 configuration [4]. Starting with the activated 4-keto glycoside 4a via enolization at C-3 (5), the 3-uloside 6 is obtained. Formation of a Schiff base with pyridoxamine phosphate leads to 7. This undergoes a 1,4-elimination of water favored by a six-membered transition state furnishing a conjugated system. The resulting 2,3-olefin sugar derivative of enimine structure 8 is reduced by NADH and protonated at C-2. Following cleavage of the 2-deoxy azomethine,

287

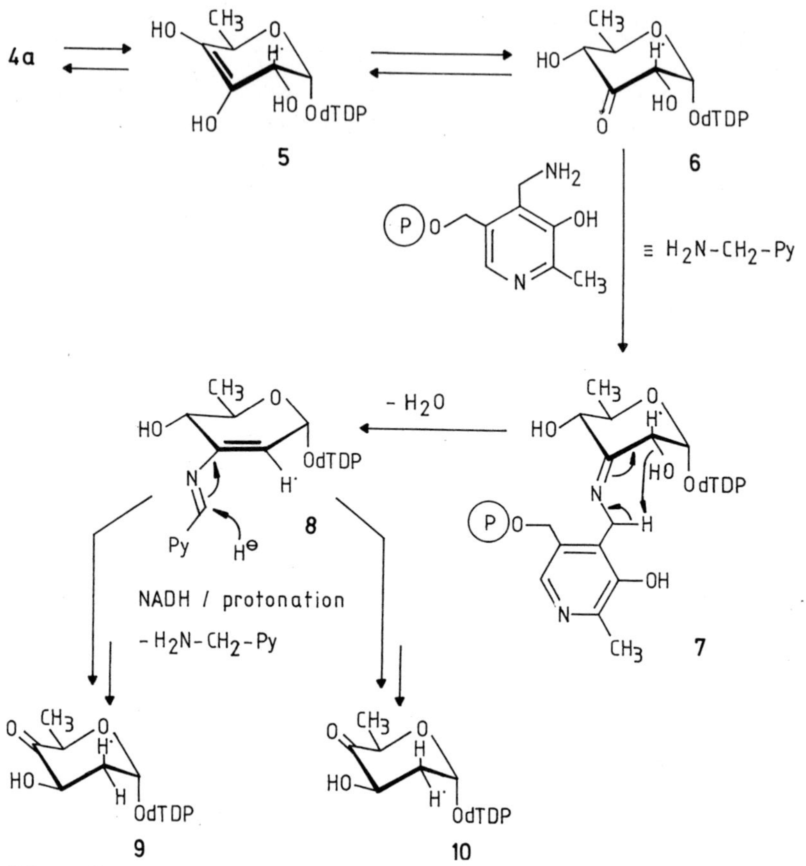

Scheme 2

the 3-keto derivative enolizes to the more stable 4-uloside. Mechanistic studies proved the formation of the differently labelled 2,6-dideoxy-4-keto-D-*threo* glycosides **9** and **10**.

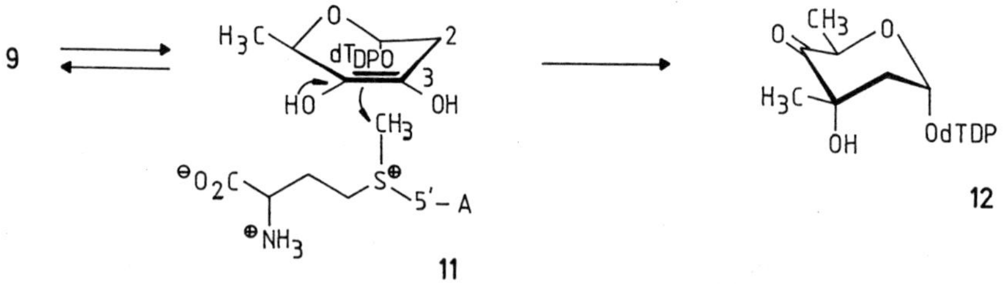

Scheme 3

The methyl-branch is introduced at the C-3 position of tautomer **11** of the keton **9** by an electrophilic attack with active methione (*S*-methyl-5'-adenosyl methionine). Thus the 2,6-dideoxy-4-keto-3-C-methyl-D-*erythro* glycoside **12** results which after reduction furnishes, e.g., D-mycarose (cf. also Ref. [5]).

288

3 Glycosylation of 2-Deoxy Sugars

The basic concepts of glycosylation have been known for more than eighty years. Nevertheless, the selective formation of a full acetal still represents one of the major challenges in carbohydrate chemistry. Within the recent decade, a number of attractive approaches for the glycosylation of simple alcohols as well as more complex aglycons including sugar derivatives have been developed [6, 7]. In all cases a high stereoselectivity is desired, and it seems, simple transfer of a procedure used for a certain sugar series does not necessarily apply to other isomeric series.

High stereoselectivity in glycoside synthesis is of major concern. Furthermore, the high prices of certain aglycones or sugars should be considered, and often only a single isomer is needed for the biologically active form of a pharmaceutical. Glycosylation procedures, which are presently applied in the normal sugar series, make use of a neighboring group adjacent to the anomeric center either by means of a real anchimeric assistance or by operation of steric influence [6, 7].

Evidently the particular problems in the chemical synthesis of 2-deoxy sugar glycosides are aligned with the missing neighboring group, and are also associated with the enhanced lability of their glycosyl halides. This implies a special handling of known tools in carbohydrate chemistry or the development of new techniques.

3.1 Classical Techniques

A typical Koenigs-Knorr reaction may outline the problems mentioned above. Treatment of the α- or β-glycosyl halide (the former being slightly more stable owing

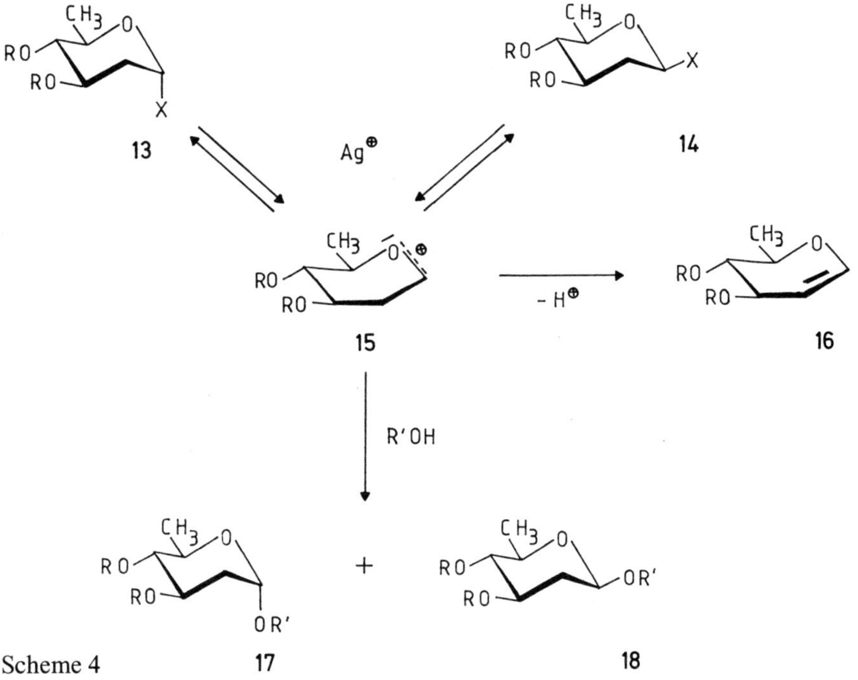

Scheme 4

to the anomeric effect) in the 2,6-dideoxy-D-*arabino*-series (**13**) or (**14**) with an alcohol in the presence of a silver promoter is supposed to proceed via an oxocarbenium ion intermediate (**15**). After nucleophilic attack of the alcohol, both the protonated precursors are formed, which after release of the proton give the α- and β-glycosides **17** and **18**. Mostly the former prevails, probably again by the operation of an anomeric effect. The intermediate **15** may be also stabilized by subsequent loss of the 2-H which leads to the glycal (**16**). This is an often observed by-product in 2-deoxy sugar glycosylation thus decreasing the yields of the wanted glycosides.

By glycosylation of **19** with the halide **20** following Helferich conditions (HgBr$_2$, HgO), a typical but acceptable amount of the desired α-glycoside **21** (40%) of a 2,6-dideoxy sugar in the L-series was obtained which represents the A–B disaccharide unit of aclacinomycin A [8].

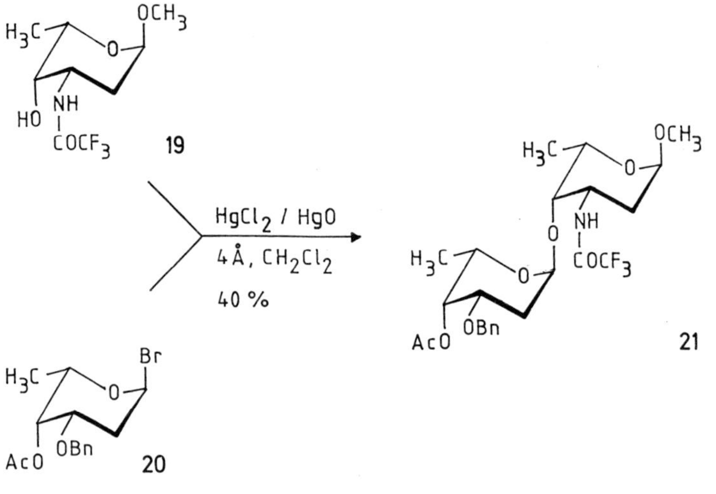

Scheme 5

Obviously there is a need for more generally valid and more stereoselective procedures.

3.2 Formation of 2-Deoxy-α-glycosides

There is no steric guidance for an approaching nucleophile in this series because neighboring groups are lacking. The following example will outline the general aspect to this concept:

Treatment of the glycal **22** with an electrophile E$^+$ leads to intermediate **23** which will be favored owing to the operating of an inverse anomeric effect. Nucleophilic attack from the opposite side affords the α-glycosidic linkage and *trans*-configurated bonds in **24**. This approach was developed and often proved valid in several cases of α-L- and α-D-oligosaccharide formation using *N*-iodsuccinimide as suitable iodonium donor [9].

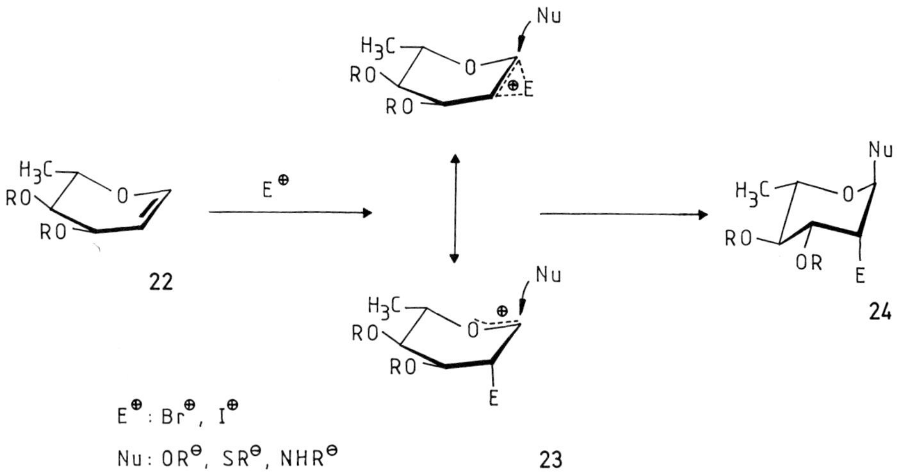

Scheme 6

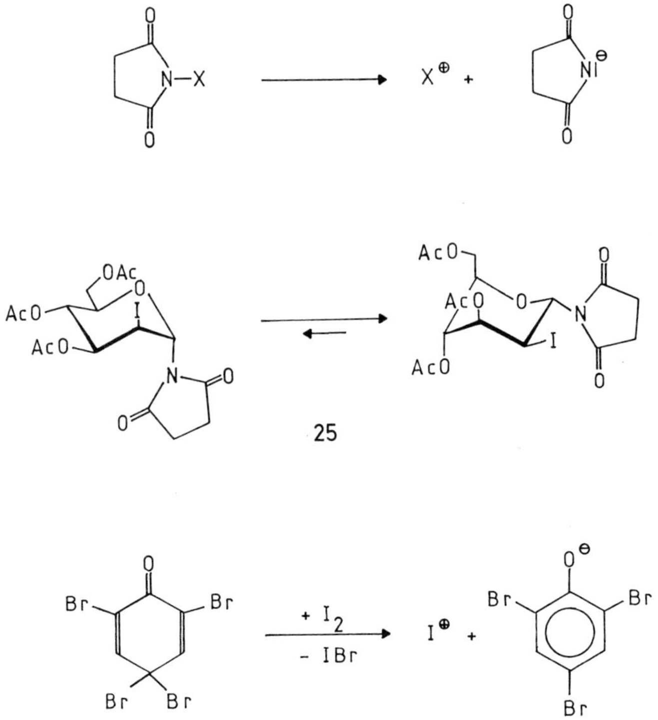

Scheme 7

The succinimide anion obtained by the NIS heterolysis is frequently observed to compete with a less potent nucleophile in NIS glycosylations. This leads to glycosyl succinimides like **25** in the 2-deoxy-D-*arabino* case [10] which, owing to the marked

291

inverse anomeric effect of that group, adopts predominantly an inverted $^1C_4(D)$ chair conformation.

A modification of the NIS procedure [11] makes use of 2,4,4,6-tetrabromo-2,5-hexadienone (TBCO) [12]. In the presence of iodine, TBCO generates I^+ and the non-nucleophilic 2,4,6-tribromophenolate which does not interfere with any other nucleophile.

The 2-deoxy function may be regenerated from the 2-halo derivatives by reduction with tributyl stannic hydride, $NiCl_2/NaBH_4$, or hydrogen/palladium on charcoal. The latter should be carried out in the presence of base in order to quench the resulting acid. Another method following the same concept is the oxyselenation-deselenation of glycals [13]:

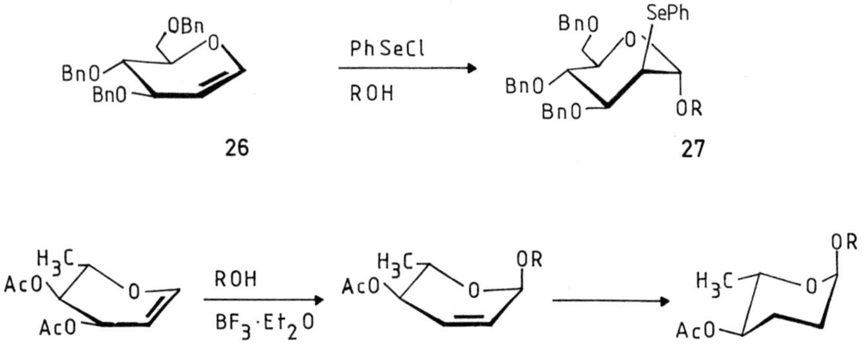

Scheme 8

Phenylselenyl chloride adds to the double bond of the glycal 26 to give a phenyl-selenoxonium ion which is *trans*-opened by an approaching nucleophile to the glycoside 27. Other electrophiles have also been used like *p*-toluenesulfonic acid, bromine, or iodine. Acid conditions, however, may cause destruction of labile 2-deoxy sugar products and the value of such reagents will be diminished by this evident disadvantage.

The Lewis acid-mediated allylic rearrangement of glycals like 28 to glycosides 29 known as the Ferrier-reaction is well known in carbohydrate chemistry [14–16]. It yields predominantly α-configurated hex-2-enopyranosides (cf. 29), either in the D-or L-sugar series. These may be hydrogenated to glycosides like the 2,3,6-trideoxy-species 30.

3.3 Formation of 2-Deoxy-β-glycosides

As mentioned above, there is particular need for β-stereospecific glycosylation in the 2-deoxy sugar series. Wiesner et al. [17] have developed a method which starts with 4-(*p*-methoxy)-benzoyl-3-methylurethane digitoxose 31. On treatment with the aglycon and *p*-toluene sulfonic acid, the β-glycoside 32 is obtained in 83% with an β:α-ratio of 7:1.

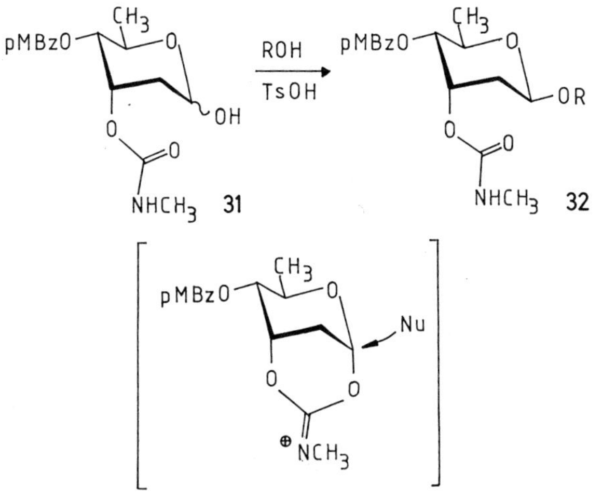

31 32

33

R = digitoxigenin-3-OH
(digi)

Scheme 9

In this case the substituent at C-3, is believed to account for the stereoselectivity by formation of the charged intermediate **33**. Similar to this case is the $HgCl_2$-promoted glycosylation [18] of a corresponding thioglycoside like **35** with the aglycon, component **34**. This leads predominantly to the β, 1→4-disaccharide **36**.

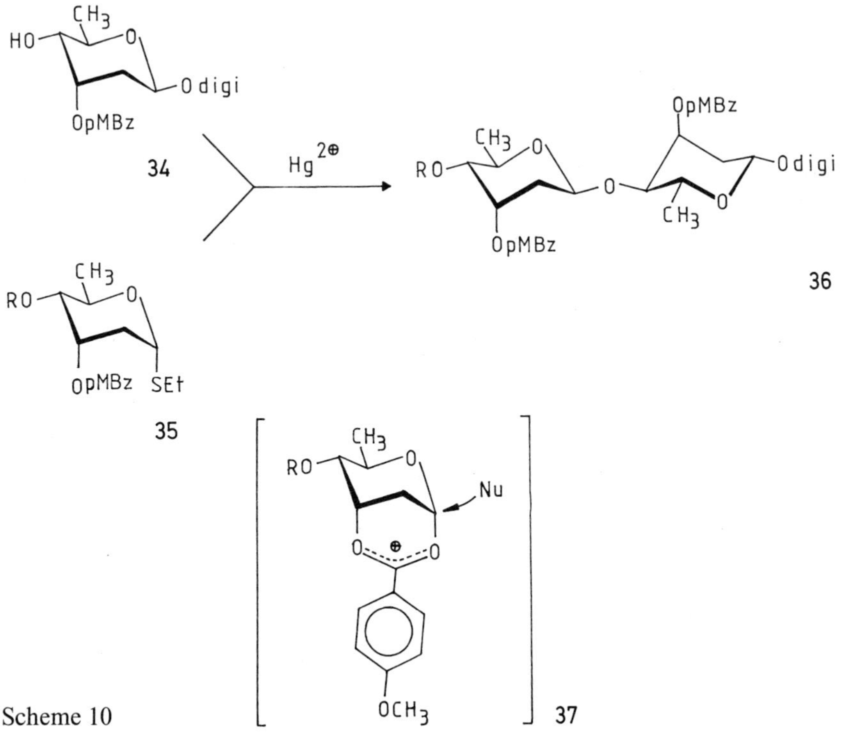

34 35 36 37

Scheme 10

The glycosylation is supposed to operate via a 1,3-O-(p-methoxy)-benzoxoniumion **37**. This approach may be characterized as a 1,3-anchimeric assisted process in which the configuration at C-3 is responsible for the anomeric orientation.

In recent studies, acceptable yields of β-configurated 2-deoxy glycosides by application of the NIS reaction to uronic ester glycals like **38** have been reported [19]. By NMR studies the equilibrium conformation of the glycal **38** is shown to be shifted towards the inverted half-chair or half-boat conformation $[^5H_4(D)]$.

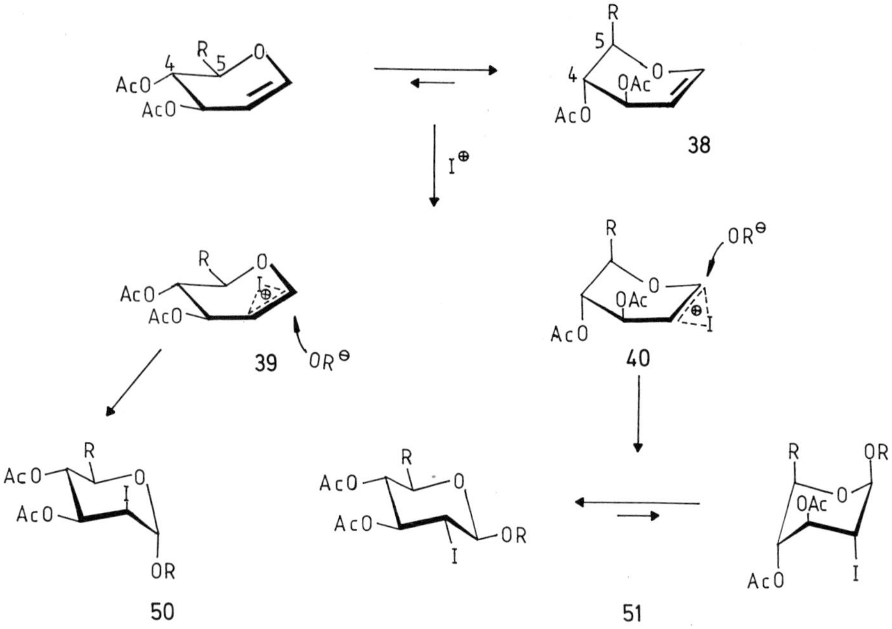

Scheme 11

$R = COOCH_3, CONH_2, CN$

Owing to the Curtin-Hammett principle, the ratio of the iodonium intermediates **39** and **40** does not reflect the distribution of the conformers in **38** $(^5H_4:{}^4H_5 = 95:5)$. Thus, both the 2-deoxy-2-iodo-α-mannoside **50** and the 2-deoxy-2-iodo-β-glucoside **51** are formed in almost equimolar amounts. Similar results are observed in glycals with nitrile or carboxamide functions [11].

Another procedure known as the dibromomethyl-methyl ether method (DBE-method) [20] leads to 2-deoxy-2-bromo-β- and in some cases to 2-deoxy-2-bromo-α-glycosides. As with the NIS reaction, this represents an indirect synthesis of 2-deoxy sugars, because the C-2 halide substituent may be cleaved in a subsequent step.

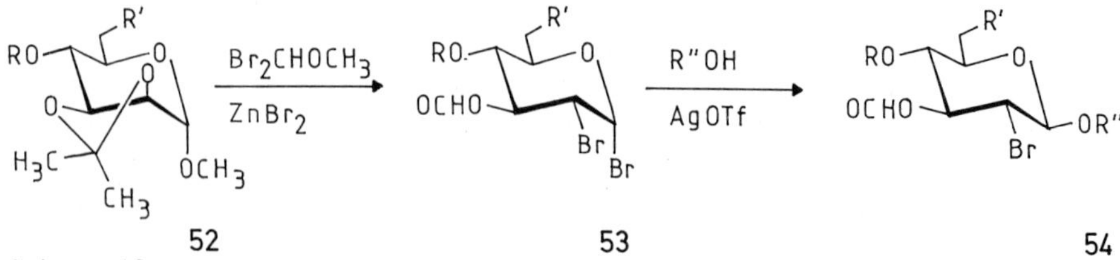

Scheme 12

294

For instance, starting with a 2,3-di-*O*-isopropylidene-α-D-mannopyranoside **52**, an attack by dibromomethyl-methyl ether (DBE) yields the 2,6-dideoxy-2-bromo-3-*O*-formyl-α-D-glucopyranosyl bromide **53**. In the presence of a silver salt, predominantly β-glycosides like **54** are formed [21–23]. Little is known about the mechanistic aspects; however, the stereoselectivity favors a 1,2-bromonium ion intermediate.

Recently, a stereospecific 1,2-migration was shown [24] to occur by treatment of various 2-hydroxy sugar derivatives like glycosides, acetates, azides, and thioglycosides with an excess of diethylamino sulfur trifluoride (DAST). Products of this migration are glycosyl fluorides (both α and β anomers) reportedly carrying the previously anomeric substituent in an inverted configuration at C-2. Apart from its application potential in the normal sugar series it also allows the preparation of α- and β-2-deoxy glycosides when starting from phenylthio glycosides.

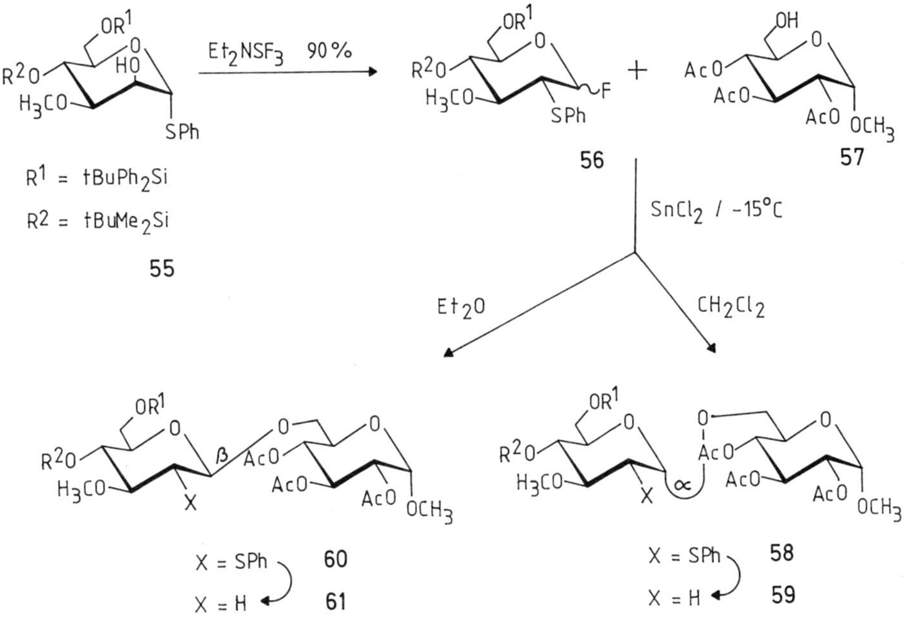

Scheme 13

The manno thioglycoside **55** on treatment with DAST give the 2-deoxy-2-phenyl-thio-glucosyl fluoride as anomeric mixture **56**. On glycosylation with the 6-hydroxy-glucoside **57** in dichloromethane and the presence of SnCl$_2$, the α,1→6 disaccharide **58** is obtained exclusively (92%), which on reduction with Raney-Nickel yields the 2′-deoxy derivative **59**.

In contrast, with ether as the solvent and otherwise the same conditions, **56** and **57** were condensed to give again exclusively (92%) the β,1→6-linked derivative **66** which in turn could be reduced similarly to **61**.

The plausible but somewhat speculative attempts to interpret the stereochemical outcome resembles the previously reported stereoselectivities in glycosylations of glycosyl fluorides [25] many of which could not be confirmed in other studies [26, 27].

4 Anthracycline Antibiotics

Malignant carcinomas are defined by pathologic transformations of the genetic information and cell proliferation. Such cancer is characterized by unregulated high mitotic activity leading to tumors and metastases. Obviously, surgery is indicated only in cases of localizable tumors. X-irradiation therefore represents an additional therapy. Several antitumor therapeutics like the anthracyclines can exert specific effects on the genetic material. Generalized neoplasms like cancers of blood and lymphatic tissues can be treated [28, 29]. Owing to their excellent therapeutic width, anthracyclins are particularly suited for such applications [30]. Anthracyclines belong to a group of antibiotic compounds which occur as intermediates in the metabolism of several *Streptomyces* species. Their aglycon consists of a tricyclic quinoide chromophor with a functionalized cyclohexane moiety attached [30]. The various substitution patterns are outlined in Scheme 14.

4, 6, 11: H, OH, OCH_3

10: H, OH, CO_2CH_3

13: X = H_2,O

14: H, OH

Scheme 14

Adriamycin is one of the best-studied members for the interaction with human cell tissues and genetic material. Its activity is believed to be based on the formation of intercalation complexes with the DNA. This results in a change of its helical shape and causes the total inhibition of transcription which terminates the ability of cell separation [31]. Obviously, all fast-proliferating tissues are damaged by anthracyclines. Side effects like bone marrow depression can also be understood by such a mechanism. In cooperation with a disarrangement of the plasmatic calcium content, this NADPH-dependent redox cycle causes a large number of radical reactions which effect a high cumulative cardiotoxicity [32]. Another parameter may also be the inhibitory effect on cardiac guanylate cyclase, which can be altered by substitution of the methyl ketone side chain by long-chain hydrocarbons [33].

Thus, the development of new anthracycline antibiotics is of interest in which the therapeutical width is enhanced by decreasing toxicity and increasing specificity. Several screening methods are presently available in clinical tests. One is carried out by measurement of the survival rate of mice, induced with P 388 leukemia carcinoma [30, 32]. Other methods are based on in-vitro tests; either the 50% inhibitory concentration (IC_{50}) of nucleosomal RNA synthesis is measured or the growth of tumor cell cultures like HeLa is observed [34].

In all cases, the sugar residues are attached to the 7-hydroxy group and this is the point of interest to carbohydrate chemistry. Until present, only the 3-amino-2,3,6-trideoxy sugars daunosamine (**62**) and rhodosamine (**63**) were found to be α-glycosidically linked to this position. The other saccharides 2-deoxy-L-fucose (**64**), L-cinerulose (**65**), and L-amicetose (**66**) occur in the oligosaccharide chains of the class-II anthracycline antibiotics.

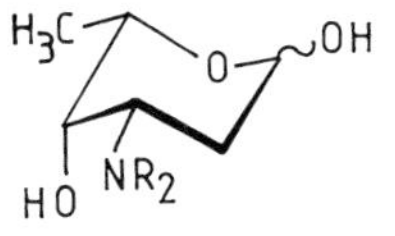

Daunosamine   R = H   62

Rhodosamine   R = CH$_3$ 63

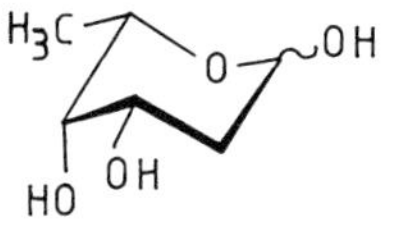

2-Deoxy-L-fucose   64

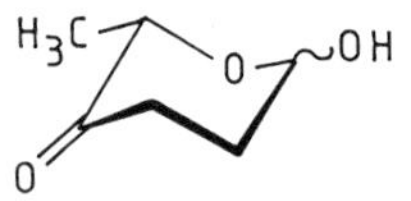

L-Cinerulose   65

L-Amicetose   66

Scheme 15

Daunorubicin (**67**) found independently by Di Marco et al. [35] and Du Bost et al. [36] was the first antileucemic agent in this group [37]. Adriamycin (**68**), discovered by Arcamone et al. [38] in 1969, marked an advantage in the development of chemo-therapeutics. It is presently one of the most valuable substances in clinical application [39]. Modifications carried out both in the aglycone and the saccharide residue are still of main interest.

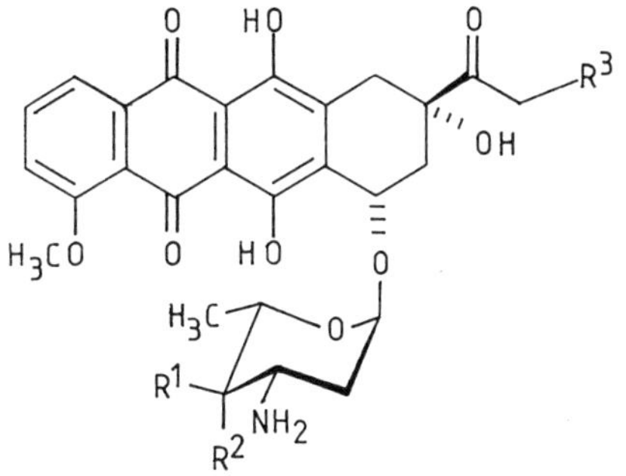

R^1 = H, R^2 = OH, R^3 = H   67

R^1 = H, R^2 = OH, R^3 = OH   68

R^1 = OH, R^2 = H, R^3 = H   69

Scheme 16

Certain *N*-alkylation products [40] of daunosamine show a decreased ability for intercalation and eliminated the inhibitory action on cardiac guanylate cyclase as mentioned above [33]. The epimeric C-4' derivative (**69**), the sugar of which is called *epi*-daunosamine, extended the medical value to adriamycin-resistant carcinoma [41].

Oligosaccharide anthracyclines like aclacinomycin (**70**), dehydroaclacinomycin (**71**) [42, 44] and marcellomycin (**72**) [45] are known as "class-II" anthracyclins [45].

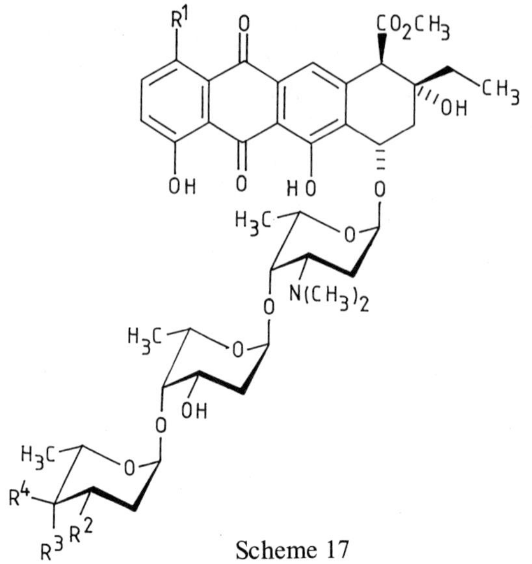

Joachim Thiem and Werner Klaffke

$R^1 = R^2 = H, R^3 + R^4 = O$ 70

$R^1 = R^2 = R^3 = H, R^4 = OH$ 71

$R^1 = R^2 = R^3 = OH, R^4 = H$ 72

Scheme 17

4.1 Synthesis of Class I Sugars and Glycosides

Anthracyclines isolated from *streptomyces* show a 2,3,6-trideoxy-3-amino-L-*lyxo*-configurated sugar α-attached to the aglycon. Therefore, the first interest centered on the synthesis of aminodeoxy sugars in the L-*lyxo* series, e.g., daunosamine [46–50]. Several new glycosides were prepared [51] in order to evaluate structure-activity relationships.

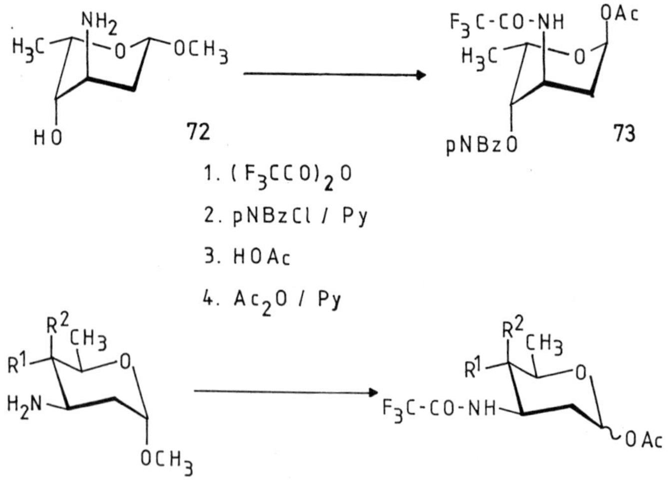

1. $(F_3CCO)_2O$
2. pNBzCl / Py
3. HOAc
4. Ac_2O / Py

$R^1 = OH, R^2 = H$ 74

$R^1 = H, R^2 = OH$ 75

$R^1 = OpNBz, R^2 = H$ 76

$R^1 = H, R^2 = OpNBz$ 77

Scheme 18

The glycosides **72**, **74**, and **75** are usually available by nucleophilic substitution reactions at C-3 of precursors, furnished with good leaving groups, like methanesulfonyloxy and trifluormethanesulfonyloxy substituents. Alternatively, oxirane ring opening of 2,3-anhydro sugars may be applied. Usually the 4,6-O-benzylidene group is opened by Hullar-Hanessian cleavage, followed by formation of an exocyclic glycal in the presence of silver fluoride. A final hydrogenation step affords the 2,3,6-trideoxy-3-amino-β-L-(**72**) or α-D-glycosides (**74–75**). Selective N- and O-protection, and acetolysis followed by acetylation leads to the glycosyl precursors **73**, **76** and **77**.

In order to simplify these preparations, Boivin et al. [52] made use of the allylic rearrangement reaction (Ferrier type) which was one of the first accesses to 3-amino glycals. By use of azide as the nucleophile, the L-*arabino* configurated glycal **78**

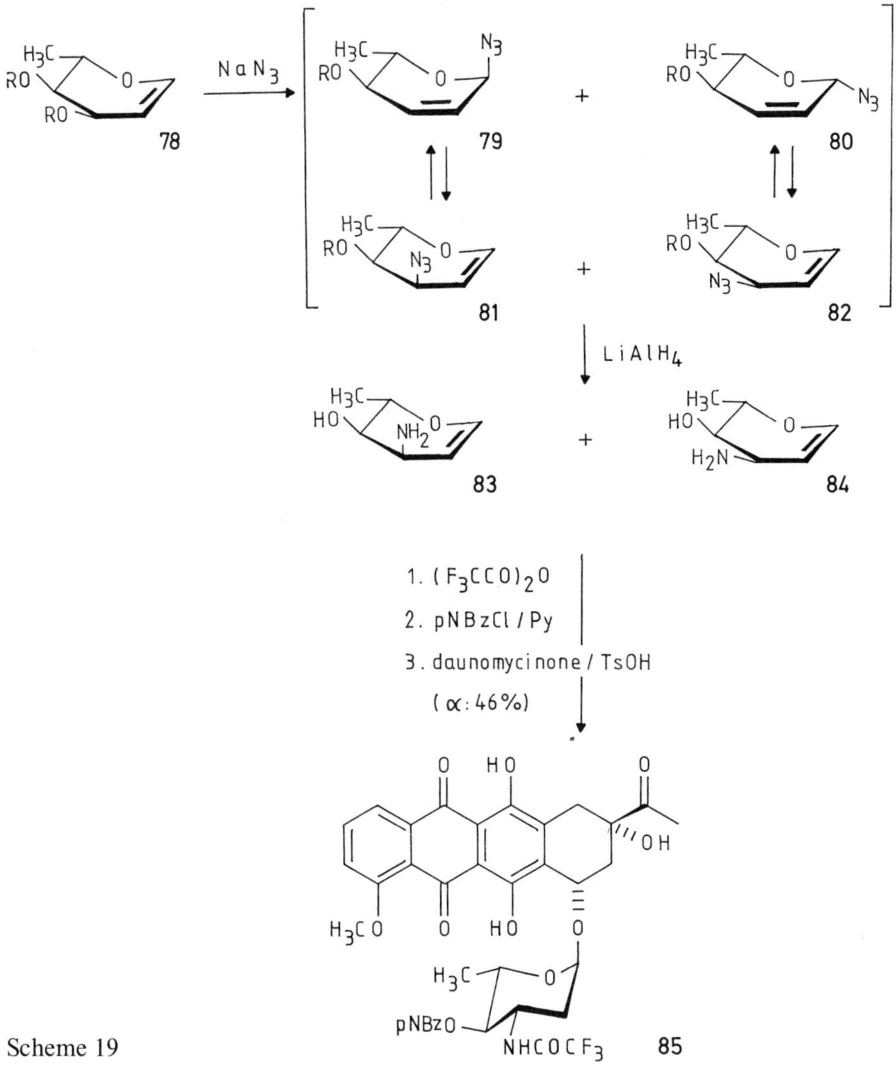

Scheme 19

299

(R = Bz) leads to four products. Both the α- and β-hex-2-enopyranosyl azides **79** and **80** undergo a [3.3]-sigmatropic hetero-Cope rearrangement and stereospecifically give the 3-azido glycals **81** and **82**, respectively. Unfortunately, the desired compound **82** shows the lower yield in this mixture. Subsequent reduction with $LiAlH_4$ leads to both the 3-amino glycals **83** and **84**. After separation, the latter is *N*- and *O*-acylated, and on treatment with daunomycinone in the presence of toluenesulfonic acid the 4′-epidaunosaminide **85** is obtained.

Heyns et al. [53] started with di-*O*-acetyl-L-rhamnal (**28**) which after reaction with sodium azide in the presence of boron trifluoride etherate as Lewis catalyst gave again the four products **79**–**82** (R = Ac). Subsequent quenching with NIS and the aglycone led to ristosamine and acosamine glycosides **86** and **87**.

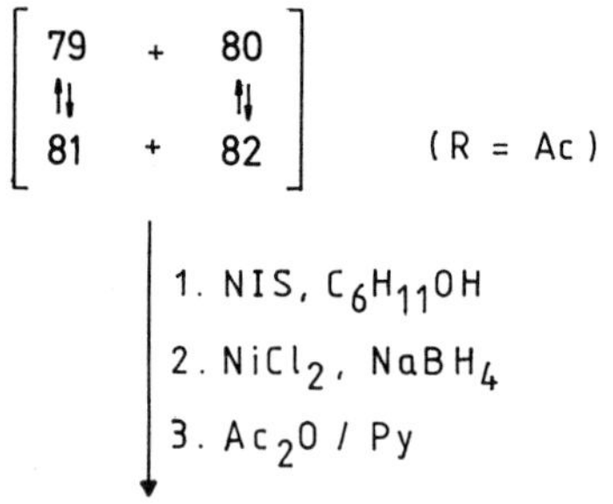
79 + 80
81 + 82
(R = Ac)
1. NIS, C6H11OH
2. NiCl2, NaBH4
3. Ac2O / Py

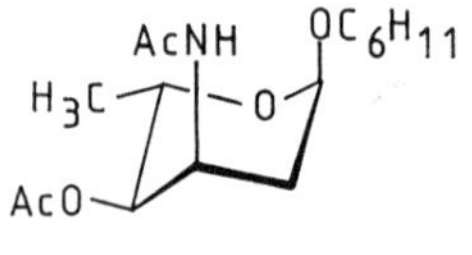
AcNH
OC6H11
H3C
O
AcO

86

+

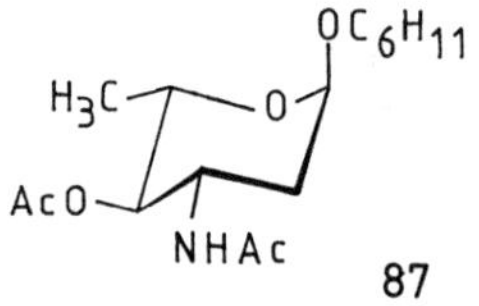
OC6H11
H3C
O
AcO
NHAc

87

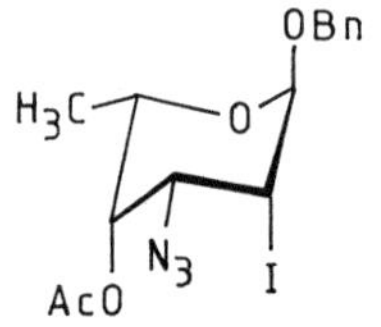
OBn
H3C
O
N3
I
AcO

88

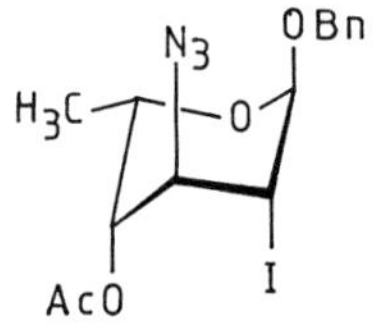
N3
OBn
H3C
O
I
AcO

89

Scheme 20

A corresponding sequence from di-*O*-acetyl-fucal gives a mixture of the precursors of benzyl daunosaminide **88** and the epimeric compound **89** [54]. By this NIS quenching method, enhanced yields of the equatorial 3-azido compounds are obtained.

Most recently reported glycosylations with daunosamine and related 3-deoxy-3-amino epimers start with acetyl daunosamine [55] or silvertriflate and anomeric chlorides [56]. By use of trimethylsilylmethanesulfonate (TMSOTf), 1,4-di-*O*-p-nitrobenzoylated daunosamines **91** could be stereospecifically attached to (+)-4-

demethoxydaunorubicinone **90** to give **92**. This very promising approach to α-gly-cosides is carried out very smoothly at low temperatures (−15 °C) [57].

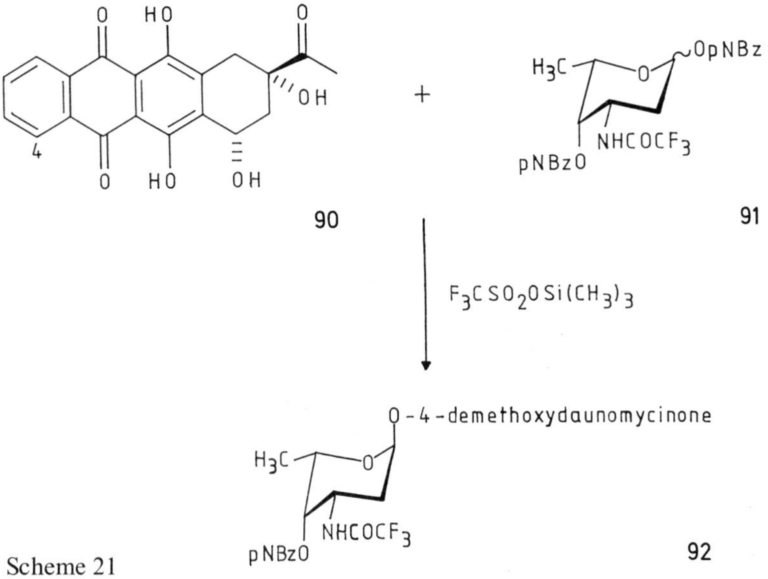

Scheme 21

An interesting synthesis of a methyl-branched anthracycline has been reported [58]. The methyl-branched 3-amino-*ribo* derivative **93** was opened by Hullar-Hanessian reaction to **94**, transformed into the exocyclic glycal **95**, inverted at C-5 and reacted to the glycal precursor **96**. Its treatment with the aglycone and *p*-toluenesulfonic acid gave the derivative **97**.

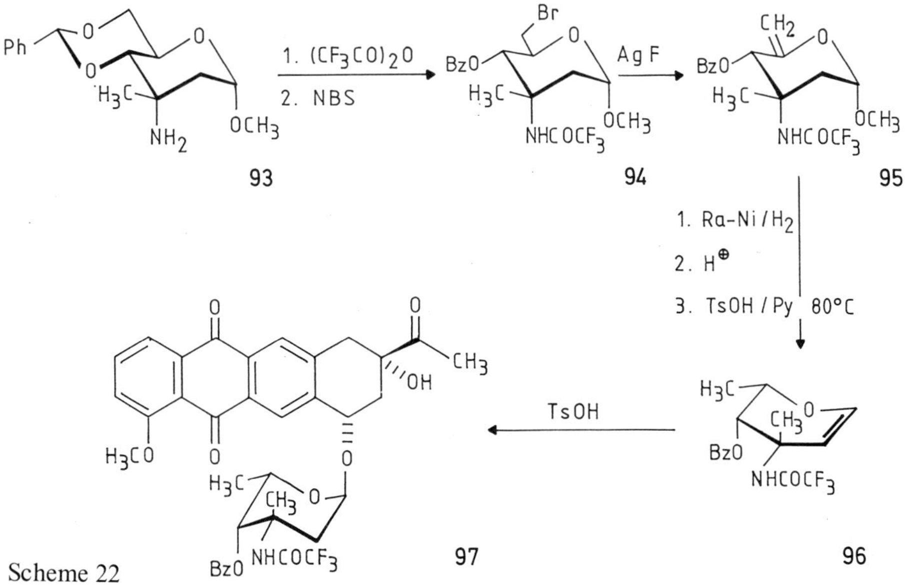

Scheme 22

Also 2'-halo-3'-hydroxy derivatives of various anthracyclinones show a high activity against P 388 mouse leukemia in certain in-vivo tests. A compound like **98** has been prepared in the 2-deoxy-2-iodo-α-L-*manno*-series by Horton et al. [59, 60] starting from di-*O*-acetyl-L-rhamnal (**28**). Similarly, Thiem et al. have also successfully prepared the tetracenomycinone-C glycoside **99** [61]. From 4-O-acetyl-3-O-(*p*-methoxy)-benzyl-L-fucal (**102**) the glycoside derivatives **103** and **104** in the *talo*-series were obtained.

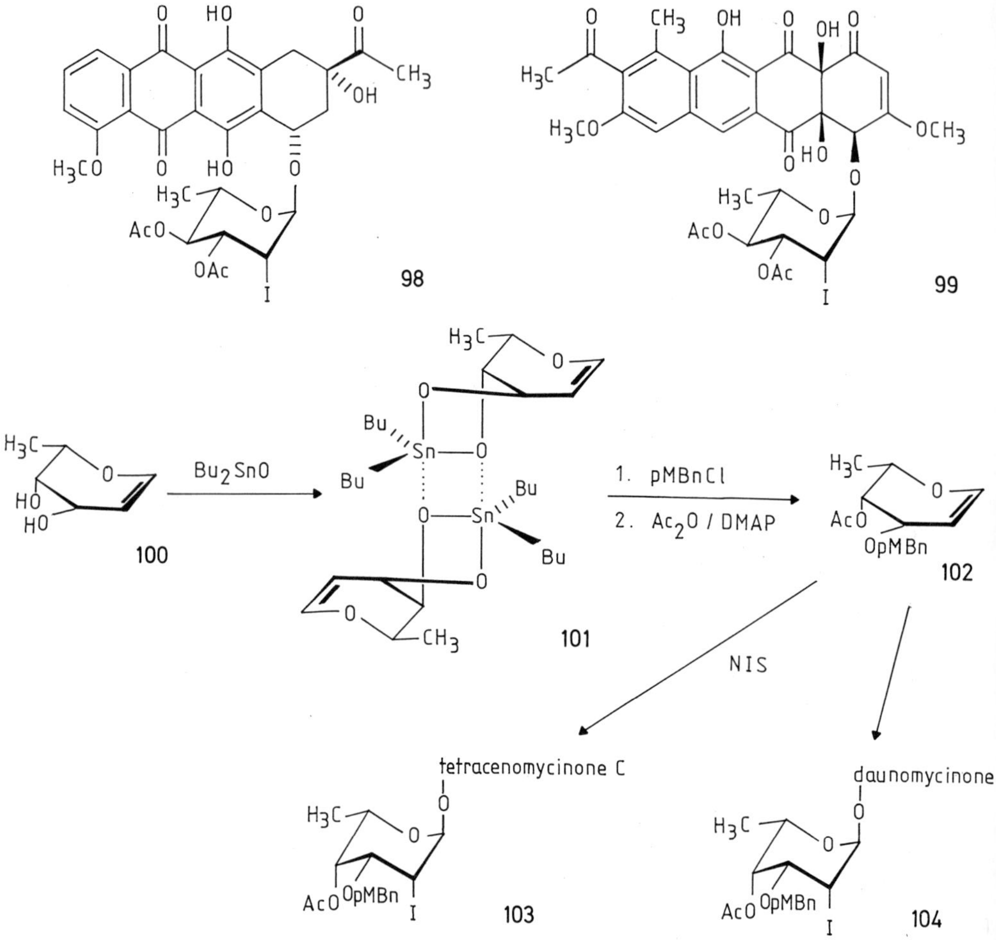

Scheme 23

Starting with L-fucal (**100**), the stannylidene derivative **101** is prepared. As supposed by ^{119}Sn-NMR studies [62], the dimeric species is cleaved preferentially at the apical position. Thus, the subsequent opening by *p*-methoxybenzyl chloride affords the equatorial benzyl ether which, following Steglich acetylation, gives the glycal **102**. Its NIS glycosylation with tetracenomycinone-C as well as daunomycinone leads to the glycosides **103** and **104**, respectively [63].

302

By use of the Ferrier reaction with boron trifluoride catalysis, the formation of both the epimeric α-linked hex-2-enopyranosides **105** [64] and **106** [65] is achieved.

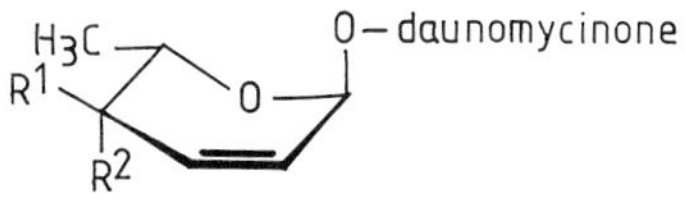

R¹ = OAc, R² = H   105

R¹ = H, R² = OAc   106

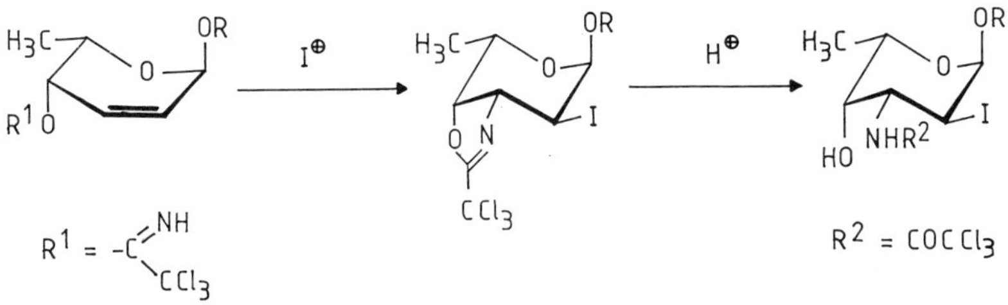

107   108   109

Scheme 24

Fraser-Reid [66] as well as Cardillio et al. [67] have developed the oxazoline procedure by which 4-*O*-trichloroacetimido-hex-2-enopyranosides **107** are treated with an iodonium source [(I⁺Coll)ClO₄⁻ or NIS [64]] to give the 3,4-oxazoline precursors **103** which are opened to the aminosugar glycosides **109**.

4.2 Synthesis of Class II Oligosaccharides and Glycosides

One of the first disaccharides prepared in this group was published by El Khadem et al. [68].

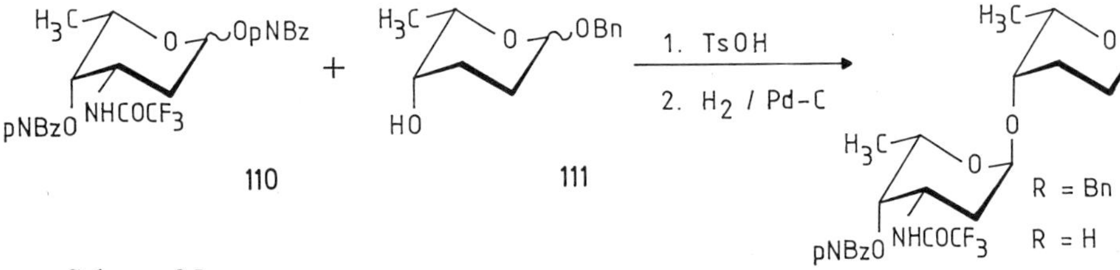

110   111   R = Bn   112

                                                    R = H   113

Scheme 25

On acid-catalyzed glycosylation of **110** with the rhodinoside **111**, the disaccharide **112** is formed. This, after hydrogenolytic cleavage of the benzyl ether gives the 2,3,6-trideoxy-4-*O*-[2,3,6-trideoxy-*N*-trifluoroacetyl-4-*O*-*p*-nitrobenzoyl-α-L-*lyxo*-hexopyranosyl]-L-*threo*-hexopyranose (**113**) in 20% yield.

It became evident rather early that an axial 4-OH group is less nucleophilic than the equatorial one. Attempts for a direct glycosylation at 4-OH furnished low yields, unless there is some flexibility in the saccharide conformation. Therefore, a new synthetic strategy is requested by which the glycosylation step is preponed to the equatorial 4-hydroxy group of the corresponding D-sugar derivative. As depicted in Scheme 26, the model saccharide **114** shows the correctly, equatorially oriented hydroxy group in the $^4C_1$ chair conformation. Its synthetic equivalent **115** with the exocyclic enol ether structure bears the 6-deoxy functionality for the L-series.

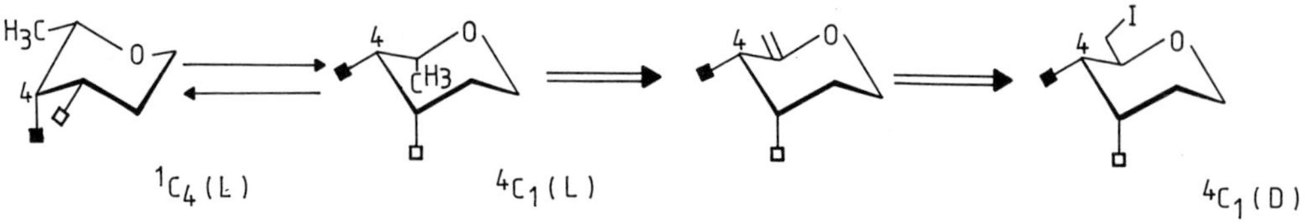

$^1C_4$ (L) $^4C_1$ (L) $^4C_1$ (D)

114   **115**   **116**

Scheme 26

Another retrosynthetic step leads to the 6-deoxy-6-iodo derivative **116**. Throughout this sequence the relative configuration between positions 3 and 4 remains constant, however, altogether the D-*ribo* componed **116** will be transformed into the L-*lyxo* derivative **114**.

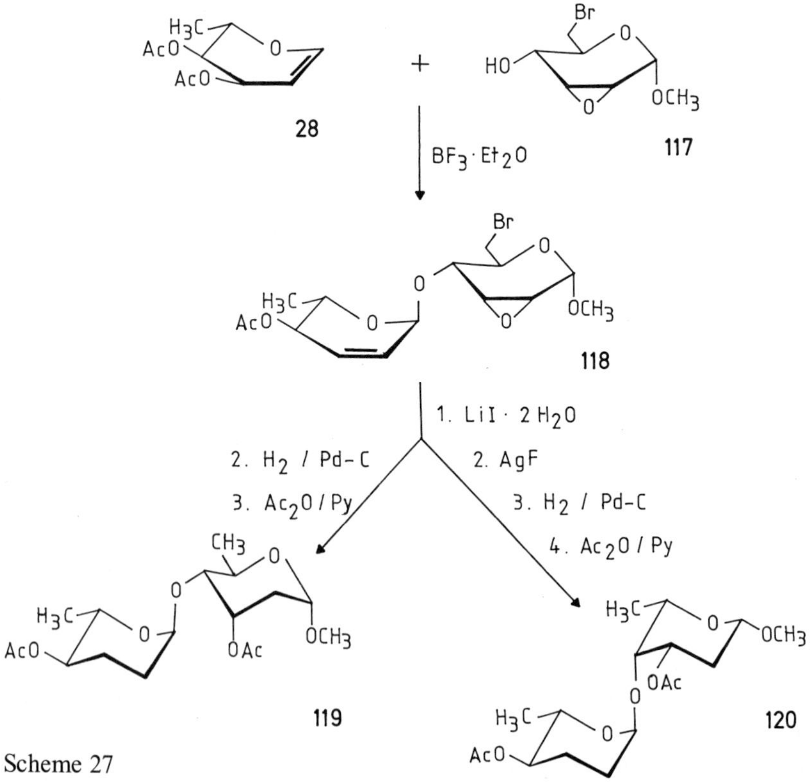

Scheme 27

By Ferrier glycosylation of di-O-acetyl-L-rhamnal (28) with the epoxide derivative 117 the disaccharide 118 is obtained. This is *trans*-opened by lithium iodide according to Fürst-Plattner's rule. Upon different ways, two interesting products are obtained, either the original C-B disaccharide of aclacinomycin A 120 or its C-5 epimer 119 which contains one sugar in the D-configuration [69].

Starting from methyl 2-deoxy-4,6-O-benzylidene-3-O-mesyl-α-D-*arabino*-hexopyranoside (121), a S$_N$2 reaction by use of sodium azide in HMPT at 140 °C and NBS-cleavage gives compound 122. Following transglycosylation with benzyl alcohol and deprotection by Zémplen transesterification to 123, acid-catalysed glycosylation with tri-O-acetyl-2-deoxy-L-fucose 124 affords the disaccharide 125. Upon treatment with silver fluoride, subsequent hydrogenolytic reduction and acetylation the disaccharide derivative 126 is received, which represents the A-B moiety of aclacinomycin A. After a hydrogenolytic deprotection, a further acid-catalyzed glycosylation with daunomycinone is performed to give 127 in 9 % yield [70].

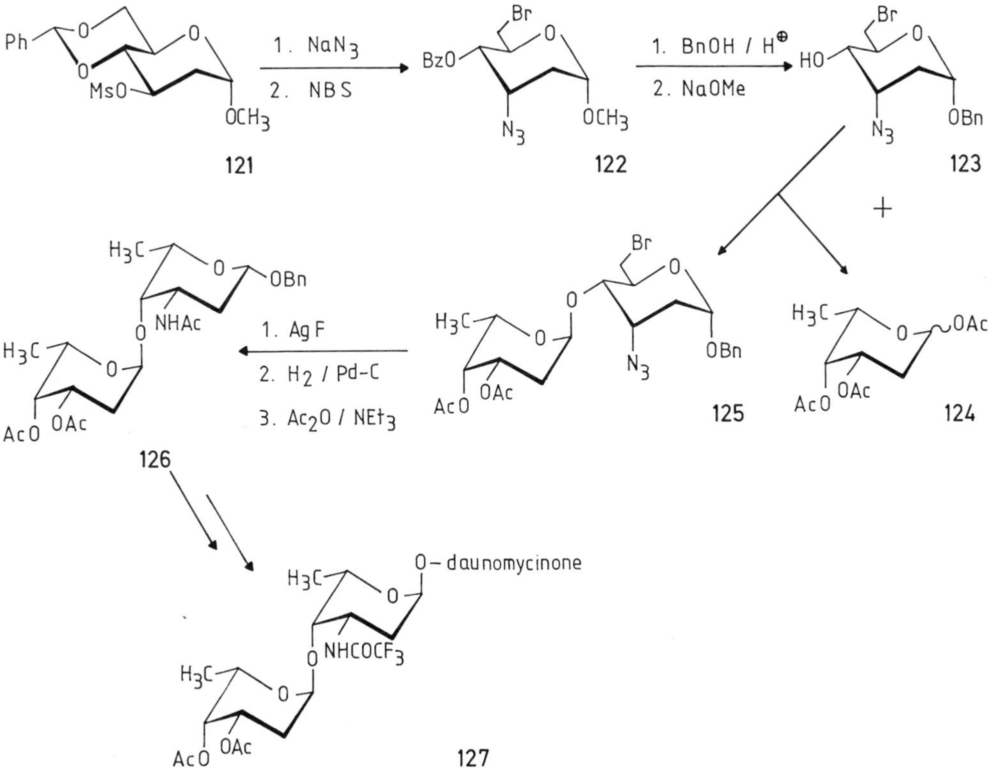

Scheme 28

Further studies by Monneret's group [8, 71] furnished the trisaccharide portions of marcellomycin 135 and dihydroaclacinomycin 136.

Both Monneret syntheses are quite similar to the disaccharide way mentioned above, except for a final NIS glycosylation step, by which the terminal saccharide C is attached. Glycosylation of the 2-deoxy fucosyl bromide 130 with the daunosaminide

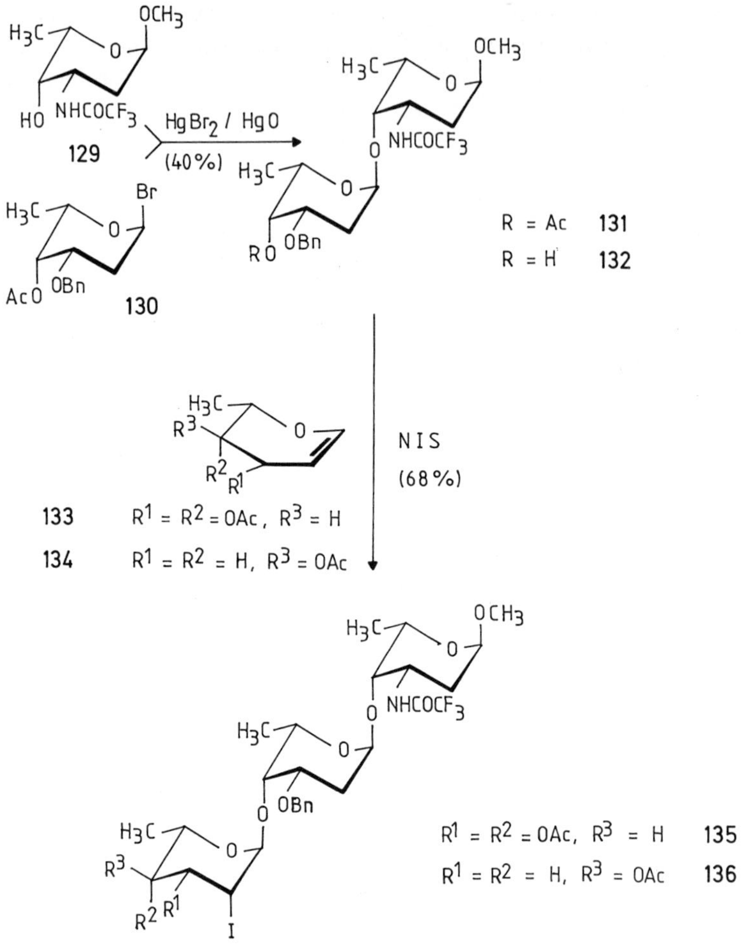

Scheme 29

129 under Helferich conditions yields the disaccharide 131 (40 %). Following deacetylation to 132, an NIS reaction with either di-O-acetyl-L-fucal (133) or O-acetyl-L-amicetal (134) gives the trisaccharide derivatives 135 (68 %) or 136, respectively.

By consequent use of the NIS-alkoxygenation procedure combined with the retrosynthetic approach (Scheme 26), Thiem et al. [64, 72] prepared the trisaccharide 143. The D-ribo-configurated 3-azido glycoside 137 is attached to the glycal 138 by NIS to give the disaccharide 139. Treatment of the hex-2-enopyranoside 141 with lithium aluminum hydride leads to the L-amicetal 134. Its NIS glycosylation with 140 affords the trisaccharide precursor 142. This is deiodinated by radical reduction with tributyl stannic hydride at both the C-2′ and C-2″ positions. The deblocked compound is iodinated at C-6 by the use of triphenyl phosphine/NIS [73, 74] and reduced to give an exocyclic glycal by means of a smooth elimination step with diazabicyclo-[5.3.0]undec-7-ene (DBU), which is much better in yield than the one with silver fluoride described above. Final hydrogenation with inversion of the unit A and acetylation gives the sugar portion **143** of dihydroaclacinomycin A.

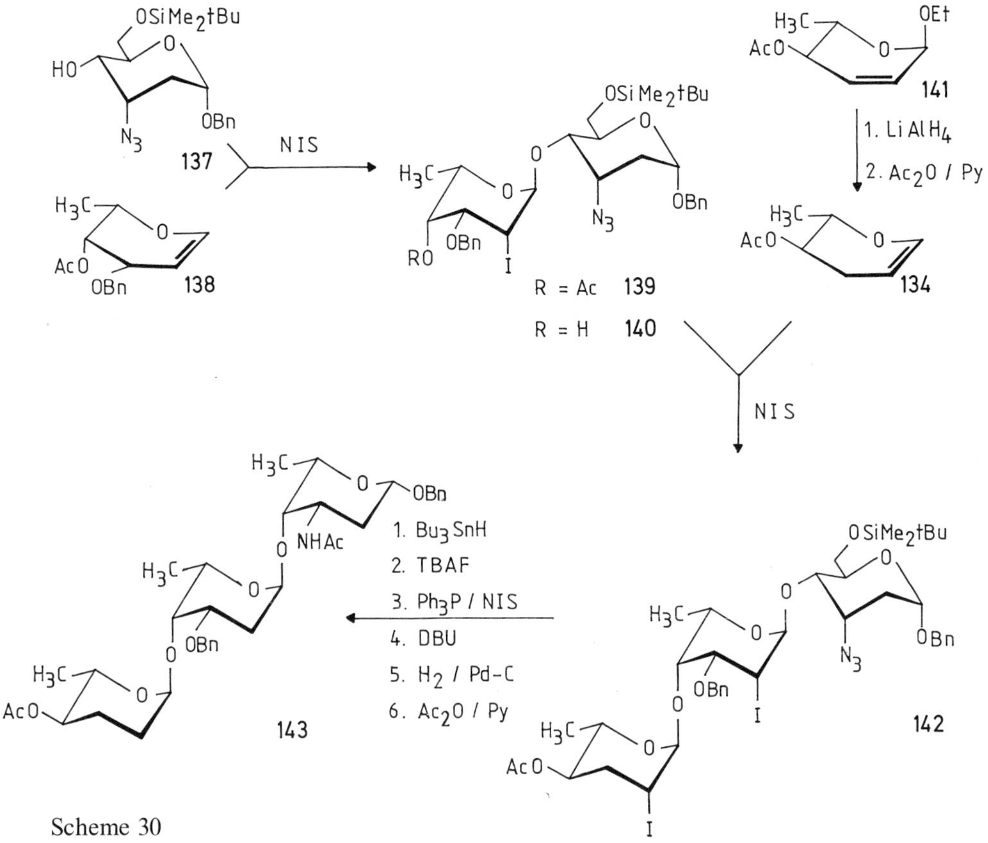

Scheme 30

5 Aureolic Acids

Aureolic acids are a group of tetrahydroanthracene oligosaccharide antibiotics the name of which relates to their characteristic golden appearance. The most prominent members chromomycin A_3 (**144**), olivomycin A (**145**), and mithramycin (**146**) represent potent cytostatics which, even though they are extremely toxic, enjoy selected clinical application in the treatment of malignant tumors [30]. Mithramycin, however, is the member of this list which is approved for clinical use in the U.S. Recent developments led to modified derivatives which display improved activities; all these control cancer-induced hypercalcemia and calciurea. Their cytostatic activities are assumed to result from strong and selective inhibitions of the DNA-dependent RNA synthesis. The mechanism, however, seems to differ from that of the anthracycline intercalation. The inhibition of the DNA-dependent RNA polymerase requires the presence of Mg^{2+} ions and an interaction occurs predominantly at guanosine-rich regions [76].

307

There is another important difference concerning the therapeutic width. The *dosis curativa minima* (ED) is very similar to the *dosis tolerata maxima* (LD). This results in a therapeutic index, defined as the quotient of LD_{50} and ED_{50} close to 1. The effect is attributed to the digitalis-like shape of the oligosaccharide side chains, which cause cardiotoxicity. Nevertheless, the RNA synthesis inhibitory effect is attributed to the carbohydrate moieties. The influence of aglycon modifications in comparison to those in the sugar residue is rather small.

Earlier studies proved the structure of the almost similar tetrahydrocenone aglycons in **144–146** with the chiral five-carbon side chain at C-3 as well as those of the individual

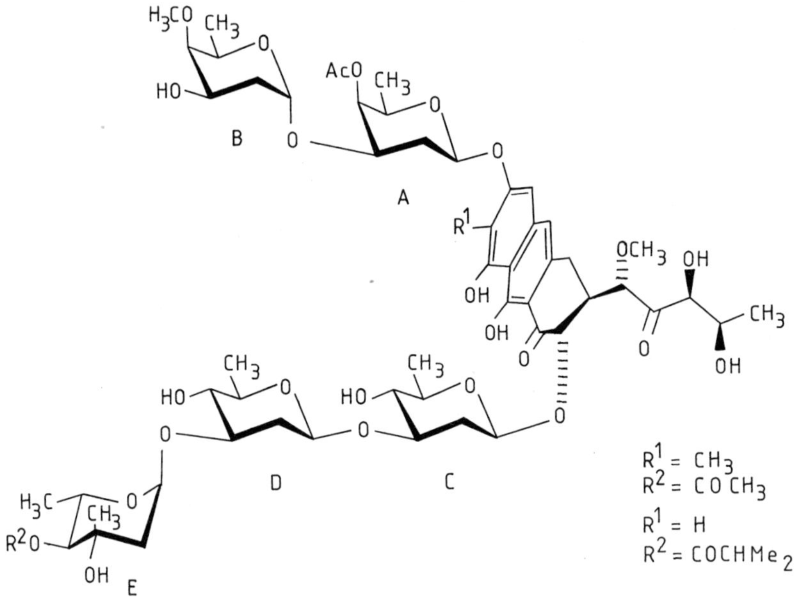

$$R^1 = CH_3$$
$$R^2 = CO\,CH_3 \qquad 144$$

$$R^1 = H$$
$$R^2 = COCHMe_2 \qquad 145$$

$$R^1 = CH_3 \qquad 146$$

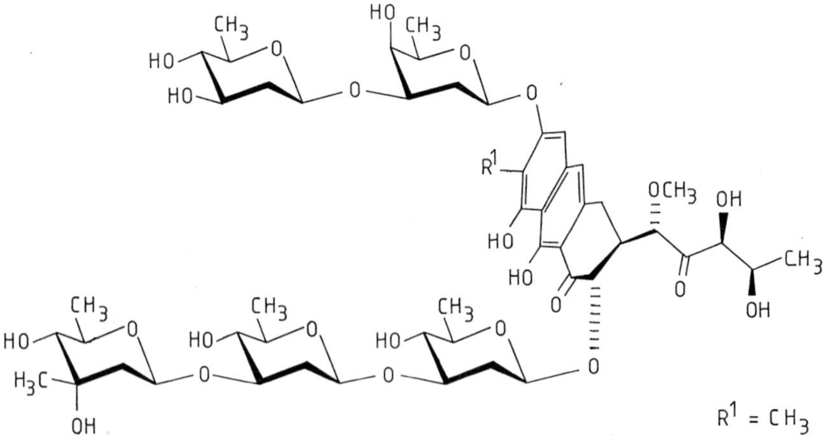

Scheme 31

monosaccharides [77, 78]. This interglycosidic linkage to oligosaccharide units was deduced by applying Klyne's rule, but this was not convincing throughout [79]. The complete sugar sequence as well as the direction and type of their interglycosidic linkages were assigned by extended NMR-spectroscopy and supported by syntheses [80–83]. There are only minor deviations between chromomycin A_3 (**144**) and olivo-mycin A (**145**) although they are produced by different *Streptomyces* strains. In both these compounds, a differently substituted $\alpha,1 \rightarrow 3$-linked bis-2,6-dideoxy-*lyxo*-unit B-A is attached to the phenolic site at C-6. Their E-D-C trisaccharide shows a terminal 3-C-methyl-branched sugar, L-olivomycose E, linked via an $\alpha,1 \rightarrow 3$-bond to the $\beta,1 \rightarrow 3$ connected dimeric olivosyl-olivose D-C. In mithramycin (**146**) the B-sugar is a 2,6-dideoxy-arabino unit attached to A via a $\beta,1 \rightarrow 3$ linkage. In the E-D-C trisaccharide part, only $\beta,1 \rightarrow 3$ linkages occur. The D-C unit is similar as before, and the terminal sugar again is of the 3-C methyl-branched type, but this time it is D-mycarose.

Starting from methyl 2,6-di-*O*-acetyl-3,4-*O*-isopropylidene-α-D-galactopyranoside (**147**), photolytic deoxygenation at C-2 and C-6 leads to the D-*lyxo* compound. Cleavage of the isopropylidene ring with trifluormethane sulfonic acid to **148** and monoacetylation with *N*-acetyl imidazole gives both the acetates (**149**) and (**150**), useful for further synthesis. Boron trifluoride-mediated methylation of **149** occurs with diazomethane regioselectively at C-4 without acetyl migration to give **151**. The latter is transformed into the reactive 2,6-α-D-*lyxo*-hexopyranosyl bromide **152** by the trimethylsilyl halide (TMSX) procedure [84]. The stereoselective condensation step with **150** yields only the $\alpha,1 \rightarrow 3$-linked disaccharide **153** [85], and this proceeds via the well-known halide ion-catalyzed glycosylation mechanism.

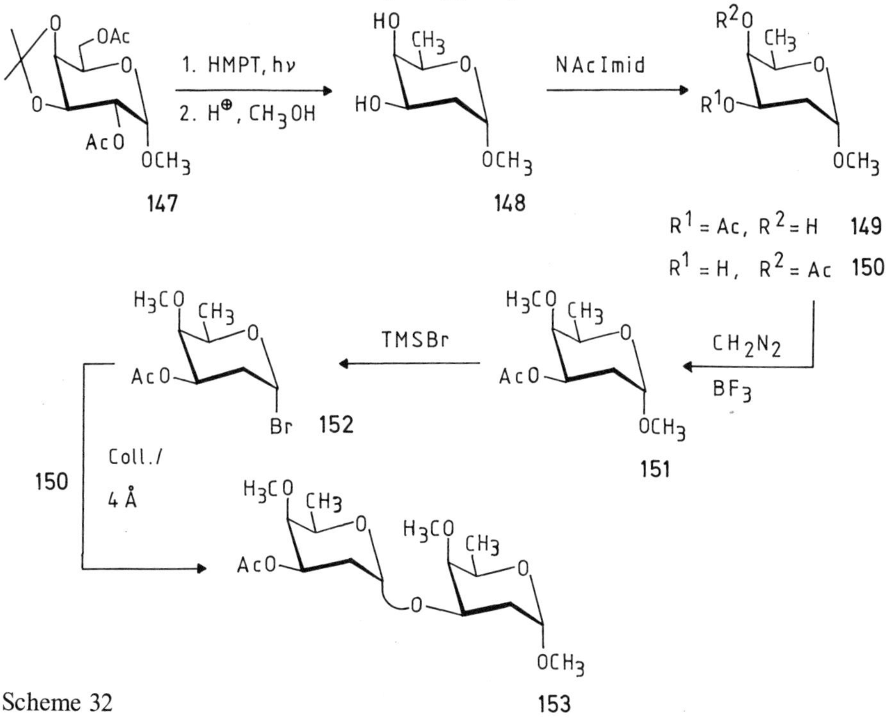

Scheme 32

Structure elucidation of the B-A disaccharide portion of mithramycin was also carried out by synthesis and ^{1}H- and ^{13}C-NMR comparison with the original compound. As it turned out later, the D-*arabino*-β,1 → 3-D-*lyxo* configuration was shown to be correct [80, 81]. Upon treatment of di-*O*-acetyl-D-rhamnal (154) with NIS in the presence of water, the 2,6-dideoxy-2-iodo-hexopyranose is formed. Reductive cleavage of the 2-iodo function by hydrogenolysis and acetylation in pyridine gives the 2,6-dideoxy sugar. The glycosylation is carried out after application of the TMSX-approach via the bromide 155, and subsequent glycosylation in the presence of silver triflate with 150 as the reducing unit to give the α- and β,1→3-linked derivatives 156 and 157. Under carefully optimized conditions, a yield of approx. 60% and an α:β ratio of 2:1 is obtained [80, 81]. Obviously, this more-or-less classical approach is tedious and not as stereospecific as wanted.

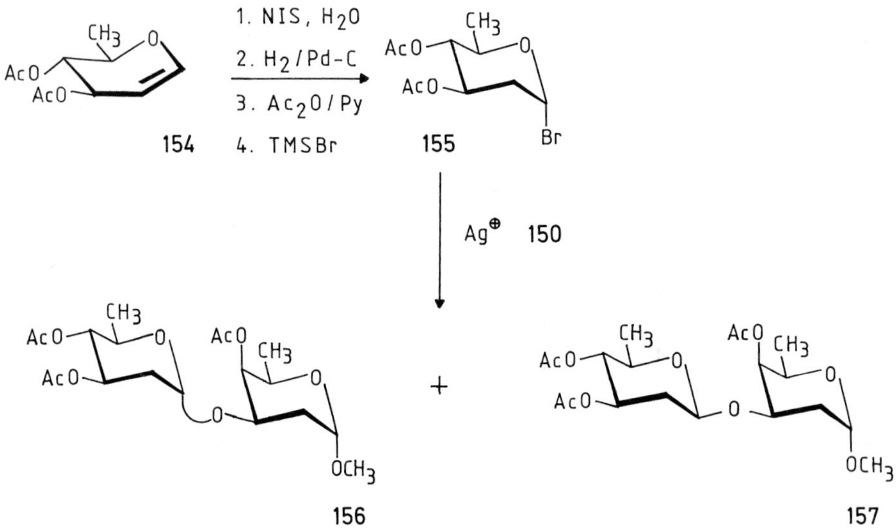

Scheme 33

Both chromomycin A$_3$ (144) and olivomycin A (145) are further furnished with a C-3 methyl-branched sugar, namely L-olivomycose, which is α-glycosidically linked to the 3 position of the sugar unit D. For this synthetic purpose the NIS reaction is ideally suited, provided a high yielding procedure for the precursor glycal is at hand. Following Klemers approach [86], treatment of methyl 2,3-*O*-benzylidene-α-L-rhamnopyranoside (158) with methyl lithium gives the L-olivomycal (160) and L-mycaral (161) in a 7:2 ratio. After abstraction of the axial H-3, an electrocyclic ring-opening sequence is terminated by the loss of the anomeric methoxy group. This leads to the intermediate 3-keto glycal 159, the carbonyl function of which is attacked immediately by another equivalent of methyl lithium. A previous method [87] for their preparation applied an allylic oxidation of unblocked L-rhamnal with manganese dioxide, and following a similar nucleophilic attack at the carbonyl site with methyl lithium, a 7:3 ratio of 160 and 161 is obtained.

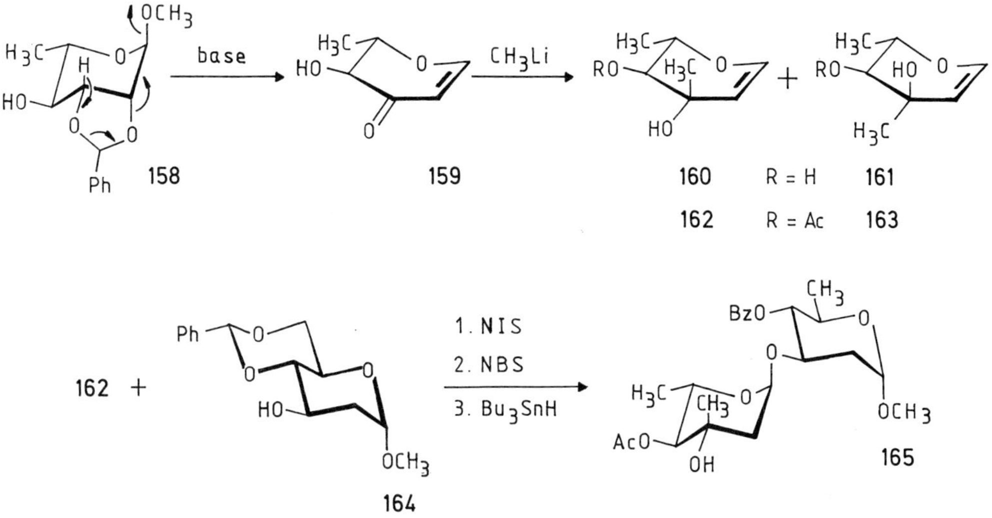

Scheme 34

By attachment of *O*-acetyl-L-olivomycal (**162**) to the methyl 4,6-*O*-benzylidene glycoside (**164**), the α,1→3-linked disaccharide is obtained. By Hullar-Hanessian cleavage and a subsequent radical reduction of both the 2'- and 6-halo functions with Bu₃SnH, the E-D moiety **165** is obtained [88].

For the synthesis of the D-C portion, two different concepts were followed either by modification of laminaribiose (**166**) [89] or by a stereospecific β,1→3-glycosylation [20]. Laminaran is isolated from seaweeds or from *Poria cocos Wolf*, degraded by selective acetolysis, and the lower oligomers separated by preparative HPLC [90]. Following acetylation, the heptaacetyl laminaribiosyl bromide is prepared and transformed into the disaccharide glycal **167** by the classical approach in 93 % yield. The 2-deoxy-2-iodo-α-glycoside is formed by application of the NIS procedure; after deprotection and subsequent 4,6-*O*-benzylidenation, the precursor **168** for the radical formation of the 6,6'-dibromo-6,6'-dideoxy derivative is at hand. This compound may be further reduced to methyl-3-*O*-(β-D-chinovosyl)-α-D-olivoside (**169**).

Much more critical, however, is the introduction of the 2'-deoxy function. This can be achieved after selective 3'-*O*-benzoylation to **170**, xanthation at the 2' position to **171** and reductive cleavage (Bu₃SnH) to **172** following Barton's method [91]. Further radical ring opening by the Hullar-Hanessian procedure and final reduction leads to the D-C disaccharide **173** [92].

The stereospecific glycosylation makes use of the dibromomethyl methyl ether (DBE) method [20]. Starting from methyl mannoside **174**, the compound **175** is obtained, the dioxolane ring-opening reaction of which with DBE gives a 2,6-dideoxy glucosyl bromide **176**. This is β-glycosylated with benzyl alcohol and the formyl-protecting group cleaved to give the β-benzyl glycoside **177**. The glycosylation is promoted by silver triflate and optimized reaction conditions give the α- (**178**) and the β-disaccharide **179** in 92 % yield and an α:β-ratio of 1:7.

The precursor **179** is deblocked at C-3' and reduced with Bu₃SnH to give **181**. By NIS reaction the olivomycal **162** could not be attached neither to compound **180**

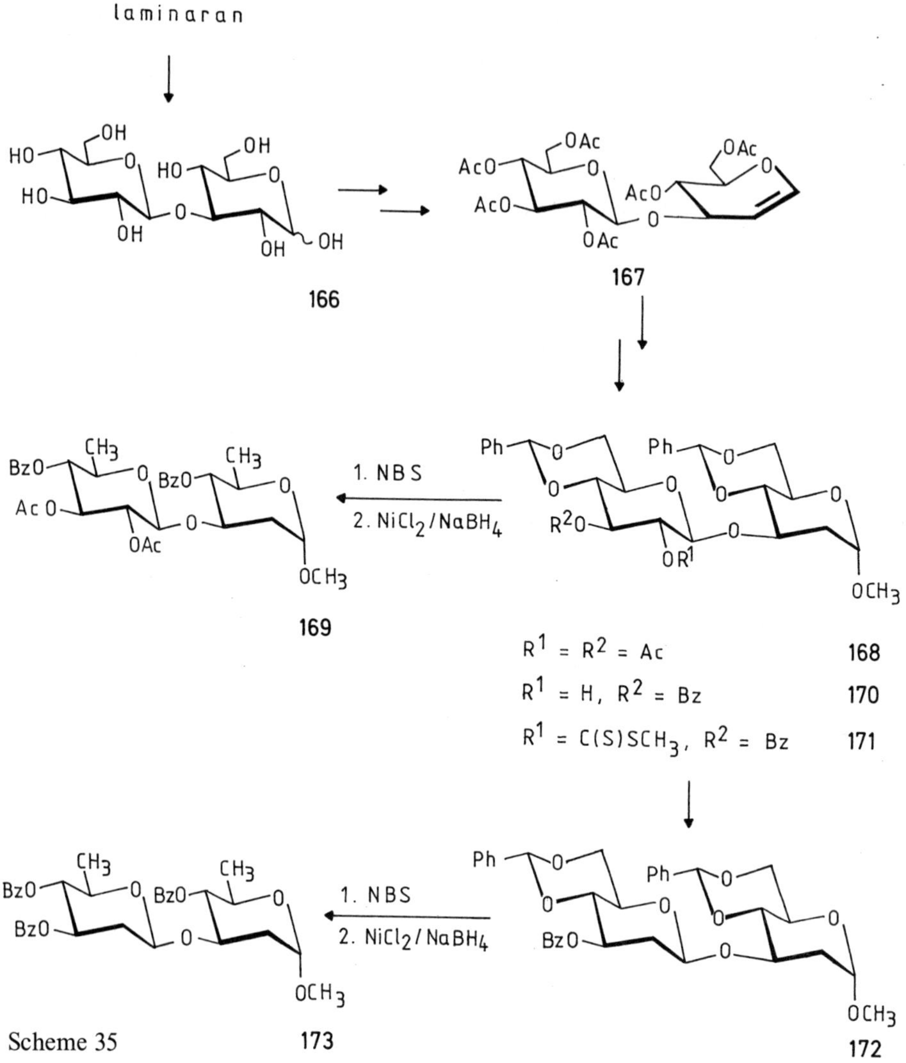

laminaran

Scheme 35

nor to **181**. However, the 3′,4′-unblocked compound **182** regio- and stereospecifically leads to the trisaccharide **183** [23]. After acetylation and deiodination the E-D-C moiety **184** of olivomycin A and chromomycin A_3 is obtained.

Mithramycin shows a completely β-linked chain of D-configurated saccharides. This requires a totally different approach for the synthesis which is also done by application of the DBE method. The previously obtained disaccharide **180** is β-glycosylated with the monosaccharide precursor **176** to give the trisaccharide **185**. After reductive debromination (Bu_3SnH), an acid deformylation deblocked the C-3″ position which is oxidized with pyridinium dichromate. Nucleophilic attack at the carbonyl group by methyl lithium affords a 1:1.2 mixture of **186** and **187** none of which is the desired compound [93]. Obviously, the methyl branch is formed exclusively in the axial way.

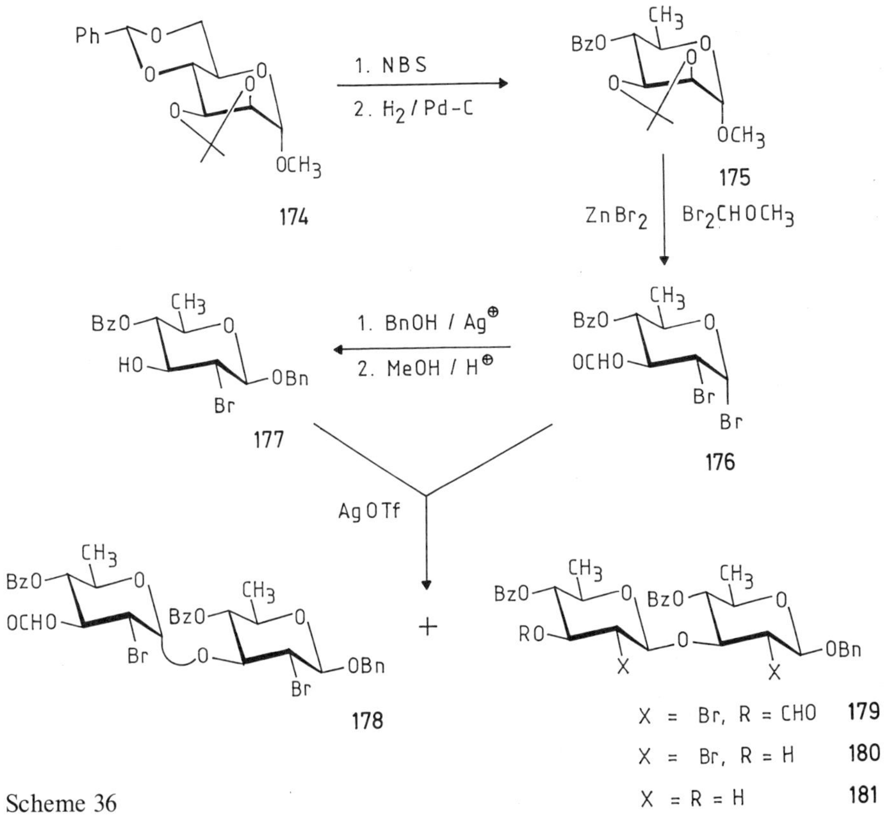

1. NBS
2. H_2 / Pd–C

174

175

$ZnBr_2$ Br_2CHOCH_3

176

1. BnOH / Ag$^{\oplus}$
2. MeOH / H$^{\oplus}$

177

AgOTf

178

+

X = Br, R = CHO   179

X = Br, R = H   180

X = R = H   181

Scheme 36

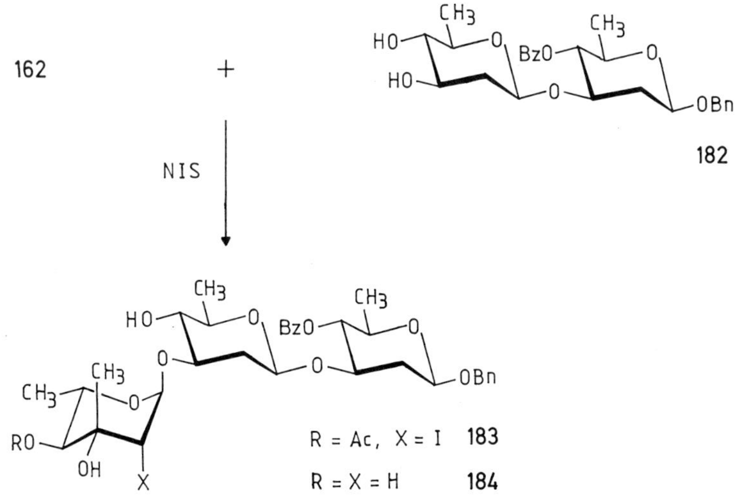

162   +

182

NIS

R = Ac, X = I   183

R = X = H   184

Scheme 37

313

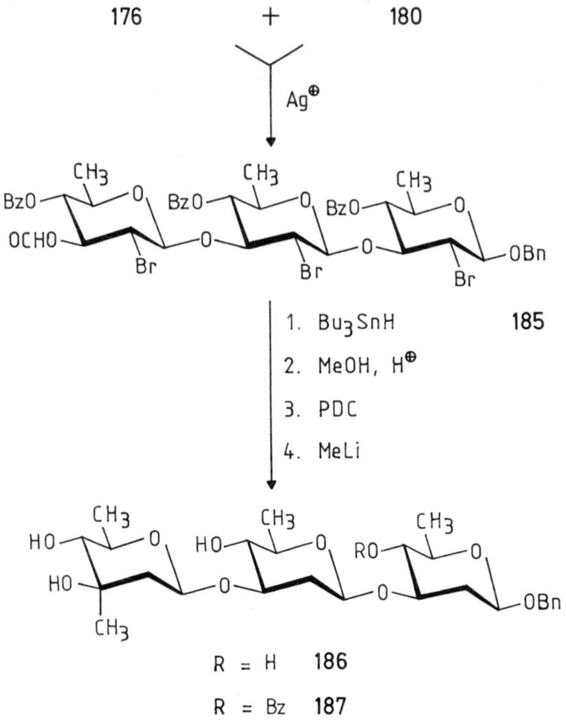
176
+
180
Ag⊕
BzO
OCHO
Br
OBn
1. Bu3SnH
2. MeOH, H⊕
3. PDC
4. MeLi
185
HO
RO
R = H 186
R = Bz 187

Scheme 38

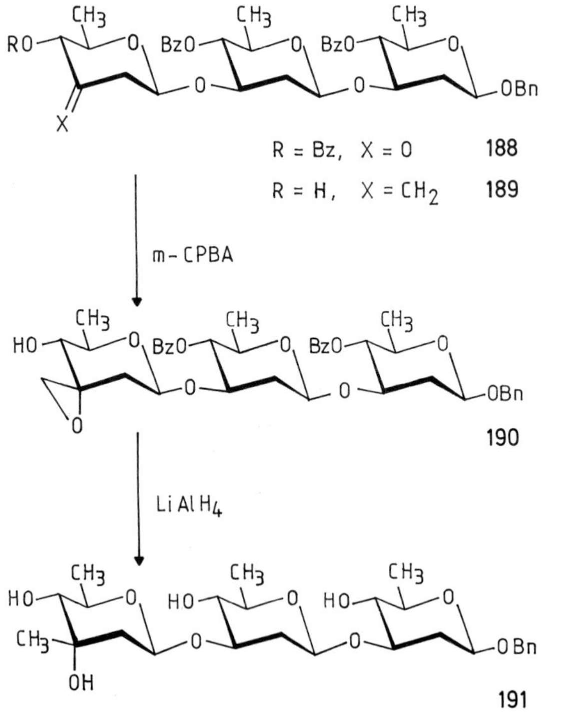
RO
BzO
OBn
X
R = Bz, X = O 188
R = H, X = CH2 189
m-CPBA
HO
190
LiAlH4
OH
191

Scheme 39

Equatorially positioned methyl-branched derivatives may be obtained by reductive cleavage of spiro epoxides [94]. Thus the Peterson olefination of **188** gives the exocyclic 3″-methylene function in **189**. By means of a Sharpless epoxidation the allylic 4″-hydroxy group should determine the chirality of the resulting epoxide. However, the Sharpless method does not show any reaction neither in a monosaccharide model system nor in this trisaccharide precursor [95]. Amazingly, the classical epoxidation with *m*-chloroperbenzoic acid is employed to give exclusively the desired (3″R) epoxide **190** in excellent yield. These results may be associated with a sufficient chiral induction of the stereochemical information at C-1″, C-4″, and C-5″. A subsequent reduction furnishes the original E-D-C trisaccharide sequence **191** of mithramycin [95, 96].

6 Cardiac Glycosides and Related Antibiotics

Cardiogenic insufficiency represents a widespread disease in modern industrial societies, which explains the particular request for appropriate therapeutic agents.

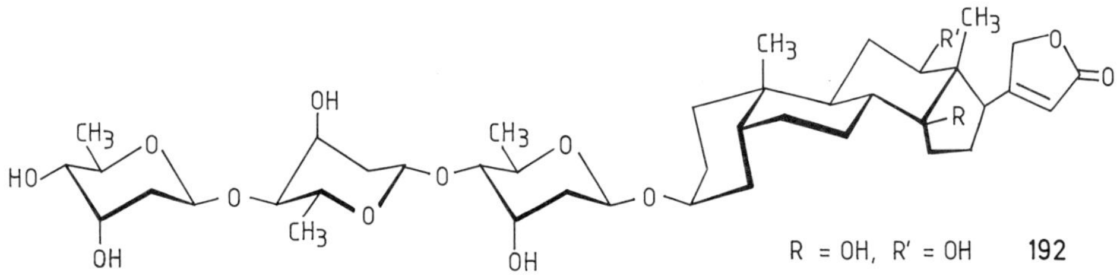

R = OH, R' = OH   192

R = OH, R' = H   193

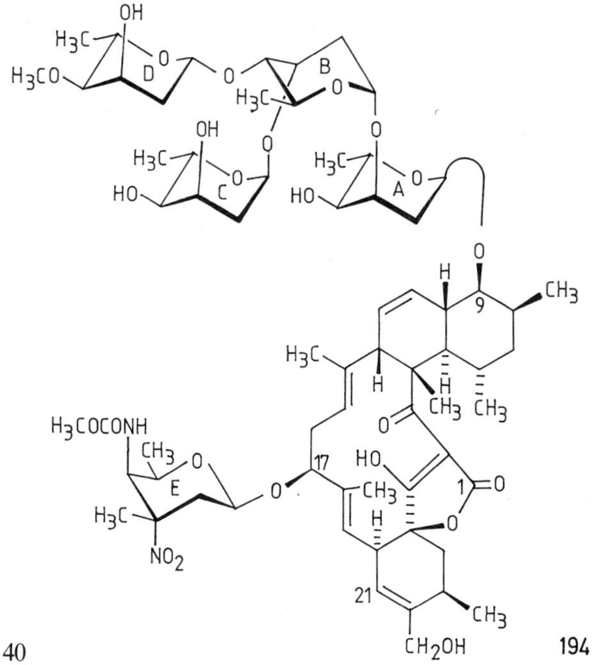

Scheme 40

194

Its treatment with cardiac glycosides remains problematic owing to their small therapeutic width. Too low doses do not show effects and enhanced doses lead to heart inactivation during the systole.

Cardiac glycosides cause a positive inotropic effect which means an increase of the cardiac beat volume by enhanced contraction ability. The reason for this is supposed to be aligned with the direct inhibition of the transport enzyme sodium/potassium-ATPase. The decrease of sodium ions enhances the calcium ion concentration, which activates the myofibrillic enzyme and inactivates proteins like tropomyocine and tropomine. Till present, a final proof for this hypothesis is lacking, the toxicity, however, is definitely aligned with these effects [97].

In general, pharmacokinetics are considerably influenced by the saccharide portion, which consist mainly of $\beta,1 \rightarrow 4$-interglycosidically linked 2,6-dideoxyhexoses of D-*ribo* configuration [98]. The typical $\beta,1 \rightarrow 4$-linked cardiac glycosides shown in Scheme 40 are digoxin (192) and digitoxin (193).

A mainly $\alpha,1 \rightarrow 3$-linked oligosaccharide type in the L-series is represented by the antibiotic kijanimycin (194), isolated in 1981 as a metabolic product of *Actinomadura kijaniata* from Kenian soil and structurally elucidated by Mallams et al. [99–101]. The sugars A, B, and C are $\alpha,1 \rightarrow 3$-linked, and at the B unit another 2,6-dideoxy L-*ribo* sugar D is $\beta,1 \rightarrow 4$-glycosidically attached. Kijanimicin belongs to the class of macrolide antibiotics which are defined as polyoxygenated 12–40-membered ring lactones, furnished commonly with deoxy or amino deoxy saccharides.

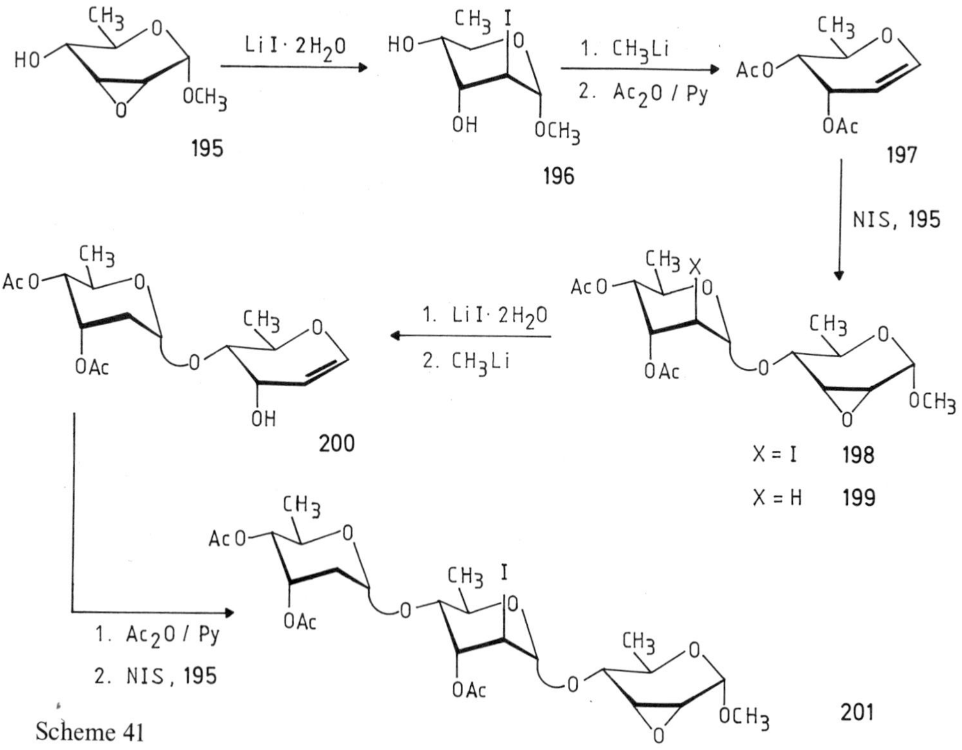

Scheme 41

Macrolides, in contrast, are bacteriostatics and fungicides by inhibition of ribosomal translation. Several members of this group cause the same effects on mitrochondrial ribosomes and are antineoplastic agents. Kijanimicin is active against anaerobic bacteria — and malaria — and it also shows cytostatic activity. Kijanimicin may be counted among the group of tetronic acids like the tetrocarcines and antherimycines, all of which contain L-digitoxose residues.

One of the first attempts to synthesize cardiac glycosides was made by Zorbach et al. in 1963 [102] by Fischer glycosylation. After treatment of digitoxigenin and unprotected digitoxose in dioxane with a saturated solution of hydrogen chloride in dichloromethane, 5.4% of the α- and 4.5% of the desired β-glycoside were isolated after chromatographic purification.

Later is was shown that syntheses using peracetylated digitoxose [103] or digitoxosyl bromide [104], the former in presence of acid, the latter with a silver salt promotor, gave only slightly enhanced yields of β-glycosides with β:a ratios up to 3:2.

A number of 1→4-linked anomerically modified oligosaccharides can be obtained by NIS glycosylations. By treatment of the 2,3-anhydro-*allo* derivative **195** with lithium iodide the *trans*-opening predominantly gives the 2-deoxy-2-iodo-*altro* glycoside **196**, which in turn can be reductively eliminated to furnish the D-digitoxal. After acetylation to **197** and glycosylation with a second molecule of *allo*-epoxide **195**, the disaccharide **198** is obtained in 72% yield.

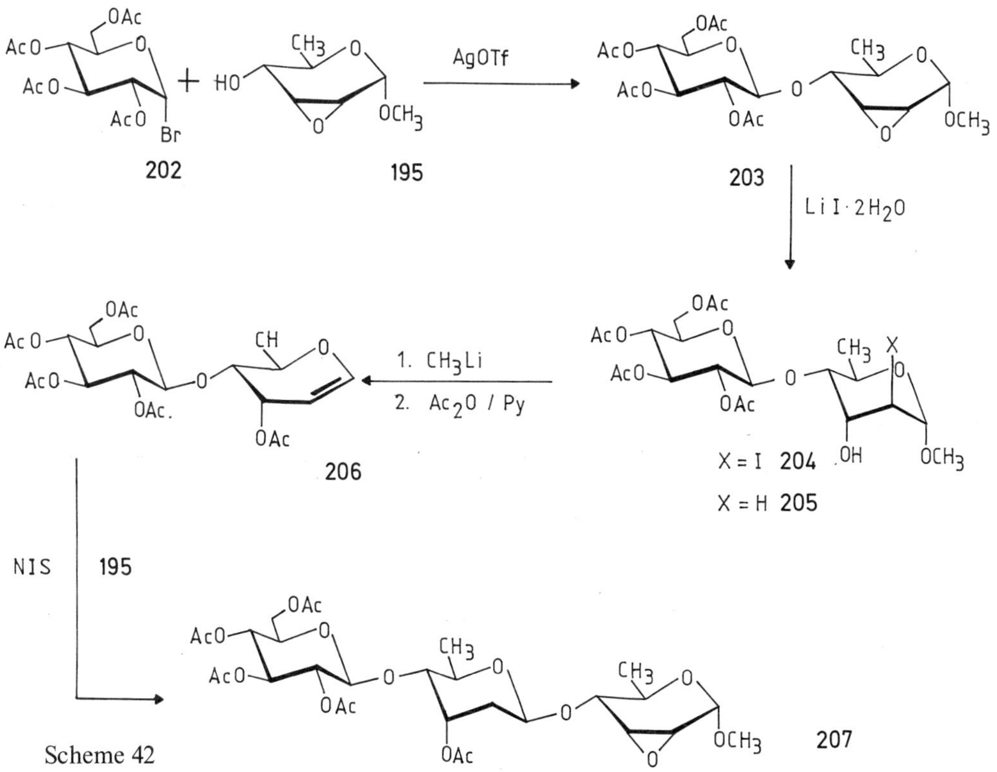

Scheme 42

Following nickel boride reduction, the disaccharide **199** results which by a similar treatment first with lithium iodide and then with methyl lithium gives the disaccharide glycal **200**. Another glycosylation with **195** and NIS leads to the trisaccharide digitoxose derivative **201** (53%) [105].

Glycosylation of methyl 2,3-anhydro-6-desoxy-α-D-allopyranoside (**195**) with acetobromoglucose **202** in the presence of silver triflate gives the β,1 → 4-linked disaccharide **203** in yields between 12% and 30% depending on the type of bases used. Nucleophilic opening of the epoxide with LiI · 2 H$_2$O in dichloromethane gives the more stable *trans*-diaxial product **204** (51%) according to the Fürst-Plattner rule predominantly. The *trans*-diequatorial, crystalline *gluco*-configurated side product was isolated in 13%. Nickel boride reduction affords the acetylated methyl digilanidobioside **205**.

After MeLi-mediated elimination and acetylation digilanidobial, the disaccharide glycal **206** is obtained. This provided a useful intermediate for the formation of the trisaccharide **207** which could be prepared by NIS glycosylation again with **195** in 29% yield [106].

The α,1→4-linked bis-digitoxoside **208** after treatment with triethyl orthoacetate and subsequent acetylation, gives both the diastereomeric orthoester intermediates **209**. Acid-catalyzed 3′,4′-orthoacetate opening leads to both the 3′,3′- and the 3,4′-diacetates **210** and **211** in the ratio 3:1. NIS glycosylation of **210** with digilanidobial **206** furnished the central α,1→4-linkage in **212**, and following reduction the tetrasaccharide species **213** results [107].

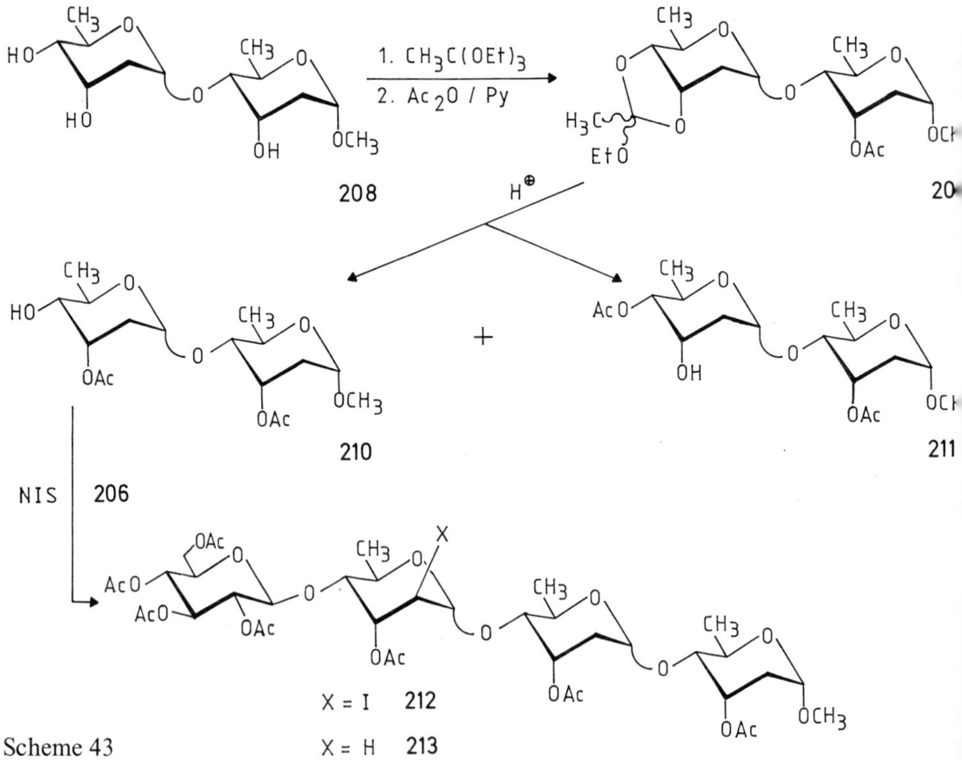

Scheme 43

X = I   **212**
X = H   **213**

318

Recently, Wiesner et al. [108] succeeded in a stereoselective β-glycosylation in the D-*ribo* series by use of an anchimeric assistance from the 3 position. The 3-*O*-*p*-methoxy-benzoyl group is supposed to form the 1,3-acyloxonium ion intermediate **215** either starting from an α- or β-configurated ethyl thioglycoside precursor (**214**).

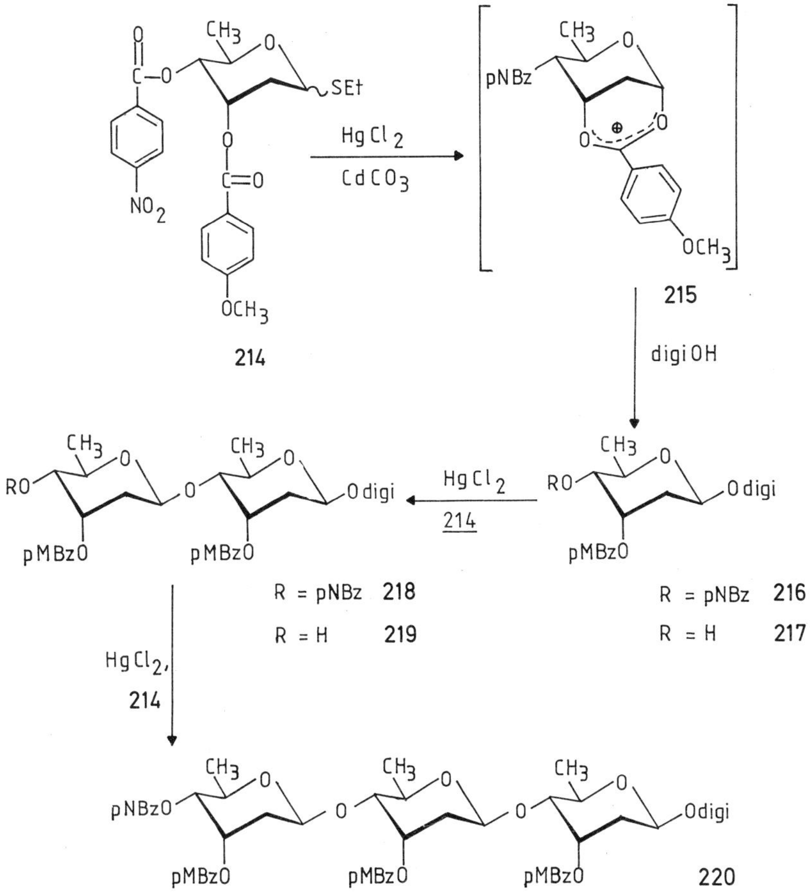

Scheme 44

These are obtained by treatment of the anomerically free digitoxose in 96% with thioethanol and *p*-toluene sulfonic acid in dichloromethane. The coupling of the thioglycoside is performed in dichloromethane, in the presence of HgCl$_2$, CdCO$_3$, and a drop of dimethylformamide. A reaction time of two days is required to give the β-glycoside **216** in 59% yield with only small amounts (2.3%) of the undesired α-anomer. After selective removal of the *p*-nitrobenzoyl protecting group at the C-4 position (by stirring with a saturated solution of ammonia in ethanol) **217** is subjected to another glycosylation with **214** under the same conditions. The disaccharide **218** is obtained. The trisaccharide synthesis is performed similarly by repeating the same procedure twice over. Deblocking of **218** affords **219** which is again glycosylated with the monosaccharide precursor **214** to give **220** in 58% yield.

Similarly, a 1,3-anchimeric assistance is thought to operate in the case of an urethane-protecting group at the C-3 position in **221**. On hydrolysis with aqueous acetic acid at 110 °C for 40 min, the anomeric mixture **222** is obtained. The coupling with a modified digitoxigenin derivative (digi*OH) **223** is catalyzed with *p*-toluene sulfonic acid in dichloromethane and benzene for 45 min. For purification and separation of the anomeric mixture **224** obtained in this step, deblocking of both the 3- and 4-hydroxy groups to **225** is necessary. Following silica gel chromatography, the mixture is obtained in 83 % yield with a β:α-ratio of 7:1. The mechanism is supposed to proceed via the charged species **227** and rather not an uncharged urethane. Further transformations are necessary to generate the β-digitoxigenin glycoside **226** from the β derivative of **225** [17].

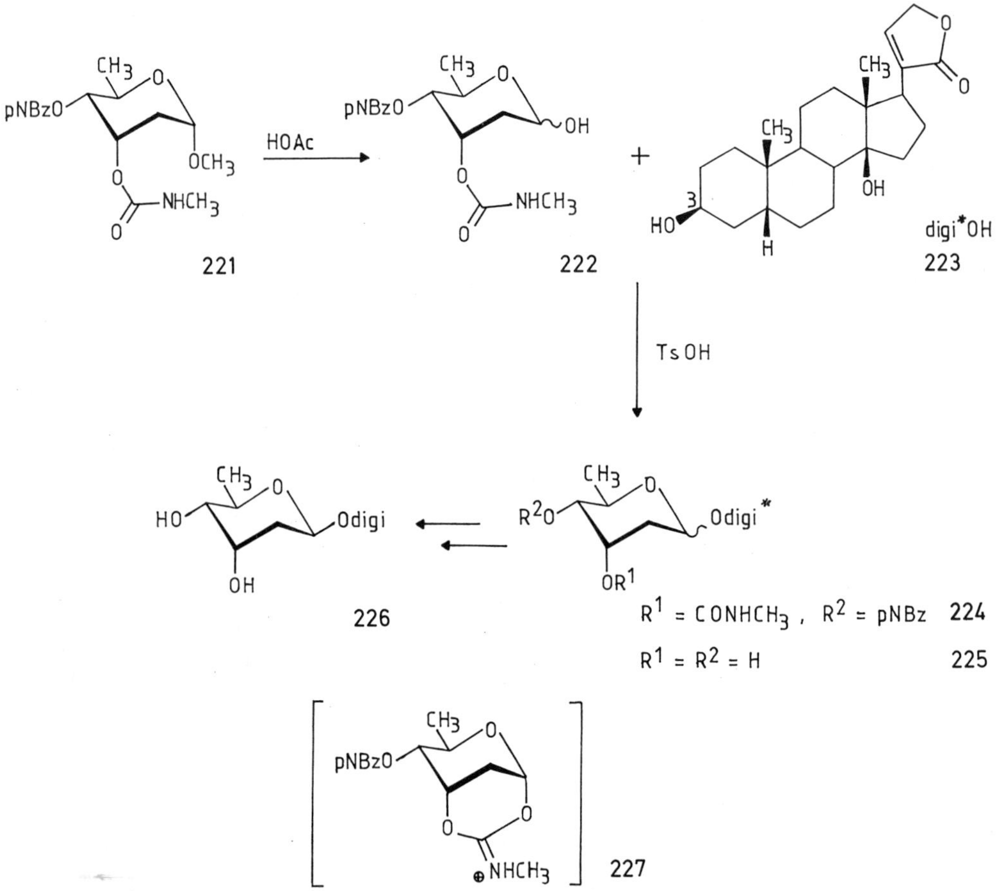

Scheme 45

The naturally occurring α,1→3-linked oligosaccharides of, e.g., kijanimicin are synthesized either in a stepwise manner or following a one-pot procedure [109]. Starting with the 4-*O*-benzyl-L-digitoxal (**228**) [10], NIS is added in solution in a half-equimolar amount. Then the same aliquote of glycal **228** and NIS is added under the same conditions. Subsequently, addition of a half-molar equivalent of NIS and benzyl alcohol quenches the oligomerization. Thus, the homogeneously substituted trisaccharide product **229** is obtained in 20% yield. A final hydrogenation step affords the natural trisaccharide derivative **230** of kijanimicin.

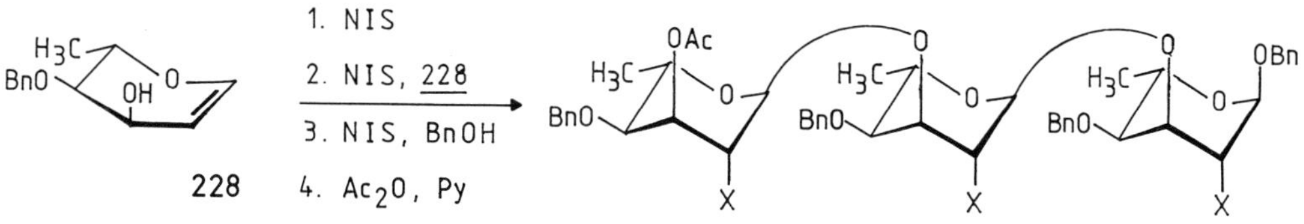

Scheme 46

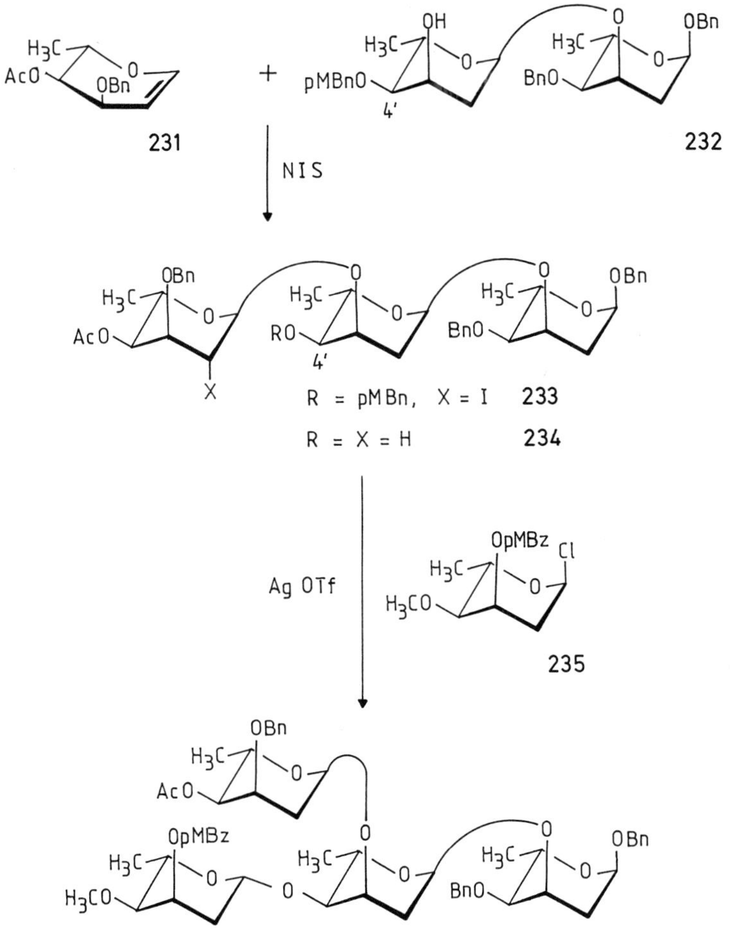

X = I   **229**
X = H   **230**

Scheme 47

In order to devise a synthesis of the tetrasaccharide, a discrimination between the 4-, 4'-, and 4''-hydroxy groups is necessary, and different protecting groups in these positions are required. By NIS-glycosylation of the 4'-O-p-methoxy-benzylated disaccharide **232** with the glycal **231**, the trisaccharide derivative **233** is obtained. Removal of the 4'-p-methoxybenzyl group by oxidative cleavage using DDQ leads to the trisaccharide **234** unprotected at the 4' position. In order to achieve a β,1→4 linkage, the in situ prepared glycosyl chloride **235** is attached to **234** in the presence of silver triflate. This synthesis comprises the first preparation of a fully protected kijanimicin tetrasaccharide **236** [109].

7 Orthosomycine Antibiotics

Another group of carbohydrate antibiotics was isolated and structurally elucidated several years ago [111]. In contrast to the compounds mentioned before, these are predominantly characterized by a larger oligosaccharide chain. In some cases benzoic acid derivatives like the dichloroisoeverninic acid (unit A in **237–243**) are part of the molecule. The most important and characteristic feature is located in some of the anomeric linkages of the oligosaccharide chain. Several 2,6-dideoxy sugars are attached via a unique interglycosidic spiro ortholactone linkage, replacing the normal acetal junctions. Members of this group are flambamycin **237** [112], the avilamycins **238** and **239** [113, 114], and the everninomycins **240–243** [115]. Another branch of orthosomycins are hygromycin B [116, 117] and several destomycins like, e.g., desto-mycin A (**244**) [118–122], furnished with an aminocyclitol aglycone. Other antibiotics like curamycin [123] are classified as members of the same group, owing to their similar biological behavior.

Flambamycin **237** and the everninomycin complex were isolated from *Streptomyces hygroscopicus* and *Micromonospora carbonacea*. Curamycin and avilamycin were obtained from culture broth of different *Streptomyces*. The biological effect of everni-nomycin, e.g., is a high in vitro activity against Gram-positive bacteria, especially penicillin-resistant strains. This fact is of particular interest because an increase of penicillin resistances was reported within the last years, owing to the extensive use of this antibiotic. Everninomycin shows no activity against Gram-negative bacteria like *E. coli* [115, 124], though it does affect Gram-positive bacteria like strepto- and staphylococci. Unless applied i.v., everninomycin and flambamycin, show no toxicity in mice. Furthermore, avilamycin advantageously has shown no mutagenicity. The antibacterial effect was shown to operate via protein synthesis perturbation [125]. By blocking of the binding site of aminoacyl-t-RNA, a ribosomal attachment is inhibited. This results in an interuption of the translation step and thus a defect in peptide processing.

In these particular compounds, the synthesis is facing three main problems: the oligosaccharide assembly in "normal" glycosidic linkages, the introduction of alkyl branches in pyranose rings, and the generation of the special anomeric orthoesters. Some of the interglycosidic bonds between normal sugars were shown to be β and thus modern Koenigs-Knorr variations were adapted for this aim.

More difficult to construct, however, are the β-linkages in the case of 2-deoxy sugar glycosides. The B-C disaccharide subunit comprises such a problem. Starting

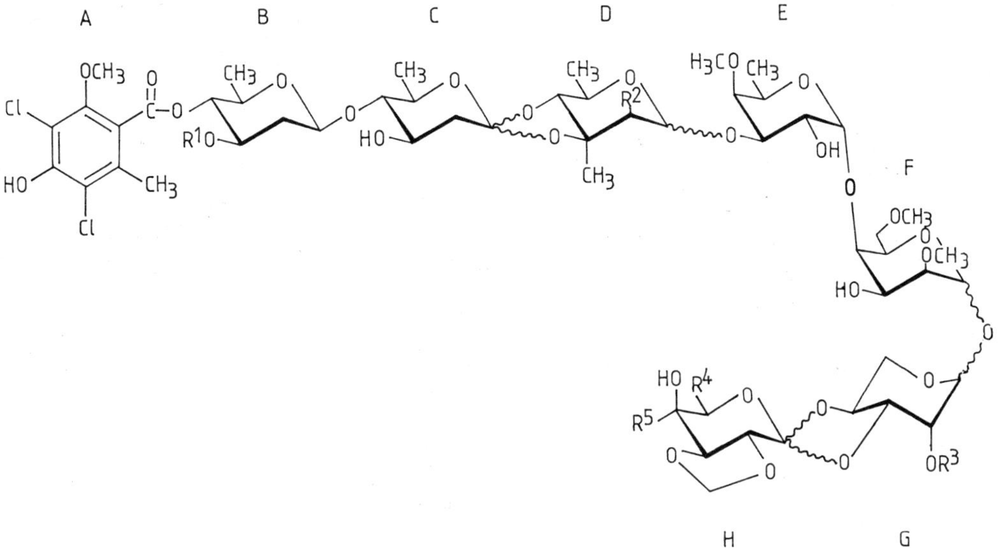

		R^1	R^2	R^3	R^4	R^5
237	Flambamycin	H	OH	$COCH(CH_3)_2$	CH_3	$COCH_3$
238	Avilamycin A	H	H	$COCH(CH_3)_2$	CH_3	$COCH_3$
239	Avilamycin C	H	H	$COCH(CH_3)_2$	CH_3	$(S)-CH(OH)CH_3$
240	Everninomicin B	E	OH	CH_3	H	$CH(OCH_3)CH_3$
241	Everninomicin C	E	H	CH_3	H	H
242	Everninomicin D	E	H	CH_3	H	$(S)-CH(OCH_3)CH_3$
243	Everninomicin 2	H	H	CH_3	H	$(S)-CH(OCH_3)CH_3$

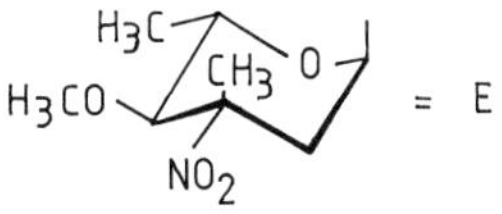

Evernitrose

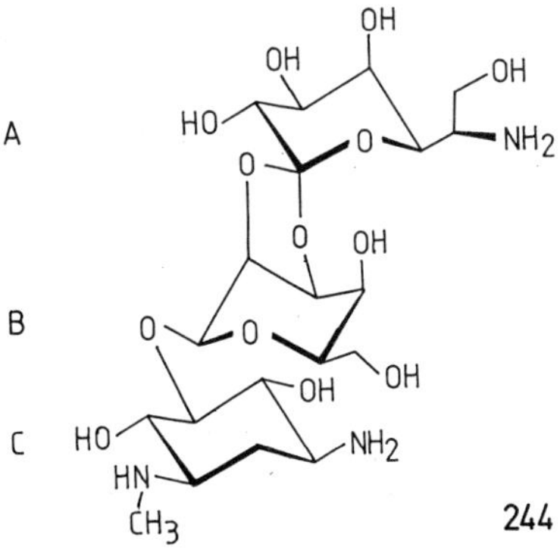

244   Scheme 48

with the 2,6-dibromo-2,6-dideoxy-glucosyl bromide **245**, glycosylation with the regioselectively blocked glycoside **246** gives the desired β,1→4-linkage in 73% yield and
a β:α-ratio of 5:1. The β-compound **247** is reductively transformed into the disaccharide **248**, which in turn can be oxidized by bromine/water to give the bamflalacton
249 [126].

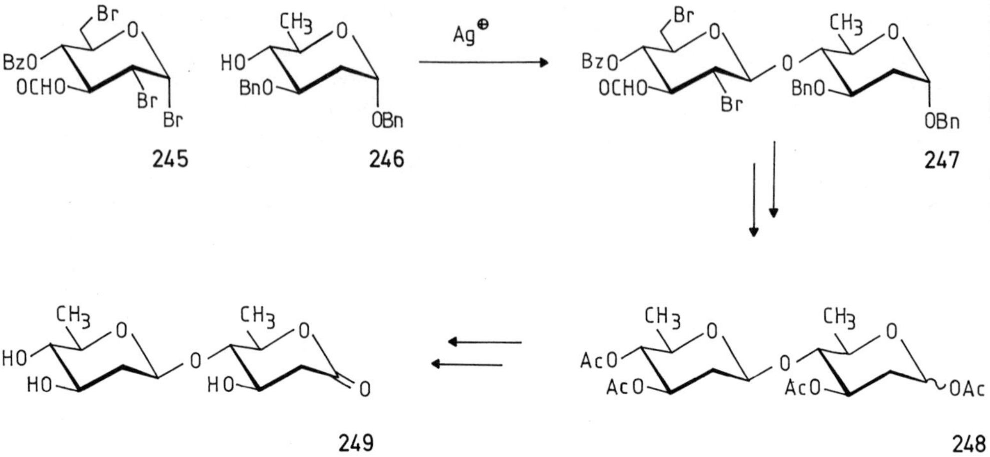

245   246   247

249   248

Scheme 49

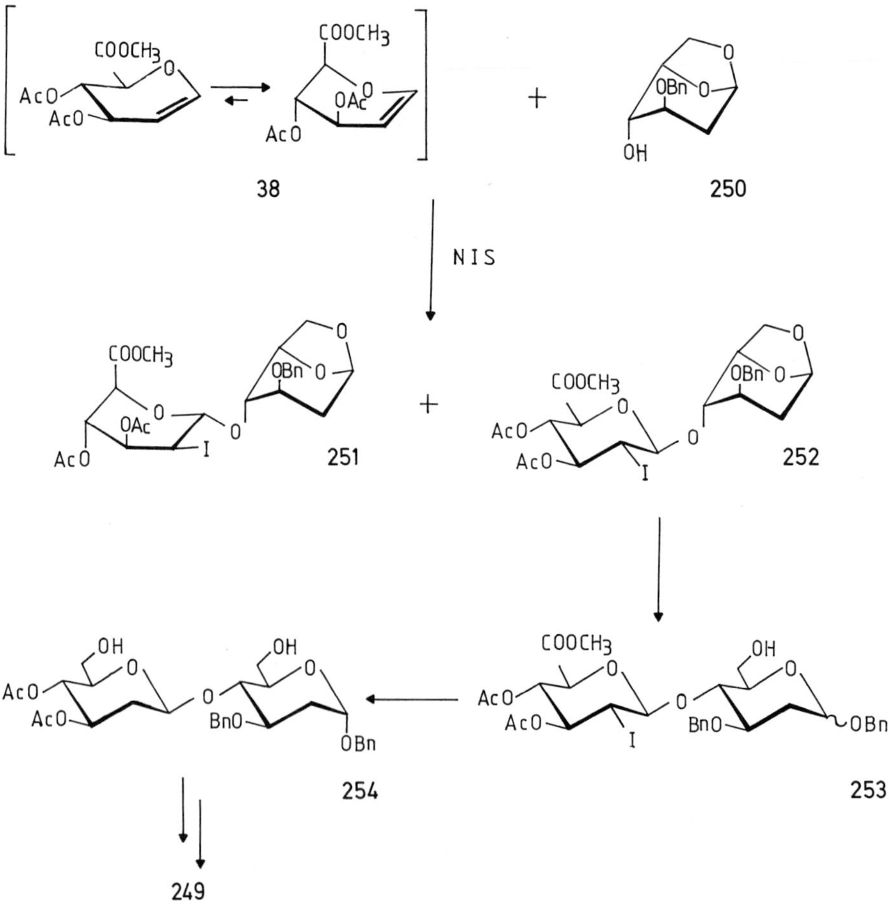

38   250

NIS

251   252

254   253

249

Scheme 50

In an alternative approach, the same terminal disaccharide B-C is prepared by the NIS reaction using the uronal **38** (R=COOCH$_3$) [19]. Here, the strict α-selectivity is modified into a α:β-ratio of approx. 1:1 [127]. The key step of this synthesis is a glycosylation of **38** as electrophilic acceptor with the nucleophilic 1,6-anhydro-3-O-benzyl-2,6-dideoxy-D-*arabino* species **250** obtained in a number of steps from levoglucosan. The iodonium intermediate obtained from **38** is opened to give a considerable amount of the desired β-glycoside **252** in addition to the α,1→4-anomer **251** (61%, α:β-ratio 1:1). Following the anhydro ring opening to **253**, reductions, and a number of further steps, the desired B-C target molecule **254** is obtained which again is transformed into bamflalactone **249** [127].

The E-F disaccharide moiety of flambamycin and its interglycosidic isomer are obtained by application of silver triflate-promoted condensations [128]. Starting with methyl α-D-mannopyranoside, the 4 and 6 positions are protected with the divalent 1,1,3,3-tetraisopropyldisiloxyl group (TIPS) to give **255**. Following the acid-mediated shift of the 8-membered towards a more stable 7-membered ring system in **256**, a blocking of both the equatorial 3 and 4 positions is achieved. Thus, the remaining C-6 and the secondary C-2 positions are methylated with MeI/Ag$_2$O (Purdie conditions) to give the 2,6-dimethyl product **257**. After cleavage of the TIPS-group by fluoride ion catalysis and careful regioselective benzylation by application of the phase transfer method, the aglycone precursor **258** (sugar unit F) is obtained. The E-compound precursor is synthesized from acetobromogalactose. The intermediate benzyl 2,3-O-benzyl-α-D-galactopyranoside (**259**) can be tosylated selectively at the primary hydroxy function, and following a hydride reduction, the fucoside **260** is obtained. Upon treatment with MeI/KOH in dimethyl formamide (Kuhn-methyla-

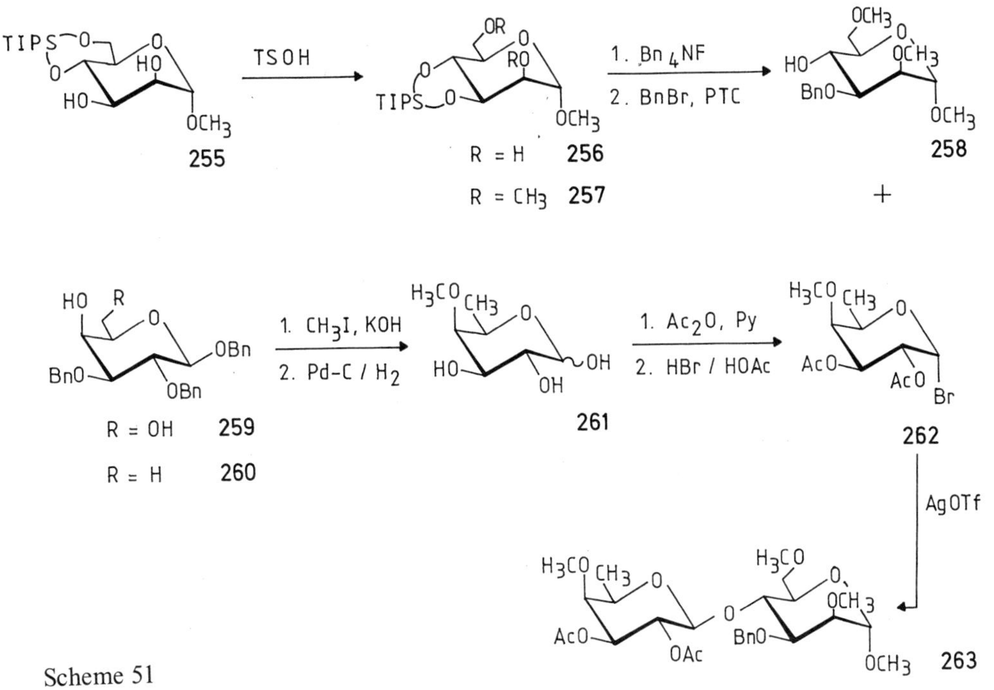

Scheme 51

tion) as well as benzyl ether hydrogenolysis, 4-O-methyl-D-fucose (261) results. By peracetylation and reaction with HBr/acetic acid, its α-bromide 262 is prepared. The glycosylation of the latter with 258 is performed in nitromethane/toluene at −78 °C to yield the modified trimethyl fucosyl-mannoside 263 (72%), the precursor of the E-F disaccharide [128].

In recent years considerable work has been done in the field of branched-chain sugars by several groups [129]. Three major methods for the introduction of a methyl-branch into ulosides (keto glycosides) shall be outlined. All of them make use of the carbonyl functionality, either at the electrophilic carbonyl center or at the nucleophilic position adjacent to the carbonyl function. A 3-uloside like, e.g., 264 may be methyl-branched by use of methyl lithium or methyl magnesium iodide to give either 265 or 266. Alternatively, 264 on Wittig olefination gives the exocyclic olefin sugar 267, which on epoxidation with m-chloroperbenzoic acid may give both the spiro epoxides 268 and 269. Their reduction leads to both the epimeric branched-chain sugars 265 or 266, respectively [129].

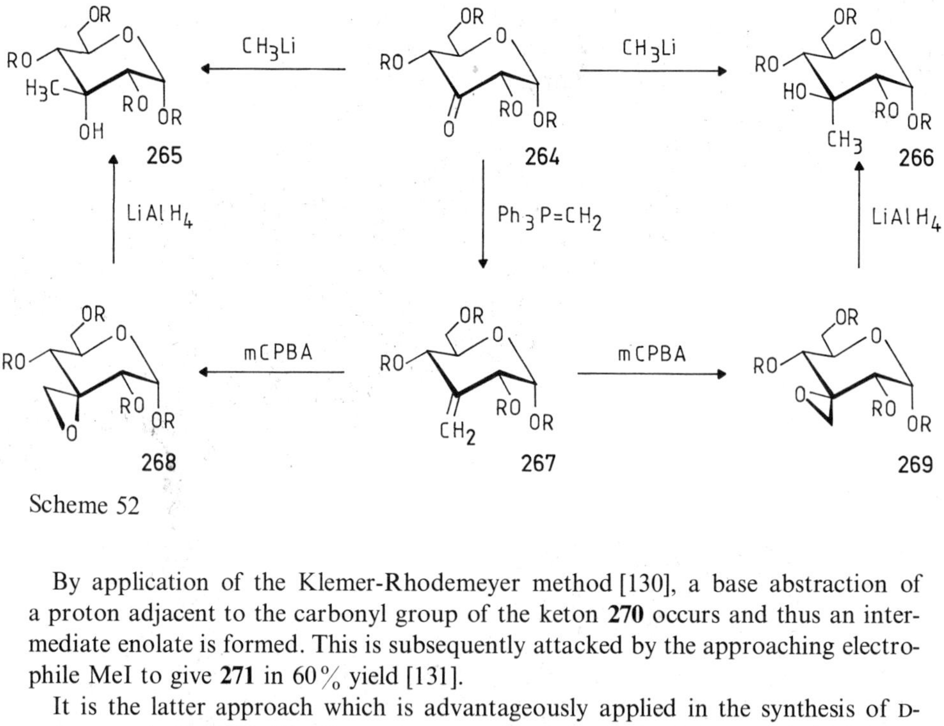

Scheme 52

By application of the Klemer-Rhodemeyer method [130], a base abstraction of a proton adjacent to the carbonyl group of the keton 270 occurs and thus an intermediate enolate is formed. This is subsequently attacked by the approaching electrophile MeI to give 271 in 60% yield [131].

It is the latter approach which is advantageously applied in the synthesis of D- and L-nogaloside [132]. The former target molecule is obtained from methyl 2,3-di-O-isopropylidene-α-D-rhamnopyranoside (272). Pyridinium dichromate oxidation is carried out to give the 4-uloside 273, the methylation of which occurs upon treatment with lithium diisopropylamine/methyl iodide to give 274 in 72% yield. In this biomimetic process only the axial methyl-branched product is generated with regioselectivity and stereospecificity. Acid-catalyzed dioxolane ring cleavage, acetylation, uloside reduction, and deacetylation yields methyl α-D-evaloside (275). The corresponding enantiomer L-nogaloside was obtained similarly [132].

326

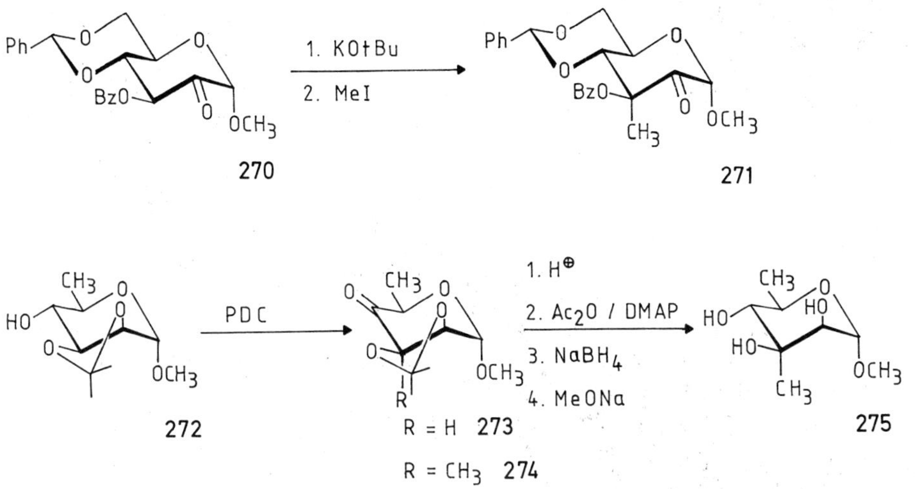

Scheme 53

The remaining central problem in the synthesis of orthosomycin glycosides is the generation of the complex anomeric spiro orthoesters. Two general approaches are leading to the desired structures, either by starting with a lactone or a glycal. Yoshimura et al. [133, 134] obtained the anomeric orthoesters starting with, e.g., the gluconolactone **276** and, e.g., the methyl-branched silyl ether-protected diol **277**. Under

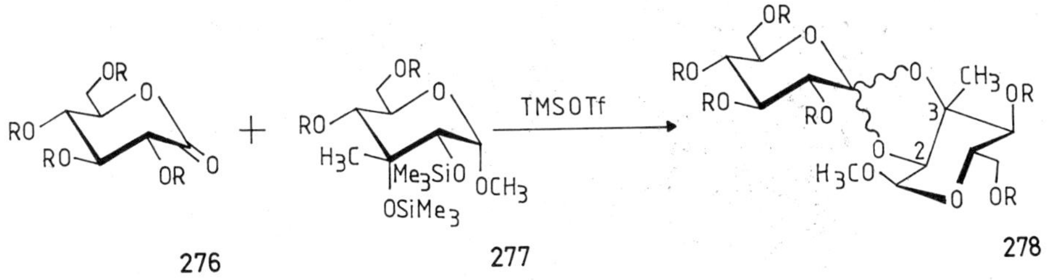

Scheme 54

trimethylsilyltrifluoromethane sulfonate (TMSOTf) catalysis in dichloromethane, the corresponding product **278** was formed. The method gives good yields if simple diols are involved; in case of more complex derivatives, the yield is considerably lower (like approx. 10% for **278**, R = Bn). This was the first method published, the mechanism of which is supposed as follows.

Employing the lactone **279**, the trimethylsilyl cation generates a positive charge at the anomeric carbon position of **280**. The corresponding triflate anion recombines with one of the silyl ethers in **281** and yields the alkoxide **282** and a molecule of catalyst. After nucleophilic attack, **283** is formed, and by a subsequent intramolecular reaction, the orthoester **284** is obtained simultaneously releasing a molecule of hexamethyldisiloxane.

Descotes et al. developed a stereoselective photocyclisation procedure of hydroxyalkyl glycosides to give spiro orthoesters [135]. By oxymercuration of the glycal **154**

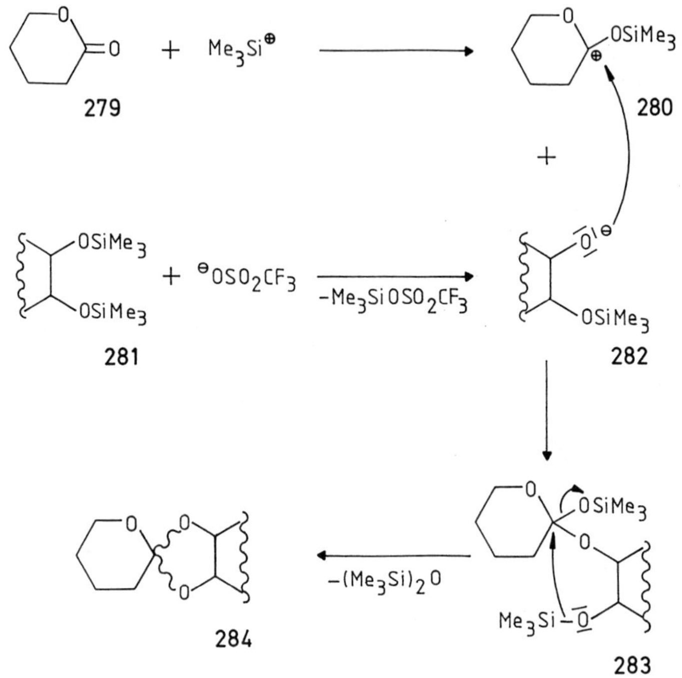

Scheme 55

with ethylene glycol as a model diol, the 2-deoxy-2-acetoxymercuri-glycoside **285** is formed as an anomeric and 2-epimeric mixture. This is reductively transformed into the 2-deoxy compound **286**. Irradiation with visible light in the presence of iodine and mercury(II)oxide gives the spiroorthoester **287** [136].

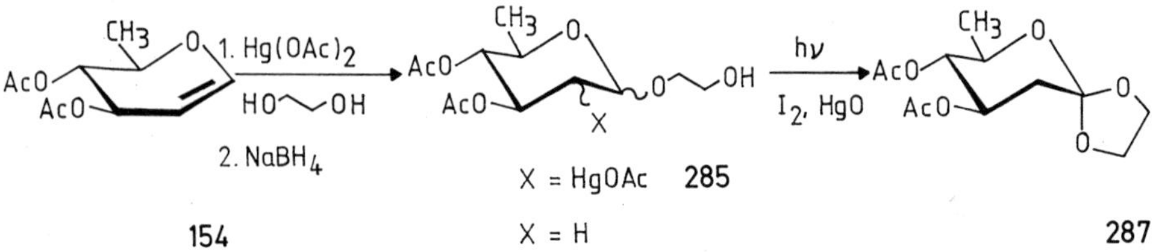

Scheme 56

Previous work of Sinaij et al. [137] established the method of diastereoselective conversion of glycals into anomeric spiro orthoesters using the oxyselenation-elimination sequence. Recently, the C-D-unit of everninomycin itself was synthesized [138]. Methyl α-D-evermicoside **289**, the D-ring precursor is prepared conventionally. Its condensation with the 3,4-di-O-benzyl-D-rhamnal (**288**) using phenyl-selenyl chloride in acetonitrile is shown to occur regiospecifically. In contrast to the expectation, glycosylation is observed via the tertiary 3-position to give the disaccharide **290**. After sodium metaperiodate oxidation, the diastereomeric mixture of phenylselen-oxides **291** is obtained. Heating to 120 °C in toluene in the presence of vinylacetate and diisopropylamine as base results in the formation of both ortholactones **292** and **293** in similar yields. Final debenzylation, followed by peracetylation, gives derivatives which are structurally assigned.

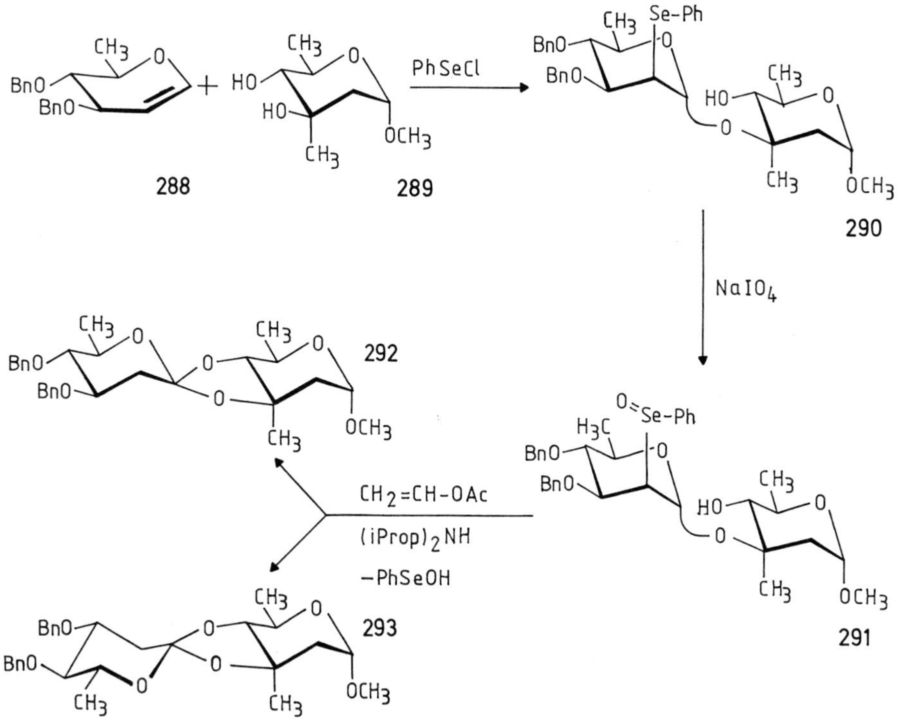

Scheme 57

8 References

1. Snipes CE, Chang C, Floss HG (1979) J Am Chem 101: 701; and literature cited therein
2. Snipes CE, Brillinger GU, Sellers L, Mascaro L, Floss HG (1977) J Biol Chem 252: 8113
3. Vanek Z, Majer J (1967) Macrolide Antibiotics, in: Gottlieb D, Shaw PD (eds) Antibotics Vol. 2, Springer, Berlin, p 154
4. Lee JJ, Lee JP, Keller PJ, Cottrell CE, Chang CJ, Zähner H, Floss HG (1986) J Antibot 39: 1123
5. Grisebach H (1968) Helv Chim Acta 51: 928
6. Paulsen H (1982) Angew Chem Int Ed Engl 21: 155
7. Schmidt RR (1986) Angew Chem Int Ed Engl 25: 212
8. Martin A, Pais M, Monneret C (1983) J Chem Soc, Chem Commun 306
9. Thiem J, Karl H, Schwentner J (1978) Synthesis 693
10. Thiem J, Köpper S, Schwentner J (1985) Liebigs Ann Chem 2135
11. Thiem J, Duckstein V (unpublished); Duckstein V (1984) Diplomarbeit Univ Münster
12. Kato T, Idinose I, Hosogai T, Kitahasra Y (1976) Chem Lett 1187
13. Jaurand G, Beau JM, Sinaij P (1981) J Chem Soc, Chem Commun 572
14. Ferrier RJ, Prasad N (1968) Chem Commun 476: (1969) J Chem Soc (C) 570
15. Thiem J, Schwentner J (1976) Tetrahedron Lett 3117

16. Ferrier J (1965) Adv Carbohydr Chem 20: 67; (1969) Adv Carbohydr Chem Biochem 24: 199; (1970) Fortsch Chem Forsch 14: 389
17. Jin H, Tsai R, Wiesner K (1983) Can J Chem 61: 2442
18. Ferrier RJ, Hay RW, Vethaviyasar N (1973) Carbohydr Res 27: 55
19. Thiem J, Ossowski P (1984) J Carbohydr Chem 3: 287
20. Bock K, Pedersen C, Thiem J (1979) Carbohydr Res 73: 85
21. Thiem J, Gerken M (1982) J Carbohydr Chem 1; (1983) 229
22. Thiem J, Gerken M, Bock K (1983) Liebigs Ann Chem 462
23. Thiem J, Gerken M (1985) J Org Chem 50: 954
24. Nicolaou KC, Ladduwaleetty T, Randall JL, Chucholowski A (1986) J Am Chem Soc 108: 2466
25. Hashimoto S, Hayashi M, Noyori R (1984) Tetrahedron Lett 25: 1379
26. Kreuzer M, Thiem J (1986) Carbohydr Res 149: 347
27. Kreuzer M (1986) Diss Univ Münster
28. Forth W, Henschler D, Rummel W (1983) Allgemeine und spezielle Pharmakologie und Toxikologie, 4. Aufl., B.I.-Wissenschafts-Verlag, Mannheim
29. Mutschler E (1981) Arzneimittelwirkungen, 4. Aufl., Wissenschaftl. Verlagsgesellschaft, Stuttgart
30. Remers WA (1979) The Chemistry of Antitumor Antibiotics, vol 1, Wiley, New York
31. Pigram WJ, Fuller W, Hamilton LD (1972) Nature new Biol 235: 17, as cited in: Füllenbach D, Nagel GA, Seeber S (eds) (1981) Adriamycin-Symposium, Ergebnisse und Aspekte, Karger, Basel
32. Crooke ST, Reich SD (eds) (1980) Anthracyclines — Current Status and Developments, Academic Press, New York
33. Lehotay DC, Levey BA, Rogerson BJ, Levey GS (1982) Cancer Treat Rep 66: 311
34. Neidle S, Waring MJ (eds) (1983) Molecular Aspects of Anticancer Drug Action, VCH, Weinheim
35. Grein A, Spalla C, Di Marco A, Canevazzi G (1963) Giorn Microbiol 11: 109
36. Du Bost M, Ganter P, Maral R, Ninet L, Pinnert S, Preu'dhomme J, Werner GH (1963) CR Acad Sci 157: 1813
37. Bernard J, Paul R, Boiron M, Jacquillat C, Maral R (eds) (1969) Rubidomycin, Springer, New York, Berlin, Heidelberg
38. Arcamone F, Francesci G, Penco S, Selva A (1969) Tetrahedron Lett 1007
39. Arcamone F (1981) Doxorubicin — Anticaner Antibiotics, Academic Press, New York
40. Acton EM, Long GL, Moser CW, Wolgemuth RL (1984) J Med Chem 27: 638
41. Arcamone F, Penco S, Vigevani A, Radaelli S, Franchi G (1975) J Med Chem 18: 703
42. Oki T et al. (1975) J Antibiot 28: 830
43. Oki T et al. (1977) J Antibiot 30: 613
44. Oki T et al. (1977) J Antibiot 30: 683
45. Nettleton JE (Jr.), Bradner WT, Busch JA, Coon AB, Moseley JE, Myllymaki RW, O'Herron FA, Schreiber RH, Volcano AL (1977) J Antibiot 30: 525
46. March JP, Mosher CW, Acton EM, Goodman L (1967) J Chem Soc, Chem Commun 973
47. Horton D, Weckerle W (1975) Carbohydr Res 44: 227
48. Pauls HW, Fraser-Reid B (1983) J Chem Soc, Chem Commun 1031
49. Cardillo G, Orena M, Sandri S, Tomasini C (1984) J Org Chem 49: 3951
50. El Khadem H, Liav A, Swartz DL (1979) Carbohydr Res 74: 345
51. Boivin J, Monneret C, Pais M (1980) Carbohydr Res 85: 223
52. Boivin J et al. (1980) Carbohydr Res 79: 193
53. Heyns K et al. (1981) Chem Ber 114: 232
54. Thiem J, Springer D (1985) Carbohydr Res 136: 325
55. Arcamone F et al (1978) Experientia 34: 1255
56. Penco S et al (1983) J Org Chem 48: 405
57. Kimura Y et al. (1984) Chem Lett 501
58. Thang TT, Imbach JL, Fizames C, Lavelle F, Ponsinet G, Olesker A, Lukacs G (1985) Carbohydr Res 135: 241
59. Horton D, Priebe W (1984) U.S. Pat. 4,427,664; (1984) Chem Abstr 100: 19221e
60. Horton D, Priebe W, Varela O (1984) Carbohydr Res 130: C1

61. Pudlo P, Thiem J (unpublished)
62. David S, Thieffry A (1981) Tetrahedron Lett 28: 2647
63. Klaffke W, Thiem J (1987) Royal Chemical Society, Carbohydrate Group Spring Meeting, Cambridge
64. Springer D (1985) Dissertation Univ Hamburg
65. Klaffke W, Thiem J (unpublished)
66. Pauls HW, Fraser-Reid B (1983) J Org Chem 48: 1392
67. Cardillio G et al (1982) J Chem Soc, Chem Commun 1308
68. El Khadem H, Liav A (1979) Carbohydr Res 74: 199
69. Thiem J, Kluge HW, Schwentner J (1980) Chem Ber 113: 3497
70. Boivin J, Monneret C, Pais M (1981) Tetrahedron 37: 4219
71. Martin A, Pais M, Monneret C (1986) Tetrahedron Lett 27: 575
72. Thiem J 1989 Am Chem Soc, Symp Ser 386: 131
73. Appel R (1975) Angew Chem, Int Ed Engl 14: 801
74. Hanessian S, Lavallee P (1976) Meth Carbohydr Chem 7: 49
75. Ward DC, Reich E, Goldberg IL (1965) Science 149: 1259
76. van Dyke M, Dervan PB (1983) Biochemistry 22: 2373
77. Miyamoto M, Morita K, Kawamatsu Y, Naguchi S, Marumoto R, Sasai M, Nohara A, Nakadaira Y, Lin YY, Nakanishi K (1966) Tetrahedron 22: 2761
78. Berlin YuA, Kolosov MN, Schenyakin MN (1966) Tetrahedron Lett 1431
79. Miyamoto M, Kawamatsu Y, Kawashima K, Shinohara M, Tanaka K, Tatsuoka S, Nakanishi K (1967) Tetrahedron 421
80. Thiem J, Schneider G (1983) Angew Chem, Int Ed Engl 22: 58
81. Thiem J, Schneider G, Sinnwell V (1986) Liebigs Ann Chem 814
82. Thiem J, Meyer B (1981) Tetrahedron 37: 551
83. Thiem J, Meyer B (1979) J Chem Soc, Perkin Trans 2: 1331
84. Thiem J, Meyer B (1980) Chem Ber 113: 3075
85. Thiem J, Meyer B (1980) Chem Ber 113: 3067
86. Klemer A, Jung G (1981) Chem Ber 114: 740
87. Thiem J, Elvers J (1979) Chem Ber 112: 818
88. Thiem J, Elvers J (1980) Chem Ber 113: 3049
89. Thiem J, Gerken M, Bock K (1983) Liebigs Ann Chem 462
90. Thiem J, Sievers A, Karl H (1977) Chromatogr 130: 305
91. Barton DHR, McCombie SW (1975) J Chem Soc, Perkin Trans I: 1574
92. Thiem J, Karl H (1980) Chem Ber 113: 3039
93. Thiem J, Gerken M (1985) J Org Chem 50: 954
93a. Thiem J, Gerken M, Schöttmer B, Weigand J (1987) Carbohydr Res 164: 327
94. Yoshimura J (1984) Adv Carbohydr Chem Biochem 42: 69
95. Schöttmer B (1987) Dissertation Univ Münster
96. Thiem J, Schöttmer B (1987) Angew Chem, Int Ed Engl 26: 555
97. Conn HF (ed) (1980) Current Therapy, Saunders, Philadelphia
98. Reichstein T, Weiss E (1962) Adv Carbohydr Chem 17: 65
99. Mallams AK, Puar MS, Rossman RR (1981) J Am Chem Soc 103: 3938
100. Mallams AK, Puar MS, Rossman RR, Mc-Pahil AT, Macfarlane RD (1981) J Am Chem Soc 103: 3940
101. Mallams AK, Puar MS, Rossman RR, Mc-Phail AT, Macfarlane RD, Stephens RL (1983) J Chem Soc, Perkin Trans I: 1497
102. Zorbach WW, Henderson N, Saeki S (1964) J Org Chem 29: 2016
103. Thiem J, Köpper S, Schwentner J (1985) Liebigs Ann Chem 2135
104. Thiem J, Köpper S (1982) Angew Chem, Int Ed Engl 21: 779
105. Thiem J, Ossowski P, Schwentner J (1980) Chem Ber 113: 955
106. Thiem J, Ossowski P (1981) Chem Ber 114: 733
107. Thiem J, Ossowski P (1983) Liebigs Ann Chem 2215
108. Wiesner K, Tsai TYR, Jin H (1985) Helv Chim Acta 68: 300
109. Köpper S (1985) Dissertation Univ Hamburg and Münster
110. Köpper S, Lundt I, Pedersen C, Thiem J (1987) Liebigs Ann Chem 531
111. Wright DE (1979) Tetrahedron 35: 1207

112. Ninet L et al. (1974) Experientia 30: 1270
113. Heilmann W et al. (1979) Helv. Chim Acta 62: 1
114. Keller-Schierlein W et al. (1979) Helv Chim Acta 62: 7
115. Weinstein JJ et al. (1964) Antimicrob Agents Chemotherapy 24
116. Mann RL, Bromer WW (1958) J Am Chem Soc 80: 2714
117. Neuss N et al. (1970) Helv Chim Acta 53: 2314
118. Shogi J et al. (1970) J Antibiotics 23: 291
119. Kondo S, Akita E, Koeke M (1966) J Antibiotics Ser A 19: 139
120. Kondo S et al. (1975) J Antibiotics Ser A 28: 79
121. Shimura M et al. (1976) Agr Biol Chem 40: 611
122. Shimura M et al. (1975) J Antibiotics 28: 83
123. Galmarim OL, Denlofen V (1961) Tetrahedron 15: 76; (1968) Experienta 24: 323
124. Saunders WE, Crow CC (1973) XIIIth Interscience Conference on Antimicrobial Agents and Chemotherapy, Abstr 139
125. Wolf H (1979) FEBS Lett 36: 181
126. Lundt I, Thiem J, Prahst A (1984) J Org Chem 49: 3063
127. Thiem J, Prahst A, Lundt I (1986) Liebigs Ann Chem 1044
128. Thiem J, Duckstein V, Prahst A, Matzke M (1987) Liebigs Ann Chem 289
129. Yoshimura J (1984) Adv Carbohydr Chem Biochem 42: 69
130. Klemer A, Rodemeyer G (1974) Chem Ber 107: 2612
131. Klemer A, Klaffke W (1987) Liebigs Ann Chem 759
132. Klemer A, Stegt H, Prahst A, Thiem J (1986) J Carbohydr Chem 5: 67
133. Yoshimura J et al (1983) Carbohydr Res 121: 187
134. Yoshimura J et al. (1983) ibid 121: 175
135. Praly JP, Descotes G, Grenier-Loustalot MF, Metras F (1984) Carbohydr Res 128: 21
136. Descotes G, Idmoumazet H, Praly JP (1984) Carbohydr Res 128: 341
137. Jaurand G, Beau JM, Sinaÿ P (1982) J Chem Soc, Chem Commun 223
138. Beau JM, Jaurand G, Esnault J, Sinaÿ P (1987) Tetrahedron Lett 28: 1105

Author Index Volumes 151—154

The volume numbers are printed in italics